69.00

FEEDBACK AND MOTOR CONTROL IN INVERTEBRATES AND VERTEBRATES

FEEDBACK AND MOTOR CONTROL IN INVERTEBRATES AND VERTEBRATES

Edited by

W.J.P. BARNES, BSc, PhD
Department of Zoology, University of Glasgow

M.H. GLADDEN, MB, BS, PhD, MRCS(Eng), LRCP(Lond), DCH
Institute of Physiology, University of Glasgow

CROOM HELM
London • Sydney • Dover, New Hampshire

©1985 W.J.P. Barnes and M.H. Gladden
Croom Helm Ltd, Provident House, Burrell Row,
Beckenham, Kent BR3 1AT
Croom Helm Australia Pty Ltd, Suite 4, 6th Floor,
64-76 Kippax Street, Surry Hills, NSW 2010, Australia

British Library Cataloguing in Publication Data
Feedback and motor control in invertebrates and
vertebrates.
 1. Animal locomotion
 I. Barnes, W.J.P. II. Gladden, M.H.
 591.1'852   QP303
ISBN 0-7099-3277-4

Croom Helm, 51 Washington Street, Dover,
New Hampshire 03820, USA

Library of Congress Cataloging in Publication Data
Main entry under title:

Feedback and motor control in invertebrates and
vertebrates.
 +Papers given ... at a symposium ... held at the
University of Glasgow from July 10th to 13th 1984 ...
organised by the Scottish Electrophysiological Society+—pref.
 Includes indexes.
 1. Locomotion—regulation—congresses.
2. Sensory-motor integration—congresses. 3. Feedback
control systems—congresses. 4. Invertebrates—
physiology—congresses. 5. Vertebrates—physiology—
congresses. I. Barnes, W.J.P., 1940–
II. Gladden, M.H., 1940– . III. Scottish
Electrophysiological Society.
QP303.F36   1986   591.1'852   85-26895
ISBN 0-7099-3277-4

Printed and bound in Great Britain by
Biddles Ltd, Guildford and King's Lynn

CONTENTS

Contributors
Preface
SECTION 1 : ORGANISATION OF MOTOR SYSTEMS 1

Chapter 1 Introduction
D.A. DORSETT 3

Chapter 2 Central feedback loops and some implications for motor control
W.J. DAVIS 13

Chapter 3 Neural control of vertebrate locomotion - central mechanisms and reflex interaction with special reference to the cat
S. GRILLNER 35

Chapter 4 Generation of behaviour: the orchestration hypothesis
G. HOYLE 57

Chapter 5 Convergence of several sensory modalities in motor control
A. TAYLOR and S. GOTTLIEB 77

Chapter 6 Feedback control of an escape behaviour
J.M. CAMHI 93

SECTION 2 : CENTRAL CONTROL OF SENSE ORGAN EXCITABILITY 113

Chapter 7 Introduction
A. PROCHAZKA 115

Chapter 8 Intrafusal muscle fibres in the cat and their motor control
I.A. BOYD 123

Chapter 9 How do crabs control their muscle receptors?
B.M.H. BUSH and A.J. CANNONE 145

Contents

SECTION 3 : AFFERENT INPUT DURING NORMAL MOVEMENTS 167

Chapter 10	Introduction A. PROCHAZKA	169
Chapter 11	What the cat's hind limb tells the cat's spinal cord G.E. LOEB	173
Chapter 12	Proprioceptive feedback and the control of cockroach walking S.N. ZILL	187

SECTION 4 : REFLEXES 209

Chapter 13	Introduction W.J.P. BARNES	211
Chapter 14	Stretch reflexes in man: the significance of tendon compliance P.M.H. RACK	217
Chapter 15	The synaptic basis for integration of local reflexes in the locust A.H.D. WATSON and M. BURROWS	231

SECTION 5 : THE CONTROL OF EQUILIBRIUM 251

Chapter 16	Introduction W.J.P. BARNES	253
Chapter 17	Control of eye-head coordination by brain stem neurones A. BERTHOZ	259
Chapter 18	Multisensory interactions in the crustacean equilibrium system D.M. NEIL	277

SECTION 6 : THE CONTROL OF MOVEMENT 299

| Chapter 19 | Introduction
E. JANKOWSKA | 301 |
| Chapter 20 | Are there central pattern generators for walking and flight in insects?
K.G. PEARSON | 307 |

Contents

Chapter 21	The role of movement-related feedback in the control of locomotion in fish and lamprey P. WALLÉN and T.L. WILLIAMS	317
Chapter 22	How locusts fly straight C.H.F. ROWELL, H. REICHERT and J.P. BACON	337
Chapter 23	Interactions of segmental and supra-segmental inputs with the spinal pattern generator of locomotion S. ROSSIGNOL and T. DREW	355
Chapter 24	Stepping reflexes and the sensory control of walking in Crustacea F. CLARAC	379

SECTION 7 : FEEDBACK AND MOTOR CONTROL IN MAN 401

Chapter 25	Introduction J.A. STEPHENS	403
Chapter 26	Proprioceptive activity from human finger muscles A.B. VALLBO	411
Chapter 27	Human long-latency stretch reflexes - a new role for the secondary ending of the muscle spindle? P.B.C. MATTHEWS	431
Chapter 28	Phase dependent step adaptations during human locomotion H. FORSSBERG	451
Chapter 29	Abnormal feedback and movement disorders in man, with particular reference to cortical myoclonus C.D. MARSDEN and J.C. ROTHWELL	465

Author index 477

Subject index 489

CONTRIBUTORS

Dr J.P. BACON, School of Biological Sciences, University of Sussex, Falmer, Brighton BN1 9QG, U.K.

Dr W.J.P. BARNES, Department of Zoology, University of Glasgow, Glasgow G12 8QQ, U.K.

Dr A. BERTHOZ, Laboratoire de Physiologie Neurosensorielle du CNRS, 15, Rue de l'École de Médecine, 75270 Paris Cedex 06, France.

Professor I.A. BOYD, Institute of Physiology, University of Glasgow, Glasgow G12 8QQ, U.K.

Dr M. BURROWS, Department of Zoology, University of Cambridge, Downing Street, Cambridge CB2 3EJ, U.K.

Dr B.M.H. BUSH, Department of Physiology, University of Bristol, Park Row, Bristol BS1 5LS, U.K.

Professor J.M. CAMHI, Department of Zoology, Hebrew University of Jerusalam, 91904 Jerusalam, Israel.

Dr A.J. CANNONE, Department of Zoology, University of the Witwatersrand, Johannesburg 2001, South Africa.

Dr F. CLARAC, Laboratoire de Neurobiologie, CNRS-Université de Bordeaux 1, Place du Docteur Bertrand Peynau, 33120 Arcachon, France.

Professor W.J. DAVIS, Thimann Laboratories, University of California, Santa Cruz, California 95064, U.S.A.

Dr D.A. DORSETT, Marine Science Laboratories, Menai Bridge, Gwynedd LL59 5EH, U.K.

Dr T. DREW, Départment de Physiologie, Université de Montréal, P.O. Box 6128, Station "A", Montreal, Quebec, Canada H3C 3J7.

Contributors

Dr H. FORSSBERG, Institutionen för Fysiologi III, Karolinska Institutet, Lidingövägen 1, 114 33 Stockholm, Sweden.

Dr S. GOTTLIEB, Sherrington School of Physiology, St Thomas's Hospital Medical School, Lambeth Palace Road, London SE1 7EH, U.K.

Professor S. GRILLNER, Institutionen för Fysiologi III, Karolinska Institutet, Lidingövägen 1, 114 33 Stockholm, Sweden.

Professor G. HOYLE, Department of Biology, University of Oregon, Eugene, Oregon 97403, U.S.A.

Dr E. JANKOWSKA, Fysiologiska Institutionen, Göteborgs Universitet, Box 33031, 400 33 Göteborg, Sweden.

Dr G.E. LOEB, Laboratory of Neural Control, National Institute of Neurological Disorders and Stroke, National Institutes of Health, Bethesda, Maryland 20205, U.S.A.

Professor C.D. MARSDEN, University Department of Neurology, Institute of Psychiatry and King's College Hospital Medical School, Denmark Hill, London SE5 8AF, U.K.

Dr P.B.C. MATTHEWS, University Laboratory of Physiology, University of Oxford, Parks Road, Oxford OX1 3PT, U.K.

Dr D.M. NEIL, Department of Zoology, University of Glasgow, Glasgow G12 8QQ, U.K.

Professor K.G. PEARSON, Department of Physiology, University of Alberta, Edmonton, Alberta, Canada T6G 2H7.

Dr A. PROCHAZKA, Sherrington School of Physiology, St Thomas's Hospital Medical School, Lambeth Palace Road, London SE1 7EH, U.K.

Professor P.M.H. RACK, Department of Physiology, University of Birmingham, Vincent Drive, Birmingham B15 2TJ, U.K.

Dr H. REICHERT, Zoologisches Institut der Universität Basel, CH-4051 Basel, Rheinsprung 9, Switzerland.

Dr S. ROSSIGNOL, Département de Physiologie, Université de Montréal, P.O. Box 6128, Station "A", Montreal, Quebec, Canada H3C 3J7.

Dr J.C. ROTHWELL, University Department of Neurology, Institute of Psychiatry and King's College Hospital Medical School, Denmark Hill, London SE5 8AF, U.K.

Contributors

Professor C.H.F. ROWELL, Zoologisches Institut der Universität Basel, CH-4051 Basel, Rheinsprung 9, Switzerland.

Professor J.A. STEPHENS, Department of Physiology, Middlesex Hospital Medical School, London W1P 7PN, U.K.

Professor A. TAYLOR, Sherrington School of Physiology, St Thomas's Hospital Medical School, Lambeth Palace Road, London SE1 7EH, U.K.

Professor A.B. VALLBO, Institutionen för Fysiologi, Umeå Universitet, 901 87 Umeå, Sweden.

Dr P. WALLÉN, Institutionen för Fysiologi III, Karolinska Institutet, Lidingövägen 1, 114 33 Stockholm, Sweden.

Dr A.H.D. WATSON, Department of Zoology, University of Cambridge, Downing Street, Cambridge, CB2 3EJ, U.K.

Dr T.L. WILLIAMS, Department of Physiology, St. George's Hospital Medical School, Cranmer Terrace, Tooting, London SW17 ORE, U.K.

Dr S.N. ZILL, Department of Anatomy, University of Colorado Medical School, 4200 East Ninth Avenue, Denver, Colorado 80262, U.S.A.

To the memory of Graham Hoyle

PREFACE

This book is a collection of papers given by invited speakers at a Symposium on 'Feedback and Motor Control', held at the University of Glasgow from July 10th to 13th 1984, which was attended by over 200 scientists from 20 countries. The Symposium was the Fourth International Symposium organised by the Scottish Electrophysiological Society (SES), and on this occasion the SES joined forces with the Society for Experimental Biology (SEB), so that the Symposium was held during the annual Summer Meeting of the SEB. A policy of the SES since its formation in 1970 has been to promote dialogue between scientists working on invertebrate and vertebrate nervous systems by holding local Scottish inter-university meetings and international symposia. Previous symposia were on 'Simple' Nervous Systems, the proceedings of which were published by Edward Arnold in 1975, Synapses published by Blackie in 1977, and Sense Organs published by Blackie in 1981.
    A major objective of this Symposium was to bring together neurobiologists working on the roles of sensory feedback in motor control in invertebrates as well as vertebrate animals and man to discuss common problems and learn about each other's research. Thus speakers were invited to give talks and write articles that were general enough to be readily understood by non-specialists. As organisers, we did, however, add that this did not mean that latest findings should not be described. Rather that they should be put into context, so providing a broad view of the current state of their field. Thus this book should be useful to senior undergraduates as well as graduate students and more senior research workers.
    Like the Symposium itself, the book consists of seven sections, each consisting of an introduction by the chairman of the session and a chapter by each of the contributing speakers. Additionally, we include a chapter by Professor Hoyle, who kindly wrote us a review in spite of being unable to participate in the Symposium itself. An initial section on

Preface

the organisation of motor systems not only provides a framework for the rest of the book, but highlights one of the current controversies within the field; whether motor control systems are hierarchically organised, a traditional view expounded here by Professor Grillner, or whether they incorporate central feedback loops making terms such as 'higher' and 'lower' centres inappropriate as claimed by Professor Davis.

This section is followed by two on the sensory input. The first of these, on the central control of sense organ excitability, provides a striking example of the frequent similarity of the motor control systems of different animal groups, for the crab thoraco-coxal muscle receptor organ and mammalian muscle spindles are almost identical in their response to perturbations, their mechanism of central control and even the function of their central connections. The technical difficulties of recording from proprioceptors during natural movement have been overcome for cats and cockroaches as described in Section 3 by Drs Loeb and Zill. For humans too, access to the axons of proprioceptors is possible, but only during movements of restricted range and velocity (see Chapter 26 by Professor Vallbo). Such studies are of particular importance for understanding the role played by 'active' receptors whose outputs can be controlled by the central nervous system.

In the following section on reflexes and their role in motor control, the complexity of even the simplest of reflexes is forcibly illustrated for both vertebrates and invertebrates by Professor Rack and Drs Watson and Burrows, the latter article demonstrating that the nervous systems of lower animals are by no means as simple as was once thought. Sections 5 and 6 cover the control of equilibrium and movement, respectively. In the former, Dr Neil describes how visual, statocyst and proprioceptive inputs interact to produce compensatory or righting responses, while Dr Berthoz demonstrates that such compensatory responses are actively inhibited during saccadic eye movements when the eyes are turned towards objects of interest. During locomotion, as in other rhythmic movements, the sensory input has a variety of roles. It interacts with a centrally generated rhythm to produce the 'normal' pattern of motor output as discussed by Professor Pearson; it can entrain the motor output as described for fish and lamprey by Drs Wallén and Williams; and it can adjust for perturbations as described for the cat by Drs Rossignol and Drew and for man by Dr Forssberg in Chapter 28. It is also involved in course maintenance as elegantly demonstrated for locust flight by Professor Rowell and Drs Reichert and Bacon, and in inter-leg coordination as described by Dr Clarac. It is interesting to note that many of the reflexes involved in these responses appear to be phase-dependent, in that they are centrally gated-out during the

Preface

phase of the locomotor cycle when such a reflex would be inappropriate.

Finally, since most of us, whether we work on invertebrates or vertebrates, are ultimately interested in man, it is appropriate that there is a section entirely devoted to feedback and motor control in man. Within this section, which covers disorders of motor control as well as studies of normal physiology, another of the current controversies within the field is highlighted. According to Professor Marsden and Dr Rothwell, the long-latency components of some human stretch reflexes involve a trans-cortical loop, while Dr Matthews argues that they are simply attributable to the slower conducting secondary spindle afferents.

Although a book of this length can never hope to be comprehensive, both the way in which authors have put their results into context and the inclusion of introductions to each section have, we believe, enabled us to present an integrated story. We have also been fortunate that so many distinguished scientists, leaders in the field, have contributed chapters to the volume. We thus feel confident that this book is not just another collection of review papers of only passing interest, but will be of lasting value in that it presents a well-argued case for studying motor control systems in a variety of different animals.

Professor Graham Hoyle, to whom this book is dedicated, died shortly after submitting his chapter to us. He was a lecturer in Zoology at the University of Glasgow from 1952-9, and a major figure in invertebrate neurobiology for many years. His contribution to this volume is typically both original and provocative. Unlike many of the other contributors to this book, neither of us was fortunate enough to have worked with him, but one of us (WJPB) does have vivid memories of lively discussions, enthusiastic lectures and of hillwalking together in the Swiss alps, during a period a few years ago when we were both visiting the University of Konstanz in West Germany.

We wish to express our gratitude to all who have contributed chapters to this book. Also, most especially, to thank Mr M. Murthy, not only for so efficiently producing the camera-ready manuscript, but also for maintaining his good humour in spite of all too frequent requests for minor changes to the text.

Glasgow   W.J.P. Barnes
July 1985   M.H. Gladden

Section 1
Organisation of Motor Systems

The Centipede was happy quite,
Until the Toad in fun,
Said, 'Pray which leg goes after which?'
And worked her mind to such a pitch,
She lay distracted in the ditch,
Considering how to run.

      Mrs. Craster

Chapter One

INTRODUCTION

D.A. Dorsett

One objective in the study of excitable systems and behaviour has been to formulate a set of principles upon which the operation of the nervous system is based, which is applicable to organisms at different levels of evolutionary complexity. This concept has conditioned our thinking in a number of ways. It implies that the genetic resources which determine the biochemical and physiological characteristics of the neural elements impose constraints on the capacity of single units to process information, and also that there may be a finite number of ways in which available units can be utilised to perform a given function. The validity of this concept gains support in the fundamental division of neurones into three major categories; sensory, motor and interneurones. The transition from simple organisms to those of greater organisational and evolutionary complexity has not been accompanied by any remarkable increase in cellular diversity, but rather by a relative increase in the number of interneurones available as information processing channels ultimately projecting to the motor centres.
In recent years the application of intracellular recording techniques to the central nervous systems of both vertebrate and invertebrate preparations has enabled us to characterise their structural and physiological properties and study the network relationships of many individual and some identifiable neurones. While these studies have undoubtedly furthered our understanding of the operation of both sensory and motor systems, it is the opportunity for access to the interneurones and analysis of central organisation that has justifiably attracted most attention.
The subject of the first section of this book is "organisation of motor systems", and it is perhaps inevitable that most contributors have approached this topic from the centralist viewpoint. However, we should remind ourselves that the "final motor pathway" is not just a means of translating a command formulated by a central nervous network into movement, but itself exhibits both plasticity and flexibility.

Peripheral organisation of motor systems

The concept of the motor unit was first introduced by Sherrington as the basic unit of motor activity. However, the realisation that vertebrate muscles could be divided into "fast" and "slow" response types and that even mammalian muscles contained fibres with different response characteristics meant that the original concept had to be considerably broadened. The size of a given motor unit varies considerably according to the animal and the muscle chosen, but it has only recently been realised that in humans the size of the motor unit may be much larger than the corresponding unit in laboratory animals (Table 1). The muscle fibres, particularly of fast motor units, tend to be scattered throughout a large area of the muscle and are all of the same physiological type. Fibres are broadly of two kinds, the "fast" and "slow twitch", the latter being used to maintain tension during posture. Gradation of force at low tension levels is thought to be mainly by recruitment of additional motor units, small slow units being recruited before larger fast units according to the "size principle" (Henneman, 1979). At intermediate force levels, adjustment is made by an increase in firing rate in the motor neurones from around 8 to 20-30 i.p.s. (Buchthal & Schmalbruch, 1980).

Amphibian skeletal muscles fall more distinctly into fast (twitch) and slow (tonic) types. In the fast fibres the end plate potential generates a propagating action potential that precedes the twitch, while tonic fibres depolarise by temporal summation of individual excitatory junctional potentials (EJPs). In fish the swimming muscles also form two broad divisions, a red external layer active during slow swimming and a white myotomal muscle consisting of fast fibres used for bursts of speed. In elasmobranchs the fast fibres are excited by propagating action potentials arising from a myoseptal end plate, a motor unit consisting of about 100 muscle fibres. In most teleosts the fast muscles show a quite different pattern of innervation from the elasmobranchs, each fibre receiving input from up to three axons arising from different spinal

TABLE 1

Size of motor unit in some mammals. (From Buchthal & Schmalbruch, 1980)

Muscle	No. of muscle fibres/unit			
	RAT	CAT	BABOON	MAN
Gastrocnemius	70	600	700	1720

roots which form multi-terminal endings along the fibre. These fibres also sustain propagating action potentials. Depending upon the type or number of axons activated, the motor units respond by graded contractions or a fast twitch. With this pattern of innervation, it is difficult to define what constitutes a motor unit (Roberts & Williamson, 1983).

In the lamprey there are estimated to be some 150-200 motor neurones per segment, each fast motor unit consisting of a central group of innervated fibres which are electrically coupled to more peripheral fibres. The slow fibre innervation resembles that of elasmobranchs (Teräväninen, 1971).

Considerable variation exists in the motor organisation of the invertebrate phyla. Generally invertebrate muscles are supplied by a small number (between 1 and 4) motor neurones which may evoke fast or slow responses from the muscle. A major difference from the vertebrate pattern is the presence of a peripheral inhibitory supply which may terminate on the muscle fibres or pre-synaptically on the terminals of the motor neurones. In general, fast motor neurones have larger somata and axon diameters than the slow type and show less tendency to repetitive or tonic discharge. They produce large EJPs which do not usually facilitate. Fast motor units are often recruited as the force requirement of a muscle is increased, in accordance with the size principle. They are often used in the execution of centrally programmed movements, while slow systems are used for postural adjustments in conjunction with proprioceptive feedback.

Invertebrate muscle fibres may also be separated into "fast" and "slow" categories on the basis of their responses, and these characteristics show a strong correlation with fine structural features such as the pattern of striation, sarcomere length, and extent of the sarcoplasmic reticulum and transverse tubular system. These two types must, however, be regarded as the extremes of a broad spectrum of functional types. In the same way, membrane responses range from large amplitude EJPs, which result in propagating action potentials, to non-propagating graded local responses. Muscles showing local responses are electrically inexcitable and show graded contractions resulting from overall depolarisation by summation of distributed EJPs.

In the simplest case the muscle may be innervated by a single axon, fast or slow responses being determined by the pattern of stimulation. In the accessory flexor muscle of the crayfish leg, the whole muscle forms one motor unit, innervated by an excitatory and an inhibitory motor neurone. The physical contraction evoked is determined by the input pattern to a heterogeneous group of muscle fibres (Angaut-Petit, 1977). In other instances the muscles are of a single functional type, such as the slow flexor/extensor system which controls abdominal posture in the crayfish. Here the superficial muscle sheets are innervated by six motor neurones, which generate junctional potentials of different amplitudes and temporal properties in the muscle fibres. The motor neurones show regional trends in their distribution within the muscle sheet and in the proportion of fibres they innervate, so that the overlapping sets formed by different combinations probably have considerable functional specificity. As Kennedy and his colleagues point out, this presents a considerable departure from our concept of the motor neurone pool, and we might add to this the "motor unit" (Kennedy et al., 1966).

It is not uncommon for arthropod muscles to contain a mixture of several fibre types which extend their functional capacity, for example their ability to develop tension quickly and maintain it. Regional specialisation within the muscle is less common but has been recorded in a number of instances (Hoyle, 1968; Hoyle & Burrows, 1973). In the basi-ischiopodite levator muscle of the 5th swimming leg of the crab, Portunus, specialised regions are further distinguished by having differently coloured fibres. Each portion of the muscle is independently served by overlapping sets of motor axons and is without an inhibitory supply. The mechanical and electrical properties range from a fast twitch and propagating spikes in the white fibres to slow tonic contractions and electrical inexcitability in the deep pink muscle. The light pink fibres are intermediate in their properties.

An intensive study of the innervation and fibre distribution of a specialised muscle may not lead to findings which represent the general condition, but the locust extensor tibiae (which in the hindleg is used for jumping) shows considerable regional distribution of fibre types (Hoyle, 1978). Features such as the dense proximal and distal innervation by the slow extensor (SETi) motor neurone and the consistent location of certain synaptic combinations of the four motor neurones are paralleled in closely related genera, suggesting the genetic regulation and strong evolutionary significance of these associations.

The operational limits of arthropod muscles are defined by the jointed exoskeleton, but such constraints do not apply to the annelids and molluscs where the muscle fibres are capable of operating over a great range of lengths. The longitudinal muscles of oligochaetes and polychaetes are obliquely striated and appear to be innervated by fast and slow axons which form distributed endings along the fibres. Support for fast and slow systems is also derived from the presence of giant-fibre mediated escape responses. Leeches are reported to have muscles of the slow type, the swimming muscles being innervated by 11-14 pairs of excitatory motor neurones in each segment and 3 pairs of inhibitors. The motor neurones produce graded contractions in strip-like fields of longitudinal muscle of varying size which slightly overlap.

There is a broad range of complexity in the motor

innervation of molluscan muscles. Fast and slow systems are found in the mantle of cephalopods and in the adductor muscles of the scallop, Pecten, associated with escape responses. Among the gastropods favoured by neurophysiologists, unstriated smooth muscles are most common, obliquely striated fibres being reported from the buccal mass, which is one of the few structures to be provided with anatomically defined muscles. The entire posterior jugalis muscle of Helisoma is innervated by a single neurone (Heyer & Kater, 1973), but in Lymnaea, Planorbis and Tritonia, muscles of the buccal mass and radula typically receive bilateral innervation from pairs of motor neurones in the buccal ganglia. Large muscle sheets are regionally innervated by several neurones which show functional differences in their discharge patterns, while in Lymnaea and Aplysia motor neurones serving symmetrical or synergistic muscles are either electrically coupled or are driven by common interneurone inputs. Neurobiologists studying the generation of rhythmic patterned outputs in the motor neurones during feeding cycles have mostly paid scant attention to their detailed distribution to the buccal musculature.

In the opisthobranch mollusc Philine, we have studied the innervation and fine structure of the two pairs of retractor muscles to the buccal mass (Dorsett & Roberts, 1980). These each consist of 130 and 180 unstriated fibres, and are innervated by two motor neurones from the buccal ganglion of its own side, and by fibres of the bilateral neurone of the opposite side (Fig. 1). Although all fibres have not been sampled, each neurone appears to contact each muscle fibre. The muscle fibres are electrically inexcitable but respond to single EJPs with a small slow twitch. The electrical and mechanical response facilitates and summates to a prolonged discharge by the motor neurone. The muscle also receives quite separate innervation from a paired neurone in the cerebral ganglia, which is not active during the feeding cycle but responds to tactile stimulation of the head.

This brief survey may serve to remind us that, in our dissection of the neural mechanisms associated with a particular behaviour pattern, it is ultimately necessary to return to a detailed study of motor neurone activity. Small differences in motor neurone output or timing, and co-activation of different neurone sets may signal some significant departure from the normal sequence, due entirely to the inherent properties of "the final motor pathway".

Patterns of motor output

It is now widely recognised, both in vertebrates and invertebrates, that innate rhythmic behaviours, such as the various forms of locomotion, feeding and respiratory movements, are generated by neural networks within the central nervous system. Although the fundamental pattern of motor output is often deceptively simple and can be reduced to a basic alternation of sets of motor neurones driving antagonist muscle groups, the natural sequence of activation often involves asymmetric motor patterns with subtle phase differences between antagonists and synergists dependent on the cycle period. These relationships are not always observed in preparations where the central nervous system is completely isolated from sensory input.

In <u>Tritonia</u> the patterned motor activity characteristic of the stereotyped escape swimming behaviour can be elicited by a short electrical stimulus to the isolated brain (Dorsett et al., 1969, 1973; Getting, 1983). In other preparations, the

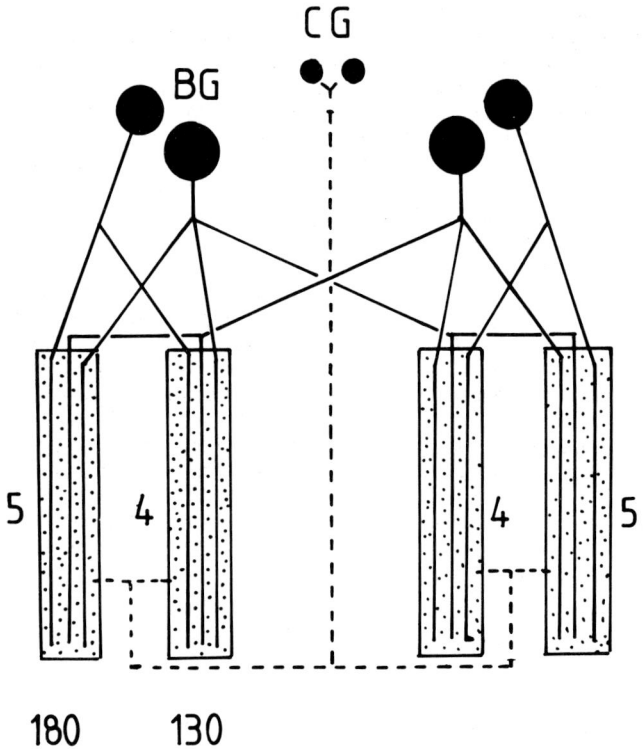

Fig. 1. Innervation of the retractor muscles (4 and 5) of the buccal mass of <u>Philine</u>. Two motor neurones in each buccal ganglion (BG) are distributed to most ipsilateral and contralateral fibres. Two cells in the cerebral ganglia (CG) also innervate the muscles, but are activated by different types of stimulus. The numbers indicate the number of fibres in each muscle.

centrally generated motor pattern following deafferentation differs considerably from the motor output of the intact system. In cockroaches, changes are observed in the discharge pattern of the elevator motor neurones (Pearson & Iles, 1970; see also Chapter 20 by Prof Pearson), while in the swimming leech the phase relationship between the dorsal and ventral muscle motor neurones is also fundamentally altered (Kristan et al., 1974).

Grillner and Zangger (1984) have described how the mesencephalic cat walking on a treadmill generates essentially similar electromyographic patterns in the hindlimb musculature before and after transection of the dorsal roots that supply that limb, although the pattern becomes more variable following the operation. However, although central pattern generating networks function in the absence of afferent information, peripheral inputs may modulate their activity at several levels. Movement-related feedback is capable of entraining the intrinsic spinal locomotory rhythm in the cat (Andersson & Grillner, 1983), the dogfish (Grillner & Wallén, 1982) and the lamprey (Grillner et al., 1983). In the cat some of this information is derived from receptors associated with the hip joint, while in the lamprey intraspinal mechanoreceptor elements (the edge cells) are ideally situated to perform this role.

Proprioceptive and cutaneous receptors also influence the efferent pattern through monosynaptic or local interneurone reflex pathways to the motor neurones. In Chapter 5, Prof Taylor and Dr Gottlieb point out that the apparent inflexibility of the monosynaptic excitatory connections made by spindle afferents is moderated by the relative amplitude of the post-synaptic response, which appears to be inversely related to the size of the motor unit. The interpolation of one or more interneurones in the reflex pathway allows much greater flexibility of response, which may be gated either by descending pathways from the brain, or by peripheral inputs acting on the interneurones. Convergence of sensory input from the periodontal and palatal receptors appears to be used to modulate the strong inhibitory action of the former on the masseter muscle when food is present in the mouth.

Gating mechanisms of this kind may also operate in invertebrates. Resistance reflexes generated in the limbs of arthropods are suppressed during centrally programmed walking sequences, but reappear accompanying unexpected movements such as a slip (Barnes, 1977). In many stereotyped escape behaviours the initial stimulus serves to trigger the response, after which sensory input plays no further part. In the crustacean tail-flip, the rapidly adapting sensory input is inhibited for the duration of the behaviour to preserve its excitability and prevent re-excitation of the escape system. In Chapter 6, Prof Camhi reveals how the sensory information channels remain operative throughout the escape response of the cockroach, but once the behaviour is initiated, the information is switched to another set of giant interneurones which control deviations from the original turn.

The organisation of pattern generating pre-motor systems discussed by Prof Davis in Chapter 2 reviews an earlier set of proposals (Davis, 1976) in the light of more recent work on the mollusc, Pleurobranchaea. The discovery of extensive feedback loops between centres which previously had been thought to have a specific function (compartmentalised function e.g. command, coordinating, oscillator networks), leads to a revised concept of distributed function where no cell or group of cells is responsible for a particular aspect of the output pattern. Thus a hierarchical pattern of organisation is replaced by one of reciprocity where many groups of cells interact through feedback loops in producing the motor output.

To many of us (see also Chapter 3 by Prof Grillner) this argument depends upon extreme interpretations of terms such as "hierarchies" and "command elements", and we would not agree that these necessarily imply the absence of feedback from levels closer to the motor output. The principle of consensus might equally involve redundancy, and the example chosen deals with a heterogeneous cell group whose operational characteristics are not precisely specified. In lower vertebrates spinal oscillating networks receive descending inhibitory and excitatory input from the brainstem, which is itself influenced by inputs from the cerebellum. The network controlling swimming in Tritonia is based on a sequential activation of two sets of interneurones and the motor neurones, while in the leech neurones 204 and 205 appear to be quite separate from the intersegmental oscillating network. All these appear to provide examples of hierarchies, if one does not seek to interpret the term in a rigid way. A similar problem arises with the use of the term "command" when seeking to apply it to neurones.

Sadly the opportunity to include a commentary on the chapter submitted by Prof Hoyle came shortly after the news of the untimely death of this charismatic personality. All those privileged to work with him will appreciate the loss this means to the community of neurobiologists. His submitted contribution to this book is a typically provocative and challenging discussion of the question of the initiation and selection of specific motor programmes by the organism. Many variable behaviours such as the several ambulatory patterns of orthopterans, songs of crickets, ingestion and regurgitation in molluscs, utilise common sets of muscles while operating in different modes. In formulating his "orchestration" hypothesis he emphasises the fundamental conservatism of neural circuitry while attempting to show how plasticity may be introduced under the influence of relevant sensory input, variation in the intensity of the central drive, or particularly by the action of neuromodulatory interneurones at critical points in the neural pathway. The prospect of medium to long term changes in excitability of specific junctions through the modulatory action of amines or peptides is now widely recognised, and both classes of compound may release repetitive coordinated sequences of neural activity associated with specific behaviours when applied topically or locally to the neuropile (e.g. Grillner, Chapter 3). We should remind ourselves that the initiation of one motor programme would normally be associated with the suppression of other, incompatible activities. Does this property of the nervous system also come within the province of neuromodulation?

At this stage we do not yet have complete understanding of even the most simple short chain circuits, when attempting to define their operation in absolute terms. This may never be realised. Each type of behaviour presents the nervous system with its own particular organisational requirements and problems, and there is no guarantee that the answers to each are in any way unique. I commend to you the comparative approach.

REFERENCES

Andersson, O. & Grillner, S. (1983). Peripheral control of the cat's step cycle. Acta physiol. scand., 118, 229-239.

Angaut-Petit, D. (1977). Mise en évidence de la complexité d'une "unité motrice" chez un Crustacé. J. Physiol., Paris, 73, 565-580.

Barnes, W.J.P. (1977). Proprioceptive influences on motor output during walking in the crayfish. J. Physiol., Paris, 73, 543-564.

Buchthal, F. & Schmalbruch, H. (1980). The motor unit of mammalian muscle. Physiol. Rev., 60, 90-142.

Davis, W.J. (1976). Organisational concepts in the central motor networks of invertebrates. In Neural Control of Locomotion. Eds. Herman, R.M., Grillner, S., Stein, P.S.G. & Stuart, D.G.. Plenum, New York, pp. 265-292.

Dorsett, D.A. & Roberts, J. (1980). A transverse tubular system and neuromuscular junctions in a molluscan unstriated muscle. Cell Tissue Res., 206, 251-260.

Dorsett, D.A., Willows, A.O.D. & Hoyle, G. (1969). Centrally generated nerve impulse sequences determining swimming behaviour in Tritonia. Nature, Lond., 224, 712-713.

Dorsett, D.A., Willows, A.O.D. & Hoyle, G. (1973). The neuronal basis of behaviour in Tritonia. IV. The central origin of a fixed action pattern demonstrated in the isolated brain. J. Neurobiol., 4, 287-300.

Getting, P.A. (1983). Neuronal control of swimming in Tritonia. Symp. Soc. exp. Biol., 37, 89-128.

Grillner, S. & Wallén, P. (1982). On peripheral control mechanisms acting on the central pattern generators for swimming in the dogfish. J. exp. Biol., 98, 1-22.

Grillner, S., Wallén, P., McClellan, A., Sigvardt, K., Williams, T. &

Feldman, J. (1983). The neural generation of locomotion in the lamprey: an incomplete account. Symp. Soc. exp. Biol., 37, 285-303.

Grillner, S. & Zangger, P. (1984). The effect of dorsal root transection on the efferent motor pattern in cat's hindlimb during locomotion. Acta physiol. scand., 120, 393-405.

Henneman, E. (1979). Functional organization of motoneuron pools: the size-principle. In Integration of the Nervous System. Eds. Asanuma, H. & Wilson, V.J.. Igaku-Shoin, Tokyo, pp. 13-24.

Heyer, C. & Kater, S.B. (1973). Neuromuscular systems in molluscs. Am. Zool., 13, 247-270.

Hoyle, G. (1968). Correlated physiological and ultrastructural studies on specialized muscles. Ia. Neuromuscular physiology of the levator of the eyestalk of Podophthalmus vigil (Weber). J. exp. Zool., 167, 471-486.

Hoyle, G. (1978). Distribution of nerve and muscle fibre types in locust jumping muscle. J. exp. Biol., 73, 205-233.

Hoyle, G. & Burrows, M. (1973). Correlated physiological and ultrastructural studies on specialized muscles. IIIa. Neuromuscular physiology of the power-stroke muscle of the swimming leg of Portunus sanguinolentus. J. exp. Zool., 185, 83-96.

Kennedy, D., Evoy, W.H. & Fields, H.L. (1966). The unit basis of some crustacean reflexes. Symp. Soc. exp. Biol., 20, 75-109.

Kristan, W.B., Stent, G. & Orr, C.A. (1974). Neuronal control of swimming in the medicinal leech. III. Impulse pattern of the motoneurons. J. comp. Physiol., 94, 155-176.

Pearson, K.G. & Iles, J.F. (1970). Discharge patterns of coxal levator and depressor motoneurones of the cockroach, Periplaneta americana. J. exp. Biol., 52, 139-165.

Roberts, B.L. & Williamson, R.M. (1983). Motor pattern formation in the dogfish spinal cord. Symp. Soc. exp. Biol., 37, 331-350.

Teräväinen, H. (1971). Anatomical and physiological studies on muscles of lamprey. J. Neurophysiol., 34, 954-973.

Chapter Two

CENTRAL FEEDBACK LOOPS AND SOME IMPLICATIONS FOR MOTOR CONTROL

W.J. Davis

The concept of feedback, as applied to motor control, usually suggests an image of sensory receptors on the periphery supplying information to the central nervous system (CNS) for the regulation of posture or movement. It is now widely recognised, however, that motor behaviour is typically generated by central networks of neurones that can operate without the aid of such "reafference". In the last decade it has also become apparent that neural feedback occurs entirely within the boundaries of the CNS, independently of sensory reafference. Such central feedback occurs because central networks of neurones are frequently organised not in linear chains or hierarchies, but rather in complex, interactive loops, with feedback from one point in the loop to another.

Based on this concept of central neural loops, a series of "principles" of central motor organisation was proposed a decade ago to account for an apparent misfit between then-current theory and data (Davis, 1976). My purpose here is to review these principles and their status in the light of investigations performed in the intervening years. I will also explore some implications of central neural loops for the control of motor output.

PRINCIPLES OF MOTOR CONTROL

The elements of any central motor system must perform four basic roles: 1) initiation, maintenance and termination of movement ("command"); 2) generation of the spatio-temporal impulse pattern that is translated by muscles into meaningful movements (pattern generation); 3) the direct control of muscle contractions that underlie movement (muscle innervation); and 4) coordination of the many individual central neurones that participate in any particular movement (coordination). Until a decade ago it was presumed that these roles were played by neurones aligned in a linear sequence, with information flowing unidirectionally through the network

Central feedback loops and some implications for motor control

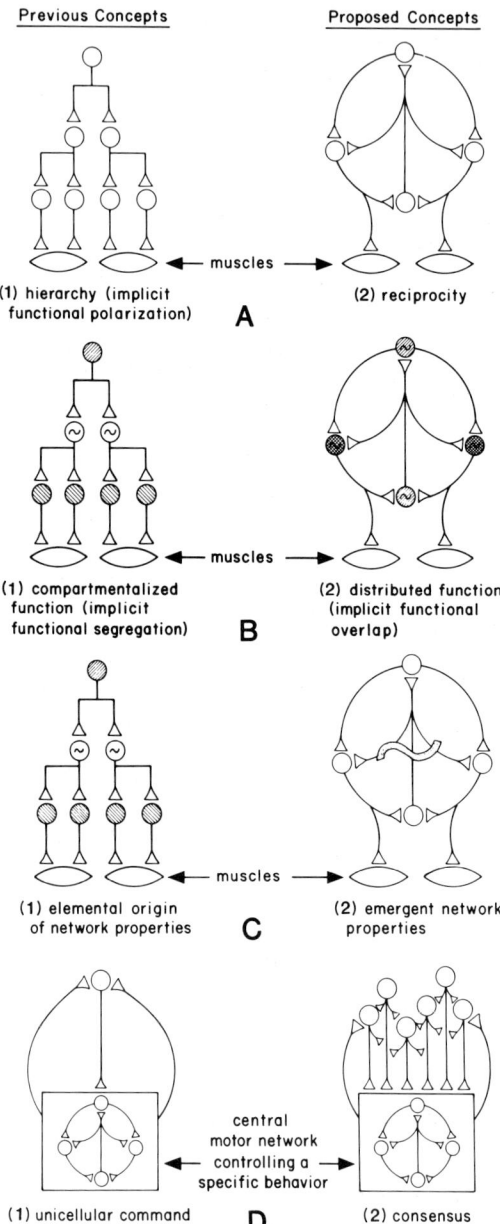

ORGANIZATIONAL CONCEPTS IN CENTRAL MOTOR NETWORKS

A
(1) hierarchy (implicit functional polarization)
← muscles →
(2) reciprocity

B
(1) compartmentalized function (implicit functional segregation)
← muscles →
(2) distributed function (implicit functional overlap)

C
(1) elemental origin of network properties
← muscles →
(2) emergent network properties

D
(1) unicellular command
← central motor network controlling a specific behavior →
(2) consensus

from sensory input to motor output (the concept of hierarchy) [Fig. 1A(1)].

As an implication of such hierarchical organisation, it was presumed further that the four functional roles listed above were played by separate and non-overlapping neurones (the concept of compartmentalised function) [Fig. 1B(1)]. The initiation of movement was thought to be accomplished by command neurones; the generation of motor patterns by oscillator neurones; the control of muscles by motor neurones; and communication between disparate elements of a functional network by coordinating neurones. By this scheme command neurones occupied the top position in motor hierarchies, with oscillator neurones next, followed by the motor neurones that innervate the respective muscles, and coordinating neurones somewhere in between.

Corollaries of these concepts included the idea that output parameters of motor networks, such as oscillation, were fully described by properties of individual neurones [the concept of elemental origin of network properties; Fig. 1C(1)], and that single neurones commanded motor behaviour [the concept of unicellular command; Fig. 1D(1)].

Among the first clues that this conceptual scheme might be inadequate in certain instances came from studies of neurones that participate in the feeding behaviour of the carnivorous gastropod mollusc Pleurobranchaea californica. The brain of this animal contains identified interneurones that send descending axons to the buccal ganglion to activate the feeding rhythm. These neurones, including the metacerebral and paracerebral neurones (Fig. 2A), are capable of initiating the feeding motor programme in intact animals and deafferented nervous systems (Gillette & Davis, 1977; Gillette et al., 1978; Croll, Kovac, Davis & Matera, 1985). The same neurones also oscillate during feeding, firing in cyclic bursts of action potentials during proboscis eversion (e.g. Fig. 2B,C). The occurrence of this oscillation in the same neurones that initiate the behaviour implies that the neurones playing the command role also receive synaptic feedback from other elements of the central motor network; i.e. that neurones previously designated as "command" neurones are incorporated into central neural loops. The concept of hierarchy is inadequate to explain these results, and hence the alternative concept of reciprocity was proposed [Fig. 1A(2)].

Fig. 1. Principles of motor organisation derived from the feeding motor system of Pleurobranchaea. Old concepts are illustrated in the left column, while revised concepts suggested are illustrated in the right column. Single neurones are indicated by a circle: axons by lines; and synapses by triangles; ⌀ indicates a command function; ⊘ a pattern-generating function; and ⊙ a motor function. (From Davis, 1976).

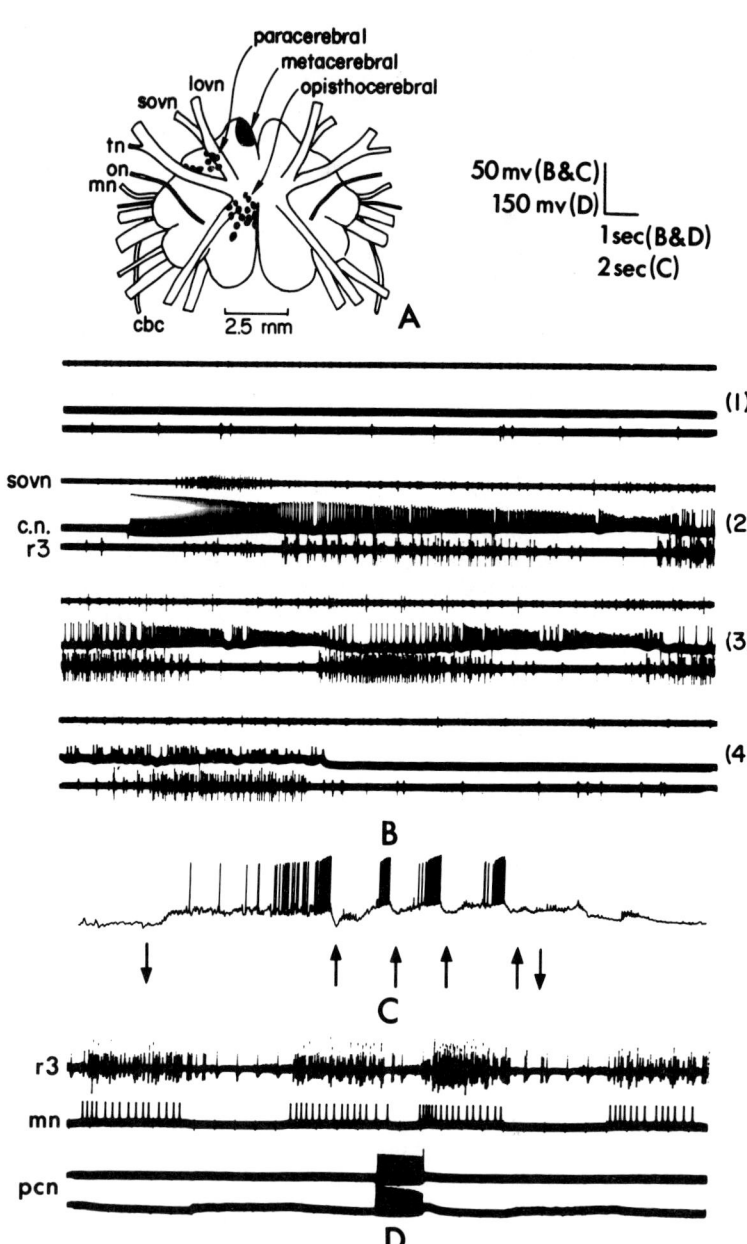

Central feedback loops and some implications for motor control

The finding that these brain neurones oscillate during the motor rhythm raised the possibility that they may participate in pattern generation. This hypothesis was tested by phase-reset experiments (Gillette et al., 1982). When bursts of activity at physiological spike frequencies are interpolated into an ongoing feeding rhythm, the phase of the rhythm is advanced (Fig. 2D) or retarded, depending upon the phase position of the interpolated burst. This result implies that the stimulated neurones either are, or have direct access to, rhythm-generating neurones. This observation, combined with the fact that these neurones normally fire in cyclic bursts, implies that they can exercise a direct effect on the timing of the motor rhythm; i.e. they participate in pattern generation.

In addition to command and pattern-generating roles, the metacerebral and paracerebral neurones send descending axons to the buccal ganglion, where they synapse with a variety of neurones that participate in feeding behaviour (Gillette et al., 1982; Kovac et al., 1982). These brain neurones thus carry timing information about the feeding rhythm between different central ganglia, and therefore may play a coordinating role. The metacerebral giant cell also sends numerous axon branches to the muscles of the buccal mass, implying a potential motor role (Gillette & Davis, 1977). This neurone may therefore play all four of the previously identified motor roles - command, pattern generation, motor

---

Fig. 2. Properties of paracerebral neurones (PCNs) in the brain of Pleurobranchaea.
A. Soma positions of all brain neurones that stain following back injection of a cerebrobuccal connective (cbc) with cobaltous chloride. Abbreviations: lovn, large oral veil nerve; sovn, small oral veil nerve; tn, tentacle nerve; on, optic nerve; mn, mouth nerve.
B. Sufficiency of a single PCN for inducing "fictive" feeding in an isolated CNS preparation. The four traces in each record [(1)-(4)] are continuous. Tonic current injection into the soma of a PCN (c.n.) starts in the second trace, and results immediately in cyclic motor output from the third buccal root (r3), which innervates radula retractor muscles. Activity in sovn usually occurs during proboscis eversion.
C. Appropriateness of a single PCN to feeding illustrated in a whole animal preparation. The trace is an intracellular recording from the soma of a single PCN during feeding behaviour. Food is applied at the downward arrow, and the PCN is immediately depolarised and begins to burst. Upward arrows signify visible bites.
D. Pattern-generating role of the PCNs, illustrated by a phase-reset experiment. Two PCNs are briefly depolarised during feeding motor output recorded from r3 and a feeding motor neuron (mn), re-setting the motor rhythm.
(A-C from Gillette et al., 1978; D from Gillette et al., 1982).

and coordinating. The other brain neurones (paracerebral neurones) play at least two motor roles (command and pattern generation) and perhaps a third (coordination). The rigid separation of functional roles into discrete neurones, previously presumed to characterise motor systems, is clearly absent in the case of these neurones. Accordingly the concept of <u>distributed</u> <u>function</u> with implicit functional overlap was proposed as an alternative to compartmentalised function [Fig. 1B(2)].

With the replacement of the concept of compartmentalised function, two of its corollaries could also be modified. First, compartmentalised function implied that the properties of a neural network, such as oscillation, might directly and fully reflect properties of individual elements within the network. An alternative that was more consistent with distributed function was that network properties emerge as a consequence of interactions between the individual elements that comprise the network. A network might oscillate, for example, even though no single neurone within it could independently oscillate. Thus the principle of <u>emergent</u> <u>network</u> <u>properties</u> was proposed in place of elemental origin of network properties [Fig. 1C(2)]. Recent evidence supports this hypothesis for <u>Pleurobranchaea</u>'s feeding system, as summarised below.

Similarly, the old concepts of compartmentalised function and elemental origin both implied that single neurones might play the role of initiating behaviour. But the new principles of distributed function and emergent properties invited an alternative interpretation, in which the function of initiating behaviour is relegated not to single neurones but to populations of neurones acting in concert. Thus the principle of <u>consensus</u> was proposed as a modification of unicellular command [Fig. 1D(2)]. According to this principle, the task of initiating behaviour is assigned not to individual neurones, but falls instead to populations of neurones necessarily acting in concert. Evidence for this principle in <u>Pleurobranchae</u>'s feeding system is summarised below.

RECENT DATA FROM PLEUROBRANCHAEA'S FEEDING SYSTEM

Since the formulation of the above four principles of motor control, continued investigations of <u>Pleurobranchaea</u>'s feeding motor system have shed additional light on how these principles are manifested. In particular, the "command" system for feeding behaviour has been studied in considerable detail (Gillette et al., 1978,1982; Kovac et al., 1982; Kovac et al., 1983a,b; Fig. 3), furnishing new data relevant to each of the principles. In this section I will briefly review the more recent data and discuss how they relate to the previously proposed principles.

The principle of reciprocity

Recent work has elaborated the central neural loops in Pleurobranchaea's feeding system at the level of identifiable neurones within the paracerebral brain neurones (Fig. 3), and between these neurones and coordinating neurones in the buccal ganglion (Fig. 4). With respect to the former, many individual neurones of the paracerebral population are reciprocally connected. For example, two sub-populations of the paracerebral neurones, the "standard" phasic paracerebral neurones (PC_ps) and the type II electrotonic neurones (ET_{II}s), are coupled by non-rectifying (reciprocal) electrical synapses (Fig. 3A). All members of the phasic paracerebral population excite a small network of oscillatory neurones, the "cyclic inhibitory neurones" (CIN) which in turn reciprocally inhibit the paracerebral neurones (Gillette et al., 1978, 1982; Fig.

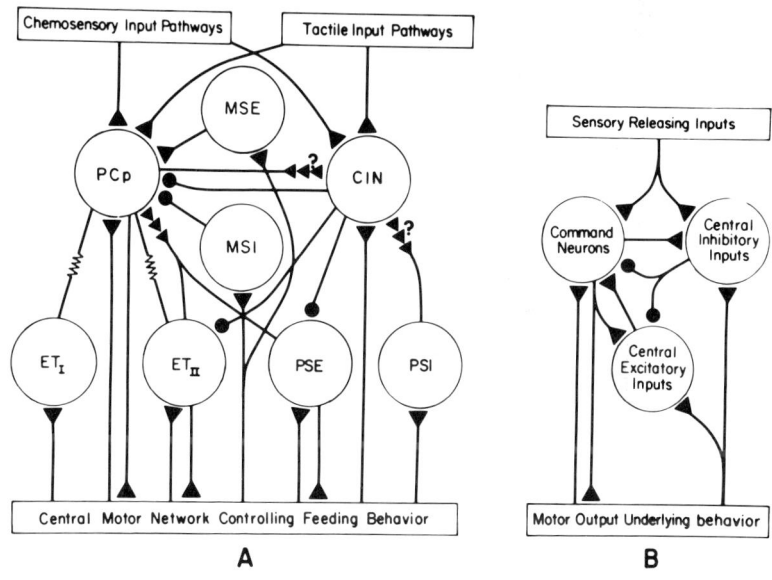

Fig. 3. Condensed summary diagrams of the synaptic organisation of Pleurobranchaea's feeding "command" system. A shows details while B shows the underlying conceptual schemata, namely excitation of feeding by command neurones, and also excitation by the command neurones of their excitatory and inhibitory inputs. Note extensive reciprocity within this central network. Abbreviations: PC_p, "standard" phasic paracerebral neurones (tonic paracerebral neurones are not shown); MSE, monosynaptic excitors; CIN, cyclic inhibitory network; MSI, monosynaptic inhibitors; ET_I, class I electrotonically coupled neurone; ET_{II}, class II electrotonically coupled neurones; PSE, polysynaptic excitors; PSI, polysynaptic inhibitors. (From Kovac et al., 1983b).

3B). These recurrent connections contribute to the rhythmic bursting of the command interneurones during feeding behaviour. Modifications of this recurrent pathway appear partially to underlie associative learning in this preparation, as discussed in more detail below.

Reciprocal connections also exist between paracerebral neurones and coordinating neurones in the buccal ganglion, namely the previously identified "corollary discharge" (CD) neurones (Davis et al., 1973; Davis et al., 1974; Gillette et al., 1978). Recently we have analysed one such interaction, between an identified CD neurone and members of the paracerebral population (Davis et al., 1984; Kovac, Matera & Davis, in preparation). Each polysynaptic excitor paracerebral neurone (PSE) in the brain forms a reciprocal, monosynaptic chemical synapse with an identified contralateral corollary

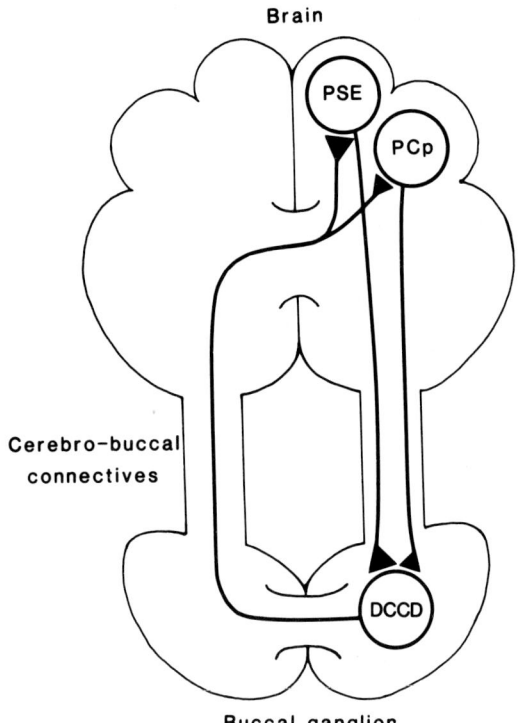

Fig. 4. Schematic diagram illustrating demonstrated reciprocal monosynaptic connections between feeding command interneurones in the brain (a PSE and a PC_p) and an identified interneurone, the "deep contralateral corollary discharge" (DCCD) neurone in the buccal ganglion. (From Kovac, Matera & Davis, in preparation).

discharge neurone in the buccal ganglion, to form a long neural loop between the brain neurones and the coordinating neurone of the buccal ganglion (Fig. 4). This loop is presumably responsible at least in part for coordinating the independent feeding oscillators demonstrated for the brain and buccal ganglion (Davis et al., 1984). This same loop also figures prominently in the cellular modifications that are induced by associative training, as will also be discussed below.

The principle of distributed function

As discussed above for both the metacerebral and paracerebral neurones, a number of functional roles is distributed across the same population of identified neurones. Such distribution of function does not necessarily imply lack of functional specialisation, however. Indeed, the eight paracerebral neurones on each side of the brain are divided into four classes each consisting of two neurones, with each individual neurone displaying a unique morphology and distinct physiological properties. Two paracerebral neurones are "tonic", meaning that they fire in continuous trains rather than in bursts, and have a weak effect in terms of their ability to elicit the feeding rhythm (Gillette et al., 1982; Kovac et al., 1982). The six remaining "phasic" paracerebral neurones include two "standard" $PC_p s$, which have an intermediate potency for eliciting the feeding rhythm; two neurones that are coupled with the standard $PC_p s$ by non-rectifying electrical synapses (type II electrotonic neurones, or $ET_{II}s$) and have a weak effect on the feeding rhythm; and two neurones that polysynaptically excite the standard $PC_p s$ and have a powerful effect on the feeding rhythm (the PSEs) (Kovac et al., 1983a). In addition a number of identified inhibitory interneurones are incorporated into the feeding command system (Kovac et al., 1983b; Fig. 3).

The principle of emergent properties

The phase-reset experiments described above indicate that the PCNs are pattern-generating neurones; and yet the pattern can continue even when one or two PCNs are silenced by hyperpolarisation (Gillette et al., 1982). The generation of the oscillatory pattern may therefore emerge from synaptic interactions among a large population of neurones, some of which may none the less possess intrinsic oscillatory properties (Siegler et al., 1974; Gillette & Davis, 1977).

The principle of consensus

Hyperpolarisation of paracerebral neurones has shown also that no single paracerebral neurone is necessary for the initiation of feeding behaviour. Moreover, the effects of paracerebral stimulation summate algebraically to regulate the feeding rhythm (Gillette et al., 1982). In other words, the feeding

Central feedback loops and some implications for motor control

system appears to operate not by unicellular "command", but rather by consensus. The discovery of consensus in the feeding command system of Pleurobranchaea suggests that the initiation of movement is in this case a network property, rather than a function of individual neurones.

STUDIES ON OTHER INVERTEBRATE MOTOR SYSTEMS

The utility of any principle depends in part on its generality. Although the data are far from complete, a number of studies on several rhythmic invertebrate motor systems suggests that the principles of motor organisation proposed previously and reviewed above may have more general applicability (see also Davis & Kovac, 1981).

For example, the best-understood "simple" system, the lobster stomatogastric ganglion, exemplifies several of the above principles (Selverston & Miller, 1980; Eisen & Marder, 1982; Miller & Selverston, 1982a,b). In this system the P neurones, located in the commissural ganglion, both excite the pyloric rhythm and also receive powerful synaptic feedback from a pyloric neurone, the AB interneurone, with the result that the P neurone fires cyclically in phase with the motor rhythm. In the same motor system, neurones that innervate muscles, the PD neurones, also have oscillatory characteristics; the oscillation can also result from emergent properties of the entire network; and the motor output is commanded not by single neurones but rather by a population of neurones (Dando & Selverston, 1978). Therefore, all four of the principles of motor control seen in Pleurobranchaea's feeding system are manifest also in the lobster stomatogastric ganglion.

Similarly, in the swim system of the medicinal leech, a neurone termed cell 204 (Fig. 5A) not only commands cyclic swimming motor output (Fig. 5B), but also oscillates in phase with the swimming motor output (Fig. 5C), indicating that it is incorporated into a loop that includes elements of the motor network (Weeks & Kristan, 1978). Another neurone in the swim system that also exhibits command properties, namely cell 205, also contributes to the generation of the swim pattern, as shown by phase-reset experiments (Weeks, 1980).

The swim system of the gastropod mollusc Tritonia contains a neurone, termed C2 (Fig. 6A), with properties similar to cell 205 of the leech swim system. This neurone is capable of initiating cyclic swim activity when it is depolarised (Fig. 6B; Getting, 1977). The same neurone fires in cyclic bursts that are phase locked with the normal swim rhythm (Fig. 6C), and contributes to pattern generation in the swim system, as demonstrated by phase reset experiments (Fig. 6D; Getting et al., 1980; Lennard et al., 1980).

Recent experiments on locust walking (Kien, 1983; Kien &

Central feedback loops and some implications for motor control

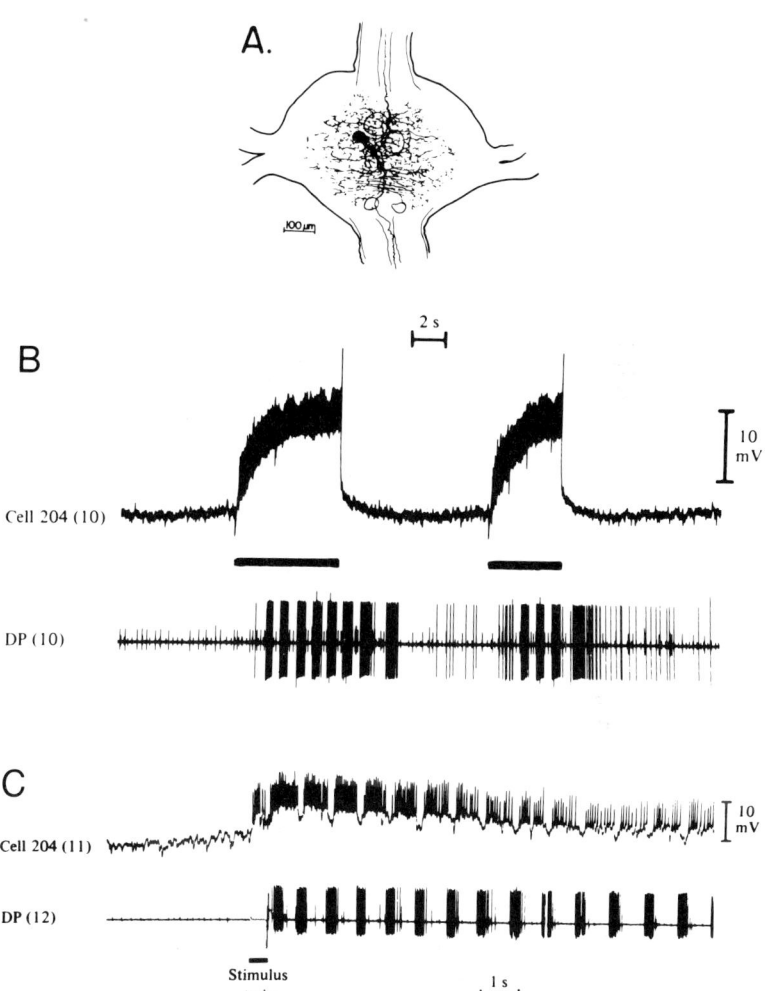

Fig. 5. Properties of cell 204 in the leech. A. The morphology of cell 204 as revealed by injection of horseradish peroxidase. B. "Command" effects of cell 204 on the motor rhythm recorded from an efferent nerve [DP(10)]. Cell 204 is depolarised tonically in two episodes, causing cyclic motor output in both cases. C. Activity of cell 204 during a "natural" swim episode induced by nerve stimulation. (From Weeks & Kristan, 1978).

Central feedback loops and some implications for motor control

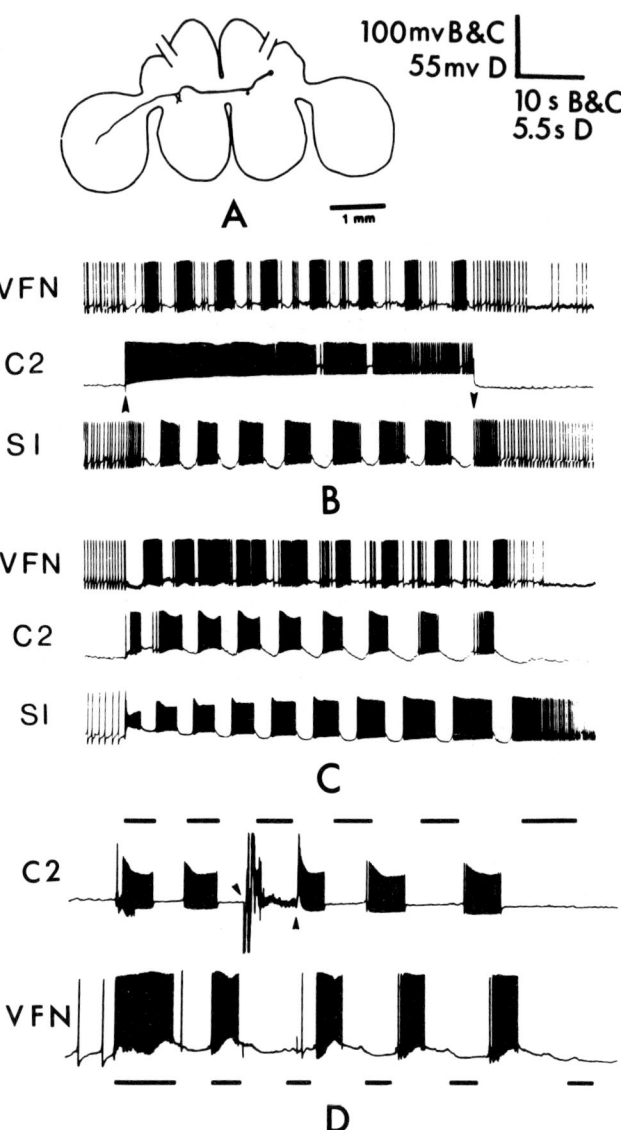

Fig. 6. Properties of C2 neurone in the brain of the mollusc Tritonia. A. Gross morphology of cell C2 as revealed by Lucifer Yellow injection. B. "Command" effect of cell C2 on the swim rhythm. Depolarisation of C2 occurred between arrows. C. Activity in C2 during a "natural" swim induced by nerve stimulation. D. Re-set of the swim rhythm by brief

Central feedback loops and some implications for motor control

Williams, 1983) suggest that the principle of command by consensus also operates here. Localised extracellular stimulation of anterior connectives has revealed 73 loci capable of reproducibly eliciting one form or another of walking. On this basis Kien has suggested that walking is normally initiated by many descending neurones acting in consensus and sending their "recommendations" for a specific form of walking to pattern generators in lower ganglia.

As these findings from other motor systems illustrate, the principles of motor organisation observed in Pleurobranchaea's feeding motor system appear to be general ones. Certain behavioural requirements may be better suited to reciprocal arrangement of the corresponding motor elements than others, however. In particular, central neural loops may characterise cyclic but not tonic motor systems. Especially in the case of rapid, non-rhythmic movements such as escape responses or prey capture, hierarchical central organisation might impart beneficial response speed. Even in the case of rhythmic behaviours, central loops may be most advantageous for relatively slow cycle times (e.g. <1Hz), where the resultant delays imposed by the loops would not conflict with the need for rapid repetition frequencies. For analogous or related reasons, central loops may be favoured in the case of variable-frequency rhythmic movements, behaviours requiring substantial modulation from sensory feedback, and especially plastic behaviours; i.e. ones that are readily modifiable by experience (Table 1). Whether the same principles apply in tonic or postural motor systems, and whether vertebrate motor systems are organised similarly, remain to be critically examined.

IMPLICATIONS OF NEURAL LOOPS

The "command" neurone concept

In terms of functional implications, the most important principle to emerge from our studies of Pleurobranchaea's feeding system is that of reciprocity. In contrast to old hierarchical views of the organisation of motor systems, in which the flow of information through the network was polarised (unidirectional), the organisation of central motor networks into loops permits reciprocal (bidirectional or multidirectional) information flow among the elements of the network. Among other implications, this organisational feature makes motor networks more challenging to study and comprehend,

---

hyperpolarisation of C2 (arrows in middle of record). VFN, ventral flexion neurones; S1, swim interneurone. (A and D from Getting et al., 1980; B and C from Getting, 1977).

since every element is in principle interactively coupled with every other one, as in an interdependent ecosystem.

The finding of reciprocity in motor systems has special implications for the definition of a command neuron. Three criteria have been proposed for establishing the command function of an individual neurone (Davis, 1977; Kupfermann & Weiss, 1978; Croll, Kovac, Davis & Matera, 1985): <u>sufficiency, appropriateness</u> and <u>necessity</u>. Sufficiency is demonstrated by stimulating a putative command interneurone and showing that it releases a behaviour; appropriateness is demonstrated by recording from a putative command interneurone and showing that it fires as expected during the performance of the behaviour; and necessity is demonstrated by showing that silencing the putative command neurone eliminates or at least alters the behaviour or its corresponding motor programme.

Given the reciprocal organisation of motor systems, however, the sufficiency criterion becomes ambiguous. Stimulating a neurone may evoke a behaviour not because the neurone normally plays a command role, but only because it is connected with a neurone that does. The appropriateness criterion is likewise rendered ambiguous, since a neurone may fire in the appropriate pattern only because it is reciprocally connected to other elements of the motor system. Finally, in any system where several neurones share the role of initiating behaviour (consensus), no single neurone is likely to be necessary to this role. Therefore, the necessity criterion is likewise of limited use.

In view of the principle of reciprocity and the related principles, new criteria are needed to establish the

Table 1. Conditions postulated to favour (left column) and not favour (right column) the organisation of central motor systems into loops.

POSTULATED CONDITIONS FAVOURING CENTRAL LOOPS	POSTULATED CONDITIONS NOT FAVOURING CENTRAL LOOPS
1. Rhythmicity in motor output	1. Tonicity in motor output
2. Slow cycle periods (<1 Hz)	2. Fast cycle periods (>1 Hz)
3. Variable frequency movements	3. Fixed frequency movements
4. High degree of sensory modulation	4. Low degree of sensory modulation
5. High degree of plasticity	5. Low degree of plasticity

functional roles of central elements. The concept of command is still valid, in as much as certain central neurones must play the initiating role, but the way we identify such neurones requires modification. Two new criteria suggest themselves. First, any central neurone playing a causal role in initiating behaviour must be the first to fire in association with that behaviour (temporal primacy). Second, any such neurone must preferentially receive input from the sensory systems that normally elicit the behaviour (privileged access). These criteria, however, could apply equally well to the pattern-generating function. Indeed, the command and pattern-generating roles could be one in the same. New criteria are required to distinguish these roles as well.

The hierarchical organisation of behaviour

The principle of reciprocity has implications beyond the definition of a command element. According to the old view, for example, motor behaviour originated in central nervous "centres" and was then imposed on neurones belonging to lower levels and finally on motor neurones. This hierarchical view of motor organisation has been resurrected in the case of Pleurobranchaea's feeding motor system by Cohan and Mpitsos (1983), who proposed that the feeding pattern originates from a single "master" oscillator in the buccal ganglion. We have found, however, that independent neural oscillators for feeding exist in the brain, in addition to those demonstrated previously in the buccal ganglion (Davis et al., 1984). Coordinated feeding motor output is accomplished by the coupling of these independent oscillators, which in turn depends upon reciprocal connections between them. An analogous decentralised organisation characterises several other motor systems in both invertebrates and vertebrates, and appears to be a general organisational rule for motor systems (for a review, see Davis et al., 1984).

Redundancy and specialisation

The wide distribution of the command and pattern generating functions among a population of central neurones implies considerable functional overlap or redundancy. A likely effect of such redundancy in command systems is increased gradation of the control of the system's output. The purpose of the specialisation within Pleurobranchaea's command population, however, is unclear at present. It has been shown that all elements of the command system elicit ingestion rather than egestion or other buccal motor patterns (Croll, Davis & Kovac, 1985; Croll, Kovac & Davis, 1985; Croll, Kovac, Davis & Matera, 1985). Therefore the documented specialisation within the paracerebral population is apparently not related to the production of qualitatively different motor programmes. It is possible that each command element elicits a subtly different version of the ingestion motor programme having a discrete functional role or capacity, but as yet there is no evidence for such an arrangement.

Cellular mechanisms of behavioural plasticity
The organisation of motor systems into loops increases the number of synaptic loci at which the activity of the network could in principle be modified by experience. Experiments have now shown that multiple synaptic loci within the feeding system are in fact modified by associative training. Food avoidance conditioning causes facilitated inhibition and suppressed excitation at the level of the paracerebral neurones (Davis & Gillette, 1978; Davis et al., 1983). The increased inhibition results from modifications involving the cyclic inhibitory network (CIN) mentioned above (Davis et al., 1983; Kovac et al., 1985).

The decrease in synaptic excitation of the paracerebral neurones following training entails the pathway described above involving the PSE via the CD neurones. Prior to training, intracellular stimulation of a PSE causes a long-latency polysynaptic chemical excitation of the PC_p at a relatively fixed mean spike threshold frequency of about 18Hz (Fig. 7A). After training (but not control procedures), the spike threshold frequency for the same polysynaptic response nearly doubles, to a mean of about 29Hz (Fig. 7B). In contrast, the mean spike threshold frequency for the PSE-to-PC_p response in control animals (unpaired food and shock) is not significantly different from trained animals (Fig. 7C). Similarly, the mean threshold is unchanged for animals that have been food satiated, thereby reducing their feeding motivation by a non-associative mechanism (Fig. 7D). Therefore, somewhere in the identified neural loop from the PSE to the PC_p, learning is manifested as a reduction in the functional impact of action potentials, resulting in reduced excitation of the command interneurones following training. The biophysical and molecular bases of these changes is currently under investigation.

As a consequence of these findings, food avoidance learning in Pleurobranchaea can now be interpreted in terms of modifications of pre-existing neural loops (Fig. 3) within the corresponding motor system (Fig. 8). Specifically, associative training alters particular pathways within these motor loops, such that the flow of information is re-routed from mainly excitatory to mainly inhibitory pathways. Each synapse in the loop represents a localised target at which the transfer function or gain around the entire network can be asserted. The data briefly summarised here thus suggest that many of the synaptic loci available because of the loop organisation of Pleurobranchaea's feeding network are in fact utilised to alter behaviour on the basis of associative training, resulting in a widespread distribution of the physiological effects of learning throughout the affected motor network.

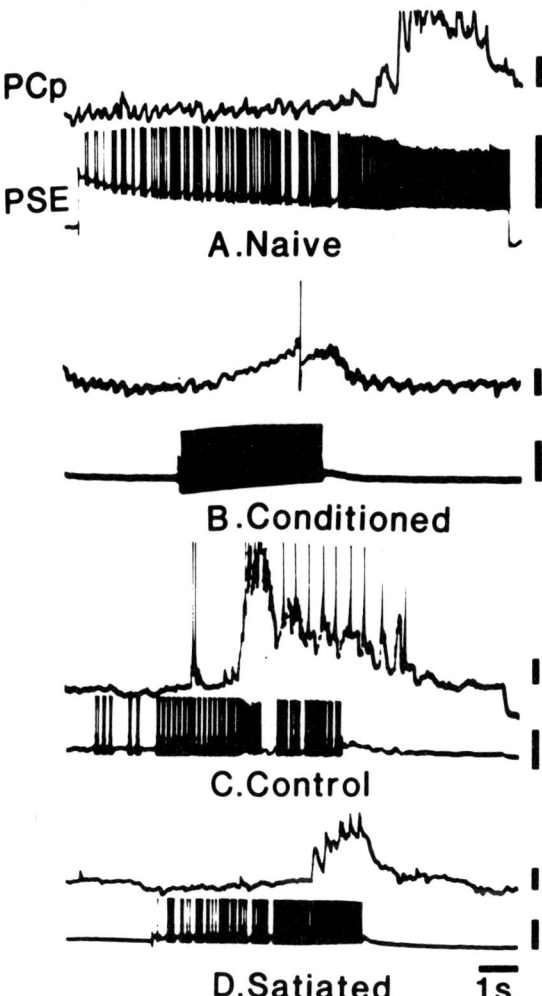

Fig. 7. Effect of associative training on the efficacy of the PSE-CD-PC$_p$ pathway. In each set of records a single PSE has been stimulated intracellularly at the spike threshold frequency necessary to elicit the polysynaptic response from the PC$_p$, in an isolated brain preparation taken from the following categories of specimens. A, naive; B, conditioned; C, control (explicitly unpaired CS/US); D, food satiated. (From Kovac et al., 1985).

CONCLUSIONS

The principles discussed in this paper have developed in two phases: initial formulation (Davis, 1976), and a preliminary test of generality (Davis & Kovac, 1981; this paper). A third phase could well be to define the limits of generality of these principles, i.e. to test the speculations formalised in Table 1, and to develop their theoretical and practical implications. Do these principles apply to tonic as well as cyclic motor systems? With respect to the latter, which characteristics of rhythmic motor systems favour these principles and which do not? And what are the implications for the function of the nervous system? Some of these implications are identified earlier (Davis, 1976) and discussed further here. Now that central neural loops are known to occur more widely in motor systems, the time would seem ripe for a more formal and detailed consideration of their broader implications for the neural sciences and animal behaviour.

SUMMARY

Recent data from several invertebrate motor systems illustrate that the central neural apparatus subserving rhythmic movements is organised into reciprocal loops rather than

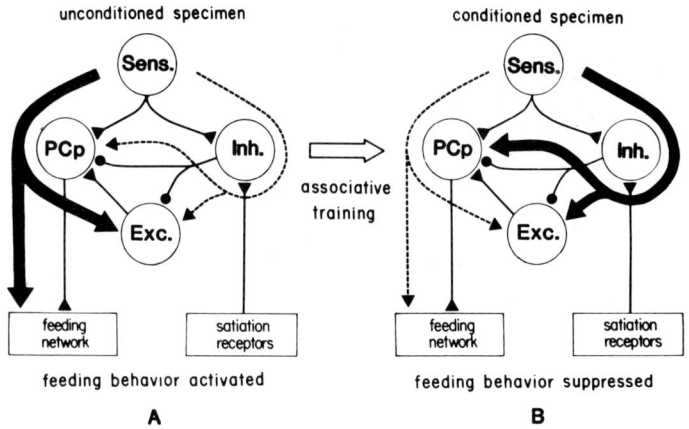

Fig. 8. Schematic summary of the neural changes underlying associative learning in the feeding neural network of <u>Pleurobranchaea</u>. Food avoidance learning entails a functional modification of identified neural loops, resulting in the re-routing of the flow of information about food (heavy arrows) from predominantly excitatory to inhibitory pathways. (From Davis et al., 1983).

linear hierarchies. Several "principles" of motor organisation follow from this finding, including the distribution of motor function over populations of neurones (with the implicit investment of several functions into single neurones), the origin of network properties from synaptic interactions within the network, and the initiation of behaviour by consensus rather than unicellular command. These principles carry several implications for motor control. In particular, existing criteria for the "command" neurone are outmoded by the documentation of central neural loops. The organisation of motor systems into loops also furnishes multiple synaptic loci at which learning can manifest itself, a possibility that is realised in the feeding motor system of the mollusc Pleurobranchaea.

Acknowledgements

Work from our laboratory has been supported by NIH Research Grants NS-09050 and MH-23254, NSF Grants BNS-8110235 and GZ 3120, and DoD Grant DAAG29-83-G-0071. This paper was written during the tenure of an Alexander von Humboldt Senior Scientist Award. I am grateful to the Humboldt Foundation for generous support and to my German hosts, Prof H. Markl (University of Konstanz) and Prof F. Huber (Max-Planck Institute, Seewiesen), for warm hospitality, stimulating environments and excellent facilities. I also thank Prof Huber for critically reading and discussing the manuscript.

REFERENCES

Cohan, C.S. & Mpitsos, G.J. (1983). The generation of rhythmic activity in a distributed motor system. J. exp. Biol., 102, 25-42.

Croll, R.P., Davis, W.J. & Kovac, M.P. (1985). Neural mechanisms of motor program switching in the mollusc Pleurobranchaea. I. Central motor programs underlying ingestion, egestion and the "neutral" rhythm(s). J. Neurosci., 5, 48-55.

Croll, R.P., Kovac, M.P. & Davis, W.J. (1985). Neural mechanisms of motor program switching in the mollusc Pleurobranchaea. II. Role of the ventral white cell, anterior ventral, and B3 buccal neurons. J. Neurosci., 5, 56-63.

Croll, R.P., Kovac, M.P., Davis, W.J. & Matera, E.M. (1985). Neural mechanisms of motor program switching in the mollusc Pleurobranchaea. III. Role of the paracerebral neurons and other identified brain neurons. J. Neurosci., 5, 64-71.

Dando, M.R. & Selverston, A.I. (1972). Command fibers from the supraoesophageal to the stomatogastric ganglion in Panulirus argus. J. comp. Physiol., 78, 138-175.

Davis, W.J. (1976). Organizational concepts in the central motor networks of invertebrates. In Neural Control of Locomotion. Eds. Herman, R.M., Grillner, S., Stein, P.S.G. & Stuart, D.G.. Plenum, New York, pp. 265-292.

Davis, W.J. (1977). The command neuron. In Identified Neurons and Behavior of Arthropods. Ed. Hoyle, G.. Plenum, New York, pp. 293-305.

Davis, W.J. & Gillette, R. (1978). Neural correlate of behavioral plasticity in command neurons of Pleurobranchaea. Science, N.Y., 199, 801-804.

Davis, W.J., Gillette, R., Kovac, M.P., Croll, R.P. & Matera, E.M. (1983). Organization of synaptic inputs to paracerebral feeding command interneurons of Pleurobranchaea californica. III. Modifications induced by experience. J. Neurophysiol., 49, 1557-1572.

Davis, W.J. & Kovac, M.P. (1981). The command neuron and the organization of movement. Trends Neurosci., 4, 73-76.

Davis, W.J., Kovac, M.P., Croll, R.P. & Matera, E.M. (1984). Brain oscillator(s) underlying rhythmic cerebral and buccal motor output in the mollusc, Pleurobranchaea californica. J. exp. Biol., 110, 1-15.

Davis, W.J., Mpitsos, G.J., Siegler, M.V.S., Pinneo, J.M. & Davis, K.B. (1974). Neuronal substrates of behavioral hierarchies and associative learning in Pleurobranchaea. Am. Zool., 14, 1037-1050.

Davis, W.J., Siegler, M.V.S. & Mpitsos, G.J. (1973). Distributed neuronal oscillators and efference copy in the feeding system of Pleurobranchaea. J. Neurophysiol., 36, 258-274.

Eisen, J.S. & Marder, E. (1982). Mechanisms underlying pattern generation in lobster stomatogastric ganglion as determined by selective inactivation of identified neurons. III. Synaptic connections of electrically coupled pyloric neurons. J. Neurophysiol., 48, 1392-1415.

Getting, P.A. (1977). Neuronal organization of escape swimming in Tritonia. J. comp. Physiol., 121, 325-342.

Getting, P.A., Lennard, P.R. & Hume, R.I. (1980). Central pattern generator mediating swimming in Tritonia. I. Identification and synaptic interactions. J. Neurophysiol., 44, 151-164.

Gillette, R. & Davis, W.J. (1977). The role of the metacerebral giant neuron in the feeding behavior of Pleurobranchaea. J. comp. Physiol., 116, 129-159.

Gillette, R., Kovac, M.P. & Davis, W.J. (1978). Command neurons in Pleurobranchaea receive synaptic feedback from the motor network they excite. Science, N.Y., 199, 798-801.

Gillette, R., Kovac, M.P. & Davis, W.J. (1982). Control of feeding motor output by paracerebral neurons in the brain of Pleurobranchaea californica. J. Neurophysiol., 47, 885-908.

Kien, J. (1983). The initiation and maintenance of walking in the locust: an alternative to the command concept. Proc. R. Soc. B, 219, 137-174.

Kien, J. & Williams, M. (1983). Morphology of neurons in locust brain and suboesophageal ganglion involved in initiation and maintenance of walking. Proc. R. Soc. B, 219, 175-192.

Kovac, M.P., Davis, W.J., Matera, E.M. & Gillette, R. (1982). Functional and structural correlates of cell size in paracerebral neurons of Pleurobranchaea californica. J. Neurophysiol., 47, 909-927.

Kovac, M.P., Davis, W.J., Matera, E.M. & Croll, R.P. (1983a). Organization of synaptic inputs to paracerebral feeding command interneurons of Pleurobranchaea californica. I. Excitatory inputs. J. Neurophysiol., 49, 1517-1538.

Kovac, M.P., Davis, W.J., Matera, E.M. & Croll, R.P. (1983b). Organization of synaptic inputs to paracerebral feeding command interneurons of

Pleurobranchaea californica. II. Inhibitory inputs. J. Neurophysiol., 49, 1539-1556.

Kovac, M.P., Davis, W.J., Matera, E.M., Morielli, A. & Croll, R.P. (1985). Learning: Neural analysis in the isolated brain of a previously trained mollusc, Pleurobranchaea californica. Brain Res., 331, 275-284.

Kupfermann, I. & Weiss, K.R. (1978). The command neuron concept. Behav. & Brain Sci., 1, 3-39.

Lennard, P.R., Getting, P.A. & Hume, R.I. (1980). Central pattern generator mediating swimming in Tritonia. II. Initiation, maintenance and termination. J. Neurophysiol., 44, 165-173.

Miller, J.P. & Selverston, A.I. (1982a). Mechanisms underlying pattern generation in lobster stomatogastric ganglion as determined by selective inactivation of identified neurons. II. Oscillatory properties of pyloric neurons. J. Neurophysiol., 48, 1378-1391.

Miller, J.P. & Selverston, A.I. (1982b). Mechanisms underlying pattern generation in lobster stomatogastric ganglion as determined by selective inactivation of identified neurons. IV. Network properties of pyloric system. J. Neurophysiol., 48, 1416-1432.

Selverston, A.I. & Miller, J.P. (1980). Mechanisms underlying pattern generation in lobster stomatogastric ganglion as determined by selective inactivation of identified neurons. I. Pyloric system. J. Neurophysiol., 44, 1102-1121.

Siegler, M.V.S., Mpitsos, G.J. & Davis, W.J. (1974). Motor organization and generation of rhythmic feeding output in the buccal ganglion of Pleurobranchaea. J. Neurophysiol., 37, 1173-1196.

Weeks, J. (1980). The roles of identified interneurons in initiating and generating the swimming motor pattern of leeches. Ph.D. Thesis, University of California at San Diego.

Weeks, J. & Kristan, W.B. (1978). Initiation, maintenance and modulation of swimming in the medicinal leech by the activity of a single neuron. J. exp. Biol., 77, 71-88.

Chapter Three

NEURAL CONTROL OF VERTEBRATE LOCOMOTION - CENTRAL MECHANISMS AND REFLEX INTERACTION WITH SPECIAL REFERENCE TO THE CAT

S. Grillner

How the nervous systems of vertebrates make their respective bodies swim or walk is the subject of this paper. Our knowledge has increased markedly over the last decades. The movements of a walking mammal in its natural habitat depend upon a precise adaptation of each step to the terrain and the overall goal of the animal. Each species generates its characteristic type of propulsive locomotor movements, here referred to as the basic locomotor synergy. This article deals with how this synergy is generated and, in particular, with the interaction between sensory and central elements of the control system in cat. The complex mechanisms underlying the precise movements necessary to place the foot on a predetermined spot in each step are only now starting to become unravelled (see Chapter 28 by Dr Forssberg) and will not be dealt with here.

If animals like the cat are made decorticate, they still exhibit a very complex behavioural repertoire, particularly if the decortication is performed early in life (Bjursten et al., 1976, see Grillner & Wallén, 1985). They display goal directed behaviour, seek food, eat, chew, locomote, and avoid obstacles in the environment while doing so. The situation is strikingly different in decerebrate cats in which a transection of the central nervous system has been made in the rostral mesencephalon or caudal diencephalon. Such decerebrate cats can initiate walking movements over ground spontaneously (Bard & Macht, 1958), but their ability to avoid obstacles is very reduced, and they would attempt to walk through a wall in their way rather than turn around as do intact or decorticate cats.

Shik, Orlovsky and Severin made the important discovery that stimulation of specific circumscribed brain areas (see Fig. 1) in decerebrate cats with decerebrate rigidity could induce well coordinated locomotion over ground or on a treadmill (see Shik & Orlovsky, 1976; Stein, 1978; Grillner, 1981). Increasing the strength of the stimulation changes the speed of locomotion from slow walk to trot and gallop; i.e. a very uncomplicated type of stimulus (e.g. 20Hz, 8µA, 1ms pulses) can induce a complex pattern of behaviour in which the four limbs are coordinated with each other and with the different parts of the spine. Moreover, the individual joints in each limb and the different muscle groups are coordinated with each other, each in its specific way. Brainstem locomotor areas have been found in the corresponding regions of the brain in other tetrapods (primates, rat, turtle) and in fish and lamprey (for references see Grillner & Wallén, 1985). These brainstem areas project at least partially via relays to the spinal cord (some of which are noradrenergic) by way of its ventral quadrant (Jordan, 1983, cf. Grillner, 1981). The mesencephalic locomotor region is at least partially comprised of neurones within the caudal pole of the nucleus cuneiforme, and neurones in this region receive an input from the different output nuclei of the basal ganglia, i.e. the entopeduncular (globus pallidus), the subthalamic and the pars reticulata of the substantia nigra (Garcia-Rill et al., 1983; see Grillner & Wallén, 1985). This is presumably significant in relation to the normal initiation of locomotion, which occurs in a seemingly normal way in animals which have their basal ganglia intact but lack cortex (see Grillner & Wallén, 1985).

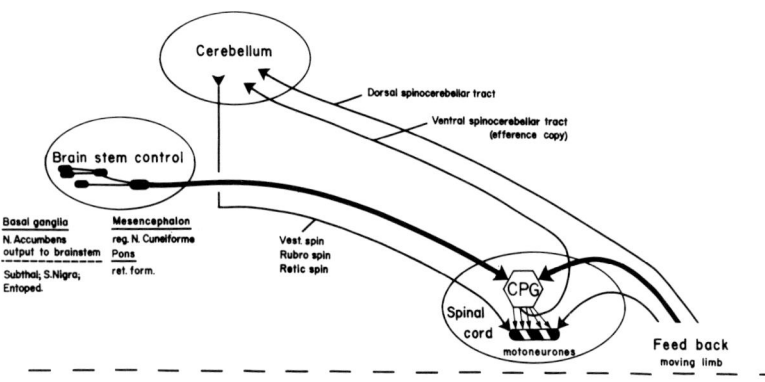

Fig. 1. Schematic diagram of the control system for locomotion in vertebrates, based mainly on data from mammals. Abbreviations: CPG, central pattern generator; Entoped., entopeduncular nucleus; N., Nucleus; ret. form., reticular formation; Retic spin, Reticulo-spinal tract; Rubro spin, Rubro-spinal tract; S. Nigra, Substantia Nigra; Subthal, Subthalamic nucleus; Vest spin, Vestibulo-spinal tract. See text for details.

Neural control of vertebrate locomotion

CEREBELLUM IN LOCOMOTION

The basic locomotor synergy can be generated after a decerebellation, but the quality of the movement has deteriorated somewhat. For instance, foot placement on the ground is not as smooth, interlimb coordination is less accurate and equilibrium control less perfect (see Shik & Orlovsky, 1976; Orlovsky & Shik, 1976). Orlovsky and his colleagues have in a series of outstanding papers shown that the cerebellum in each step cycle receives information via the different direct and indirect spinocerebellar pathways, not only about the execution of the actual ongoing movements, but also "efference copy" information concerning the commands issued from the central pattern generators (see below) to different muscle groups. These cerebellar signals are used to activate the different bulbospinal pathways subserving the cerebellum. Different fast vestibulo-, rubro- and reticulo-spinal pathways are phasically active in different parts of the step cycle and contribute to perfecting each step (Udo et al., 1980; Arshavsky et al., 1984). Analogous results have recently been reported in the dogfish (Paul & Roberts, 1984).

THE EFFECT OF DEPRIVATION OF SENSORY INPUT ON THE MOTOR PATTERN DURING LOCOMOTION

The view that vertebrate rhythmic movements were generated exclusively via some sort of reflex chain arrangement was rejected early in this century (e.g. Brown, 1911, see Grillner, 1981). The next level of hypothesis was that the neural centres provided a very simple motor output such as, for example, alternating activity in flexor and extensor motor neurones during locomotion; the reflex input arising in that limb during the movement would add to and substract from that central pattern to generate the complex motor output characteristic of the intact cat (Lundberg, 1969). That this hypothesis could also be rejected was shown by comparing the motor output during locomotion (electromyography, movement) before and after the dorsal roots supplying the limb had been transected (Grillner & Zangger, 1975, 1984). After such a dorsal root transection, the complex timing arrangements of the different muscle groups remained, thus demonstrating that the central network is able to produce such a pattern (Fig. 2). As is made clear below, it is important that essential features of this pattern are also left when decerebrate or decorticate animals have been rendered motionless by curarisation (Perret & Cabelguen, 1980) and that spinal curarised animals retain complex features of the motor pattern (Grillner & Zangger, 1979, cf. also O'Donovan et al., 1982). It is relevant to note here that the presumed ventral root afferents in the cat lumbar ventral root do not appear to enter the spinal cord as was assumed earlier (Risling & Hildebrand, 1982).

A complex locomotor output has also been demonstrated in birds, reptiles, amphibians, fish and lamprey after deprivation of sensory input (see Grillner & Wallén, 1985). In the lamprey, a detailed comparison between activity in intact and spinal swimming lampreys and the motor pattern produced by the isolated spinal cord in vitro has shown that the timing characteristics can remain unchanged (Wallén & Williams, 1984). As there are intraspinal mechanoreceptors in lampreys (Grillner et al., 1984), which provide feedback to the locomotor pattern generator (see Chapter 21 by Drs Wallén & Williams), a dorsal root transection would not be sufficient to determine which components of the motor pattern can be generated without sensory feedback; a motionless (e.g. curarised) preparation is required. Whether other vertebrates utilise intraspinal mechanoreceptors has yet to be investigated, but for the time being results in lower vertebrates which have been obtained in moving animals after a dorsal root transection (e.g. von Holst, 1935) cannot be viewed as conclusive in this respect.

Although the motor pattern produced after sensory deprivation of one type or the other can be surprisingly normal, it is much less stable and it can break down to

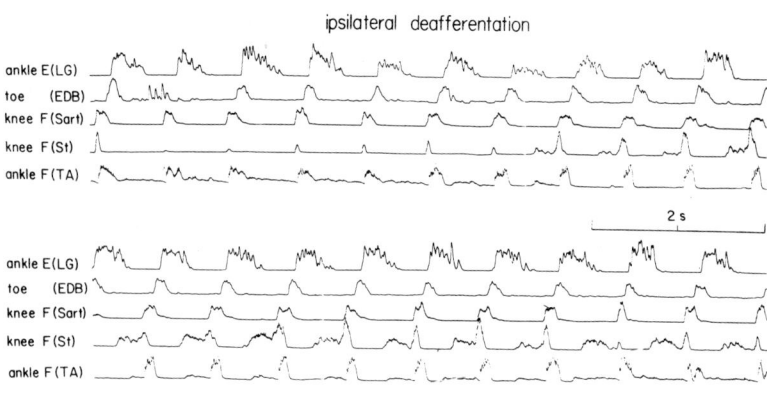

Fig. 2. The effect of dorsal root transection of all afferents supplying the hindlimb of a decerebrate cat. Rectified and filtered EMG pattern recorded after ipsilateral dorsal root transection. The lower set of traces is a direct continuation of the upper panel. Note the variation of semitendinosus (St) bursts throughout the recording. Abbreviations: E, extensor; F, flexor; LG, lateral gastrocnemius; EDB, extensor digitorum brevis; Sart, sartorius; TA, tibialis anterior. (From Grillner & Zangger, 1984).

Neural control of vertebrate locomotion

produce quite abnormal patterns for shorter or longer periods of time (Grillner & Zangger, 1975, 1984; Wallén, 1980). Under normal conditions, sensory information is of paramount importance to provide adjustments of each step cycle (see below). The motor pattern produced by the central nervous system, although detailed, should be viewed, at least in the case reviewed here, as providing a rough motor template of some sort of standard step cycle. Normal control depends on both centre and periphery, not one or the other (see below and Grillner, 1975, 1981; Grillner & Wallén, 1985).

LOCOMOTOR CAPACITY OF THE SPINAL CORD

After a spinal transection the ability of lamprey, dogfish, eel, turtle, chicken and cat to produce locomotor-like movements and motor patterns has been studied in quite some detail (see Grillner, 1981; Grillner & Wallén, 1985; Chapter 21 by Drs Wallén & Williams). In general the spinal preparations can be said to generate a replica of most aspects of the basic locomotor movements.

A low spinal cat can be made to perform walking movements on a treadmill which adapt to the treadmill speed within limits. The movement trajectories in hip, knee, ankle and foot joints of each limb resemble those of the intact cat (Fig. 3) at the corresponding speed, and as the treadmill speed changes the duration of the different components of the step cycle change in the same way as that of the intact cat (Grillner, 1973; Forssberg, Grillner & Halbertsma, 1980; Halbertsma, 1983). The electromyographic pattern of activity is also similar. As the speed of the treadmill increases, the alternating limb movement characteristic of walking and trotting can abruptly change to more simultaneous movements (Fig. 4) as in a gallop (Forssberg, Grillner, Halbertsma & Rossignol, 1980). Coordination with the spine can also be retained (Zomlefer et al., 1984). There are, on the other hand, important deficiencies; first, the level of activity cannot be controlled from the brain and thus the speed range is much smaller than under intact conditions; second, equilibrium control is deficient and the cat tends to fall over if no aid is given by the experimenter; third, the propulsive component is smaller in the spinal preparation; fourth, the motor pattern is more susceptible to fatigue and other factors which may induce a gradual or complete break down of the pattern; fifth, it has been our impression that a daily "training" on the treadmill is important for maintaining an adequate locomotor pattern in the spinal cat. Spinal cats can be made to walk even if the spinal cord transection is made before the cat has actually started to use its limbs for walking (Forssberg, Grillner & Halbertsma, 1980; Forssberg, Griller, Halbertsma & Rossignol, 1980; Smith et al., 1982).

Adult spinal cats can also be made to perform walking movements (Julien et al., 1982), directly after the transection if noradrenergic agonists are administered (Budakova, 1973; Forssberg & Grillner, 1973).

If an acute transection is made at a high spinal level, all four limbs can be made to perform coordinated movements as in walking. Hence there is clearly a spinal potential for coordination of hind- and forelimbs (Miller & van der Meche,

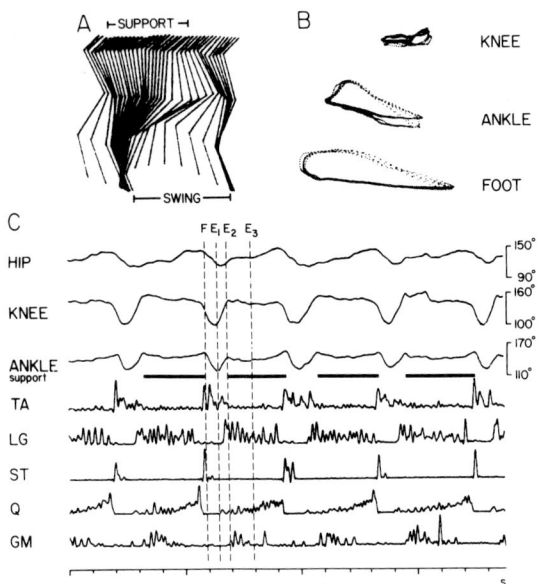

Fig. 3. Locomotion in chronic spinal cat. Movements and muscular activity during slow walk (0.20m s^{-1}). In A the forward movement of the right limb is presented as stick figures throughout the step cycle at 18ms intervals. Swing and support phases are indicated. In B the knee, ankle and foot trajectories are displayed at 6ms intervals during 6 consecutive step cycles. The positions are related to the position of the pelvis, which is fixed in the horizontal direction. In C the angular movements of the hip, knee and ankle are displayed together with the simultaneously recorded EMGs of the tibialis anterior (TA), lateral gastrocnemius (LG), semitendinosus (ST), Quadriceps (Q) and gluteus medius (GM). The EMGs are filtered and rectified. The onset of flexion (F) and the different extension phases (E_1, E_2 & E_3) are indicated in the angular movements of the ankle (Phillipson, 1905). The support phases are also indicated by schematic blocks. In the time scale at the bottom large lines represent seconds and small ones 200ms. (From Forssberg, Grillner & Halbertsma, 1980).

1976). A detailed study of movements and motor patterns has unfortunately not been carried out to allow an estimate of the quality of the movement pattern.

SENSORY CONTROL OF THE STEP CYCLE

The very fact that a low spinal cat can adapt to the speed of the treadmill is sufficient to conclude that sensory information can influence the spinal network producing the motor pattern. As described above, the central pattern generator can operate without sensory feedback, but when sensory information is available, the motor pattern becomes coordinated to the movements.

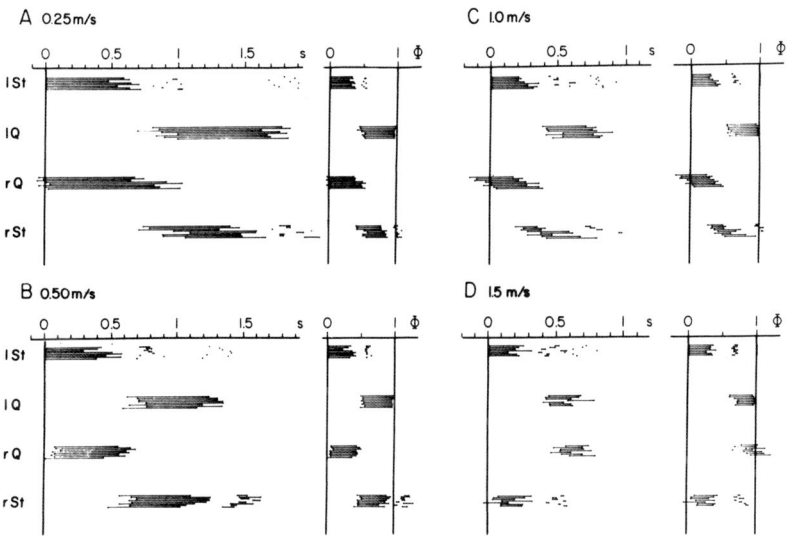

Fig. 4. Schematical representation of the muscular activity from a chronic spinal cat during treadmill locomotion at different speeds. The duration of the EMG bursts of left semitendinosus (lSt), left quadriceps (lQ), right semitendinosus (rSt) and right quadriceps (rQ) are measured from Mingograph recordings on a graphic tablet (digitiser) connected to a desk calculator (HP9830) and to a plotter. They are drawn as bars in real time on the left (note time scale at the top), and normalised to the cycle duration on the right ($\emptyset$ = phase). The bursts of the four muscles from several consecutive step cycles are plotted beneath each other. A cycle starts with the onset of activity of lSt. The end is marked with a single dot. A, B and C show alternating gaits at increased speeds. D shows a non-alternating pattern as occurs during gallop. (From Forssberg, Grillner & Halbertsma, 1980).

41

If one of the hindlimbs of a low spinal walking cat is stopped in the first half of the support phase, activity in all extensor muscles of the limb continues tonically until the limb is released, when it resumes walking in coordination with the contralateral limb (Fig. 5; Grillner & Rossignol, 1978). These effects are mainly due to the position of the most proximal joint (the hip). When the hip joint is slowly extended, the limb will suddenly initiate a flexion phase at approximately the joint angle at which this would occur in normal walking. All types of receptors influenced by hip position could possibly contribute to these effects.

In order to analyse these effects in greater detail, we utilised a more reduced preparation of an acute spinal and curarised cat in which fictive locomotion was induced pharmacologically. The motor pattern was recorded in peripheral nerve filaments, while the entire limb including the skin was denervated except for the hip joint and the small muscles around the hip. Under these conditions the fictive motor pattern could be recorded at rest and the effect of applied hip movements examined (Andersson & Grillner, 1983). Fig. 6A shows the response to hip movements of gradually increasing frequency, while Fig. 6B illustrates the effects of decreasing frequencies of applied movement. In both cases there is initially a 1:1 entrainment, though subsequently the pattern of coordination becomes either 2:1 or 1:2. Fig. 7 shows stable entrainment in response to applied frequencies of 0.5, 0.6 and 0.7Hz of fictive locomotion of resting frequency 0.55Hz. It should be noted that flexors are active during induced hip flexion, extensors during induced hip extension, and that proximal as well as distal joints are effectively entrained. Also, the efferent motor bursts retain their characteristic locomotor features even under such externally induced entrainment (Fig. 6). As predicted when two oscillators (spinal CPG and external hip movement) interact, the phase relationship varies over the frequency region where a 1:1 relationship occurs. For example, Fig. 7 shows that the motor bursts occur earlier in the movement cycle at 0.5Hz than at 0.7Hz (Andersson & Grillner, 1983).

To test if the central network has a differential sensitivity to the induced movement, we used the same experimental arrangement as described above, and applied short ramp movements of the hip in different parts of the fictive step cycle. The effect of applied flexion movements on bursts of the ankle flexor muscle, tibialis anterior, are illustrated in Fig. 8, while in Fig. 9, the effects of applied extension movements are plotted in the form of a phase response curve in which the differences in cycle time or burst duration are plotted against the phase of the fictive movement cycle at which the ramp stimulus was applied. As Fig. 9 shows, extension movements applied early in the locomotor cycle (at a phase of 0.2 - 0.3) produce a marked shortening of the cycle

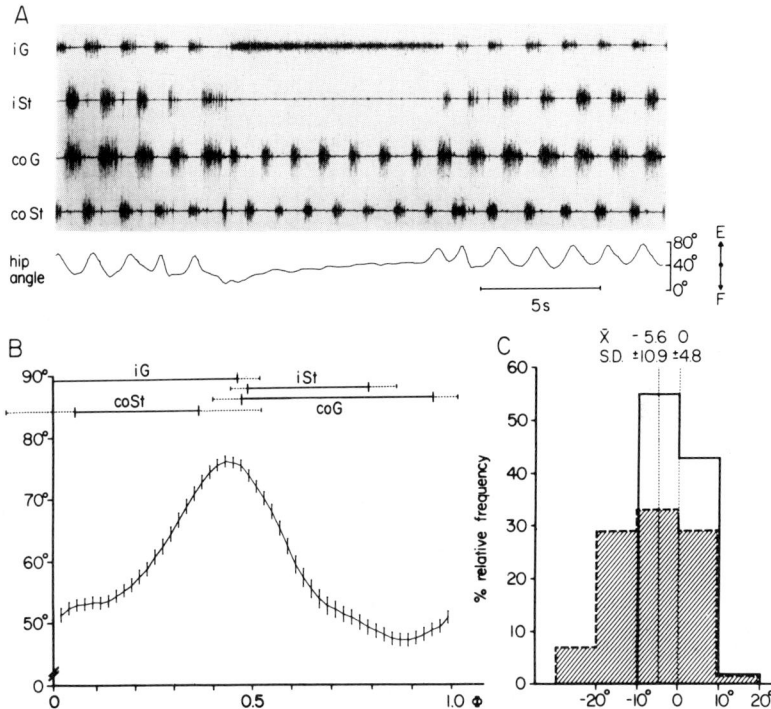

Fig. 5. EMG and hip movements during lift-off responses in chronic spinal cats.
A. Ipsilateral (i) and contralateral (co) gastrocnemius (G) and semitendinosus (St) EMG recordings. During the period where the ipsilateral limb is held, iG is tonically active and iSt is silent while the contralateral leg walks. The continuous measurement of hip angle was retraced with the help of a computer to transform the cosine output to an angle output.
B. Average hip angle and EMG during 40 normal walking cycles normalised to 1 from the onset of iG activity. Vertical lines on hip angle graph and dotted horizontal lines on the EMG bars indicate standard deviations.
C. Hip angle at initiation of swing. The hip angles corresponding to the onset of St activity (beginning of limb flexion) were measured in 100 normal walking cycles and 48 lift-off responses. These angles are plotted as histograms as a percentage of their relative frequency. $0°$ on the histogram is the mean hip angle of onset of St activity during normal locomotion. Open histogram, normal locomotion; hatched histogram, list-off responses. During lift-off responses, the onset of St activity began at a mean hip angle of $-5.6° \pm 10.9°$ with respect to its onset during normal locomotion.
(From Grillner & Rossignol, 1978).

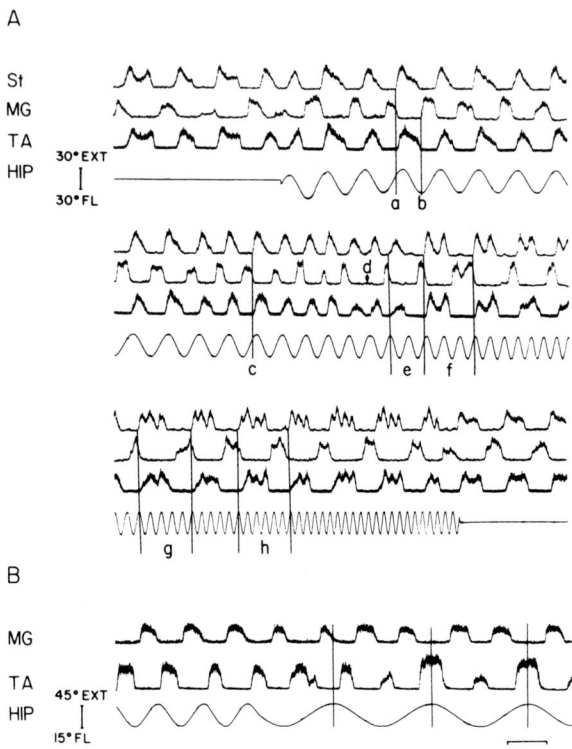

Fig. 6. Entrainment of fictive locomotor activity by sinusoidal hip movements of different frequencies in a spinal preparation.
A. Movement frequencies faster than the resting rate. Rectified and filtered nerve filament activity to semitendinosus (St), gastrocnemius (MG) and tibialis anterior (TA) recorded together with associated position of the hip joint (30° flexion - 30° extension). Continuous recording over 3 panels. When the hip is moved in the flexion direction, flexors (St and TA) are active (see a), while during extension movements there is extensor activity (MG, see b). When the movement frequency is increased, the extensor bursts are the first to fail (see d). Then the rhythm enters a 2:1 pattern (see e), with two movement cycles for every burst activity cycle. Later 3:1, 5:1 and 6:1 patterns occur (see f, g and h).
B. Movement frequencies slower than the resting rate. Recordings of MG and TA as in A, together with hip movements (15° flexion -45° extension). For movement cycles of a duration less than around 12s a 1:1 entrainment occurs (resting burst duration 9s). Later the rhythm follows a 1:2 pattern (one movement cycle for every two burst activity cycles) as indicated by the vertical bars. Preparation with iliopsoas muscle cut.
(From Andersson & Grillner, 1983).

duration, while the same stimulus causes a prolongation of the step cycle when it is applied later (at a phase of 0.9 - 1.0). Such a phase response curve can be expected to result in phase locking between applied movement and CPG with a phase

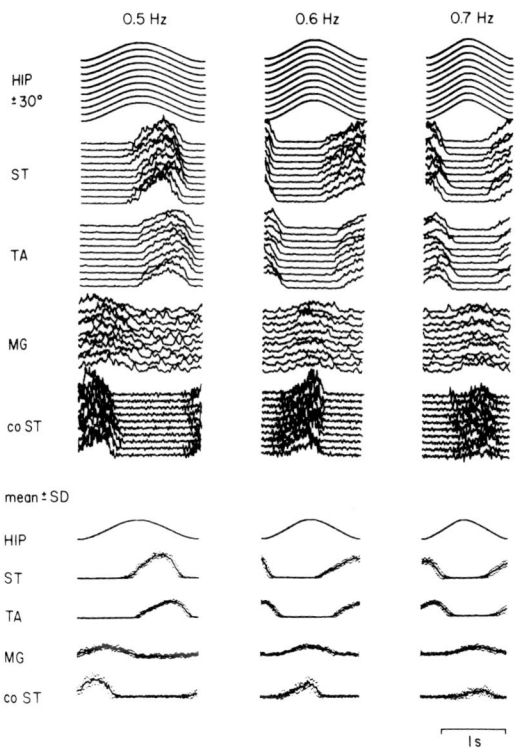

Fig. 7. Entrainment of the fictive locomotor rhythm at three different frequencies of applied hip movement (from left to right 0.5, 0.6 and 0.7Hz), in a preparation with a fast and stable resting rhythm (0.55 Hz). Each panel shows 10 consecutive movement cycles of applied hip movement (30° flexion - 30° extension), together with the associated activity in semitendinosus (ST), tibialis anterior (TA), gastrocnemius (MG) and contralateral semitendinosus (coST). Mean and standard deviation (SD) curves obtained from these cycles are plotted below. Note the very good reproducibility of the bursts at a given movement frequency. The typical lag between ST and TA is clear and the burst amplitude for the contralateral flexor (coST) is attenuated to a higher degree at the higher frequency (0.7Hz) compared to ST, TA and MG. Preparation with skin over leg and abdomen removed. (From Andersson & Grillner, 1983).

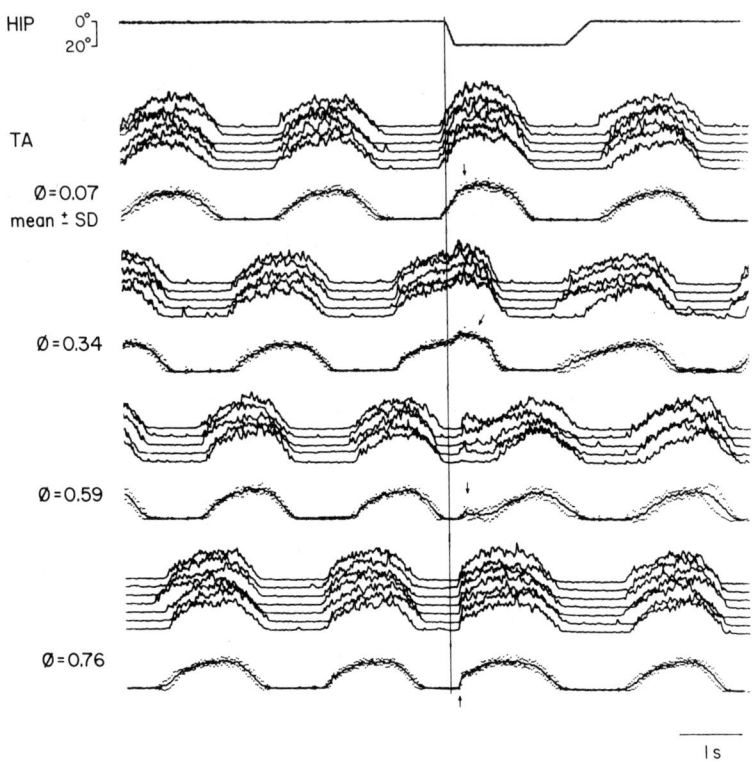

Fig. 8. Ramp movements of the hip applied in the flexion direction during fictive locomotion in an acute spinal cat. Ramp movements (0–20° flexion, upper trace), with an angular velocity of $130°\,s^{-1}$, are applied at four different phases of the spontaneously generated bursts of tibialis anterior (TA) activity (values $\emptyset$=0.07, 0.34, 0.59 and 0.76). Each group of records shows for each phase value several consecutive bursts of rectified and filtered neurograms from the nerve to TA, together with associated mean and standard deviation (SD) curves obtained from the illustrated bursts. A ramp starting near the beginning of the TA-burst ($\emptyset$=0.07) slightly increases the amplitude, and shortens the duration of the burst (see arrow and compare with preceding burst). Moving the stimulus further forward in the TA-cycle to the middle of the burst ($\emptyset$=0.34) reinforces and prolongs the burst (see arrow). If the ramp occurs in the interburst interval, two different responses may be seen. Early in the interval ($\emptyset$=0.59) an extra burst can be evoked, but later ($\emptyset$=0.76) the next TA-burst occurs earlier and resets the rhythm (see arrows). (From Andersson & Grillner, 1981).

relationship that varies with the frequency of imposed movement (Andersson & Grillner, 1981, 1983).

In addition to these effects, Pearson and Duysens (1976) and Duysens and Pearson (1980) have shown that a heavy load on the ankle extensors applied during the middle of the support phase will prevent the initiation of limb flexion, and conversely the unloading of the limb at the end of the support phase will promote a limb flexion. Also, Rossignol and Gauthier (1980) have shown that muscle receptors from the distal part of the limb (tibialis anterior) can affect the pattern generator.

SIGNIFICANCE OF THE PERIPHERAL EFFECTS ON THE CPG

Adaptation of the movements during normal movements
As an animal walks over the complicated terrain of its natural habitat, the demands on the limb will vary markedly from cycle to cycle and peripheral controls will act to optimise the amplitude of each support phase.

It is clear that a position signal from the most proximal joint is a more useful indication of how far the step has proceeded than the joint angle trajectories of the distal joints. The type of control discussed above and illustrated in Figs. 5-9 will result in regulation of the amplitude of limb extension. This will prevent limb flexion occurring too early during the support phase which would be inefficient, and also an excessively long limb extension during which the limb would not be providing any real support or propulsive force.

Effects during turning and sudden acceleration and deceleration
When the animal changes its direction during locomotion, the forelimbs are rotated and the body curved (see Grillner, 1981). As a consequence the outer limbs have to cover a somewhat longer distance than the inner limbs (compare walking in a narrow circle). The adaptations of the step cycle in this situation have been studied in decerebrate (Kulagin & Shik, 1970) and intact cats (Halbertsma, 1983) by using an experimental situation where the left and the right pairs of limbs walk on separate treadmill belts (c.f. also Chapter 24 by Dr Clarac for equivalent experiments on lobsters). The speed of these two belts can be changed independently to induce a mismatch between the two sides, allowing a detailed study of the interaction between the left and right limbs. With a moderate mismatch in belt speed, fore- and hindlimbs maintain a strict alternation, but the support phase becomes longer on the "slow" inner side and shorter on the "fast" outer side. All phases of the step cycle turn out to be somewhat modifiable in relation to the step cycle duration. With large discrepancies in speed the animal can change to a

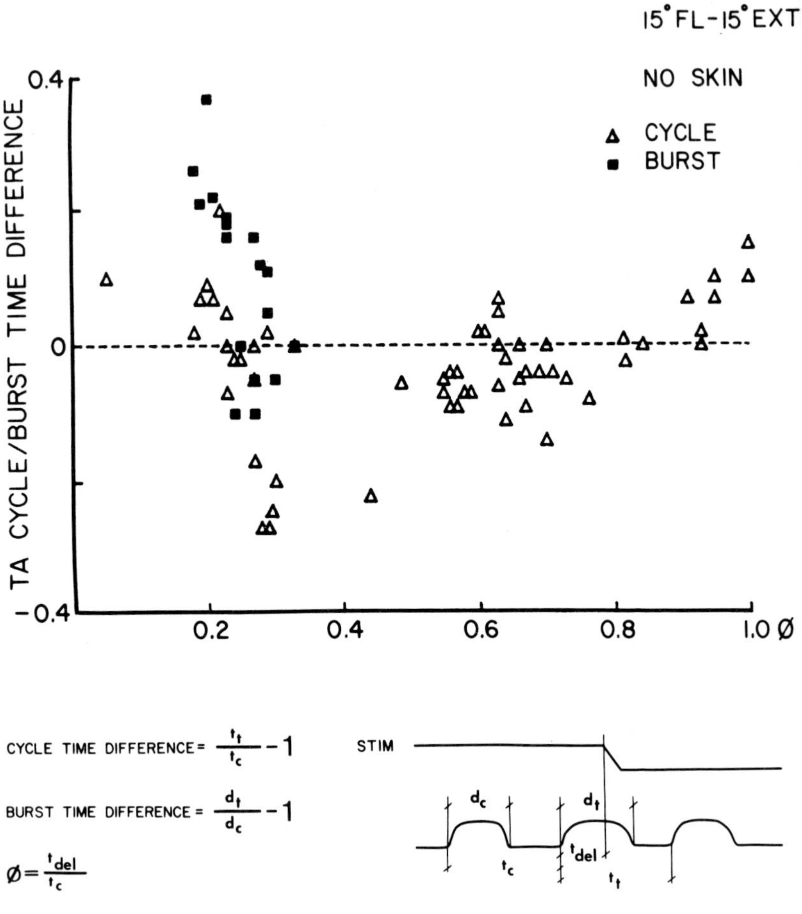

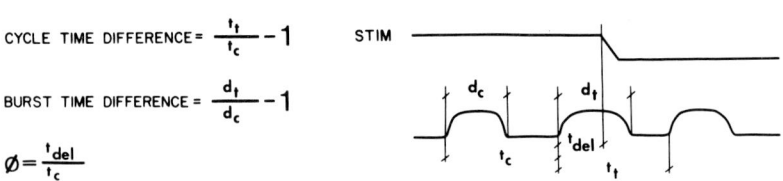

Fig. 9. Phase response curves for ramp movements of the hip applied during fictive locomotion in an acute spinal cat. The limb has been skinned (see text). Effects of extension ramps (15° flexion - 15° extension at an angular velocity of $105°\ s^{-1}$) on both cycle time (open triangles) and tibialis anterior (TA) burst duration (closed squares). As indicated below the graph, differences in cycle time (or burst duration) between test and control cycles are plotted against the phase of the cycle at which the ramp stimulus was applied. Note that stimuli applied near the end of the cycle cause a prolongation of the "step cycle", but that there is a sudden switch to a marked shortening of the cycle length at a phase value of approximately 0.2 - 0.3.

2:1 coordination with a very specific type of coordination (Halbertsma, 1983).

If a low spinal animal is tested under the same conditions (Fig. 10), it turns out that virtually identical patterns of coordination are observed (Forssberg, Grillner,

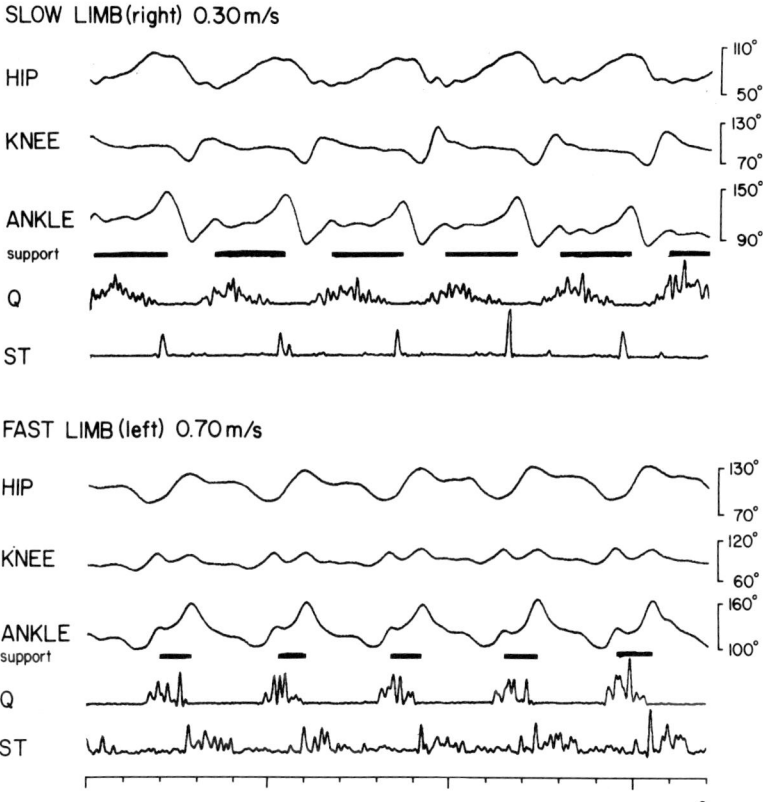

Fig. 10. Locomotion of a chronic spinal cat on a "split belt" treadmill showing 1:1 coordination between left and right limbs. The angular movement of the hip, knee and ankle, and the synchronously recorded, rectified and filtered EMGs of quadriceps (Q) and semitendinosus (ST) are shown for both hindlimbs. The support phases are indicated with bars. The belt on the right side was driven at 0.30m s^{-1}, that on the left at 0.70m s^{-1}. Angular calibrations are to the right of each curve and the time calibration is on the bottom in seconds (large bars) with 200ms subdivisions (small bars). (From Forssberg, Grillner, Halbertsma & Rossignol, 1980).

Halbertsma & Rossignol, 1980). It follows that the type of interlimb coordination used during turning can be generated by the spinal CPGs together with their powerful peripheral input. Higher centres are apparently not required for these secondary adjustments of the step cycle, which occur automatically via the spinal circuits as the body is bent to one side. In other words, the control strategy for this special adaptation of the locomotor movement synergies needs only to depend upon spinal feedback occurring during locomotion.

The control system discussed above is presumably also used during rapid accelerations and decelerations to optimise limb extension in each step cycle, since this occurs in spinal cats as the speed of the treadmill is accelerated (cf. Grillner, 1973; Forssberg, Grillner & Halbertsma, 1980). It is conceivable that it also plays a role in uphill and downhill walking, when the motor pattern is in addition somewhat changed. For example, there is an increase in the early E_1 hip extensor-knee flexor burst (Andersson, Forssberg & Grillner, unpublished).

<u>Phase dependent corrections of the movement trajectories resulting from skin stimuli during locomotion</u>
If an object impedes the forward movement of the limb during the swing phase, the resulting skin stimulus will activate a short latency reflex path which will result in the limb being lifted over the object. However, if the identical stimulus is applied during the support phase, it will instead result in enhanced extensor activity. These phase dependent responses result from a gating of the reflex by the spinal CPG (Forssberg et al., 1977; Forssberg, 1979; see also Chapter 23 by Dr Rossignol). These responses appear meaningful from a behavioural point of view as, for instance, a short latency flexion during midstance could be disastrous. Corresponding phase dependent reflexes have been demonstrated in fish (Grillner et al., 1977; Wallén, 1980).

TERMINOLOGY AND CONCEPTS

The general neural organisation underlying locomotion appears to be surprisingly similar in different types of vertebrates (see Stein, 1978; Grillner, 1981) and in invertebrates (e.g. Davis, 1976; Barnes, 1977; Pearson et al., 1983; see also Chapter 24 by Dr Clarac), even if specific modifications naturally occur depending on the type of locomotion, species etc. Let me first summarise some of these organisational principles as applied to cat locomotion.

1. Locomotor movements can be initiated and maintained from specific brainstem regions.

Neural control of vertebrate locomotion

2. A central network of neurones can generate a detailed motor output without any movement related sensory information.

3. A significant part of the detailed pattern generation can be accomplished at the spinal level, with tetrapods having independent centres for each limb (and presumably a further subdivision, Grillner, 1981, 1982).

4. Each limb network can interact with the other limb networks in several different ways. A pair of limbs can alternate or work in synchrony. These different modes can produce the coordination underlying walking, trotting, pacing and galloping.

5. A powerful peripheral feedback acts on the central network for each limb, which can change the frequency and control the duration of different parts of the step cycle.

6. Reflex effects directed to the motor neurones can be phasically gated in each movement cycle.

7. Activity in spinocerebellar pathways (partially efference copy) during locomotion modulate cerebellar neurones and cause phasic activity in descending pathways. This activity will modulate the motor neurone output.

It is convenient to have terms to apply to the different parts of the control system, provided that each term is as clear and unambiguous as possible and, most important, that it is adequate rather than misleading. Let us consider the terminology which is in current use and some of the objections made.

Pattern generators

The spinal cord with its afferents can produce a locomotor pattern. It thus must contain pattern generating circuitry. Both central elements within the spinal cord and also sensory elements, which can modify the duration of different parts of the step cycle, are important. I have referred to the central part as a central pattern generator or CPG and the peripheral part as a sensory or feedback part. The term CPG in my own writing is synonymous with the central network of the control system and does not imply that the CPG should necessarily produce every single detail of the intact pattern nor that it should not be affected by feedback (see Grillner, 1977).

The term oscillator, later often replaced by central pattern generator (CPG), was used by Wilson (1964, locust) and, for example, Davis (1973, swimmeret) in schemes in which the oscillator provided all details of the output, while sensory feedback acted at the motor neurone level but did not affect the oscillator itself. Pearson (see Chapter 20 and Pearson et al., 1983) has recently demonstrated that the motor

51

pattern of the deafferented locust is not as complete as was once thought, and that it is influenced by the wing hinge receptors (similar to hip control in cat). As the system functions as one whole pattern generator rather than as separate central and peripheral parts, Pearson feels that it is misleading to use the term CPG when there is additional sensory regulation (cf. also Wilson's early usage of the term oscillator in locust flight). In each instance in which one has separate parts of a control system, it is a potential risk for misunderstanding when terminological subdivisions are made, as for instance has been the case with left and right hemisphere functions. In the present case the term CPG has been used for a long time (see Delcomyn, 1980) in a context that allows the possibility of peripheral regulation of different phases of the locomotor cycle. I feel therefore quite comfortable in using these terms and believe also that most readers will not be seriously misled. An alternative description which stresses the unity of the two parts of the pattern generator would be (1) "the central part of the pattern generator (CPPG)" and (2) "the peripheral part of the pattern generator (PPPG)".

Coordinating neurones

The different limb pattern generators (as defined above) can be coordinated in different ways (as in walk, trot etc.). The neurones which achieve this coordination have a coordinating function and have been called "coordinating neurones" (Stein, 1971, 1978). MacMillan, Altman and Kien (1983) have recently shown that the coordinating neurones originally described in the swimmeret system (Stein, 1971, 1974) may have sensory and/or motor functions and they therefore object to the usage of this term. However, in most systems it has become clear that individual neurones may have several functions. Therefore, it appears justified to continue to use the term coordinating neurones for neurones that contribute to the coordination between pattern generators, even if these neurones simultaneously may be part of, for instance, the motor neurone pool or the CPG.

Neuronal hierarchy

Davis (see Chapter 2 or 1976) feels that the hierarchical invertebrate scheme (e.g. Davis, 1973) has to be abandoned and replaced by a new scheme because, for instance, some of the cells described as command neurones have been found to receive phasic feedback from the CPG. Davis's description of this scheme appears to me as an extreme caricature of the actual control schemes most workers conceived of in the early seventies. The old notion is said to consist of one command neurone, an oscillator constituted by one cell and motor neurones. This is not correct. Wilson and Waldron (1967), for instance, spent much time modelling oscillator networks and very few researchers thought that there would be only one command neurone (see Kupferman & Weiss, 1978 and commentaries; Grillner, 1978). The notion at that time held that at least several neurones could act to initiate and maintain a rhythmic motor pattern and that CPGs consisted of neuronal networks (although neurones with pacemaker properties sometimes contributed to the rhythm e.g. Selverston, 1976). A much finer mesh of knowledge has now accumulated and it is known for instance that sensory signals interact with the CPG and that higher centres may be affected by feedback. It appears to me that we have slowly developed the old notions of the control system to something more complex and real. At least that is how it is perceived from my vertebrate horizon. The hierarchical concepts of the early seventies have been much improved upon and remain useful. Only the present caricature of the view held in the early seventies should be abandoned, once and for all.

Acknowledgement
The dedicated help of Mrs I. Klingebrant is gratefully acknowledged as is support from the Swedish Medical Research Council (3026), Karolinska institutets fonder and Magnus Bergvalls stiftelse.

REFERENCES

Andersson, O., Forssberg, H., Grillner, S. & Wallén, P. (1981). Peripheral feedback mechanisms acting on the central pattern generators for locomotion in fish and cat. Can. J. Physiol. & Pharmacol., 59, 713-726.
Andersson, O. & Grillner, S. (1981). Peripheral control of the cat's step cycle. I. Phase dependent effects of ramp-movements of the hip during "fictive locomotion". Acta physiol. scand., 113, 89-101.
Andersson, O. & Grillner, S. (1983). Peripheral control of the cat's step cycle. II. Entrainment of the central pattern generators for locomotion by sinusoidal hip movements during "fictive locomotion". Acta physiol. scand., 118, 229-239.
Arshavsky, Yu. I., Gelfand, I.M. & Orlovsky, G.N. (1983). The cerebellum and control of rhythmical movements. Trends Neurosci., 6, 417-422.
Bard, P. & Macht, M. (1958). The behaviour of chronically decerebrate cats. In Neurological Basis of Behaviour. Eds. Wolstenholme, G.E.W. & O'Connor, C.M. Churchill, London, pp. 55-75.
Barnes, W.J.P. (1977). Proprioceptive influences on motor output during walking in the crayfish. J. Physiol., Paris, 73, 543-564.
Bjursten, L.-M., Norrsell, K. & Norrsell, U. (1976). Behavioural repertory of cat without cerebral cortex from infancy. Expl. Brain Res., 25, 115-130.
Brown, T.G. (1911). The intrinsic factors in the act of progression in the mammal. Proc. R. Soc. B, 84, 308-319.
Budakova, N. (1973). Stepping movements in the spinal cat due to DOPA administration. Fiziol. Zh. SSSR, 59, 1190-1198.
Davis, W.J. (1973). Neuronal organization and ontogeny in the lobster

swimmeret sysem. In Control of Posture and Locomotion. Eds. Stein, R.B., Pearson, K.G., Smith, R.S. & Redford, J.B.. Plenum, New York, pp. 437-455.

Davis, W.J. (1976). Organizational concepts in the central motor networks of invertebrates. In Neural Control of Locomotion. Eds. Herman, R.M., Grillner, S., Stein, P.S.G. & Stuart, D.G.. Plenum, New York. pp. 265-292.

Delcomyn, F. (1980). Neural basis of rhythmic behavior in animals. Science, N.Y., 210, 492-498.

Duysens, J. & Pearson, K.G. (1980). Inhibition of flexor burst generation by loading ankle extensor muscles in walking cats. Brain Res., 187, 321-332.

Eidelberg, E., Walden, J.G. & Nguyen, L.H. (1981). Locomotor control in macaque monkeys. Brain, 104, 647-664.

English, A.W. (1980). Interlimb coordination during stepping in the cat: effects of dorsal column section. J. Neurophysiol., 44, 270-279.

Forssberg, H. (1979). "The stumbling correction reaction" - a phase dependent compensatory reaction during locomotion. J. Neurophysiol., 42, 936-953.

Forssberg, H. & Grillner, S. (1973). The locomotion of the acute spinal cat injected with clonidine i.v. Brain Res., 50, 184-186.

Forssberg, H., Grillner, S. & Halbertsma, J. (1980). The locomotion of the low spinal cat. I. Coordination within a hindlimb. Acta physiol scand., 108, 269-281.

Forssberg, H., Grillner, S., Halbertsma, J. & Rossignol, S. (1980). The locomotion of the low spinal cat. II. Interlimb coordination. Acta physiol. scand., 108, 283-295.

Forssberg, H., Grillner, S. & Rossignol, S. (1977). Phasic gain control of reflexes from the dorsum of the paw during spinal locomotion. Brain Res., 132, 121-139.

Garcia-Rill, E., Skinner, R.D., Jackson, M.B. & Smith, M.M. (1983). Connections of the mesencephalic locomotor region (MLR). II. Afferents and efferents. Brain Res. Bull., 10, 63-71.

Grillner, S. (1973). Locomotion in the spinal cat. In Control of Posture and Locomotion. Eds. Stein, R.B., Smith, R.S. & Redford, J.B.. Plenum, New York, pp. 515-535.

Grillner, S. (1975). Locomotion in vertebrates: central mechanisms and reflex interaction. Physiol. Rev., 55, 247-304.

Grillner, S. (1977). On the neural control of movement - a comparison of different basic rhythmic behaviors. In Function and Formation of Neural Systems. Ed. Stent, G.S. Berlin, Dahlem Konferenzen, pp. 197-224.

Grillner, S. (1978). Command neurones or central program controlling system? Behav. & Brain Sci., 1, 23-25.

Grillner, S. (1981). Control of locomotion in bipeds, tetrapods and fish. In Handbook of Physiology, Section 1, The Nervous System, Vol. 2, Motor Control. Ed. Brooks, V.B.. American Physiological Society, Bethesda, pp.1179-1236.

Grillner, S. (1982). Possible analogies in the control of innate motor acts and the production of sound in speech. In Speech Motor Control, Vol. 36. Eds. Grillner, S., Lindblom, B., Lubker, J. & Persson, A.. Pergamon, Oxford & New York, pp. 217-229.

Grillner, S. & Rossignol, S. (1978). On the initiation of the swing phase of

locomotion in chronic spinal cats. Brain Res., 146, 269-277.
Grillner, S., Rossignol, S. & Wallén, P. (1977). The adaptation of a reflex response to the ongoing phase of locomotion in fish. Expl. Brain Res., 30, 1-11.
Grillner, S. & Wallén, P. (1982). On peripheral control mechanisms acting on the central pattern generators for swimming in the dogfish. J. exp. Biol., 98, 1-22.
Grillner, S. & Wallén, P. (1985). Central pattern generators for locomotion with special reference to vertebrates. Annu. Rev. Neurosci., 8, 233-261.
Grillner, S., Williams, T. & Lagerbäck, P-A. (1984). The edge cell, a possible intraspinal mechanoreceptor. Science, N.Y., 223, 500-503.
Grillner, S. & Zangger, P. (1975). How detailed is the central pattern generator for locomotion? Brain Res., 88, 367-371.
Grillner, S. & Zangger, P. (1979). On the central generation of locomotion in the low spinal cat. Expl. Brain Res., 34, 241-261.
Grillner, S. & Zangger, P. (1984). The effect of dorsal root transection on the efferent motor pattern in the cat's hindlimb during locomotion. Acta physiol. scand., 120, 393-405.
Halbertsma, J.M. (1983). The stride cycle of the cat: the modelling of locomotion by computarized analysis of automatic recordings. Acta physiol. scand., Suppl. 521.
Holst, E. von (1935). Erregungsbildung und Erregungsleitung im Fischrückenmark. Pflügers Arch. ges. Physiol., 235, 345-359.
Jordan, L. (1983). Factors determining motoneuron rhythmicity during fictive locomotion. Symp. Soc. exp. Biol., 37, 423-444.
Julien, C., Barbeau, H. & Rossignol, S. (1982). Gain changes in cutaneous reflexes during locomotion in the adult chronic spinal cat. Soc. Neurosci. Abstr., 8, 168.
Kulagin, A.S. & Shik, M.L. (1970). Interaction of symmetrical limbs during controlled locomotion. Biophysics, 15, 171-178.
Kupfermann, I. & Weiss, K.R. (1978). The command neuron concept. Behav. & Brain Sci., 1, 3-39.
Lundberg, A. (1969). Reflex control of stepping. The Nansen Memorial Lecture V. Universitetsforlaget, Oslo, pp. 1-42.
MacMillan, D.L., Altman, J.S. & Kien, J. (1983). Intersegmental coordination in the crayfish swimmeret system reconsidered. J. exp. Zool., 228, 157-162.
Miller, S. & van der Meche, F.G.A. (1976). Coordinated stepping of all four limbs in the high spinal cat. Brain Res., 109, 395-398.
O'Donovan, M.J., Pinter, M.J., Dum, R.P. & Burke, R.E. (1982). Action of FDL and FHL muscles in intact cats: functional dissociation between anatomical synergists. J. Neurophysiol., 47, 1126-1143.
Orlovsky, G.N. & Shik, M.L. (1976). Control of locomotion: a neurophysiological analysis of the cat locomotor system. In International Review of Physiology, Vol. 10, Neurophysiology, II. Ed. Porter, R.. University Park Press, Baltimore, pp. 281-317.
Paul, D.H. & Roberts, B.L. (1984). Inputs to the cerebellum of the dogfish Scyliorhinus canicula during behavioural movement. In Abstracts of the International Symposium on Feedback and Motor Control. University of Glasgow, p. 61.
Pearson, K.G. & Duysens, J. (1976). Function of segmental reflexes in the

control of stepping in cockroaches and cats. In Neural Control of Locomotion. Eds. Herman, R.M., Grillner, S., Stein, P.S.G. & Stuart, D.G.. Plenum, New York, pp. 519-537.

Pearson, K.G., Reye, D.N. & Robertson, R.M. (1983). Phase-dependent influences of wing stretch receptors on flight rhythm in the locust. J. Neurophysiol., 49, 1168-1181.

Perret, C. & Cabelguen, J-M. (1980). Main characteristics of the hindlimb locomotor cycle in the decorticate cat with special reference to bifunctional muscles. Brain Res., 187, 333-352.

Phillipson, M. (1905). L'autonomie et la centralisation dans le système nerveux des animaux. Trav. Lab. Physiol. Inst. Solvay, 7, 1-208.

Risling, M. & Hildebrand, C. (1982). Occurrence of unmyelinated axon profiles at distal, middle and proximal levels in the ventral root L7 of cats and kittens. J. Neurol. Sci., 56, 219-231.

Rossignol, S. & Gauthier, L. (1980). An analysis of mechanisms controlling the reversal of crossed spinal reflexes. Brain Res., 182, 31-45.

Selverston, A.I. (1976). Neural mechanisms for rhythmic motor pattern generation in a simple system. In Neural Control of Locomotion. Eds. Herman, R.M., Grillner, S., Stein, P.S.G. & Stuart, D.G.. Plenum, New York, pp. 377-399.

Shik, M.L. (1966). Control of walking and running by means of electrical stimulation of the mid-brain. Biophysics, 11, 756-765.

Shik, M.L. & Orlovsky, G.N. (1976). Neurophysiology of locomotor automatism. Physiol. Rev., 56, 465-501.

Smith, J.L., Smith, L., Zernicke, R.F. & Hoy, M. (1982). Locomotion in exercised and nonexercised cats cordotomized at 2 or 12 weeks of age. Expl. Neurol., 76, 393-413.

Stein, P.S.G. (1971). Intersegmental coordination of swimmeret motoneuron activity in crayfish. J. Neurophysiol., 34, 310-318.

Stein, P.S.G. (1974). Neural control of interappendage phase during locomotion. Am. Zool., 14, 1003-1016.

Stein, P.S.G. (1978). Motor systems, with special reference to the control of locomotion. Annu. Rev. Neurosci., 1, 61-81.

Udo, M., Matsukawa, K., Kamei, H. & Oda, Y. (1980). Cerebellar control of locomotion: effects of cooling cerebellar intermediate cortex in high decerebrate and awake walking cats. J. Neurophysiol., 44, 119-134.

Wallén, P. (1980). On the mechanisms of a phase dependent reflex occurring during locomotion in dogfish. Expl. Brain Res., 39, 193-202.

Wallén, P. & Williams, T.L. (1984). Fictive locomotion in the lamprey spinal cord in vitro compared with swimming in the intact and spinal animal. J. Physiol., 347, 225-239.

Wilson, D.M. (1964). The origin of the flight motor command in grasshoppers. In Neural Theory and Modeling. Ed. Reiss, R.F.. University Press, Stanford, pp. 331-345.

Wilson, D.M. & Waldron, I. (1968). Models for the generation of the motor output pattern in flying locusts. Proc. IEEE, 56, 1058-1064.

Zomlefer, M.R., Provencher, J., Blanchette, G. & Rossignol, S. (1984). Electromyographic study of lumbar back muscles during locomotion in acute high decerebrate and in low spinal cats. Brain Res., 290, 249-260.

Chapter Four

GENERATION OF BEHAVIOUR: THE ORCHESTRATION HYPOTHESIS

G. Hoyle

The papers presented in this volume address detailed questions concerning the role of proprioceptive feedback in motor control. Proprioceptors are of supreme importance in the coordination and control of ongoing behaviour. Also, tonic input from them is a major contributor to the 'stimulationsorgane' effect (von Buddenbrock, 1952), whereby tonic sensory input contributes to general arousal in awake animals. Animals do not act only as reflex machines responding to inputs, they must first generate their own activities. Circadian rhythmicity is the simplest and most compelling symptom of the tendency to innate behaviour generation, much of which is commonly produced periodically. To Nobel Laureate ethologists Konrad Lorenz and Niko Tinbergen we owe the first clear realisation that much natural animal behaviour is an expression of the cyclically-occurring tendencies of inherited central nervous system motor programme generators (MPG) to drive coordinated action. The action could generally, but not always, be seen to have functional significance, at least when the whole context of the behaviour was followed in its natural setting. Many behaviours were so stereotyped they could be labelled "fixed action patterns" (FAP).

The tendency to display any given FAP could be related to a variety of external factors, often ones peculiar to the species. Specific sensory inputs, especially visual 'signs', had quantifiable 'releasing' actions in triggering specific FAPs or parts of stereotyped behaviour sequences. In his classic synthesis of the mechanisms underlying the generation of controlled movements, Konrad Lorenz (1950) used a simple hydraulic model. Recently he re-iterated the basic conceptual value of this model (Lorenz, 1981).

For scientists interested in proprioceptors some very basic questions were raised by the ethologists. How do the MPGs for FAPs utilise proprioceptors? Are the behaviours programmed as direct motor commands to skeletal muscles? Or are they produced by an indirect route? One method would be for the programme to be achieved by biassing the

Generation of behaviour: the orchestration hypothesis

proprioceptors, in a manner first suggested by Merton (1953). The intrinsic servo-mechanisms of the stretch-receptor loops exciting α-efferents mono-synaptically, and themselves controlled via γ-efferents, would provide automatic load-compensation and guarantee smooth operation. Alternatively, did the CNS have an in-built 'comparator' which could match incoming sensory signals with an inherited 'sensory tape' sequence of inputs appropriate to the behaviour? This might enable the computation of commands to motor neurones which would adjust the output to that associated with the desired behaviour.

It was almost heretical to ignore the proprioceptors, but was it possible that the CNS contained inherited motor scores which enabled them to command at least some complex behaviours without reference to proprioceptive information? This, and the other questions raised above, could not be tackled easily in any animal, but arthropods looked promising subjects. They have proprioceptors that can be centrally controlled just like vertebrates, and their motor systems are so simple that the output is not only readily recordable in freely-moving intact animals, but is also easily quantified.

During the nine pleasant years I spent at the University of Glasgow one of the first things I did was to implant long, very fine, flexible recording wires into locust muscles, to record the detailed patterns associated with whatever might appear in the entire behavioural repertoire in the laboratory. The leads did not seem to hamper the insects too much: they had complete freedom within a 60cm cube. Eventually I accumulated a few miles of paper trace and several thousand examples of each of several different classes of movements. A great many of the movements within a class were visually/photographically identical. However, the electrical records were not. In fact the differences were sometimes so drastic that it was difficult to believe that the movements could possibly have been remotely similar had the visual record not shown otherwise.

These observations are still confounding to me. They fit neither of the extremes of available interpretation of routine motor acts: the turning on of a pure CNS motor programme on one hand, nor a chain reflex sequence on the other. Since that time, numerous examples from a variety of phyla have been brought forward to show that a strictly central MPG can be a reality. In the mollusc <u>Tritonia</u>, which has a complex escape swim FAP, Dorsett, Willows and I first recorded intracellularly from the nerve cells generating the swim (Dorsett et al., 1969). We then removed the brain from the animal, placed electrodes in the same neurones, and applied a brief burst of electric pulses to a chemosensory nerve. Following the burst, precisely the same motor pattern occurred in the same neurones, as during an escape swim. In the calling song of a cricket (Bentley, 1971) the motor output is

Generation of behaviour: the orchestration hypothesis

precisely repeated down to the level of single impulses even when isolated from sensory input, while the only influence of proprioceptive input in the control of courtship behaviour in a grasshopper is to maintain the time lag between the stridulatory movements of the two hindlegs (Elsner & Hirth, 1978).

However, the above are one end of a spectrum, which do no more than show that nervous systems can, and do, produce some behaviours without any significant reference to proprioception. They make it easy to understand that a specific stimulus can evoke a specific behaviour because there is a specific neural circuit for the production of that behaviour. But they leave open the question of how the circuit is turned on. A male cricket, even a headless one (Kutsch & Otto, 1972), is capable of singing spontaneously. On the Lorenz model this will occur because "reaction specific energy" for courtship song, which spontaneously accumulates slowly, has filled its holding tank to the brim. In this condition it forces its way out into the nervous system, hits the courtship song MPG, and causes the appropriate motor output, perhaps without reference to proprioceptive information, or perhaps utilising it.

The cricket can produce two other distinct songs using exactly the same muscles: aggression and courtship, plus various transitions. The grasshopper can also produce different songs and, since unlike the cricket it uses its legs, the same muscles can produce walking, hopping, climbing, kicking and mounting also. How are the same motor neurones used in different behaviours? Are there different overall MPG neural circuits for each? If not, how do the trigger-level programmes link to the common motor neurones? What is it that is changing slowly in the circuits which corresponds to their "reaction-specific energy"?

Emergence of a hypothesis

These were important questions about which I had hardly any ideas until after I began working with the primitive stenopelmatid insect Hemideina, the 'weta', in New Zealand. This insect has a powerful circadian rhythm in which it is vigorously active only at night. But in its daytime condition it can be awakened by prodding. The awakening occurs gradually and is accompanied by progressively-increasing muscular responses. Muscular contractions turned out to be very weak unless the insect was first awakened. The natural potentiation was enormous, at least several-fold. Eventually I found that the potentiation is largely occurring at neuromuscular junctions (Hoyle, 1984). The effect could be mimicked simply by injecting a very low concentration (ca. 10^{-8} M) of octopamine in saline into the body fluid. Octopamine is synthesised in locusts by a single dorsal unpaired median neurone (termed DUMETi) having paired axons which pass out in the left and right crural nerves, and terminate in the extensor tibiae muscles.

There is a homologue of DUMETi (Hoyle, 1978) in wetas, which when directly stimulated causes strong potentiation of peripheral transmission. This occurs first, before a condition develops in which the potentiated transmission by slow-type motor neurones leads to a catch-like tension which the insect uses well (Hoyle & Field, 1983a,b). DUMETi is one of a family of sibling neurones, derived from a single neuroblast (Goodman & Spitzer, 1979; Goodman et al., 1980), not all of which leave the ganglion. Even those which do supply muscles also have central terminals in the neuropile. These presumably release octopamine at specific sites within the neuropile. If octopamine is such a potent potentiater of peripheral excitatory transmission, perhaps it does likewise in the CNS? However, octopamine injected into the body cavity did not produce any specific posture, as it does in lobsters (Livingstone et al., 1980) and crayfish (Glanzman & Krasne, 1983), apart from generally increasing all motor activity.

The insect nervous system is surrounded by a perineurium, able to regulate the passage of ions across it (Hoyle, 1952; Treherne, 1974). The cells that make up this sheath could be keeping octopamine out. Whether or not it is kept out, the highly-complex neuropile is likely to be strictly compartmentalised with regard to octopamine, if that is the principal modulator substance. The metathoracic ganglion is responsible for the generation of several behaviours: various postures, walking, marching, hopping, jumping, kicking, stridulation in many species, and flight. Several of these behaviours utilise common muscles, but with different intensities of activation and different time relations. If these behaviours are to be governed by a common modulator substance, three-dimensional compartmentalisation for its activation and release sites is essential.

Thinking about this has led to some speculation, which I shall formulate as an <u>orchestration hypothesis</u>. There are three basically different ways in which a neuromodulator might be involved in behaviour generation. These are summarised diagrammatically in Fig. 1. The diagrams assume a specific neural circuit to be the motor programme generator (MPG) behaviour, which is symbolised by very simple reciprocal inhibition of antagonists excited by a common driver neurone. The circuit is assumed to be able to produce sufficiently strong output to cause movement when a suprathreshold concentration of modulator has been released in its vicinity. In the <u>independent</u> mode, there is no direct coupling between the neurone(s) commanding the behaviour (C), and the modulator neurone(s)(M). In the <u>coupled</u> mode, the command is delivered simultaneously to the MPG and the modulator(s). In the <u>dependent</u> mode, the command acts only on the modulator. The three possible modes are not mutually exclusive, and mixtures

POSSIBLE MODES OF MODULATION

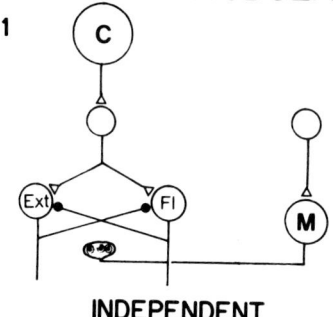

INDEPENDENT

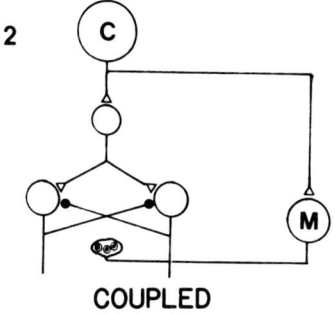

COUPLED

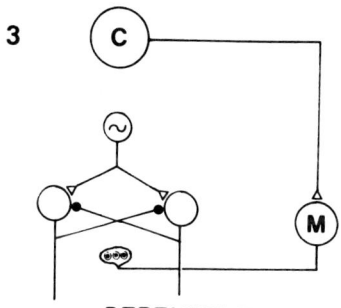

DEPENDENT

Fig. 1. Diagrams illustrating three possible modes of involvement of modulator neurones (M) in the generation of behaviour in relation to driving or "commanding" (C) interneurones. The simple reciprocal inhibitory neural circuit causing alternation in output is intended to symbolise a central motor programme generator (MPG). Oscillatoriness ($\sim$) or arhythmic circuit activity may be either intrinsic to neurone(s), or a circuit property, and either ongoing or initiated by C and/or M. (From Sombati & Hoyle, 1984b).

Generation of behaviour: the orchestration hypothesis

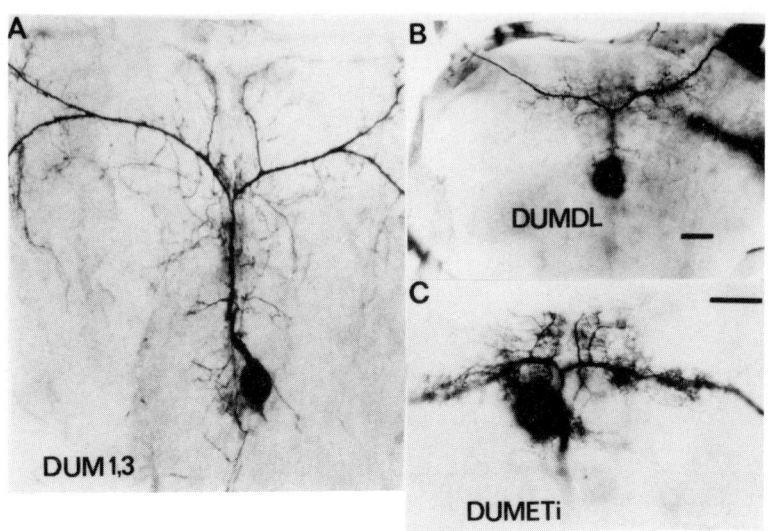

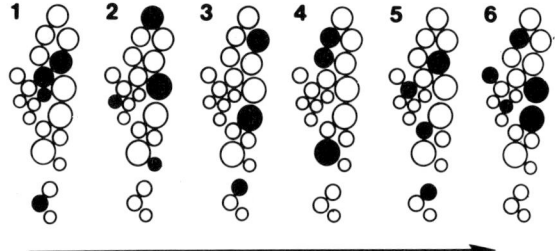

Fig. 2. Central dendritic fields of some characteristic DUM neurones as seen after direct cobalt injection, and diagram illustrating the orchestration hypothesis. A. DUM 1,3 (with 2 pairs of axons leaving in nerves 1 and 3) of the cockroach Periplaneta americana. (From Arikawa et al., 1984). B. DUMDL (with 1 pair of axons leaving in nerve 1 to supply the dorsal longitudinal flight muscles) of locust Schistocerca americana. C. DUMETi (with a pair of axons leaving in nerve 5 to supply the extensor tibiae muscle) of S. americana. (B & C from Sombati & Hoyle, 1984b). D. Diagram illustrating the basis for possible involvement of a cluster of modulator neurones in the generation of a variety of behaviours using some common muscles, according to the orchestration hypothesis. The cluster is based upon known octopaminergic

Generation of behaviour: the orchestration hypothesis

could occur.

In a nervous system in which there are many modulator neurones, the hypothesis proposes that they will be excited in an orderly manner in order to generate specific behaviours. The particular behaviour occurring will be determined by the combination of modulator neurones excited, regardless of mode. For the locust metathoracic ganglion, in which the modulators are the octopaminergic DUM neurones, hypothetical examples of different combinations would be along the lines diagrammed in Fig. 2. A complex behaviour sequence will be determined by progressive activation of a series of clusters such as those illustrated in 1 though 6 in this figure.

Neuroanatomical correlates of the orchestration hypothesis

A consequence of compartments for modulator action is a relatively short range for action. It would be considerably greater than that of a transmitter, but strictly within the CNS or muscles, and not as great as for a neurohormone. In the skeletal muscles there is only a small number of terminals of DUM neurones, much smaller than the number of neuromuscular junctions to be served. We do not have good quantitative data yet, but it is of the order of 1 to 100. Let us adopt this figure, for it is probably a reasonable one for central nervous system modulation as well. 100 input synapses to a given neurone occur, in an average region of locust neuropile, in a volume of about $120 \mu m^3$. This figure based on our unpublished electron micrographs, and is compatible with published data of Watson and Burrows (1982). The distance from the release site to the margin of the range will therefore be about $3 \mu m$. To get to the limit of its effective range the modulator would still have to diffuse along an intercellular pathway no more than 10nm wide, passing by glial cytoplasm capable of taking it up and destroying it en route. The effective range within the neuropile cannot be more than about $2-3 \mu m$ from a given modulator synapse.

The terminals of each DUM neurone present a different three-dimensional profile in the neuropile, although there is a small amount of overlap for some. At the regions of overlap different DUM neurones will affect the same synapses, which may be important because, as we have seen, different behaviours utilise common muscles. Nevertheless, the field of modulatory influence of a DUM neurone must be similar to the

---

modulatory neurones of locust metathoracic ganglia. It can be thought of as an orchestra capable of playing a variety of notes or chords, each with variable intensity. The ultimate neural product is movement, i.e. simple behaviour. A sequence of notes or chords is to be equated with a series of movements, or complex behaviour. (From Sombati & Hoyle, 1984b).

Generation of behaviour: the orchestration hypothesis

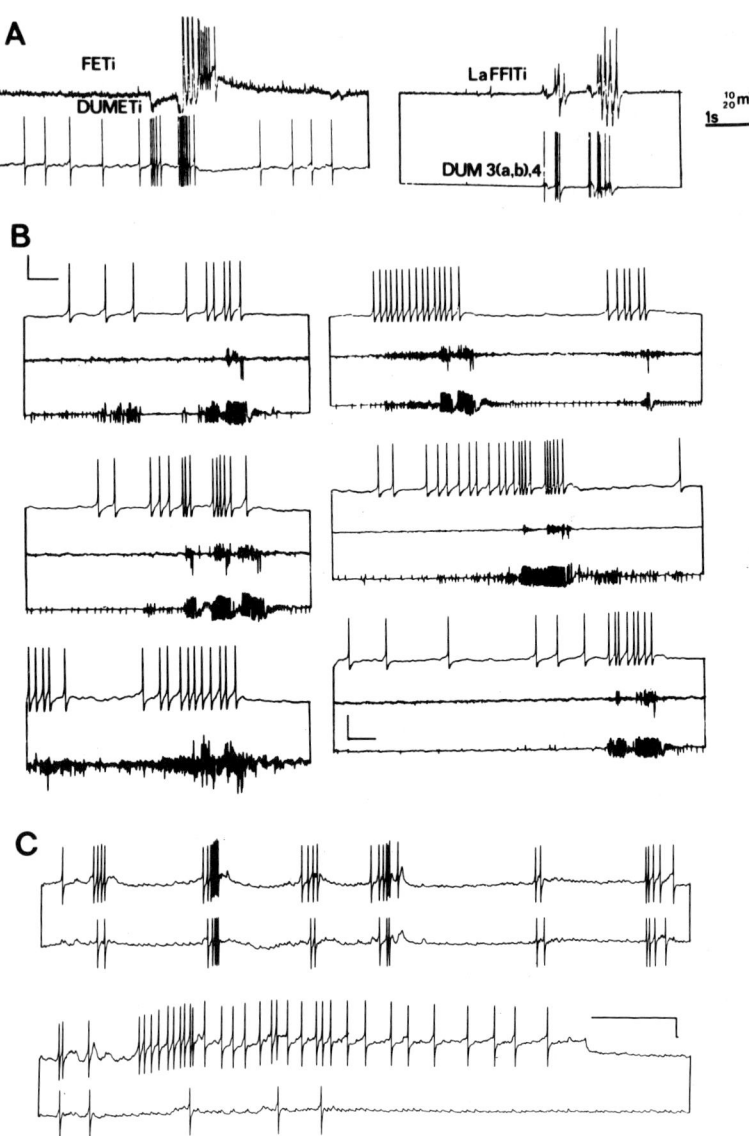

Fig. 3. Examples of intracellular activity in dorsal unpaired median (octopaminergic) neurones in preparations permitting limited behaviour. A. With intracellular recordings from identified motor neurones, as indicated (FETi, fast extensor tibiae; LaFFlTi, lateral fast flexor tibiae). (From Sombati & Hoyle, 1984a). B. With myograms recorded from

three-dimensional geometry of its terminals, plus a 3μm halo. Each much have a distinctive zone of influence. Summation of effects must occur in overlap regions when two or more DUM neurones are active at the same time.

By exciting different individual DUM neurones, or different combinations of DUM neurones, the organism would produce a wide variety of modulatory actions. These could be used to fine-tune different behaviours adaptively. They might also be used to generate behaviour if a sufficiently high local concentration of modulator were to be directly excitatory, or if there is ongoing neural activity which is subthreshold for production of movements but which can be potentiated by the modulator to a suprathreshold level. One theory of behaviour generation proposes that each MPG is running continually, like an automobile engine idling with the clutch disengaged, and that overt activity occurs when action in the CNS couples it to motor neurones, i.e. engages the clutch. This is definitely not the case for the escape swim behaviour of Tritonia, but for one insect behaviour such a mechanism has been found to be operative.

The behaviour is oviposition digging in female locusts. The neural circuitry for the rhythmic digging is complete and functional in larval forms, and is even coupled to motor neurones, because there is appropriate rhythmic motor output. However, this is too weak to produce any actual movement. When the time comes for egg-laying to occur, the digging motor bursts are intensified and at the same time neuromuscular transmission is potentiated (Thompson, 1982), a nice example of combined central plus peripheral modulation in action.

There is no sharp disparity between a clutch-engaging mechanism and initiation from rest. The circuit is there in both, and a small positive bias will turn it on even when silent. A clutch-like link to motor output might still be needed to start the behaviour. The physical layout of the locust metathoracic ganglion makes it ideal for testing the orchestration hypothesis. At least limited behaviour, certainly all forms of leg movements, and also fictive flight, can be observed and measured in preparations which permit a wide range of experimental manipulations. The latter include intracellular recording from motor neurones, interneurones and modulator neurones during execution of the behaviour, as well as iontophoretically-controlled local release into selected regions of neuropile.

To test the possibility that modular DUM neurones can be

---

flexor tibiae (middle traces) and extensor tibiae (bottom traces) metathoracic leg muscles. (From Hoyle & Dagan, 1978). C. Dual intracellular recordings from pairs of DUM neurones. Some inputs are shared, others are independent. (From Hoyle & Dagan, 1978).

Generation of behaviour: the orchestration hypothesis

generators of behaviour requires three kinds of experiment: 1) Recording from many DUM neurones during the behaviour to show that specific individuals are active during specific behaviours. 2) Demonstration that controlled local release of the modulator octopamine is followed by specific movements. 3) Evocation of specific behaviour by direct stimulation of appropriate DUM neurones.

Dr Sompong Sombati and I (Sombati, 1983; Sombati & Hoyle, 1984a,b,c) have made preliminary studies on each of these, as described below.

Correlation of spontaneous and evoked behaviour with activity in DUM neurones

We have confirmed the finding of Hoyle and Dagan (1978) that a burst of impulses occurs in one or another DUM neurone immediately before a movement (Fig. 3). There can be simultaneous activity in pairs (our current experience is from two intracellular electrodes at the same time), or in one but not another, although there is a lot of common synaptic input.

Behaviour evoked by controlled release of octopamine into specific areas of neuropile

Bathing the metathoracic ganglion in even very high concentrations (10^{-4} M) of octopamine does not evoke any specific behaviour. Pressure injection of 10^{-6} M octopamine into large areas of neuropile using pipettes of 2μm tip diameter causes long-lasting, but highly-erratic movements. A very fine-tipped iontophoretic ultra micropipette with resistance greater than 40MΩ when filled with 0.5M partially-neutralised DL-octopamine hydrochloride, having a tip diameter of about 0.2μm, was used to release octopamine into specific regions of neuropile. A transmission electron micrograph of a sample region of neuropile containing both modulator and ordinary excitatory synapses is shown in Fig. 4 to provide a sense of the scale involved. Superimposed is a silhouette of the tip of a sample iontophoretic electrode at the same magnification, made using a scanning electron microscope.

Explorations using the octopamine iontophoresis pipette were guided by a three-dimensional model of the ganglion made from serial sections after intracellular dye-filling of neurones identified physiologically. When octopamine was released into a site where there are known excitatory inputs to the flexor motor neurones of one side, these were depolarised weakly in a dose-dependent manner (Fig. 5). Provided the general level of excitation was fairly high at the time, one or more of the nine flexor motor neurones was directly excited, in an order which followed the size principle. At an exactly equivalent location on the other side, the contralateral flexors were excited.

Shifting the electrode in any direction by no more than 5μm away from the centre of the zone at which a maximal

Generation of behaviour: the orchestration hypothesis

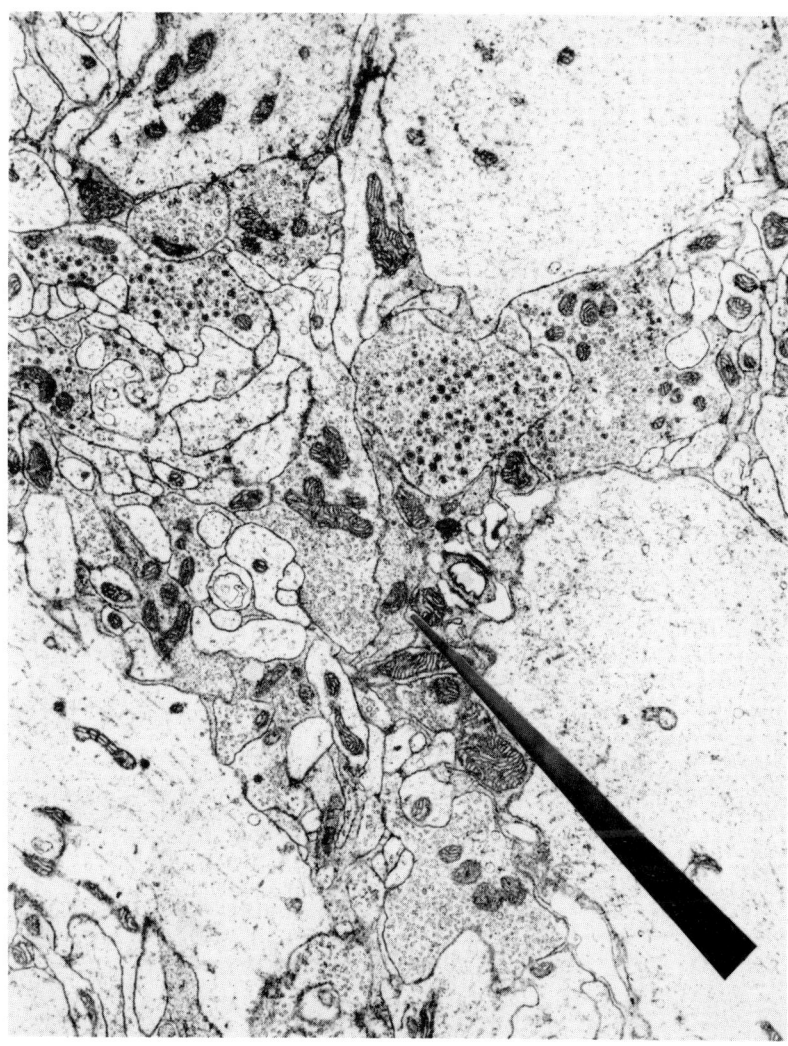

Fig. 4. Electron micrograph of locust neuropile in a region in which there are excitatory inputs to motor neurones and also terminals of modulator neurones containing dense-core vesicles. Superimposed, at the same magnification, is a silhouette of a micropipette used for iontophoretic injection of octopamine into the neuropile. (From Somabati & Hoyle, 1984a; transmission electron micrograph by Dr P.A. McNeil, silhouette, a scanning electron micrograph, by E. Schabtach).

depolarisation occurred was followed by loss of the response. No other site in the entire ganglion could be found at which a comparable simple flexion could be obtained. However, one other site was located at which an even more dramatic result occurred. At this site there was strong flexion that was followed by a prolonged sequence of extension/flexion movements (Fig. 6). If the whole leg was free to move, the movements were seen to be stepping movements, at a frequency of 2-3 steps per second, equal to fast walking. The duration of the bout, but neither the frequency nor the intensity of the movements, is dose-dependent. Similar responses occur on the contralateral side to iontophoresis of octopamine into the equivalent region. The effective radius about the optimal site was about $3\mu m$. Equivalent release at no other region evoked similar movements.

Two other sites were also found, each with its equivalent on the other side of the ganglion, at which dramatic effects were obtained. At each of these the highly characteristic rhythmical movements of fictive flight occurred (Fig. 7). The movements were bilateral. Only the duration was dose-dependent, the frequency being constant, at a mean interval of 0.43 ± 0.05s. The briefest bout of fictive flight lasted 20s and the longest an amazing 25 min, the latter in response to iontophoretic current passed for 1.5 min.

Once a bout of either stepping or fictive flight had ceased, repeat of the injection, using neither greater current strength nor a longer time, was followed by a comparable bout. We saw no sign of desensitisation and obtained as many as 30 successive bouts before any deterioration was apparent.

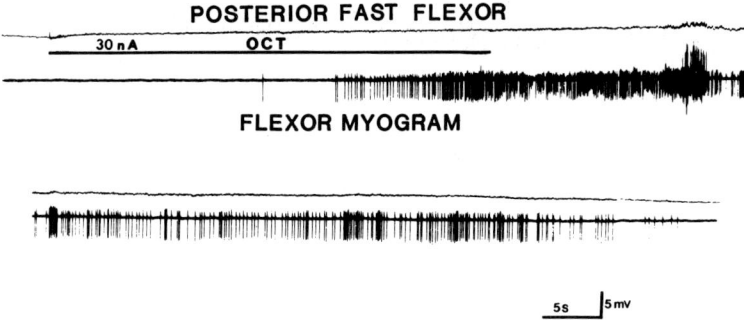

Fig. 5. Depolarisation of flexor motor neurones by octopamine iontophoresis (OCT) at the site shown in Fig. 4. The octopamine also sensitises incoming excitatory synapses which, in combination with the depolarisation, leads to early spiking of a slow flexor as indicated in the myogram obtained from the flexor tibiae muscle. This response is confined to a very narrowly-defined site. (From Sombati & Hoyle, 1984b).

Generation of behaviour: the orchestration hypothesis

<u>Evocation</u> <u>of</u> <u>behaviour</u> <u>by</u> <u>direct</u> <u>stimulation</u> <u>of</u> <u>DUM</u> <u>neurones</u>
The simplest situation in principle would be for a single DUM neurone to promote a single movement, and one would expect the movement to involve the muscle(s) innervated, if any. DUMETi has now been excited strongly in several species by a number of investigators, but none has ever reported obtaining an extension response, only potentiation of evoked ones. Kicking and jumping do not occur without activity in DUMETi, and may not be possible without it, but are not elicited unless other stimuli are present. Stimulation of other single DUM neurones have likewise not been found to cause movement. This indicated that, if the postulated orchestration mechanism were indeed operating, at least two DUM neurones must be fired either together or in overlapping bursts, in order that summation would occur and a movement be evoked. If this were the case though, there should be some occasions when the preparation had a high background activity, with some DUM neurones firing sporadically, when an evoked burst in one would exceed the threshold.

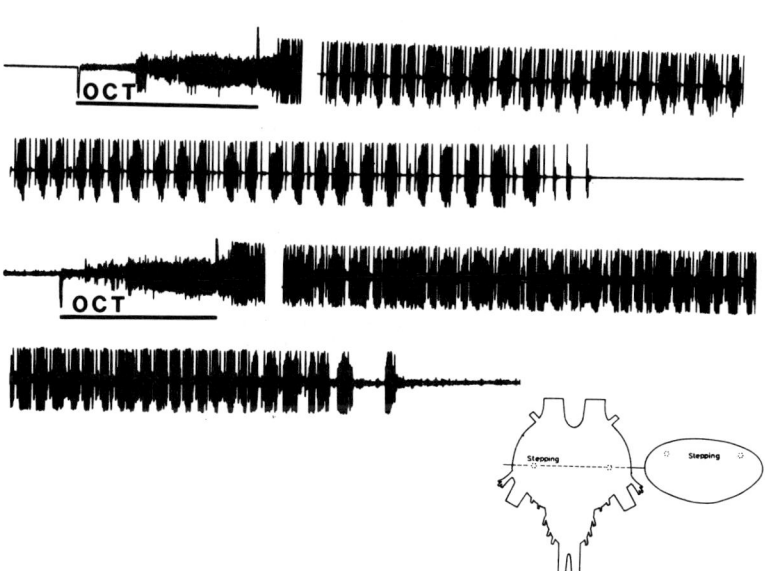

Fig. 6. Initiation of strong flexion, followed by rhythmic tibial extension/flexion movements, similar to those of stepping, as indicated by the myogram obtained from the flexor tibiae muscle. Octopamine (OCT) was iontophoretially injected at the site shown in the inset. The duration of the response is dose-dependent. (From Sombati & Hoyle, 1984b).

Generation of behaviour: the orchestration hypothesis

Such occasions have occurred during experiments by Sombati (Sombati & Hoyle, 1984b) (Fig. 8), although they can never be interpreted unequivocally. For satisfactory tests, we must await the results of multiple intracellular recordings from different DUM neurones. Only in this way will we know which ones are active in relation to specific behaviours. If they are indeed activated by commands in clusters rather than singly, the entire cluster will have to be penetrated with electrodes at the same time, and the neurones excited together, in order to simulate their natural involvement.

At this time we do not know which neurones are modulated nor in what manner they are affected physiologically, for long-lasting bouts of specific behaviour to be initiated. The three regions, two for flight and one for stepping, are ones at which there is a lot of neuropile associated with non-spiking interneurones (Siegler & Burrows, 1979). The one thing of which we are confident is that the regions are as prescribed as are the locations of identified neurone somata

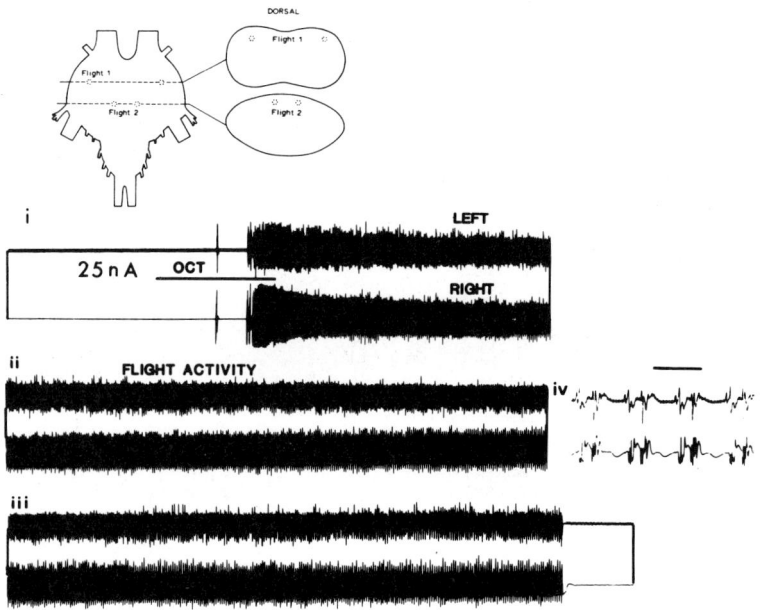

Fig. 7. Continuous record of flight muscle activity from wing depressor muscles (fictive flight) elicited from a preparation by iontophoresis of octopamine (OCT) at one of the four sites indicated in the inset above. The duration of a bout is dose-dependent, but the frequency and amplitude of individual movements are fixed. Calibration bar: i to iii, 3.6s; iv, 0.09s. (From Sombati & Hoyle, 1984b).

Generation of behaviour: the orchestration hypothesis

and the three-dimensional characteristics of their neurites and dendritic patterns. They are presumably regions where octopamine-releasing terminals of modulator neurones are concentrated close to synapses of the MPG circuits for walking and flight respectively. The sites are small, and never easy to locate. With any given preparation, one only has a few opportunities to locate them before the local neuropile has been irreparably damaged. However, the probability of success improves with experience. We have been able to elicit the behaviour sufficiently routinely to use iontophoretic release of octopamine as a tool to initiate the behaviour while using an intracellular electrode probe to locate unknown interneurones involved.

DISCUSSION

In my opinion many currently-available data are quite strongly supportive of the orchestration hypothesis, but I am well aware that many will disagree. However, unlike many branches of science, neuroethology is suffering from a paucity of hypotheses which might lead to the emergence of general principles. We have a surfeit of unrelated data, and nothing to guide us rationally in our pursuit of further facts. Under such circumstances any theory is better than none. At least the orchestration hypothesis is open to critical experimental testing. The only problem is the sheer technical difficulty of the experiments.

The hypothesis may well be global, above a minimum evolutionary level of nervous system organisation. Certainly it could apply to vertebrates, especially mammals including man, which have a formidable array of instruments in their orchestra. What a locust may do with its octopaminergic neurones is like a small string ensemble. Man has the entire

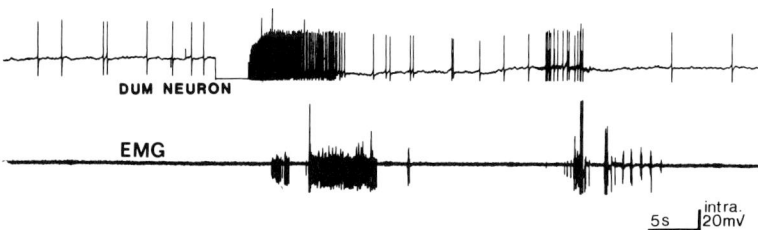

Fig. 8. Myographical activity from the metathoracic flexor tibiae muscle, initiated by a burst of impulses in an unidentified DUM neurone. The first burst was elicited at the break from hyperpolarising current. The second burst occurred as a response to touch. (From Sombati & Hoyle, 1984b).

Generation of behaviour: the orchestration hypothesis

Philadelphia orchestra plus organ and other supplements! Vertebrates have also evolved the trick of combining central and peripheral release of the same neuroactive substances. As a result of widespread neurohormonal release of some of these, which can be as different in their individual effects as those produced by adrenaline and the β-endorphins, the possibility of summed effects of release into CNS compartments, as envisaged by the orchestration hypothesis, should surely be seriously considered.

Octopamine is released globally into insect haemolymph in association with any disturbance (Davenport & Evans, 1984), and systematic changes in octopamine level doubtless play a major role in circadian rhythmicity. But this octopamine, although potentiating neuromuscular transmission and promoting the mobilisation of blood sugar (Downer, 1980), does not cause behaviour. However, as we have seen, the same substance released discretely into a small volume of neuropile evokes a long-lasting bout of a specific behaviour. In mammals, the focal release of ubiquitous, also globally-released, neuroactive substances such as noradrenaline into hypothalamic compartments reliably and routinely causes very different, discrete behaviours to occur, depending on both the nature of the substances and the regions into which they are released.

Explanation of ethological concepts

Konrad Lorenz's "reaction-specific energy" conjured up the image of slow synthesis of a neuroactive substance accumulating in modulator neurones until they were ready to burst. Octopamine could serve as such a substance for locusts. Until its concentration was above a minimal level, complex behaviour (i.e. above the reflex level) could not, according to the orchestration hypothesis, occur at all. The intensity of a behaviour would depend on the amount of octopamine accumulated in the specific DUM neurones associated with its execution since the last performance of the behaviour. Drive and mood, or readiness to perform a given behaviour, could also be simply related to modulator level.

Evolution of behaviour

The stenopelmatid orthopteran insect known to the Maoris as the 'weta' (night devil) has very large hindlegs like those of locusts, and it can kick extremely powerfully, but it does not jump. Its major metathoracic ganglion neurones including modulators have turned out to have exactly similar locations, anatomies and connections as those of locusts (Hoyle & Field, 1983b). Since it is at least 10^{10} years since wetas became isolated in New Zealand, it follows that the octopaminergic modulators, as well as various subtleties in the neuronal architecture, were evolved quite early and wetas are certainly using them functionally.

It is increasingly apparent that both neuronal anatomies

Generation of behaviour: the orchestration hypothesis

and circuits of most or all phyla are highly conservative, probably because so many genetic factors must come together correctly for an adaptive result to emerge. How can different behaviours be produced by a fixed circuit? When approached, instead of jumping, wetas adopt a threat posture, swinging both legs over the head to point the tibial spines at the offender and possibly making a forward kick (Fig. 9). If a female weta shows a threat response when approached by a male, she is attacked. The next time he approaches she adopts a flattened, submissive posture. The same muscles are involved as in a jump, but used very differently.

The answer to the questions of both versatility and plasticity may lie in subtleties of addressing a fixed circuit by modulator(s). A little extra octopamine released at excitatory synaptic input sites for coxal anterior rotator muscles of a weta, compared with a locust, would cause the whole hind leg base to rotate instead of to stiffen as it does

Fig. 9. Photograph of a female Hemideina femorata, the New Zealand 'weta', showing the long-maintained threat posture which it makes in response to a stimulus which would cause a locust to jump. The posture is developed by release of octopamine, combined with bursts of activity in slow extensor motor neurones, and utilises a catchlike tension which requires no impulses for its maintenance for many minutes.

in the co-contraction that precedes a jump in a locust. A simple adjustment in locale or even amount, of local release of octopamine would switch the behaviour drastically without changing the circuit. The question is raised: which came first in evolution, the swing, the kick or the jump? Variations on the theme of modulator biassing of parts of a behaviour sequence make more sense than neurone or circuit changes. The same applies to conditioning. Conditioning by association with a painful bite must suppress the overhead swing in females, but not in males, at least not if they are of equal size! Variations in both basic responses and conditioning by modulators, provides at least an alternative to variations in circuit properties. This may have been the route most often taken in the evolution of the neural mechanisms underlying behaviour.

REFERENCES

Arikawa, K., Washio, H. & Tanaka, Y. (1984). Dorsal unpaired median neurons of the cockroach metathoracic ganglion. J. Neurobiol., 15, 531-6.
Bentley, D.R. (1971). Genetic control of an insect neuronal network. Science, N.Y., 174, 1139-1141.
Buddenbrock, W. von (1952). Vergleichende Physiologie. Vol. 1. Sinnesphysiologie. Birkhauser, Basel.
Davenport, A. & Evans, P.D. (1984). Stress-induced changes in the octopamine levels of insect haemolymph. Insect Biochem., 14, 135-143.
Dorsett, D.A., Willows, A.O.D. & Hoyle, G. (1969). Centrally generated nerve impulse sequences determining swimming behaviour in Tritonia. Nature, Lond., 224, 711-712.
Downer, R.G.H. (1980). Short term hypertrehalosemia induced by octopamine in the American cockroach, Periplaneta americana L. In Neurotox '79: Insect Neurobiology and Pesticide Action. Society for Chemical Industry, London, pp. 335-339.
Elsner, N. & Hirth, C. (1978). Short- and long-term control of motor coordination in a stridulating grasshopper. Naturwissenschaften, 65, 160-161.
Glantzman, D.L. & Krasne, F.B. (1983). Serotonin and octopamine have opposite modulatory effects on the crayfish's lateral giant escape reaction. J. Neurosci., 3, 2263-2269.
Goodman, C.S., Pearson, K.G. & Spitzer, N.C. (1980). Electrical excitability: a spectrum of properties in the progeny of a single embryonic neuroblast. Proc. natn. Acad. Sci. U.S.A., 77, 1676-1680.
Goodman, C.S. & Spitzer, N.C. (1979). Embryonic development of identified neurones: differentiation from neuroblast to neurone. Nature, Lond., 280, 208-214.
Hoyle, G. (1952). High blood potassium in insects in relation to nerve conduction. Nature. Lond., 169, 281-282.
Hoyle, G. (1978). The dorsal, unpaired, median neurons of the locust metathoracic ganglion. J. Neurobiol., 9, 43-57.
Hoyle, G. (1984). Neuromuscular transmission in a primitive insect:

modulation by octopamine, and catch-like tension. Comp. Biochem. Physiol., 77C, 219-232.
Hoyle, G. & Dagan, D. (1978). Physiological characteristics and reflex activation of DUM (octopaminergic) neurons of locust metathoracic ganglion. J. Neurobiol., 9, 59-79.
Hoyle, G. & Field, L.H. (1983a). Defense posture and leg-position learning in a primitive insect utilize catchlike tension. J. Neurobiol., 14, 235-248.
Hoyle, G. & Field, L.H. (1983b). Elicitation and abrupt termination of behaviorally significant catchlike tension in a primitive insect. J. Neurobiol., 14, 244-312.
Kutsch, W. & Otto, D. (1972). Evidence for spontaneous song production independent of head ganglia in Gryllus campestris, L.. J. comp. Physiol., 81, 115-119.
Livingstone, M.S., Harris-Warrick, R.M. & Kravitz, E.A. (1980). Serotonin and octopamine produce opposite postures in lobsters. Science, N.Y., 208, 76-79.
Lorenz, K.Z. (1950). The comparative method in studying innate behaviour patterns. Symp. Soc. exp. Biol., 4, 221-268.
Lorenz, K.Z. (1981). The Foundations of Ethology. Springer Verlag, New York.
Merton, P.A. (1953). Speculations on the servo-control of movement. In The Spinal Cord. Ed. Wolstenholme, G.E.W. Churchill, London, pp. 247-255.
Siegler, M.V.S. & Burrows, M. (1979). The morphology of local non-spiking interneurones in the metathoracic ganglion of the locust. J. comp. Neurol., 183, 121-148.
Sombati, S. (1983). Neuroethological pharmacology of octopamine in the locust. Ph.D. Thesis. University of Oregon.
Sombati, S. & Hoyle, G. (1984a). Central nervous sensitization and dishabituation of reflex action in an insect by neuromodulator octopamine. J. Neurobiol., 15, 455-480.
Sombati, S. & Hoyle, G. (1984b). Generation of specific behaviors in a locust by local release into neuropil of the natural neuromodulator octopamine. J. Neurobiol., 15, 481-506.
Sombati, S. & Hoyle, G. (1984c). Glutamatergic central nervous transmission in locusts. J. Neurobiol., 15, 507-516.
Thompson, K.J. (1982). The neural basis of the motor pattern for grasshopper oviposition digging. Ph.D. Thesis. University of Oregon.
Treherne, J.E. (1974). The environment and function of insect nerve cells. In Insect Neurobiology. Ed. Treherne, J.E. North-Holland, Amsterdam, pp. 187-244.
Watson, A.H.D. & Burrows, M. (1982). The ultrastructure of identified locust motor neurones and their synaptic relationships. J. comp. Neurol., 205, 383-397.

Chapter Five

CONVERGENCE OF SEVERAL SENSORY MODALITIES IN MOTOR CONTROL

A. Taylor and S. Gottlieb

During the course of normal body movements many sensory receptor types are stimulated. They comprise those which seem to be specially evolved to register active movement, namely the proprioceptors, and those which are additionally very sensitive to external stimuli, namely the cutaneous mechanoreceptors. The elucidation of the connections linking these various receptors to motor neurones has led to a great body of knowledge which must be built into any comprehensive scheme of movement control.

One popular approach is to recognise some pathways which are so strong that the activation of one sensory modality or of a limited group of modalities in a well defined spatiotemporal pattern produces a well defined motor output pattern, which is termed a reflex response. Many such reflexes have been described and the function of the whole motor system has sometimes been thought of as that of an assemblage of reflex responses, disturbed and provoked into action by central motor drives or external stimuli. However, many of the above sensory inputs are remarkably ineffective in producing active contractions when stimulated separately, and their significance in regulating normal movements must be seen in terms of the convergence of various sensory modalities which occurs during the course of movements. The purpose of this chapter therefore is to point out that many so-called reflexes may be better regarded as fractionated responses of a complex system which normally works by the convergence of a wide variety of sensory inputs during centrally generated motor acts. This topic is very much bound up with interneuronal mechanisms and so the first section will deal with their properties and their general significance. The next section will consider the way in which convergence of different sensory modalities on interneurones can be regarded as an application of sensory classification. Finally, convergence of two different proprioceptive modalities will be presented as a way by which feedback control can economically be made more flexible than is generally thought.

Monosynaptic and multisynaptic reflexes

Only one reflex pathway is known in vertebrates which contains no interneurones. This is the muscle stretch reflex in which spindle Group Ia afferents and to some extent Group II afferents make monosynaptic excitatory connection with homonymous and agonist motor neurones (for review see Homma, 1976). All others involve one or more interneuronal relays. Inhibitory action requires the presence of inhibitory interneurones, but beyond that the functional reason for the added complexity of additional interneurones in most reflexes is not immediately obvious. One useful consequence of routing sensory input through a relay is that signal transmission through it may be interrupted by post-synaptic inhibition of interneurones without affecting the responsiveness of target motor neurones to other excitatory inputs. Also, depression of inhibitory interneurones can block an inhibitory action without risking the unwanted firing of motor neurones which might result if the inhibition were counteracted by post-synaptic excitation delivered direct to the motor neurones. By contrast, the monosynaptic reflex is so unusual and potentially so inflexible that it is perhaps most appropriate to try to explain why it exists at all. It is notable that it is best developed in muscles which have a clear function to maintain posture against gravity. These include the extensors of the hind limbs and forelimbs in quadrupeds, and in man the lower limb extensors. In the human arm, monosynaptic reflexes are weaker, in line with the lesser task of supporting only the weight of the limb rather than that of the body. The lower jaw also is a light load relative to the strength of the muscles needed for mastication and here again the monosynaptic connections of the spindles are weak (Appenteng et al., 1978). The jaw depressor muscles never have an antigravity role and not surprisingly therefore lack a stretch reflex (indeed they seem not to be provided even with muscle spindles). The extra-ocular muscles also have no antigravity function and so, in those species in which these muscles do contain spindles, no stretch reflex connections exist (Baker & Precht, 1972). Within the muscles which do have a postural function the monosynaptic Group Ia spindle input is most strongly directed to the motor neurones innervating small, slow, non-fatiguable (S type) units which are clearly best suited to provide postural tone economically (see Burke et al., 1976). Indeed there is a well defined gradation of the size of individual Group Ia EPSPs, the size of which is inversely proportional to the mechanical strength of the muscle units innervated (Harrison & Taylor, 1981; Zengel et al., 1983). That is, the larger the motor unit, the less influential is the monosynaptic Group Ia input in recruiting it.

Does this mean that the muscle spindle afferents, so well developed for monitoring muscle length changes, have little influence over the larger motor neurones responsible for about

75% of the force-developing capacity of the muscle? It can be argued that though the Group Ia monosynaptic influence on large motor neurones is small, the effect of it in terms of gain in the stretch reflex is scaled up because of the large force production of their muscle units (Harrison, 1983). However, the other factor which must be considered is the existence of multisynaptic stretch reflexes. Simple observation shows that the full strength of the arm flexor muscles may be engaged in resisting an attempt to extend the limb forcibly, though the monosynaptic stretch reflex appears to be quite weak (Hammond et al., 1956). It is not certain that the former effect is a spinal stretch reflex; it may be thought to involve 'voluntary reaction'. However the very strong reflex spasticity of arm flexors following a clinical 'upper motor neurone' lesion is in favour of the existence of potentially strong multisynaptic stretch reflexes, supplementing the weak postural monosynaptic reflex when appropriate.

The arguments in favour of multisynaptic autogenetic excitatory pathways for Group Ia afferents were well expressed by Matthews (1972) who emphasised that the difficulty of demonstrating them by the usual techniques in anaesthetised or spinal animals is probably due to the need for facilitation of the interneurones from other, notably supraspinal, sources. The observations illustrated in Fig. 1 give some further evidence for multisynaptic Group Ia excitation of motor neurones, in this case those of jaw-closing muscles in the decerebrate cat. Intracellular records from the three different motor neurones illustrated indicate that the occurrence of late EPSPs following Group Ia stimulation is most marked in cells in which the monosynaptic EPSP is smallest. Since the monosynaptic Group Ia EPSPs are most marked in the small, slow motor neurones it follows that the multisynaptic stretch reflex is best developed in the large, fast, high threshold unit motor neurones. Thus it may be that while the monosynaptic stretch reflex is mostly relevant for postural control, the multisynaptic reflex can exert length feedback control over motor neurones recruited by stronger efforts, notably in manipulating loads. From this point of view the monosynaptic reflex is to be regarded as the simpler and less flexible control, existing largely for antigravity action which is a well defined and constant function in any particular muscle group as indicated earlier. The multisynaptic reflex is flexible by virtue of the ease and speed with which its interneurones may be switched on or off by descending influences or by convergence of appropriate peripheral input. It is worth noting that in their series of experiments on load resisting reflexes of the human hand and arm, Merton, Marsden and Morton (see e.g. Marsden, 1973) found that the responses to loading the limb movements were greatly reduced by local anaesthesia of the skin. Possibly the reduction of sensory input affected cortical areas which those authors believe are most important as part of a long loop compensating 'reflex'. The more economical suggestion is that sensory feedback from skin normally has a facilitatory effect on interneurones of a spinal multisynaptic stretch reflex.

Returning then to the control requirements of different muscle groups the following case can be made. Body weight supporting muscles have strong postural functions and strong monosynaptic reflexes. Non-weight bearing limb muscles and jaw-closing muscles have minor postural function and weak monosynaptic reflexes. They do however have important load moving functions for which flexible length feedback control is right and this is reflected in strong multisynaptic reflexes.

Extra-ocular muscles have no postural function, and since loading is constant, no displacement feedback control is needed. There are indeed no spindle reflex connections to the extra-ocular motor neurones, mono- or multisynaptic. The same is true of the jaw opening muscles.

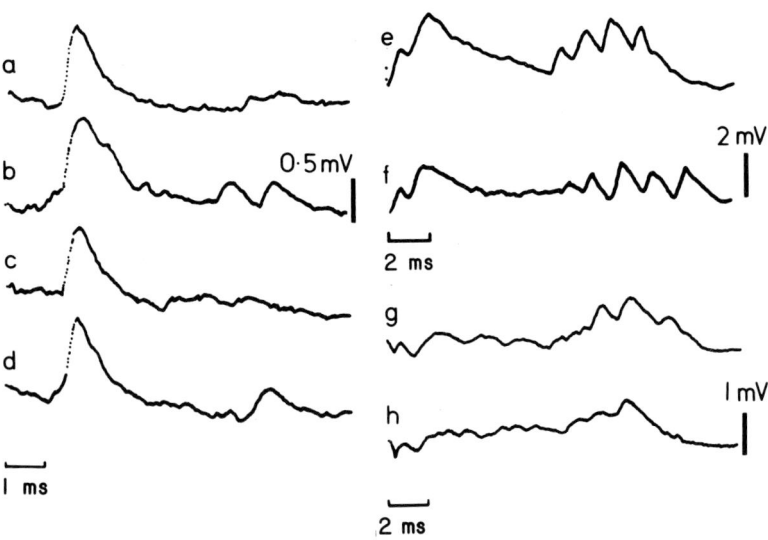

Fig. 1. Monosynaptic and multisynaptic EPSPs generated in masseter motor neurones by masseter nerve stimulation at a strength just sufficient to excite Group 1a afferents. Records a-d from a motor neurone showing a large monosynaptic EPSP; e & f from another motor neurone with a smaller EPSP, and g & h from another with very small monosynaptic EPSPs. The later, multisynaptic effects are approximately inversely related in size to the monosynaptic EPSPs. (Data obtained in collaboration with K. Appenteng).

Interneurones classify sensory inflow

Many studies have now been reported in which convergence of muscle Group 1a, 1b, joint and cutaneous afferents has been demonstrated to occur on interneurones in laminae V-VI of the spinal cord (see particularly Jankowska, 1979; Lundberg, 1979; Czarkowska et al., 1981; Jankowska et al., 1981). The projections of these interneurones are not fully established but many send collaterals to the motor nuclei where they presumably excite or inhibit motor neurones.

The functional significance of the convergence of quite different sensory modalities on interneurones which in turn project to motor neurones has not been discussed in any great detail. Matthews (1972) in discussing the Group 1a inhibitory interneurone referred to its 'integrative' capacity occurring through the interaction of several different kinds of input. Jankowska, Johannisson and Lipski (1981) express the general idea that the interneurones may be regarded as shared by two or more reflexes. The implication is that interneurones receiving Group 1a and 1b excitation for example might be made to fire by either input alone, in which case the effect is rather like an 'OR' gate. In fact the intracellular records rarely show the EPSPs from one source alone to be large enough to fire the interneurones. It seems more usual for simultaneous inputs from two or more sources to be needed to do this and so to produce a reflex effect in the motor neurones. In this case their behaviour is more like that of an 'AND' gate, and the interneurone may be said to have detected in the total sensory input the occurrence of some combination which has particular significance to the animal during a movement. This is similar to, but more specific than, the view of Lundberg, Malmgren and Schomberg (1977) who said ". . . . . but we can by no means exclude the possibility of more differentiated control systems through reflex pathways in which afferents from diverse receptors act together in different combinations". One is reminded of the principle by which orderly convergence on interneurone subpopulations is believed to account for sensory perception (see e.g. Somjen, 1972). In this scheme, greatly elaborated for the visual system, the firing of particular neurones, indicates particular features of interest by virtue of convergence from sensory elements activated by the essential components of the feature. In the same way features of importance in motor control may be detected by suitable interneuronal convergence.

Application for the hindlimb in locomotion

The existence of interneurones with excitatory input from Group 1a spindle afferents and Group 1b tendon organ afferents from a muscle means that a lengthening contraction of that muscle will be detected by the firing of these particular interneurones. Similarly, interneurones excited by Group 1a but inhibited by Group 1b would indicate stretch of an inactive muscle. Neurones with convergence of excitation from Group 1b, but inhibition from Group 1a would respond during active shortening. Illustrations of these possibilities are shown in Fig. 2. All such combinations of input to interneurones in laminae V-VI have been reported by Jankowska and her colleagues. In addition, low threshold cutaneous afferents, whose adequate stimulus is light touch, have also been found to converge on interneurones receiving muscle afferent information, by Lundberg, Malmgren and Schomberg (1977). These authors discuss the interpretation of their results in terms of either "reflexes which regulate movements

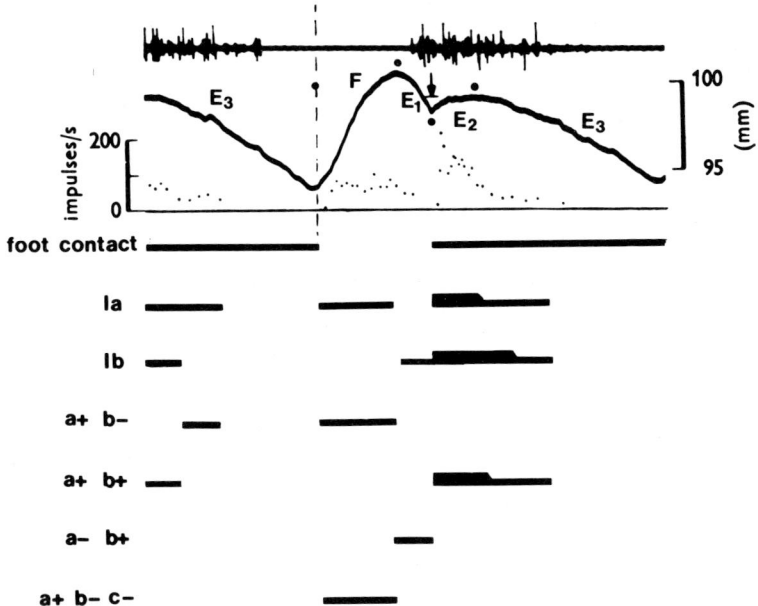

Fig. 2. Proprioceptive feedback from triceps surae of the cat during walking and its effects on interneurones. The experimental records (upper three traces) are taken with kind permission from Prochazka et al. (1977) and show from above down: e.m.g., muscle length and Group 1a afferent discharge frequency. The timing of foot contact and Groups 1a and 1b afferent firing is indicated by the horizontal bars. Also shown are the expected firing patterns of interneurones receiving excitation (+) and inhibition (-) from Groups 1a (a), 1b (b) and cutaneous (c) afferents. The assumption is made that two or more sources of excitation are required to make interneurones fire and that removal of inhibition can be taken as equivalent to excitation. The pattern of Group 1b firing is deduced from Prochazka & Wand (1980). F indicates the flexion phase and E_1, E_2 and E_3 the different extension phases of the locomotor cycle.

together . . . ." or the cutaneous input modifying the gain of a muscle reflex acting through the interneurones, and favour the former. Using the approach of the present discussion, it is suggested that convergence of touch afferents, from say the sole of the foot, with extensor Group Ib afferents would permit recognition of the situation in which force is being exerted by the extensor muscles on the foot in contact with a surface. In the above quoted experiments the specific finding was of an inhibition of motor neurones of the ankle extensor muscle, plantaris, by Group Ib afferents from its agonist, triceps surae, greatly facilitated by cutaneous input. The present interpretation would therefore be that recognition of foot contact accompanied by increasing force in triceps would lead to inhibition of plantaris. This would be quite appropriate at the onset of the stance phase in walking when triceps is yielding, but not during the second half of that phase when triceps is shortening. Thus, by the present theory, we might expect this particular group of Group Ib interneurones to have in addition excitation from triceps Group Ia afferents. The details of this sort of scheme will have to be worked out. However, a qualitative idea of some of the various categories of classification of cutaneous and muscle afferent data which would be expected to exist for controlling locomotion can be grasped from Fig. 2. It is quite easy to see how the occurrence of each different phase of locomotion could be detected. With appropriate connection of the interneurones to the motor neurone pools a mechanism would be provided for adjusting the duration of the different parts of the cycle in response to changing speed, as described by Grillner (1973) and by Stuart, Withey, Wetzel and Goslow (1973).

Application for jaw movement control

Another example of convergence of sensory modalities is provided by the control of jaw movement. The basic neural arrangement for rhythmic jaw movements appears to be quite simple, particularly in carnivores in which the jaw joint allows only for rotation about a single axis. Jaw closing muscles, temporalis and masseter, contain muscle spindles, primary and secondary endings of which make autogenetic excitatory monosynaptic connections, particularly to the motor neurones of S type units (see Appenteng et al., 1978). As mentioned above, the jaw opening muscles have no spindles. The other main proprioceptors of the jaw are the periodontal receptors (see e.g. Linden, 1978) which are very sensitive to forces on the teeth. They make disynaptic inhibitory connections with masseter and temporalis motor neurones (Kidokoro et al., 1968). The central pattern generator (Dellow & Lund, 1971) drives the opener and closer motor neurones reciprocally (Nakamura et al., 1981).

Now it will be readily accepted that use of the jaw

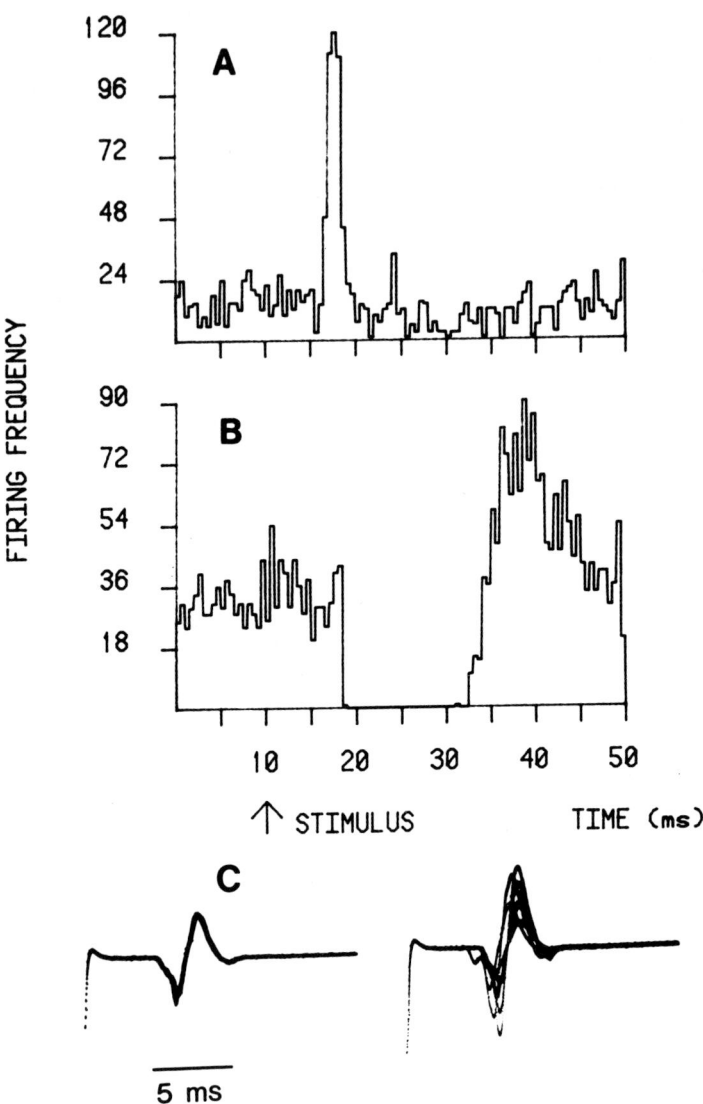

Fig. 3. Illustration of the basic proprioceptive reflexes of the jaw muscles. A: peri-stimulus time histogram (PSTH) of motor unit in masseter with minute transient stretch of jaw closing muscles. B: PSTH of the same motor unit with threshold electrical stimulation of ipsilateral inferior alveolar nerve, which contains periodontal afferents from the lower jaw. C: e.m.g. response of a jaw opening muscle, the digastric, to threshold

muscles for eating drinking, licking, vocalisation and many other purposes involves complex control likely to depend heavily on sensory feedback. However, of the various (non-nociceptor) sources of such feedback, only one (the muscle spindles) produces definite reflex contraction of closers and one other (the periodontal afferents) produces contraction of the openers (which accompanies the inhibition of closers). Fig. 3 demonstrates these responses in the cat.

Though cutaneous innervation is very rich around the mouth and is strongly activated during active jaw movements (Appenteng et al., 1982), stimulation of low threshold cutaneous afferents by electrical or natural means does not cause contraction of muscles acting on the jaw, nor does it seem to inhibit tonic contraction of closer muscles. Nevertheless, the cutaneous afferents can be shown to have potent effects on the proprioceptive reflexes. In some cases enhancing them, in others depressing them. For example in Fig. 4A reflex contraction of anterior digastric muscle is evoked by just suprathreshold stimulation of the inferior alveolar nerve (i.e. to excite periodontal afferents). In (B) the electrical stimulus is preceded by a light mechanical tap to the skin at the angle of the mouth, and this reduces the reflex response. The tap alone had no effect on masseter or digastric.

Another effect is shown in Fig. 4C to I. Inferior alveolar stimulation again produces a digastric response (C). Brief, light pressure alone on the hard palate can produce a small digastric response (D & E) and when the smaller palatal stimulus is applied with the electrical inferior alveolar nerve stimulus (F) the response is enhanced. However, when the mechanical stimulus precedes the electrical by various intervals (G,H,I) the digastric response is greatly reduced. The most probable explanation is that the low threshold cutaneous mechanoreceptors inhibit interneurones in the inferior alveolar to digastric reflex pathway. Evidence is also available of a similar effect from masseter spindle afferents.

Finally, an opposite effect is demonstrated in Fig. 5 from palatal receptors on the inhibition of masseter motor neurones by inferior alveolar stimulation. The post stimulus histogram of (A) shows the striking inhibition of a masseter motor unit by periodontal afferent stimulation. Histogram (B) shows that palatal pressure alone has no effect, but when this

---

stimulation of inferior alveolar nerve. The following record shows the effect of simultaneously tapping the ipsilateral upper canine tooth. The enhanced response with identical timing suggests that the tap excites afferents of the same type as those electrically stimulated. Chloralose anaesthetised cat.

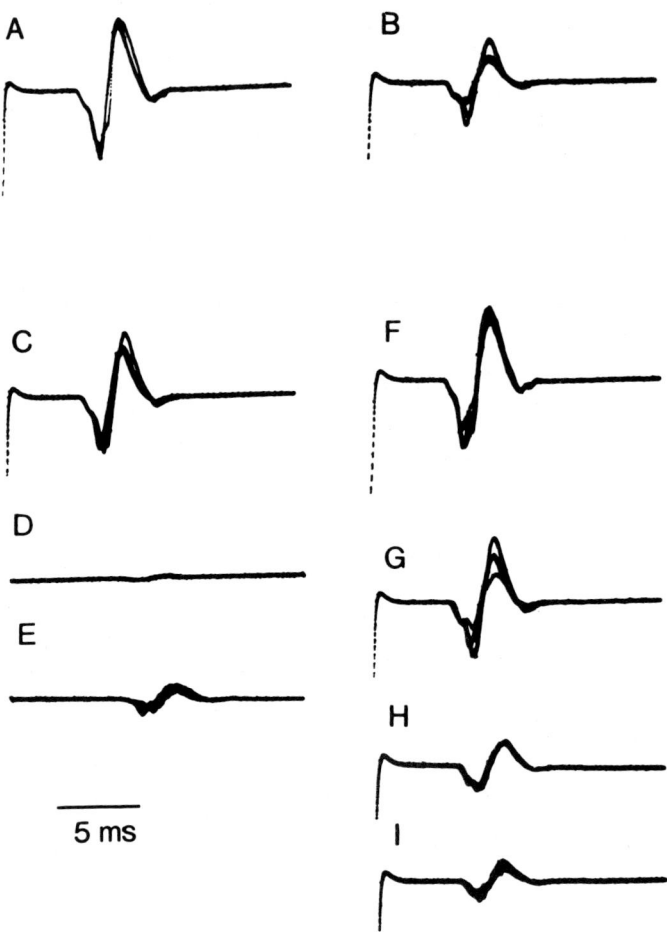

Fig. 4. Responses of the anterior digastric muscle (e.m.g.) to inferior alveolar nerve stimulation (threshold for periodontal afferents) and the effect of cutaneous and palatal mechanical stimuli. A: control, inferior alveolar stimulation alone. B: a tap to the corner of the mouth precedes the electrical stimulus by 100ms. C: control, inferior alveolar stimulation alone. D: effect of light tap and E: stronger tap to the hard palate. F: tap of D accompanies electrical stimulation. G, H, I: light tap precedes electrical stimulus by 20, 40 and 80ms. All records 5 superimposed sweeps at the same sensitivity. Chloralose anaesthetised cat.

pressure precedes the periodontal stimulus by 60ms the inhibitory effect of the latter is largely abolished (C). The most reasonable explanation for this is that the palatal afferents also cause inhibition of the periodontal to masseter interneurones.

All these effects require proper documentation by intracellular recording from the interneurones and motor neurones (for review, see Nakamura, 1980) but for the moment one general conclusion is probably quite safe. That is that there exist complex patterns of convergence from non-nociceptive cutaneous afferents with proprioceptive afferents from the jaw apparatus. These patterns of convergence may well be regarded as the basis for useful classifications of sensory data by interneurones leading to appropriate motor responses along the general line suggested for the hindlimb situation. For example, excitation of periodontal together with palatal receptors may be taken to mean that the mouth is closing on food, in which case it is reasonable to depress the inhibitory effect of periodontal afferents on masseter and their excitatory effect on digastric. Force on teeth alone means that the teeth have met without food exerting pressure on the palate and a strong inhibition of the bite is appropriate.

Proprioceptive convergence may permit switching of the controlled variable

Since jaw closing motor neurones receive negative length feedback from the muscle spindles, the excitatory drive from the pattern generator can be regarded as a command to shorten the muscle and close the jaw, and loading will be to some extent compensated by the displacement feedback (Taylor et al., 1981). This is perfectly appropriate when there is little resistance, but when hard food is encountered or the teeth come into contact, then any slight misjudgement of the command to close would be potentially damaging. At this point however firing of the periodontal receptors starts to increase (Anderson, 1976; Appenteng et al., 1982) and provides, through disynaptic inhibition of jaw closing motor neurones, negative feedback of force. Thus, since the length feedback loop has been effectively opened by arrest of movement and the force feedback loop closed, control changes automatically and smoothly from 'length servo' to 'force servo'. To some extent, this principle of convergence leading to automatic and appropriate change of control strategy may also apply in other situations such as in grasping an object. Initially the hand closes under length control with virtually no load till the object is contacted. From then on force control may be progressively asserted by Golgi tendon organ feedback via the Group 1b inhibitory interneurones. Of course cutaneous afferents will be excited from the moment of contact and they may, by convergence on Group 1b interneurones, facilitate the force feedback pathway (c.f. Marsden, 1973). The existence of

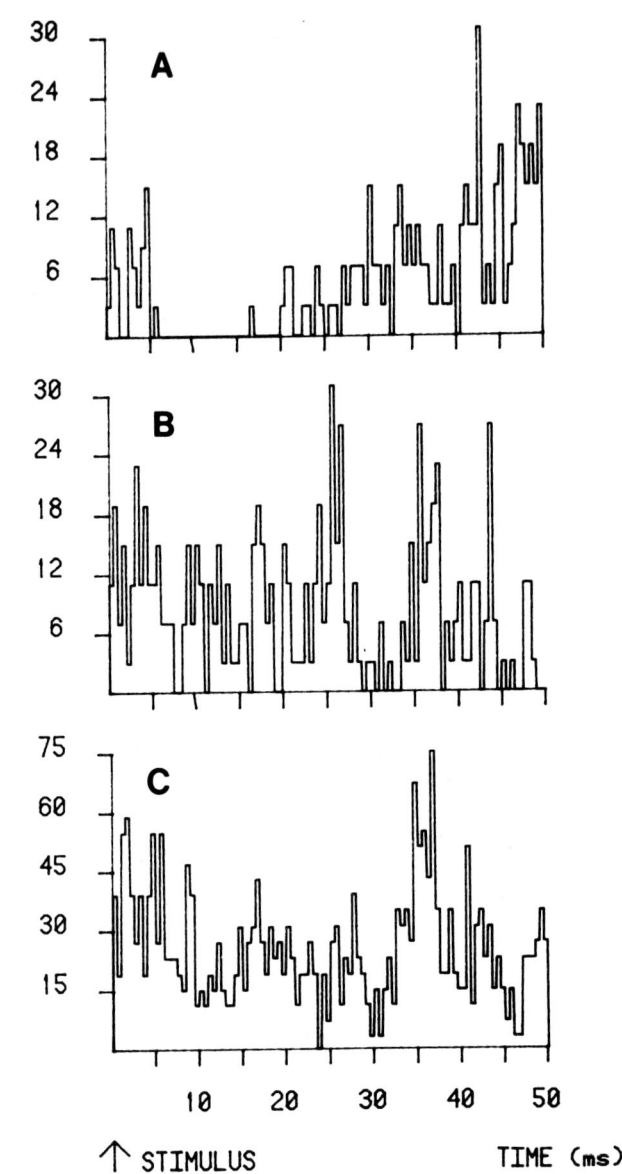

Fig. 5. Post stimulus time histograms to show interaction of palatal pressure with inhibition of masseter motor units by inferior alveolar stimulation. A: nerve stimulation alone (threshold for periodontal afferents) B: palatal tap alone. C: palatal pressure pulse precedes nerve stimulus by 60ms. Chloralose anaesthetised cat.

appropriate pathways for the human hand may be inferred from the cutaneous reflexes from low threshold afferents (Jenner & Stephens, 1982) demonstrable on a background of active contraction.

An alternative view of the interaction between tendon organ and muscle spindle signals in movement control is proposed by Houk and his collaborators (Houk et al., 1981). It is argued that simultaneous negative feedback from force detectors and displacement detectors can be represented as stiffness control. The logic of this argument is of course good, but the application of it in understanding natural movements is not easy, because the most obvious way in which it might be helpful is in setting the stiffness of the system in the face of externally applied force. Much of motor activity is directed to moving the limbs and other parts of the body often in relatively unloaded or predictably loaded situations. Once the intended displacement has been achieved the task usually changes to a rather isometric one in which displacement control or stiffness control seem to be much less appropriate descriptions of the effect of the predominant feedback (force and cutaneous) than force control. In other words, smooth and automatic changes between displacement and force control may be the consequence of the evolved design, and the intermediate situation of 'stiffness' control may be exploited only rarely.

Conclusions
No single theory can account for all the functions and consequences of multi-modal sensory convergence in motor control. We have advanced two main ideas here, namely the use of sensory classification to identify functionally significant situations, and the automatic transfer of control from one variable to another. Other possibilities of course exist, but cannot be considered in a short account. In particular, there is increasing interest in the way in which various sensory inputs generated during movement may influence fusimotor neurone activity. The consequence of this, if it should prove that the fusimotor reflexes are of significant strength, will be very complicated indeed.

For the moment the usefulness of the two ideas considered here will be measured in terms of what new observations they may stimulate. One question which clearly needs further examination is the strength of the converging drive from different sources on to interneurones. Estimates of this are needed to decide whether they have an 'OR' function or an 'AND' function as proposed. Another problem is the distribution of excitatory and inhibitory effects from the interneurones to motor neurones. The elucidation of the anatomy of the pathways must be accompanied wherever possible by estimates of their strength.

Acknowledgements

The technical assistance of Mr R.B. Chandler and Mr M.A. Bosley is gratefully acknowledged. The work was supported by Medical Research Council Grants G8111091 NA and G8311950 N.

REFERENCES

Anderson, D.J. (1976). The incidence of tooth contacts in normal mastication and the part they play in guiding the final stage of mandibular closure. In Mastication. Eds. Anderson, D.J. & Matthews, B.. John Wright, Bristol, pp. 237-241.

Appenteng, K., Lund, J.P. & Séguin, J.J. (1982). Behaviour of cutaneous mechano-receptors recorded in mandibular division of Gasserian ganglion of the rabbit during movements of the lower jaw. J. Neurophysiol., 47, 151-166.

Appenteng, K., O'Donovan, M.J., Somjen, G., Stephens, J.A. & Taylor, A. (1978). The projection of jaw elevator muscle spindle afferents to fifth nerve motoneurones in the cat. J. Physiol., 279, 409-423.

Baker, R. & Precht, W. (1972). Electrophysiological properties of trochlear motoneurones as revealed by IV nerve stimulation. Expl. Brain Res., 14, 124-157.

Burke, R.E., Rymer, W.Z. & Walsh, J.V. (1976). Relative strength of synaptic input from short latency pathways to motor units of defined type in cat medial gastrocnemius. J. Neurophysiol., 39, 447-458.

Czarkowska, J., Jankowska, E. & Sybirska, E. (1981). Common interneurones in reflex pathways from group 1a and 1b afferents of knee flexors and extensors in the cat. J. Physiol., 310, 367-380.

Dellow, P.G. & Lund, J.P. (1971). Evidence for central timing of rhythmical mastication. J. Physiol., 215, 1-13.

Grillner, S. (1973). Locomotion in the spinal cat. In Control of Posture and Locomotion. Eds. Stein, R.B., Pearson, K.G., Smith, R.S. & Redford, J.B. Plenum, New York, pp. 515-535.

Hammond, P.H., Merton, P.A. & Sutton, G.G. (1956). Nervous gradation of muscular contraction. Br. med. Bull., 12, 214-218.

Harrison, P.J. (1983). The relationship between the distribution of motor unit mechanical properties and the forces due to recruitment and to rate coding for the generation of muscle force. Brain Res., 264, 311-315.

Harrison, P.J. & Taylor, A. (1981). Individual excitatory post-synaptic potentials due to muscle spindle 1a afferents in cat triceps surae motoneurones. J. Physiol., 312, 455-470.

Homma, S. (Ed.). (1976). Understanding the Stretch Reflex. Prog. Brain Res., 44.

Houk, J.C., Crago, P.E. & Rymer, W.Z. (1981). Function of the spindle dynamic response in stiffness regulation - a predictive mechanism provided by non-linear feedback. In Muscle Receptors and Movement. Eds. Taylor, A. & Prochazka, A. MacMillan, London, pp. 299-310.

Jankowska, E. (1979). New observations on neuronal organisation of reflexes from tendon organ afferents and their relation to reflexes evoked from muscle spindle afferents. Prog. Brain Res., 50, 29-36.

Jankowska, E., Johannisson, T. & Lipski, J. (1981). Common interneurones in reflex pathways from group 1a and 1b afferents of ankle extensors in the cat. J. Physiol., 310, 381-402.

Jenner, J.R. & Stephens, J.A. (1982). Cutaneous reflex responses and their central nervous pathways studied in man. J. Physiol., 333, 404-419.

Kidokoro, Y., Kubota, K., Shuto, S. & Sumino, R. (1968). Reflex organisation of cat masticatory muscles. J. Neurophysiol., 31, 695-708.

Linden, R.W.A. (1978). Properties of intraoral mechanoreceptors represented in the mesencephalic nucleus of the fifth nerve in the cat. J. Physiol., 279, 395-408.

Lundberg, A. (1979). Multisensory control of spinal reflex pathways. Prog. Brain Res., 50, 11-28.

Lundberg, A., Malmgren, K. & Schomberg, E.D. (1977). Cutaneous facilitation of transmission in reflex pathways from 1b afferents to motoneurones. J. Physiol., 265, 763-780.

Marsden, C.D. (1973). Servo control, the stretch reflex and movement in man. In New Developments in Electromyography and Clinical Neurophysiology, Vol. 3. Ed. Desmedt, J.E. Karger, Basel, pp. 375-382.

Matthews, P.B.C. (1972). Mammalian Muscle Receptors and their Central Actions. Edward Arnold, London, pp. 357-361.

Nakamura, Y. (1980). Brainstem neuronal mechanisms controlling the trigeminal motoneuron activity. In Spinal and Supraspinal Mechanisms of Voluntary Motor Control and Locomotion. Ed. Desmedt, J.E. Karger, Basel, pp. 181-202.

Nakamura, Y., Kubo, Y., Nozaki, S., Enomoto, S. & Katoh, M. (1981). Brain stem control of masticatory rhythm. In Oral-Facial Sensory and Motor Functions. Eds. Kawamura, Y. & Dubner, R. Quintessence, Tokyo, pp. 37-44.

Prochazka, A. & Wand, P. (1980). Tendon organ discharge during voluntary movements in cats. J. Physiol., 303, 385-390.

Prochazka, A., Westerman, R.A. & Ziccone, S.P. (1977). 1a afferent activity during a variety of movements in the cat. J. Physiol., 268, 423-448.

Somjen, G. (1972). Sensory Coding in the Mammalian Nervous System. Appleton-Century-Crofts, New York.

Stuart, D.G., Withey, T.P., Wetzel, M.C. & Goslow, G.E. (1973). Time constraints for inter-limb co-ordination in the cat during unrestrained locomotion. In Control of Posture and Locomotion. Eds. Stein, R.B., Pearson, K.G., Smith, R.B. and Redford, J.B.. Plenum Press, New York, pp. 537-560.

Taylor, A., Appenteng, K. & Morimoto, T. (1981). Proprioceptive input from the jaw muscles and its influence on lapping, chewing and posture. Can. J. Physiol. & Pharmacol., 59, 636-644.

Zengel, J.E., Reid, S.A., Sypert, G.W. & Munson, J.B. (1983). Presynaptic inhibition, EPSP amplitude and motor unit types in triceps surae motoneurones in the cat. J. Neurophysiol., 49, 922-931.

Chapter Six

FEEDBACK CONTROL OF AN ESCAPE BEHAVIOUR

J.M. Camhi

The neuronal basis of escape behaviours has been a favoured subject of neurophysiological investigation for almost fifty years. Among the advantages of studying such systems are the relatively great stereotypy of the movements, simplifying behavioural descriptions, and the large axonal diameter of the participating neurones, permitting ease of experimental analysis. Considerable behavioural and neurophysiological insights have been acquired from studies of escape systems in earthworms (Drewes, 1984), crickets (Murphey, 1981), cockroaches (Camhi, 1980), crayfish (Wine & Krasne, 1982), and teleost fish (Eaton, 1983).

In the cockroach, cricket and teleost fish the escape movements can be subdivided into an initial rapid turn away from the stimulus, followed by translatory locomotion. In crayfish, the first movement is either a forward or a backward dart, depending upon whether the initiating stimulus comes from, respectively, the posterior or anterior direction. The behavioural latency of escape behaviour is generally very short (5ms in goldfish; Eaton, 1983; 14ms in cockroach; Camhi & Nolen, 1981; 10-12ms in crayfish; Wine & Krasne, 1982) and the subsequent translatory component begins soon thereafter. In general, the turning directions are not very precisely directed away from the stimulus. Rather, the animal appears to sacrifice finely tuned accuracy for life-saving speed.

Both the very rapid execution of a behaviour and directional inaccuracy are properties suggestive of an open loop organisation of information processing. For this reason, it has often been suggested that escape behaviours may generally operate under open-loop conditions, with regard to potential sources of both exteroceptive and proprioceptive feedback. In support of this idea, in several escape circuits, the sensory channels that initiate the behaviour are strongly inhibited by central circuits the moment the escape movements begin (Kennedy et al., 1974; Murphey & Palka, 1974; Russell, 1976; Daley & Delcomyn, 1980a; Camhi & Nolen, 1981).

In this paper, however, I shall present evidence

Feedback control of an escape behaviour

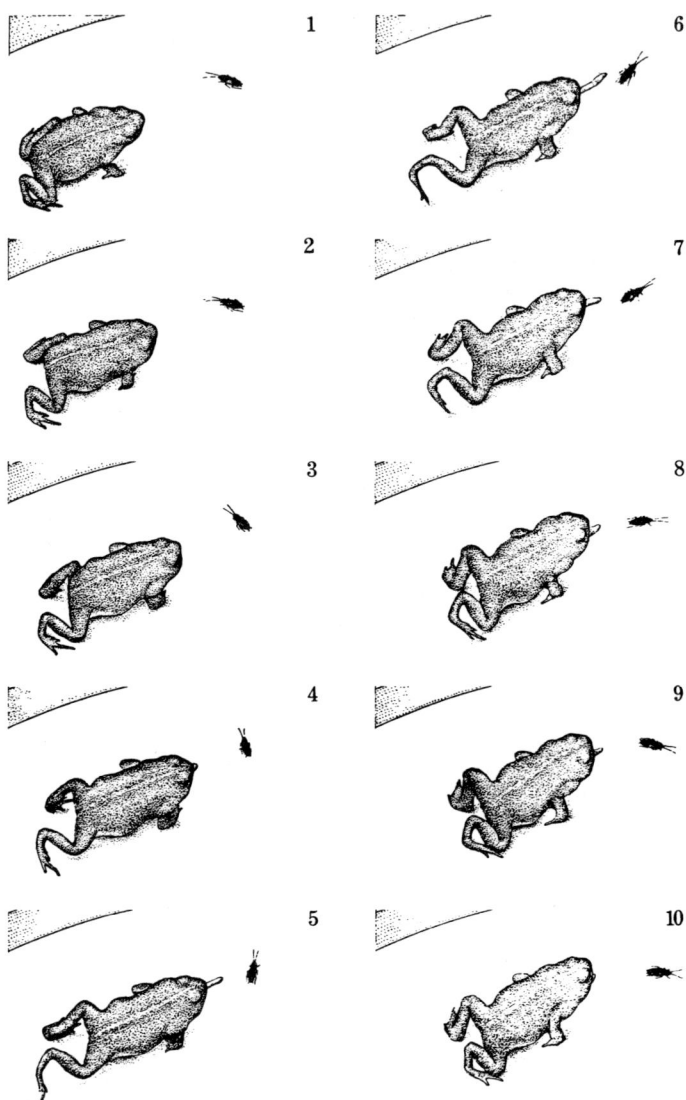

Fig. 1. The escape of a cockroach from the strike of a toad. Drawings made from a cine sequence. Interval between frames is 16ms. By frame 2, the toad has begun to lunge forward, and by frame 3, the cockroach has already begun to turn away. This turn causes the toad to miss its target.

suggesting that exteroceptive reafferent feedback received during the escape behaviour of the cockroach may play a prominent role in the control of the ongoing behaviour. Moreover, there may be a separate central pathway of giant interneurones, aside from those that initiate the behaviour, designed in part to convey these feedback signals during the escape. Thus, the motor system that controls turning and running behaviour may receive interneuronal inputs from two parallel channels, from which it must separately extract directional information.

Cockroach escape behaviour: potential feedback cues
The cockroach Periplaneta americana responds to the approach of a predator, such as a toad, with a rapid turning and then running behaviour (Fig. 1). By a series of sensory ablation and covering experiments, it has been found that the major sensory receptors evoking this behaviour are columns of wind-receptive hairs located on a pair of posterior appendages called cerci (Figs. 2 & 3). The major stimulus is the gentle, but rapidly accelerating, wind gust made by the approaching toad (Camhi et al., 1978; Camhi, 1980). As the cockroach turns away from an approaching toad or from a wind puffer simulating the toad, the insect's body pivots about its most posterior region (Fig. 4). This has the advantage of rotating the cockroach's front end sufficiently quickly and far away from the predator to permit a high frequency of successful escapes (Camhi et al., 1978). It is this front end of the body that most often encounters an attack, since the toad's hunting strategy (and that of many other nocturnal predators) is to sit and wait for its prey to walk toward it. An encounter thus requires that the cockroach be approaching the toad.

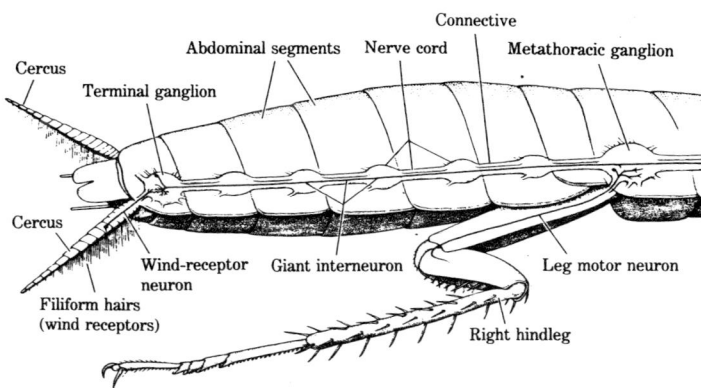

Fig. 2. Some of the neurones in the escape system of the cockroach. The sketch shows one cercal sensory cell (whose cell body sits at the base of its wind-receptive hair), one giant interneurone (GI) and one leg motor neurone.

95

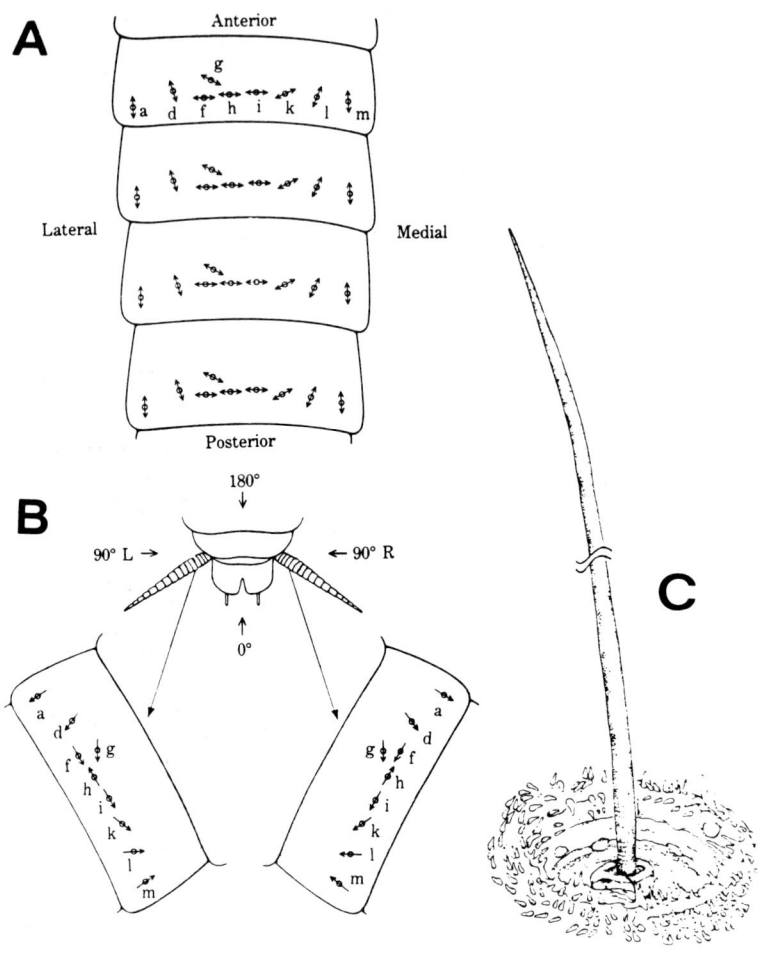

Fig. 3. Organisation of the wind-receptive hairs on the cerci. A. Underside of four of the 19 segments from an adult cercus. Each circle shows the position of a wind-receptive hair. Columns a, d, f, h, etc. extend along the cercal length. Double-headed arrows show the two directions in which each hair is most easily bent by the wind. B. A single segment from each cercus. The single-headed arrows show the optimal wind direction for the sensory cell of each hair. All hairs of a single column have the same optimal wind direction. C. A single wind-receptive hair, drawn to scale. At the hatch mark, 3/4 of the length of the hair has been omitted. Actual length is 0.5–1.0 mm.

A further consequence of the cockroach's rotational response is that this produces a rotational relative wind, which must comprise a portion of the sensory environment of the cercal hairs during the execution of the turn. Suppose, for instance, that a predator strikes from the left side, initiating a turn to the right. Because of the angles of the two cerci (Fig. 5), a right turn will produce a relative wind effectively from $150°$ left over the left cercus, and from $30°$ left over the right cercus; or an average wind angle of $90°$ left. If this relative wind were detected, over and above the initiating wind stimulus from the toad, it would help to drive a continuing turn to the right, and would thus serve a positive feedback function. However, it is possible to calculate, from the known dimensions and turning speed of the cockroach (Camhi & Tom, 1978), that the speed of this wind is only about 0.05m s^{-1}. This is much less than the peak wind speed delivered from the toad, about $1.5-2 \text{m s}^{-1}$, which continues to flow over the cerci throughout most or all of the turn (Camhi et al., 1978). The toad's wind, then, would be expected to completely swamp out this rotational relative wind.

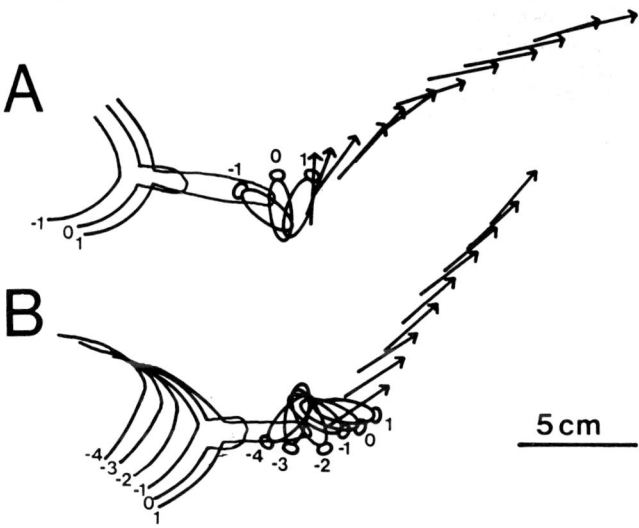

Fig. 4. Close-up view of two strikes by toads at cockroaches. Shown are the outlines of the toad's head and tongue and of the cockroach's body. Profiles are derived from filmed frames preceding and including the frame showing maximal extension of the toad's tongue. Corresponding numbers on toad and cockroach indicate corresponding times. The frame on which the toad's tongue is first visible is labelled "0". Following this, the positions of the cockroach's body on successive frames during its run are shown by the arrows.

In fact, it may be especially to the cockroach's advantage that this rotational wind, as detected by the wind receptors, is kept small by virtue of these receptors being located close to the pivot point of the turn. As will be shown below, the cockroach appears to be susceptible to wind stimulation from a predator during the course of the turn. Thus, decreasing the cockroach's sensing of the self-generated rotational wind must decrease the signal-to-noise problem in detecting the continuing wind from the predator. This may help to explain the otherwise strange observation that evolution has placed at the posterior end of the cockroach's body the wind-sensitive system used to detect predatory attacks that apparently come most often from the front. If the wind receptors were on the head of a 3cm long cockroach, the rotational relative wind they would experience during a turn would reach speeds of about 1.5m s^{-1}, and would thus likely interfere with the detection of a predator's wind of similar speed.

Let us consider the wind signal received from the toad, as sensed by the cockroach <u>during</u> its escape behaviour. The insect's initial turn away from the toad will cause the angle of this wind, as detected by the cockroach, to decrease toward $0°$ (i.e. wind from behind). Soon thereafter, however, the cockroach begins to run forward, though usually with some slight persistent turning. At about the time this run begins, the wind speed from the toad's strike begins to decrease, and the cockroach also begins to move away from the site of the

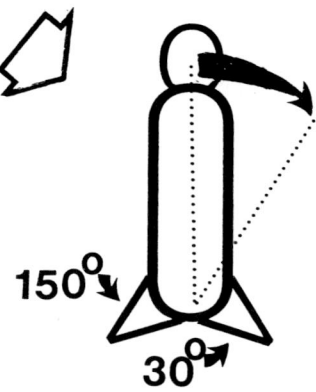

Fig. 5. Relative wind produced over the cerci by the turn of a cockroach to the right. Large open arrow-direction of predatory strike initiating the turn. Large filled arrow - direction of turn about a pivot point near the cerci (dotted lines). Arrows to cerci - directions of resulting wind. These wind directions do not take into account the wind produced by the predator's strike.

Feedback control of an escape behaviour

attack, so that the speed of the toad's wind detectable by the cockroach is markedly reduced. In its place, there is now a new source of wind, namely, the self-generated wind produced by the cockroach's forward running. Its speed must equal the insect's forward running speed (about 0.8 m/sec; Delcomyn, 1971) and its direction would be 180° (i.e. head-on), at least once all turning has ceased.

From these considerations, one can see that the wind directions experienced by a cockroach executing an escape behaviour will be composed of two sequential components. As graphed in Fig. 6, these will be first a decrease from the initial angle of the toad's approach to near zero degrees, and then a more or less steady 180° wind. Distortions from the graphed angles resulting from the cockroach's leg movements and from atmospheric winds have been shown to be small (Camhi & Nolen, 1981). Both the decrease in wind angle toward 0° and the subsequent 180° wind comprise potential feedback cues available to the cockroach. In the next section, I will examine whether these feedback cues are actually used by the escaping cockroach.

Is feedback wind information used in the escape turn?
The decrease of the wind angle during the turning component of a cockroach's escape could potentially inform the insect whether its turn is proceeding properly, and could indicate when the proper amount of turning has been achieved. Does the cockroach actually use this information? In fact, it has been shown that feedback information is not used in the selection

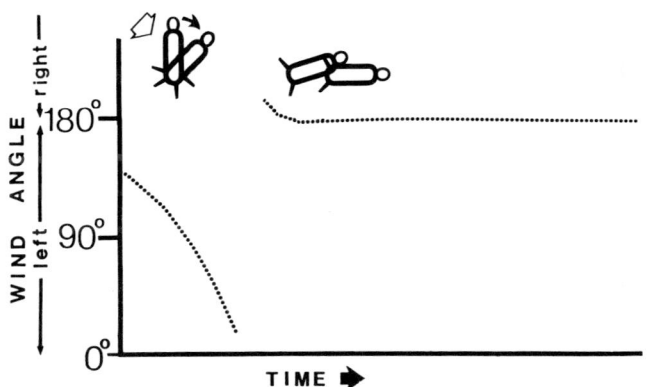

Fig. 6. Sequence of wind directions experienced by a cockroach turning to the right, away from a predator attacking from the left-front. Owing to the insect's turn, the initial wind angle of roughly 150° left will decrease toward 0°. Then, as the run begins, the continuing turn to the right should cause a relative wind from slightly to the right of head-on, followed by head-on wind (180°).

99

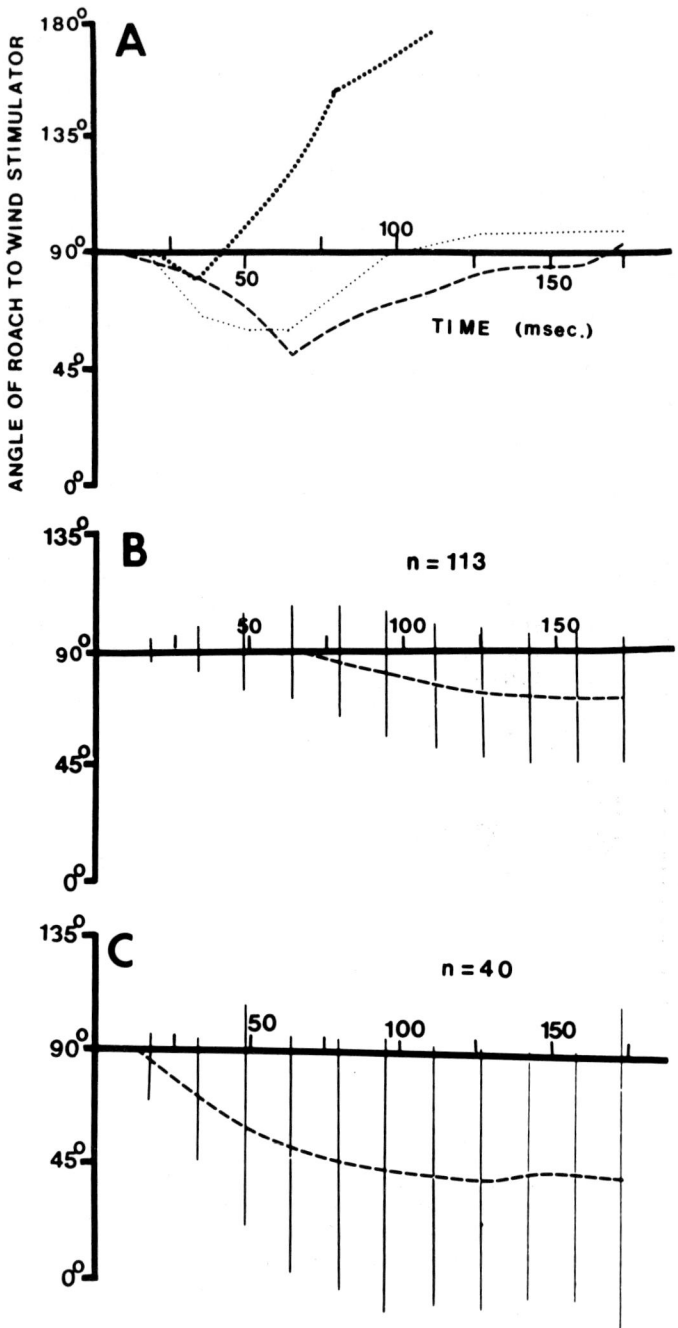

Feedback control of an escape behaviour

of the initial direction of turning (Camhi & Tom, 1978). That is, the insect does not turn randomly left or right and then use its reading of the changing wind direction (or of any other feedback cue) to correct those turns that are of improper direction. But this does not rule out the possibility of subsequent guidance by the changing wind angle, once the turn has been initiated.

To test this possibility I used a controlled wind puff to evoke a turn by a cockroach, and then quickly reversed the puff's direction, at a given predetermined moment during the turn. The initial puff was delivered from roughly $90°$ on either side (range $60°$ to $120°$). The angle with respect to the cockroach of the reversed puff (which depended upon the amount of turning that had been completed at the time of this puff's arrival) ranged from $60°$ to $160°$, on the side opposite the first puff. The wind speed of the reversed puff was no greater than that of the initial puff, and was usually less. Thus, instead of receiving a wind signal that steadily decreased in angle from $90°$ toward $0°$, these cockroaches received a signal that suddenly changed to exactly the opposite direction. If the behaviour, once initiated, operates under open-loop conditions, the insect should not respond to the second puff; but if it operates under closed loop conditions, it should respond to the second puff with a reversal of turning direction.

Fig. 7A graphs the turning responses from three trials, each showing a reversal of direction following the second puff. 85% of the trials (17 of 20) showed such reversals within the first 100ms. By contrast, on trials lacking a second puff, only 36% (12 of 33) showed any reversal of direction at all during the first 100ms. This difference is highly significant ($p < 0.01$; X^2 test). Thus, the cockroach's escape circuitry remains accessible to wind inputs during the turn.

Fig. 7. Turning responses of cockroaches to wind puffs. Data derived from cine sequences. Wind puffs, which began at time 0, were delivered from a controlled wind stimulator located at $90°$ ($\pm 30°$) to either side of the animal. Responses to wind from the two sides are presented together. Cockroach's angle with respect to the stimulator is normalised for all trials to $90°$ at time 0. A decrease in the angle of the insect represents a turn away from the stimulator; an increase represents a turn toward the stimulator. A. Three individual trials on which the initial wind puff was followed by a reversed puff in the early part of the turn. On each trial shown the reversed wind direction was followed by a reversed turning direction. B. Turns by running cockroaches away from a wind puff. Mean ± S.D. of angles are shown for 113 trials. C. Turns by standing cockroaches away from a wind puff. Mean ± S.D. of angles are shown for 40 trials.

Feedback control of an escape behaviour

This result suggests that the turning behaviour may be controlled in part by a negative feedback loop with regard to wind inputs from the toad (Fig. 8). Crucial to this interpretation is the fact that a single wind puff from a large angle (i.e. near the front) evokes a large turn away, and a single puff from a small angle (i.e. near the rear) evokes a small turn away (Camhi & Tom, 1978). Since the wind angle becomes smaller during the turn (Fig. 6), it should thus evoke even smaller turns, until finally the cockroach has achieved a nearly complete rotation away from the wind source. At this point the wind angle is close to $0°$, and thus evokes no further turn. Even though the average angle of turning is less than fully away from the wind stimulus, this feedback loop presumably is still operating, and would help to stabilise the behaviour such that the final wind angle is something close to $0°$. A further test of this idea would be to compare the sizes of turns of normal cockroaches vs. those in which the wind stimulus is permanently "clamped" at $90°$. This experiment, unfortunately, appears technically unfeasible.

Does a feedback loop operate, using wind angle as its input, during the second, running, portion of the escape behaviour? For instance, suppose that the running cockroach made a sudden yawing movement of its body to the left. While continuing to run in a constant direction, its body is now pointed somewhat leftward. The relative wind would now be coming from the right. Does the cockroach respond to such a wind from the right by swivelling its body back to the right, to restore the proper yaw attitude?

To test this, I have presented running cockroaches with wind puffs from one side. The puffs were generated from behind a screen wall next to which the insect ran on the ground. The actual wind direction experienced during the puff would be the vectorial sum of the relative wind ($180°$ with a speed equal to the running speed) and the puff itself ($90°$ with a speed that increases and then decreases during the course of the puff). The direction, then, was somewhere between $90°$ and $180°$.

Fig. 7B shows that running cockroaches turned not toward the wind stimulus, but rather away from it. In this regard

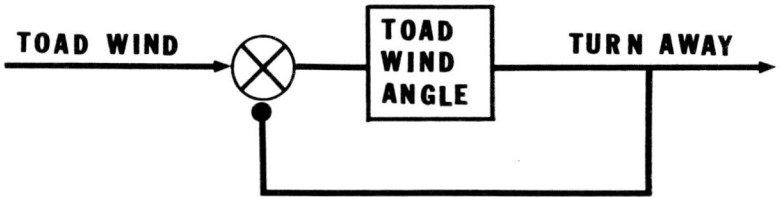

Fig. 8. Suggested reafferent negative feedback loop during a cockroach's turn away from a predator. Explanation in text.

they resembled standing cockroaches, (Fig. 7C) except that the turns during running began later and were of smaller mean angle. The response to wind during running, then, is not a case of negative feedback. Rather, at first glance, this looks like a case of positive feedback, in that a small deviation from 180° caused by a small yaw, would cause the cockroach to turn its body away from the direction needed to restore proper yaw attitude, thus causing an even larger yaw. This would lead to an instability of angular orientation during running, which apparently does not occur. Rather, the cockroach appears to regard the puff received during running as an aversive stimulus, just like a puff received while he is standing still. This implies, however, that the insect can discriminate the wind from a toad received during running from the wind produced by the running itself. How this discrimination might occur is not known. One potential cue, wind acceleration, used to discriminate a predator's wind from the relative wind during slow walking, apparently cannot help during running. Running causes high wind accelerations which would confuse this discrimination (Plummer & Camhi, 1981).

The three behavioural reactions to wind that are shown in Fig. 7, that to a single wind puff delivered to a standing cockroach (Fig. 7C), that of a cockroach executing a turn away from a reversed puff (Fig. 7A) and that of a running cockroach (Fig. 7B) all share the property of directing the animal away from the wind source. However, whereas the wind stimulus during standing (Fig. 7C) and that during running (Fig. 7B) should be regarded as exafferent stimuli, that during turning may best be regarded as a reafferent negative feedback signal, which in the experiment is made to come from the wrong direction. Although a wind puff given during standing and that during running appear to be regarded similarly by the cockroach, the problems of encoding these two wind puffs and decoding them by the motor systems would seem quite different. In the standing situation, there is much less background wind than during running, when the relative wind itself, as well as gusts made by the moving legs, must complicate the problem of signal-to-noise detection. Secondly, the movements required of the legs to execute a turn may be quite different depending upon whether those legs are already engaged in running or are stationary, or whether they are engaged in a turn. I will now examine the possibility that information on wind direction may be delivered to the motor system controlling the legs along two parallel pathways of interneurones, one that is employed when the cockroach is standing, and the other when he is running.

Delivery of directional information to the motor circuits controlling the legs

The cercal wind receptor cells send their axons to the last abdominal ganglion, where they activate a group of

Feedback control of an escape behaviour

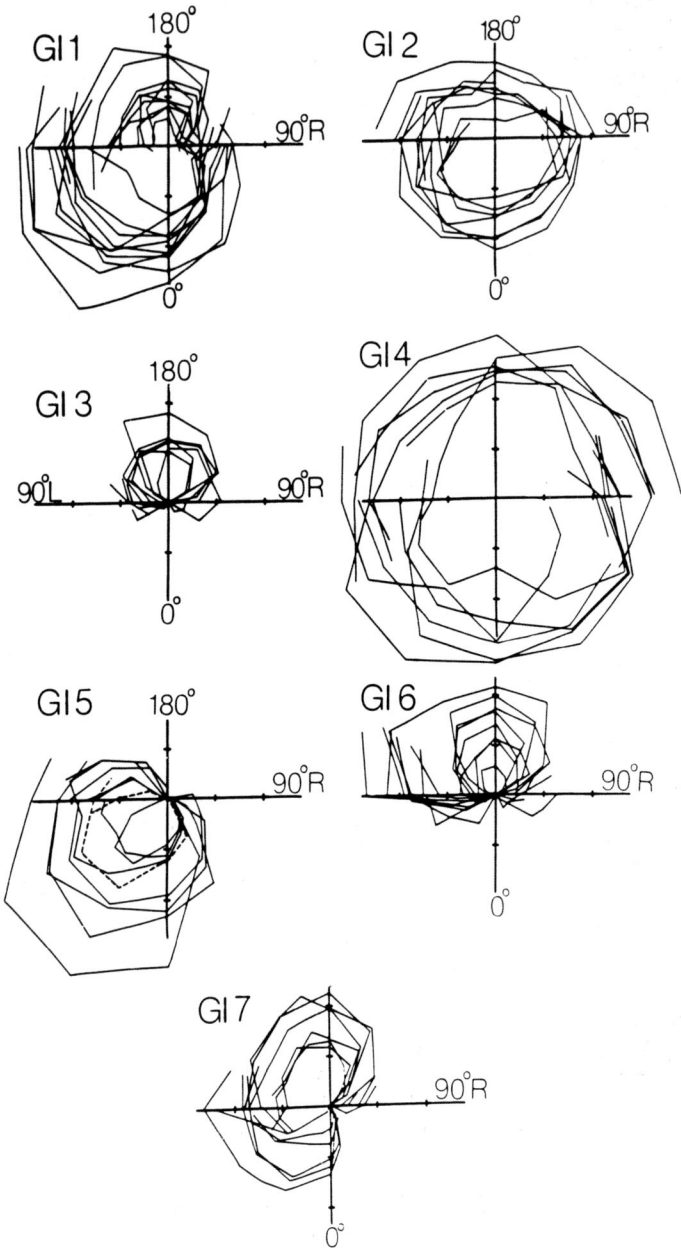

individually identified giant interneurones (GIs) (Fig. 2). These GIs are subdivided spatially into two groups, the ventrals (Nos 1 through 4) whose axons run in a ventral tract through the nerve cord, and the dorsals (Nos 5 through 7), whose axons run in a dorsal tract. Each group appears on both sides of the nerve cord. The synapses from cercal afferents to GIs are thought to be monosynaptic, though definitive proof is lacking. The GIs ascend to the brain (Spira et al., 1969), passing along the way through the three thoracic ganglia, which contain the leg motor neurones. Both the dorsal and the ventral GIs have been shown to excite leg motor neurones (Ritzmann & Camhi, 1978; Ritzmann, 1981; Ritzmann & Pollack, 1981). Although the details of the GI-to-motor connections have not been worked out, there would appear to be little in the way of complex circuitry between them from a consideration of the timing of information flow. Action potentials in the GIs can initiate action potentials in leg motor neurones in as little as one millisecond (Camhi & Nolen, 1981).

The cercal afferents, as a group, contain a code for wind direction (Fig. 3). All the hairs of one cercal column are most easily deflected by wind within either of two opposite directions, and the sensory cell under each of these hairs responds only to one of these directions. Different columns of hairs have different best wind directions, so that as a group the hair columns of both cerci encode all horizontal directions of wind.

By virtue of the specific excitatory and inhibitory connections that the cercal afferents make with different GIs (Daley, 1982), these GIs also contain a code for wind direction. The directional responses of the seven left GIs are shown in Fig. 9. The seven right GIs give mirror image responses. This GI code for wind direction must involve information contained in more than just one GI, since the response of any GI to wind depends both upon the wind's direction and upon its intensity (Westin et al., 1977). However, by comparing the impulse trains among the different GIs, wind intensity is factored out as a cue, and there remains only wind direction.

Among the ventral group of GIs, a comparison of the

Fig. 9. Directional responses of the seven left GIs to wind stimuli. (The seven right GIs, not shown here, give mirror image responses). Each line on any of the seven sets of axes represents the number of action potentials of a single cell in response to wind from different directions within the horizontal plane. (Hatch marks on the axes represent 5 action potentials). Thus, for instance, left GI 1 responds preferentially to wind from the left. (The mismatch of the curves at approximately $90°$ left and $90°$ right results from the fact that, for technical reasons, smaller wind stimuli were used from front angles).

impulse trains in the left vs. the right GI 1 could inform the motor system as to whether the wind is from the left or right, and thus whether to programme, respectively, a right or a left turn. Such a role for this GI has been revealed through experiments in which the left GI 1 is killed by intracellular pronase injection, the animal sealed up, and the behaviour tested on the next day. The results show that wind from the left produces a significant increase in the number of "incorrect" turns, that is, turns towards the left (Comer, 1981, and in preparation). Similar injections of GIs 2 or 3, both of which code symmetrically with regard to the two sides of the body (Fig. 9) or injections into the extracellular space around the GIs, produced no significant change in turning behaviour. Thus the GIs 1 appear to be involved in coding the wind direction, and instructing the motor system which direction of turn to execute.

The percentage of incorrect turns after killing left GI 1, though significant, was fairly small, only 25%. However, after killing both left GIs 1 and 2, this percentage of incorrect turns increased to 56%. Whether killing three or all four of the left ventral GIs in the same animal would further increase the percentage of incorrect turns remains to be tested. Nevertheless, it seems clear that the ventral group of GIs plays a major role in instructing the motor system for the legs which direction of turn to execute. This key role in the initial phase of the behaviour is consistent with the fact that GIs 1, 2 and 3 have the greatest axonal diameters, and thus the highest conduction velocities, of any interneurones in the nerve cord. They are also activated earliest by the cercal afferents. In sum, in response to a wind puff, they deliver their first action potentials to the thoracic ganglia about 5ms before any of the dorsal GIs (Westin et al., 1977).

Not only the ventral GIs, but also the dorsal group contains a code for wind direction. For instance, a comparison of the spike trains of the left vs. the right GI 7 could indicate whether the wind is from the left or right (Fig. 9). Thus, the dorsal group could also instruct the motor circuits for the legs which direction of turn to execute. However, it seems unlikely that the dorsal GIs are employed in this manner with regard to <u>initiating</u> the turn, in view of the timing considerations already mentioned; namely, that they require 5ms more than the ventral GIs (out of a total behavioural latency of only 14ms) to deliver their first action potentials to the thoracic ganglia. However, they could in principle help to mediate a response to a wind puff that arrives during a turn (Fig. 7A) or during a run (Fig. 7B).

This later involvement of the dorsal GIs seems even more likely in view of the fact that as soon as a tethered cockroach begins to make running movements, its ventral GIs receive strong inhibition and its dorsals received strong excitation. Both of these effects derive from a corollary

Feedback control of an escape behaviour

discharge from the thoracic ganglia (Daley & Delcomyn, 1980a,b). The inhibition of the ventrals constitutes a central negative feedback loop, and the excitation of the dorsals a central positive feedback loop. The positive loop may help to maintain running, once initiated, in a way similar to positive feedback loops in other rhythmic systems (Selverston, 1976; Gillette et al., 1978; Weeks, 1982). Whether these positive and negative feedback loops are activated during the initial, turning phase of escape, or only when the run commences, has not yet been determined.

It seems likely, then, though not yet proven, that during the course of an escape behaviour, it is first the ventral GIs that drive the motor centres for the legs, and then the dorsal GIs. This is consistent with the known synaptic interactions within the system which, together with the known temporal properties, suggest a shift from ventral GI to dorsal GI control, as sketched in Fig. 10. A similar switch from initial control by the interneurones of greatest axonal diameter, to control by smaller interneurones, has been found in the crayfish escape system (Wine & Krasne, 1982).

Whether or not it is true that the dorsal GIs take over from the ventrals when the turn or the run begins, the motor circuit for the legs, once it extracts the needed directional information from the GIs in response to a puff during a run, must carry out the turn in a different manner than if the action began from a standing position. With all six legs

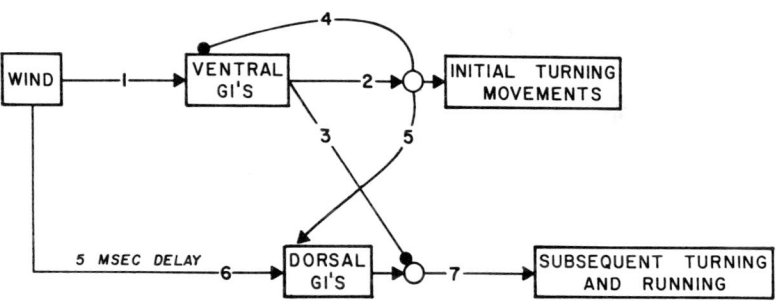

Fig. 10. Known timing and synaptic interactions involving the ventral and dorsal GIs in the cockroach escape behaviour. Wind evokes action potentials in the ventrals before the dorsals (Westin et al., 1977). While these ventrals are active, they inhibit the outputs of the dorsals (Ritzmann, 1981; Ritzmann & Pollack, 1981). This is shown by line 3 on the sketch. The initial turning movements give rise to inhibition of the ventrals (line 4) and excitation of the dorsals (line 5) (Daley & Delcomyn, 1980a,b). The inhibition of the ventrals (line 4) and the consequent disinhibition of the dorsals (line 3), together with the simultaneous activation of the dorsals (line 5), should assist a shift from ventral control to dorsal control of the escape behaviour.

already engaged in a highly coordinated set of rapid movements, the need will be either to cancel these movements and initiate a new motor programme or to superimpose upon these movements a turning component. The details of these leg motions have not yet been studied.

The leg motor circuitry also faces different problems in extracting directional information from the ventral GIs during standing than from the dorsals during turning or running. During standing, the ventrals are silent until the wind signal arrives, so that the motor circuits face little signal-to-noise problem in extracting the needed information. During running, the dorsals are highly active, owing both to the corollary discharge excitation, and presumably to the sensory activation by the relative wind created by the running itself. The problem faced by the motor system, then, would be to detect a shift in the normally balanced, ongoing activity of the left vs. the right dorsal GIs. (And yet, as mentioned above, this shift would need to be discriminated from that which would be caused by a yaw). The ventral GIs, then, are designed to carry a small brief signal to the motor system against a silent background, whereas the dorsal GIs carry a signal consisting of a shift in their ongoing activity. The fact that the behavioural latency of the turning response during running is about 50ms longer than during standing (Fig. 7B, C) may reflect in part this need for additional processing time in the face of ongoing activity in the dorsal GIs. Whether the ventral and dorsal GIs really do operate in sequence as suggested here is currently under experimental study, through pronase injections of ventral or dorsal GIs followed by analysis of turning behaviour during running.

DISCUSSION

Escape reactions have often been regarded as behaviours under open loop control. Consistent with this view are the several known cases where pronounced inhibition is brought to bear on the sensory pathways that have initiated the behaviour, just after the movements have begun. Among the escape systems showing such inhibition is that of the cockroach (Daley & Delcomyn, 1980a). Specifically, the ventral GIs, which play a major role in initiating the behaviour, are strongly inhibited when running begins. The dorsal GIs, however, are not inhibited during running. Rather they are excited, making them more responsive to wind puffs while the insect is running than while standing. It is not yet known whether these effects on the GIs begin during the turn, or only later, during the run. In any case, the enhanced wind-sensitivity of the dorsal GIs renders them good candidates for mediators of ongoing running behaviour in response to ongoing wind stimulation, and of turning in response to puffs received during running (and perhaps during turning).

I have shown in this chapter that the cercal wind receptors receive, during a normal turning and running behaviour, a complex angular sequence of wind stimulation. Moreover, as Fig. 7A shows, deviations from this normal sequence lead to changes in the insect's turning behaviour. I have indicated that the response to a reversed wind puff arriving during a turn suggests that the turn is executed under closed loop conditions; experimentally reversing the wind direction has altered the reafferent feedback loop, and the animal's response shows that it senses this alteration. However, this does not prove that the normal decrease in wind angle toward $0°$ is detected by the cockroach and is used for closed loop control as sketched in Fig. 8. Whether this decreasing wind angle is actually detected remains to be tested. However, both the wind velocity and the size of the angular change are well within the known coding capabilities of the system (Westin et al., 1977). At least it is clear that the escape behaviour is alterable by wind inputs during both the turning and the running portions of an escape behaviour.

Another category of behaviour which was thought until recently to utilise relatively little reafferent feedback is rhythmic movements, such as those in invertebrate locomotion. A case in point is locust flight, which was thought not to employ feedback from the wings for phasic adjustments of the wingbeat (Wilson, 1966). However, closer inspection showed both behaviourally and physiologically demonstrable phasic feedback loops (Wendler, 1974; Burrows, 1975). More recently, even the wind gusts produced by the wings have been shown to provide useful feedback information for phasic control of the wingbeat (Horsmann et al., 1983). Indeed, there is increasing reason to believe that reafferent feedback plays a significant role in the control of most or all categories of behaviour. Neither the speed of the movements nor their directional or other properties should be used as a priori argument against the involvement of such feedback.

As for the implications for central motor circuits, one begins to view those of the cockroach as significantly more complex than previously thought. If turning away from a wind stimulus involves three different forms of leg movement (one for a standing animal, a second for a turning, and a third for a running animal), then each time the insect turns away from a puff, its circuitry must decide which of the three programmes to execute. And if the information can arrive at the leg motor circuits via two (or more) possible interneuronal pathways, each with a different form of directional code and a vastly different signal-to-noise ratio, the motor circuits must have two or more ways of reading out and decoding the arriving information. The fact that all of this GI-to-motor readout process can be carried out in an interval as brief as a millisecond (Camhi & Nolen, 1981) makes this little piece of motor machinery a quite remarkable device. This device, with its individually identified motor neurones and GI inputs, awaits a thorough cellular investigation.

Acknowledgements

I thank Hanah Sassoon for technical assistance and Ori Sassoon for writing of computer programmes used in the behavioural analysis. This work was supported by NIH grant NS09083.

REFERENCES

Burrows, M. (1975). Monosynaptic connexions between wing stretch receptors and flight motor neurons of the locust. J. exp. Biol., 62, 189–219.

Camhi, J.M. (1980). The escape system of the cockroach. Scient. Am., 243, 158–172.

Camhi, J.M. & Nolen, T.G. (1981). Properties of the escape system of cockroaches during walking. J. comp. Physiol., 142, 339–346.

Camhi, J.M. & Tom, W. (1978). The escape behavior of the cockroach Periplaneta americana. I. Turning responses to wind puffs. J. comp. Physiol., 128, 193–201.

Camhi, J.M., Tom, W. & Volman, S. (1978). The escape behavior of the cockroach Periplaneta americana. II. Detection of natural predators by air displacement. J. comp. Physiol., 128, 203–212.

Comer, C. (1981). Effects of killing single giant interneurons on the escape behaviour of the cockroach, Periplaneta americana. Soc. Neurosci. Abstr., 7, 160.

Daley, D.L. (1982). Neural basis of wind-receptive fields of cockroach giant interneurons. Brain Res., 238, 211–216.

Daley, D.L. & Delcomyn, F. (1980a). Modulation of the excitability of cockroach giant interneurons during walking. I. Simultaneous excitation and inhibition. J. comp. Physiol., 138, 231–239.

Daley, D.L. & Delcomyn, F. (1980b). Modulation of the excitability of cockroach giant interneurons during walking. II. Central and peripheral components. J. comp. Physiol., 138, 241–251.

Delcomyn, F. (1971). The locomotion of the cockroach Periplaneta americana. J. exp. Biol., 54, 445–452.

Drewes, C. (1984). Escape reflexes in earthworms and other annelids. In Neural Mechanisms of Startle Behavior. Ed. Eaton, R.C. Plenum Press, New York, pp. 43–91.

Eaton, R.C. & Hackett, J.T. (1984). The role of the Mauthner cell in fast-starts involving escape in teleost fishes. In Neural Mechanisms of Startle Behavior. Ed. Eaton, R.C. Plenum Press, New York, pp. 213–266.

Gillette, R., Kovak, M.P. & Davis, W.J. (1978). Command neurons in Pleurobranchaea receive synaptic feedback from the motor network they excite. Science, N.Y., 199, 798–801.

Horsmann, U., Heinzel, H.-G. & Wendler, G. (1983). The phasic influence of self-generated air current modulations on the locust flight motor. J. comp. Physiol., 150, 427–438.

Kennedy, D., Calabrese, R.L. & Wine, J.J. (1974). Presynaptic inhibition: primary afferent depolarization in crayfish neurons. Science, N.Y., 186,

451-454.

Murphey, R.K. (1981). The structure and development of a somatotopic map in crickets: the cercal afferent projection. Devl Biol., 88, 236-246.

Murphey, R.K. & Palka, J. (1974). Efferent control of cricket giant fibers. Nature, Lond., 248, 249-251.

Plummer, M.R. & Camhi, J.M. (1981). Discrimination of sensory signals from noise in the escape system of the cockroach: the role of wind acceleration. J. comp. Physiol., 142, 337-357.

Ritzmann, R.E. (1981). Motor responses to paired stimulation of giant interneurons in the cockroach. II. The ventral interneurons. J. comp. Physiol., 143, 71-80.

Ritzmann, R.E. & Camhi, J.M. (1978). Excitation of leg motor neurons by giant interneurons in the cockroach, Periplaneta americana. J. comp. Physiol., 125, 305-316.

Ritzmann, R.E. & Pollack, A.J. (1981). Motor responses to paired stimulation of giant interneurons in the cockroach. I. The dorsal interneurons. J. comp. Physiol., 143, 61-70.

Russell, L.J. (1976). Central inhibition of lateral line input in the medulla of the goldfish by neurons which control body movements. J. comp. Physiol., 111, 335-358.

Selverston, A.I. (1976). Neuronal mechanisms for rhythmic motor pattern generation in a simple system. In Neural Control of Locomotion. Eds. Herman, R.M., Grillner, S., Stein, P.S.G. & Stuart, D.G.. Plenum Press, New York, pp. 317-399.

Spira, M.E., Parnas, I. & Bergman, F. (1969). Organization of the giant axons of the cockroach, Periplaneta americana. J. exp. Biol., 50, 615-627.

Weeks, J.C. (1982). Segmental specialization of a leech swim-initiating interneuron (cell 205). J. Neurosci., 2, 972-985.

Wendler, G. (1974). The influence of proprioceptive feedback on locust flight coordination. J. comp. Physiol., 88, 173-200.

Westin, J., Langberg, J.J. & Camhi, J.M. (1977). Responses of giant interneurons of the cockroach Periplaneta americana to wind puffs of different directions and velocities. J. comp. Physiol., 121, 307-324.

Wilson, D.M. (1966). Central nervous mechanisms for the generation of rhythmic behavior in arthropods. Symp. Soc. exp. Biol., 20, 199-228.

Wine, J.J. & Krasne, F.B. (1982). The cellular organization of crayfish escape behavior. In The Biology of Crustacea, Vol. 4, Neural Integration and Behavior. Eds. Sandeman, D.C. & Atwood, H.L.. Academic Press, New York, pp. 242-292.

Section 2
Central Control of Sense Organ Excitability

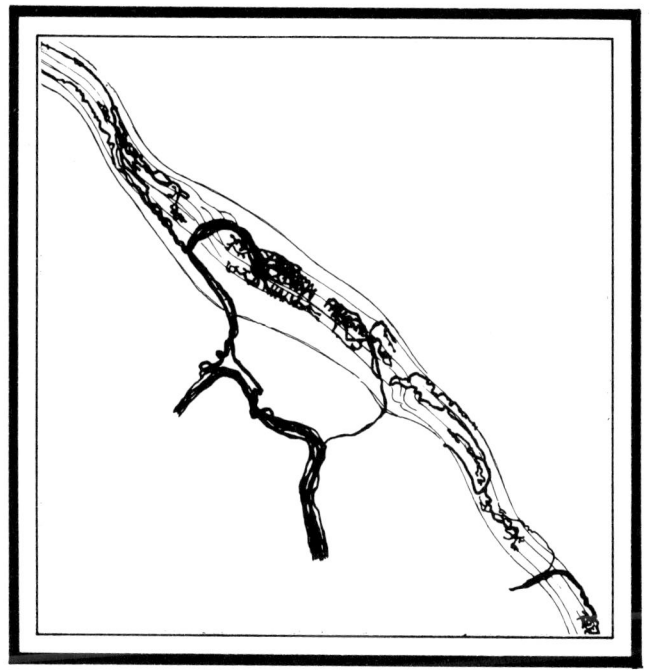

Chapter Seven

INTRODUCTION

A. Prochazka

It is commonplace to explain convergence in the separate evolution of complex organs in vertebrates and invertebrates by arguing that the range of possible biological solutions to a given task is limited. On this view, vertebrates and invertebrates have independently evolved legs and wings (rather than, say, wheels and propellors) because legs and wings are among the few workable biological solutions to the problem of locomotion on land and in the air.

Accepting this form of explanation, it is nevertheless astonishing to note the similarities in detail in the "design" of some of the sensory receptors in animals as far removed as lobsters and humans. In the following two chapters of this section, attention is focussed on the "thoraco-coxal muscle receptor organ" (TCMRO) of the crab, and the mammalian muscle spindle. These specialised receptors (whose afferents form the largest nerve fibres in their respective nervous systems) lie within muscle, in parallel with muscle fibres, and respond to variations in muscle length. They differ from all the other limb mechanoreceptors in receiving motor innervation from the CNS. This motor input can profoundly modify the afferent responses to length variations. Many aspects of the structure and function of this innervation in crustaceans and mammals have been elucidated over the last two decades by Dr Bush, Professor Boyd and their colleagues.

WHY HAVE EFFERENT CONTROL OF FEEDBACK PARAMETERS?

The way in which the nervous systems of crabs and mammals normally wield their modulatory power over TCMROs and muscle spindles, respectively, is still a matter of interest and controversy. This is because it has not yet proved possible to record satisfactorily the activity of the motor fibres to the receptors during normal movements in awake animals. All theories about the control of these fibres are therefore based either upon recordings in reduced preparations, or, indirectly, on recordings from the afferent fibres in awake mammals (review: Prochazka & Hulliger, 1983).

There are, of course, numerous other examples of receptor organs receiving motor input from the CNS: the mammalian eye and ear, the rat olfactory bulb, and the fish lateral line organ, among others. Let us therefore briefly examine the known function of the efferent control in two of these cases. In humans, the iris muscles can change the area of the pupil by a factor of five. The pupils are observed to constrict within 3 to 4 seconds of a sudden increase in light intensity or sudden accommodation for near vision, and respond to sinusoidal changes in light intensity with a high-frequency cut-off at about 2Hz (Terdiman et al., 1969). Diffuse activation of the sympathetic nervous system leads to pupillary dilation and accommodation for far vision. The functions of the pupil are evidently to provide "automatic gain control" in the face of changes in ambient light intensity, and to link the control of "lens aperture" to accommodation. Autonomic "biassing" is likely to assume significance only under extreme conditions (Ikeda, H., personal communication).

The middle ear muscles are fast-twitch muscles, whose contraction can diminish, by an order of magnitude, the transmission of sound to the round window. Carmel and Starr (1963) showed in conscious cats that sounds of widely ranging intensity caused corresponding reflex activation of the middle ear muscles. This activation could be strongly modified by prior conditioning, which, among other things, could depend on the "significance" of the sound. The muscles were also regularly activated during vocalisation and self-generated bodily movements. The functions of efferent input to the middle ear would therefore seem to be "automatic gain control" (as for the pupil), gain suppression during movement and vocalisation, and, additionally, gain control in relation to the "significance" of expected sounds. Now let us see if some of these functions might also be shared by the efferent systems to muscle receptor organs.

ANALOGIES WITH MUSCLE RECEPTOR ORGANS

Automatic gain control

A close analogy with the pupil or middle ear muscles would require that the occurrence of large variations in muscle length would reflexly evoke alterations in the activity of the efferents to the receptor organs, reducing the afferent sensitivity, so that the resultant afferent responses would be similar to those in small movements. In the case of externally applied movements in the conscious cat, the available evidence is firmly against such a mechanism, as simulation experiments have revealed that dynamic fusimotor action is steady and

Central control of sense organ excitability

strong under these conditions, regardless of the movement amplitude (Hulliger et al., 1984). The resultant afferent responses vary over a wide range, according to the size of the movements. The strict analogy with the pupil and the middle ear muscles thus breaks down. However, taking a broader view, it could be argued that a more meaningful mechanism would be one which attempted to keep within certain limits the afferent firing rates in small and large movements generated by the animal itself (Loeb & Marks, 1985).

Gain control during self-generated movement
The suppression of sound transmission during voluntary movement (especially vocalisation) seems to make good functional sense. Again, a strict analogy in muscle receptors would require suppression of muscle afferent sensitivity during, say, loud sounds or strong cutaneous stimulation. Indeed in decerebrate cats, fast, potent, reflex effects (both excitatory and inhibitory) from cutaneous afferents onto fusimotor neurones have often been documented (but innocuous skin stimulation over much of the body in the conscious cat rarely reveals significant effects on hindlimb spindle afferents (Prochazka, 1983a; see also Taylor & Gottlieb, 1985).

What about the possibility of "standardising" the depth of afferent modulation? Many authors, including Boyd, Bush and Loeb have argued that a key function of efferents to muscle receptors is to maintain afferent responsiveness (prevent silencing or "saturation") during muscle shortening (reviewed by Matthews, 1981). In the following chapters, it will be seen that in crustacean TCMROs, this could be achieved by linking the activation of the receptor muscle "fusimotor" motor neurones (Rm1 and Rm2) with that of the "skeletomotor" promotor motor neurones. In mammals, Boyd argues that static fusimotor action to chain fibres would best serve this purpose. Linkage would tend to "standardise" the afferent discharge, because the bigger the movements, the greater would be the modulation of the receptor muscles, acting to offset the unloading and stretch of the receptor endings.

The Rm2s innervate both promotor and receptor muscles, and so, like their mammalian counterparts, the beta motor neurones, are presumably automatically co-activated with the purely skeletomotor neurones. The similarities between decapods and mammals do not stop there. Remarkably, the action of Rm fibres on the response of T-fibre afferents to trapezoidal variations in muscle length shows a striking correspondence with the action of static gamma motor neurones on Ia afferents under similar conditions (equating depolarisation in the one case with firing rate in the other: Bush, 1981; Boyd, see Chapter 8). It is worth noting in passing that by having evolved Rm1 motor neurones exclusively innervating muscle receptor organs, decapod crustaceans are

117

more advanced in this respect than lizards, snakes and amphibians, which all have to get by with the "beta" type of skeletofusimotor arrangement (Proske & Ridge, 1974).

In these latter animals, the role of the efferent supply to muscle receptor organs would indeed seem to be restricted to maintaining afferent responsiveness or "bias" during muscle contractions. However, in crustaceans, Rm2 ("beta") action on T- and S-fibre afferents is evidently much weaker than Rm1 ("gamma") action (Cannone & Bush, 1983). Similarly, in mammals, beta action in most normal movements is probably weaker than gamma action, on two grounds. First, in most spindles, gamma motor neurones are likely to outnumber beta motor neurones by at least 5 to 1. Second, although individual beta fibres, when stimulated at similar rates to gamma fibres, have comparable strengths of fusimotor action, their firing rates in real life would presumably be in the alpha-motor neurone range (0-20/s), whereas gamma fibres recorded in reduced preparations often fire tonically at rates of up to 100i.p.s. (Ellaway & Trott, 1978; Appenteng et al., 1980).

Given that the skeletomotor-linked "beta" (Rm2) action in mammals and crustaceans is likely to be weak in comparison with pure "gamma" (Rm1) action, we must now ask to what extent gamma motor neurones (or Rm1s) might themselves be strictly linked to skeletomotor activation. Here there is considerable confusion. Granit's (1979) clean, clear-cut notion of "a ubiquity of alpha-gamma linkage" as a "principle of motricity" has been replaced by a jumble of relationships between alpha and gamma control, inferred from recordings from spindle afferents in a variety of circumstances. The following list summarises the permutations suggested so far:

gammas strictly alpha-linked	.. Granit (1979), Hagbarth (1981)
variable coupling between alphas & gammas	}.. Vallbo (see Chapter 25), } Burke (1981), Roll & } Vedel (1982)
some gammas linked, others tonic	}.. Eklund et al. (1964), } Sears (1964)
dynamic gammas linked, static gammas tonic (A)	}.. Goodwin & Luschei (1975), } Murphy et al. (1984)
static gammas linked, dynamic gammas tonic (B)	}.. Grillner et al. (1969), } Gottlieb & Taylor (1983)
extensor muscles, A: flexor muscles, B.	}.. Perret & Buser (1972), } Cabelguen (1981)
"stance task group", A: "swing task group", B.	}.. Loeb & Hoffer (1981) }
"fusimotor set":variable tonic, weak linked	}.. Hulliger et al. (1984) }
alpha-gamma independence (non-specific)	}.. Schieber & Thach (1980)
Rm2s linked, Rm1s tonic + linked	.. Bush (see Chapter 9)

Central control of sense organ excitability

The overall impression given by this table is that most workers have encountered afferent behaviour which suggests some linkage of gammas (Rm2's) to skeletomotor neurones, and some independent activation. This brings us back to "gain control", and our analogy with the middle ear muscles. Activation of gamma (or Rm1) motor neurones independently of skeletomotor neurones would allow the "setting" of afferent response parameters according to the "significance" of the task at hand ("parameter control": Matthews, 1981). For well-rehearsed tasks performed under predictable, undemanding conditions, low afferent sensitivities might suffice, whereas under more demanding, unpredictable conditions, higher sensitivities could be "dialled up" to allow rapid and accurate responses to external events.

Conclusion

In the opinion of the author and his colleagues, much of the power of the gamma system is held in reserve for situations involving novelty or special effort ("fusimotor set": Prochazka, 1983b; Hulliger et al., 1984). The specialisation of gamma motor neurones into at least two, and now, as suggested by Boyd and co-workers, possibly three functionally separate categories, calls for a wider search for their role in movements of different types, and under a wide variety of conditions. The related problem of the function of crustacean Rm1 motor neurones could well be more amenable to early solution by direct recording from the receptor muscle in the freely-moving animal.

REFERENCES

Appenteng, K., Morimoto, T. & Taylor, A. (1980). Fusimotor activity in masseter nerve of the cat during reflex jaw movements. J. Physiol., 305, 415-432.
Burke, D. (1981). The activity of human muscle spindle endings in normal motor behaviour. Int. Rev. Physiol., 20, 91-136.
Bush, B.M.H. (1981). Non-impulsive stretch receptors in crustaceans. In Neurones Without Impulses. Eds. Roberts, A. & Bush, B.M.H. Society for Experimental Biology, Seminar Series 6. Cambridge University Press, Cambridge, pp. 147-176.
Cabelguen, J.M. (1981). Static and dynamic fusimotor controls in various hindlimb muscles during locomotor activity in the decorticate cat. Brain Res., 213, 83-97.
Cannone, A.J. & Bush, B.M.H. (1983). Dual reflex motor control of non-spiking crab muscle receptor. II. Reinforcement of Rm1 mediated positive feedback by dual afferent excitation of Rm2. J. comp. Physiol., 153, 309-320.
Carmel, P.W. & Starr, A. (1963). Acoustic and non-acoustic factors modifying middle-ear muscle activity in waking cats. J. Neurophysiol., 26, 598-616.

Eklund, G., Euler C. von & Rutkowski, S. (1964). Spontaneous and reflex activities of intercostal gamma motoneurones. J. Physiol., 171, 139-163.

Ellaway, P.H. & Trott, J.R. (1978). Autogenetic reflex action on to gamma motoneurones by stretch of triceps surae in the decerebrated cat. J. Physiol., 276, 49-66.

Goodwin, G.M. & Luschei, E.S. (1975). Discharge of spindle afferents from jaw-closing muscles during chewing in alert monkeys. J. Neurophysiol., 38, 560-571.

Gottlieb, S. & Taylor, A. (1983). Interpretation of fusimotor activity in cat masseter nerve during reflex jaw movements. J. Physiol., 345, 423-438.

Granit, R. (1979). Interpretation of supraspinal effects on the gamma system. Prog. Brain Res., 50, 147-154.

Grillner, S., Hongo, T. & Lund, S. (1969). Descending monosynaptic and reflex control of γ-motoneurones. Acta physiol. scand., 75, 592-613.

Hagbarth, K.-E. (1981). Fusimotor and stretch reflex functions studied in recordings from muscle spindle afferents in man. In Muscle Receptors and Movement. Eds. Taylor, A. & Prochazka, A. Macmillan, London, pp. 277-286.

Hulliger, M., Zangger, P., Prochazka, A. & Appenteng, K. (1984). Fusimotor "Set" vs α-γ linkage in voluntary movement in cats. In Proceedings of the 7th International Congress of Electromyography. Eds. Struppler, A. & Weindl, A. Springer, Berlin, pp. 57-64.

Loeb, G.E. & Hoffer, J.A. (1981). Muscle spindle function during normal and perturbed locomotion in cats. In Muscle Receptors and Movement. Eds. Taylor, A. & Prochazka, A.. Macmillan, London, pp. 219-228.

Loeb, G.E. & Marks, W.B. (1985). Optimal control principles for sensory transducers. In The Muscle Spindle. Eds. Boyd, I.A. & Gladden, M.H.. Macmillan, London. (In press).

Matthews, P.B.C. (1981). Muscle spindles: their messages and their fusimotor supply. In Handbook of Physiology, Section 1, The Nervous System, Vol. 2, Motor Control. Ed. Brooks, V.B. American Physiological Society, Bethesda, pp. 189-228.

Murphy, P.R., Stein, R.B. & Taylor, J. (1984). Phasic and tonic modulation of impulse rates in γ motoneurones during locomotion in premamillary cats. J. Neurophysiol., 52, 228-243.

Perret, C. & Buser, P. (1972). Static and dynamic fusimotor activity during locomotor movements in the cat. Brain Res., 40, 165-169.

Prochazka, A. (1983a). Reflex regulation of fusimotor neurones: discussion. In Reflex Organisation of the Spinal Cord and its Descending Control. Eds. Porter, R. & Redman, S. Roche, Canberra City, p. 25.

Prochazka, A. (1983b). The uncoupling of alpha and of static and dynamic fusimotor activity in the cat: fusimotor "set". Proc. int. Union Physiol. Sci., 15, 12.

Prochazka, A. & Hulliger, M. (1983). Muscle afferent function and its significance for motor control mechanisms during voluntary movements in cat, monkey and man. In Motor Control Mechanisms in Health and Disease. Ed. Desmedt, J.E.. Raven, New York, pp. 93-132.

Proske, U. & Ridge, R.M.A.P. (1974). Extrafusal muscle and muscle spindles in reptiles. Prog. Neurobiol., 3, 3-29.

Roll, J.P. & Vedel, J.P. (1982). Kinaesthetic role of muscle afferents in man, studied by tendon vibration and microneurography. Expl. Brain Res.,

47, 177-190.

Schieber, M.H. & Thach, W.T. (1980). Alpha-gamma dissociation during slow tracking movements of the monkey's wrist: preliminary evidence from spinal ganglion recording. Brain Res., *202*, 213-216.

Sears, T.A. (1964). Efferent discharges in alpha and fusimotor fibres of intercostal nerves of the cat. J. Physiol., *174*, 295-315.

Taylor, A. & Gottlieb, S. (1985). Proprioceptive and other reflexes of fusimotor neurones of jaw muscles. In The Mammalian Muscle Spindle. Eds. Boyd, I.A. & Gladden, M.H.. Macmillan, London. (In press).

Terdiman, J., Smith, J.D., & Stark, L. (1969). Pupil response to light and electrical stimulation. Static and dynamic characteristics. Brain Res., *16*, 288-292.

Chapter Eight

INTRAFUSAL MUSCLE FIBRES IN THE CAT AND THEIR MOTOR CONTROL

I.A. Boyd

The experiments described below were carried out on muscle spindles isolated from the tenuissimus muscle of the cat hind leg and illustrate the physiological behaviour of three types of intrafusal muscle fibre called the 'dynamic bag_1 fibre' (Db_1) the 'static bag_2 fibre' (Sb_2) and the 'nuclear chain fibres' (Ch). The same three types of intrafusal fibre are found in primate spindles, including those in man, which differ only in minor respects from those in the cat. Details of spindle morphology in rat, cat, baboon and man are available elsewhere (Barker 1974; Boyd & Smith, 1984; Boyd & Gladden, 1985).
    Most spindles have one primary sensory ending supplied by a group Ia afferent axon of large diameter, and one secondary sensory ending supplied by a medium sized group II afferent axon, though the number of secondary endings varies from zero to five. The primary ending has annulo-spirals encircling the nuclear region of each of the nuclear bag fibres and nuclear chain fibres. The adjacent secondary ending has annulo-spirals round each chain fibre, usually some sprays on the static bag_2 fibre, and sometimes some sprays on the dynamic bag_1 fibre. Thus the Ia input arises in all the intrafusal fibres whereas the II input arises primarily in the chain fibres.
    The dynamic bag_1 fibre in the cat is innervated by dynamic γ, and sometimes dynamic β, fusimotor neurones, together forming a separate 'dynamic intrafusal system' (Fig. 1). In tenuissimus spindles 7% of dynamic γ axons also innervate one nuclear chain fibre, the significance of this being unknown (Kucera, 1984). The static bag_2 fibre and chain fibres are innervated by static γ fusimotor neurones; axon branches to a single spindle may innervate the chain fibres only, the static bag_2 fibre only or both together (Fig. 1). A minority of spindles have a long chain fibre innervated by a static β axon; β fusimotor neurones innervate extrafusal as well as intrafusal muscle. Of all the fusimotor axon branches to tenuissimus spindles about 80% selectively innervate only one type of fibre, whereas the rest are non-selective in

Intrafusal muscle fibres in the cat and their motor control

distribution (Boyd, 1971; Kucera, 1985). Evidence is presented that there are two types of static γ motor neurone.

METHODS

The experiments on intrafusal tension, and on the action of succinylcholine (SCh), were conducted on completely isolated spindles, free of all extrafusal muscle, which were activated by applying graded stimuli to the whole nerve to the tenuissimus muscle, successively recruiting fusimotor axons with differing thresholds (Boyd & Ward, 1975). Experiments on the mechanical properties of intrafusal fibres were initially carried out on the same isolated nerve-muscle preparation, but latterly a preparation in which the spindle was isolated in a bath, with some of its blood supply and all of its nerve supply to the spinal cord intact, was used (Boyd et al., 1977). Individual intrafusal fibres were then observed within the spindle during repetitive stimulation of a single fusimotor axon supplying that particular type of fibre only; the axon had earlier been isolated in a 'single fibre' filament dissected from the ventral spinal roots. Trains of stimuli at a constant frequency were applied, or the frequency was increased from zero to 150Hz (Figs. 13, 14). During such stimulation the discharge from primary or secondary sensory endings was recorded in group Ia or group II afferent axons in 'single fibre' filaments of the dorsal spinal roots. The effect of a ramp stretch during tonic fusimotor stimulation was also tested (Figs. 3, 6, 7). Recordings of this nature were usually made before the muscle was removed from the cat with the stretch applied to one free end of the muscle, and then again, on the following day, when the stretch was applied to both ends of the part of the muscle containing the spindle in a bath of Krebs's solution. In the experiments in which the action of a fusimotor axon on a number of spindles in the same muscle was tested, the part of the muscle in the bath contained up to six spindles; isolation of one spindle was usually achieved and sometimes intrafusal events in two or three spindles were observed.

INTRAFUSAL TENSION

The maximum tetanic tension developed by the whole intrafusal bundle varied from 2-8mg (Table 1). In one spindle the nuclear bag fibres alone produced 5mg tension, while the chain fibres produced 3mg or less in two spindles. At the time these experiments were conducted (Boyd, 1976a) it was not known that there were two types of nuclear bag fibre. In three spindles one nuclear bag fibre was observed to contract at both poles and to extend its primary sensory spiral by 15%

Intrafusal muscle fibres in the cat and their motor control

to 22% which means that it must have been a static bag_2 fibre. A much smaller contraction was observed in the other nuclear bag fibre which must have been the dynamic bag_1 fibre, only one pole of which was activated in two of the three spindles. The tension transducer was too insensitive to record fractions of 1mg; more accurate measurements of spindle tetanic tension are now available which confirm that it is of the order of a few mg (Fukami, 1985). The wide variation in maximum tension from one spindle to another was probably due to differences in the degree to which residual extrafusal muscle, into which the intrafusal fibres were inserted, absorbed part of the tension.

If it is assumed that each spindle in a muscle containing about fifty spindles, such as a hind limb muscle in a cat, produces a tension of 5mg, then the total possible tension which all the spindles could produce if maximally activated at the same time would be 250mg. It is very unlikely that all three types of intrafusal fibre are activated maximally, and probably never at the same time. Further, much of the tension produced must be absorbed by the extrafusal muscle since spindles are very much shorter than the muscles in which they lie. Thus, the maximum tension ever developed by all the

Table 1

Maximum tension developed by intrafusal fibres in single isolated muscle spindles

Tenuissimus muscle no.	Intrafusal fibres active	Tension	Static bag_2 fibre spiral extension
1	Db_1 one pole) Sb_2 both poles)	5mg	about 15%
	Ch both poles	3mg	
2	Db_1 one pole) Sb_2 both poles) Ch both poles)	3mg	15%
3	Db_1 both poles) Sb_2 both poles) Ch both poles)	2mg	22%
4	Ch both poles	<2mg	–

spindles at the muscle tendon is probably very much less than 250mg. Since the maximum tension produced by the extrafusal muscle may be ten thousand times greater than this value, the tension developed by muscle spindles is negligible in comparison with extrafusal tension.

THE BEHAVIOUR OF INTRAFUSAL MUSCLE FIBRES

The static bag_2 fibre

Activation of static γ axons produces local contraction of the static bag_2 fibre within the capsular sleeve region at one or both poles, close to the location of the motor end-plate responsible (Fig. 1). Thus, this fibre always moves towards the equator if observed at the end of the capsular sleeve.

The contraction produces a marked extension of the Sb_2 primary sensory spiral varying from about 10% if one pole is active (Fig. 2) to 25% if both poles are activated. The resultant bias in the group Ia discharge is almost proportional to the degree of extension of the Sb_2 primary spiral (Fig. 2b). When the active fibre is stretched the sensitivity of the primary ending to the length change is

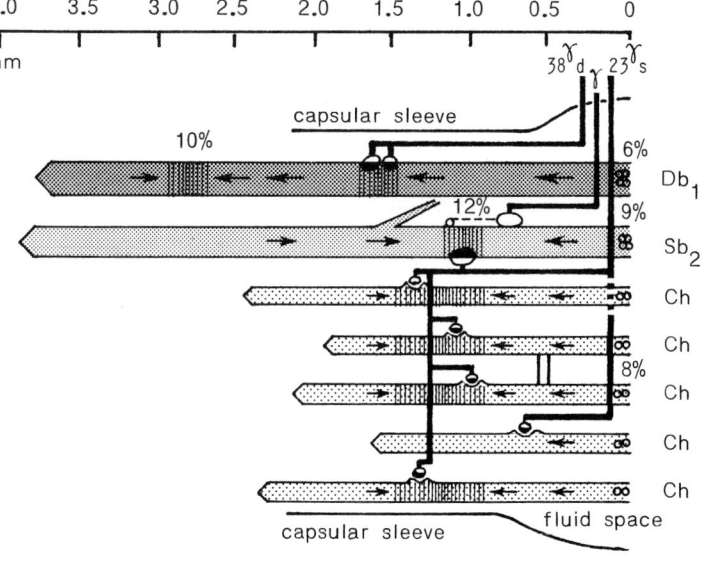

Fig. 1. Reconstruction of one pole of a muscle spindle from electron microscopy showing precise location of fusimotor nerve endings. Sarcomere convergence produced by one dynamic γ axon and one static γ axon was observed in this pole at the positions shown. Axons studied functionally are marked with conduction velocity. (From Arbuthnott et al., 1982).

Intrafusal muscle fibres in the cat and their motor control

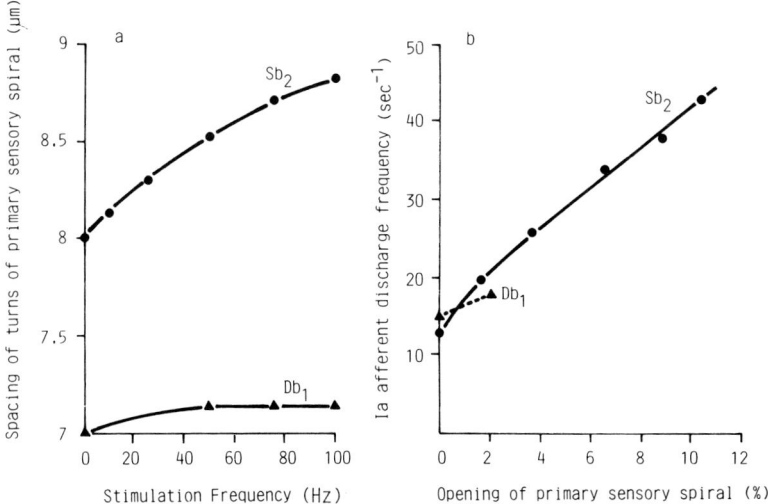

Fig. 2. Relation between fusimotor stimulation frequency, extension of primary sensory spirals round the Sb_2 fibre and the Db_1 fibre, and Ia afferent discharge frequency, in an isolated spindle during stimulation of a static γ axon and a dynamic γ axon.

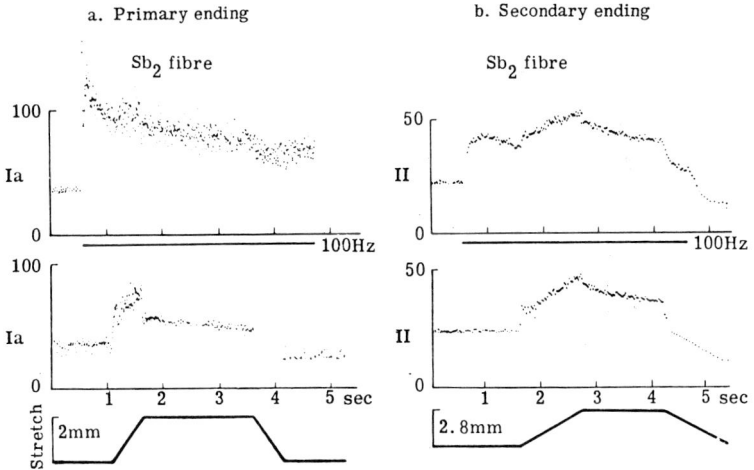

Fig. 3. Instantaneous frequencygrams showing the typical action of the Sb_2 fibre on a primary ending and a secondary ending in spindles in which contraction of this fibre only was observed. Ramp and hold stretch applied during static γ axon stimulation for the duration of the horizontal line. Responses to stretch without stimulation in middle traces. (From Boyd, 1985a).

unchanged or somewhat reduced (Fig. 3a). This results in a reduction in the 'dynamic index' (the difference between the peak discharge frequency at the end of the stretch and that 0.5 sec later, Crowe & Matthews, 1964). In fact, both the 'dynamic sensitivity', or sensitivity to the length change during the stretch itself, and the 'static sensitivity' (position sensitivity), are usually reduced. This applies to the secondary sensory ending, also (Fig. 3b), though the 'biassing' action is much smaller since the secondary ending has relatively few terminals on the static bag_2 fibre.

The dynamic bag_1 fibre

The most striking property of this fibre is that when it is stretched and held at the new length, the poles of the fibre give way and it creeps towards the spindle equator (Fig. 4). Neither the static bag_2 fibre nor the nuclear chain fibres exhibit this phenomenon. 'Intrafusal creep' in a dynamic bag_1 fibre in the absence of fusimotor stimulation is not usually as marked as that shown in Fig. 4. When a dynamic fusimotor axon is activated, a small local contraction is observed close to the site of the fusimotor end-plate usually within the capsular sleeve (Fig. 1). Contraction is not confined to this point, however, and, despite the fact that action potentials are not propagated in this fibre, convergence of sarcomeres occurs in the extracapsular region at sites where there is no end-plate (Banks et al., 1978; Arbuthnott et al., 1982). The

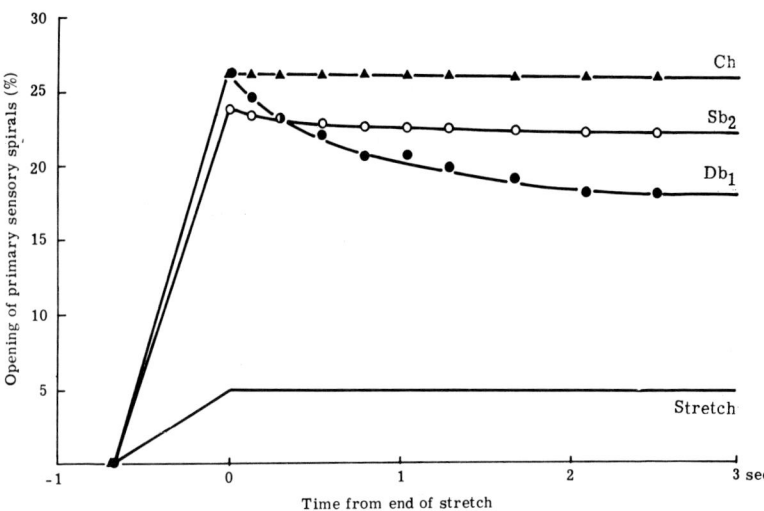

Fig. 4. 'Intrafusal creep' in the Db_1 primary sensory spiral of an inactive spindle following a ramp and hold stretch. Creep in Sb_2 or Ch fibre spirals is negligible. (From Boyd, 1976b).

fact that the fibre always moves towards the pole when observed at the end of the capsular sleeve suggests that some contraction occurs throughout the polar half of the fibre. The Db_1 primary sensory spiral is extended by a relatively small amount, less then 8% and maybe as little as 2% (Figs. 2, 5), and the bias applied to the Ia discharge is never large (Fig. 6a).

When the active dynamic bag_1 fibre is stretched, however, the stiffness of the outer part of the pole must be greatly enhanced since a much larger proportion of the stretch takes place in the primary sensory spiral (Fig. 5). This is probably due to activation of the contractile mechanism by stretch (Laporte et al., 1985; Boyd & Gladden, 1985, Part II). The result is a greatly enhanced sensitivity to the length change during the dynamic phase of stretch, which is evident as an increase in the slope of the Ia frequency record (Fig. 6a). When the fibre is held stretched a much larger degree of creep is observed than when the fibre was inactive (Fig. 5) and a corresponding phase of mechanical adaptation is seen in the Ia discharge. The resultant increase in the dynamic index indentifies the fusimotor axon as a dynamic axon. The increase in static sensitivity is much less than would have been the case had no creep occurred in the fibre.

It is not uncommon for the increase in dynamic sensitivity to be regarded as equivalent to an increase in

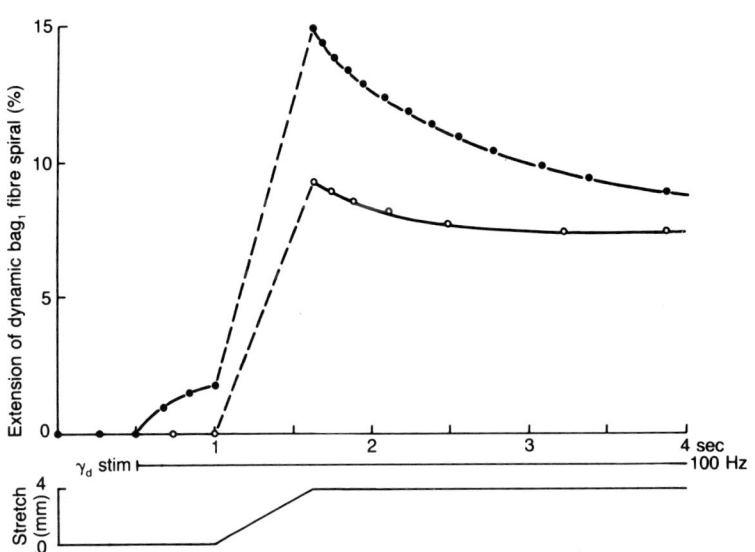

Fig. 5. Comparison of intrafusal creep in the Db_1 primary sensory spiral following a ramp stretch with (filled circles) and without (open circles) dynamic γ axon stimulation. (From Boyd et al., 1981).

velocity sensitivity. It is true that the primary sensory discharge is velocity dependent, but dynamic fusimotor activity results, mainly, in an increase in the length sensitivity of the ending under dynamic conditions which is greater than the increase in length sensitivity under static conditions (Fig. 6a). The fast rise and fall phases of the Ia response, which are velocity dependent, are little affected by dynamic fusimotor activity (Boyd, Murphy & Moss, 1985). In summary, the 'biassing' action of the dynamic bag_1 fibre is generally much less than that of the static bag_2 fibre, and the principal action of the dynamic bag_1 fibre is to make the primary ending very length sensitive during any change in length. It has a similar, but much smaller, action on the length sensitivity of any secondary ending with an appreciable number of terminals on this fibre (Fig. 6b) while its biassing action is often negligible. Many secondary endings, however, have no terminals on the dynamic bag_1 fibre and are unaffected by dynamic fusimotor activity. The properties of this fascinating intrafusal fibre are currently the subject of intensive study (Boyd & Gladden, 1985).

Nuclear chain fibres
These fibres have a relatively high fusion frequency and the individual poles twitch noticeably during low frequency stimulation of the static γ axons which supply them. Small twitch extensions of the sensory endings are produced. Unfused tetanus leads to mechanical excitation of the small primary sensory spirals round the chain fibres, and the Ia discharge is 'driven' at the fusimotor stimulation frequency (Fig. 7a). The driving action of chain fibres is particularly

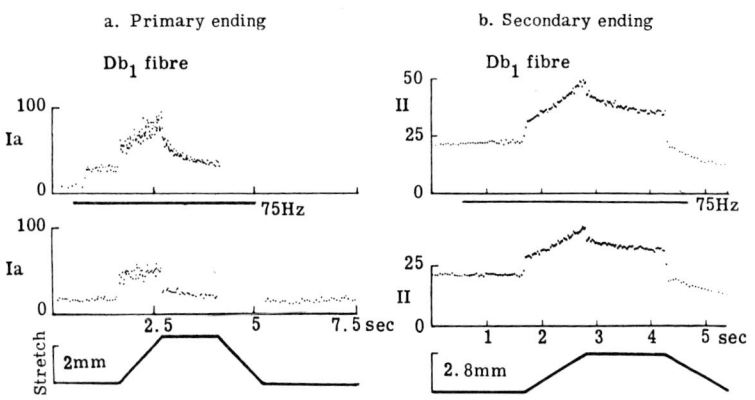

Fig. 6. Typical action of the Db_1 fibre on a primary ending and a secondary ending during a ramp stretch with and without dynamic γ axon stimulation. Details as in Fig. 3. (From Boyd, 1985a).

Intrafusal muscle fibres in the cat and their motor control

well demonstrated by the ramp stimulation test with the spindle at constant length (Fig. 13a,b). This 1:1 driving which is produced if most or all of the four or five chain fibres in any spindle pole are active, is replaced by 1:2 driving if only a few chain fibres are activated by any one axon or the response may alternate between 1:1 and 1:2 driving (Fig. 13c). The static bag_2 fibre never produces driving of the primary ending (Fig. 14a-c). When the static bag_2 fibre is activated along with most of the chain fibres by a non-selective static γ axon, however, chain fibre driving dominates the Ia response up to a fusimotor frequency of about 100Hz (Fig. 13b). Driving by a single static γ axon is most effective at muscle lengths in the middle third of the physiological range, but sometimes occurs at all muscle lengths. Two static γ axons to the two poles of the chain fibres may produce 1:1 driving when either alone produces 1:2 driving (Boyd, Murphy & Mann, 1985).

It is of particular interest to know if driving can maintain the Ia frequency at a constant value during muscle lengthening or shortening, since this would mean that the Ia input to α motor neurones would be maintained at a constant level despite changes in muscle length. If the length signal is relatively small, as is the case in the tenuissimus muscle, where there is a long length of extrafusal muscle in series with the spindles, then the length signal is quite often completely suppressed by chain fibre driving (Fig. 7a). In

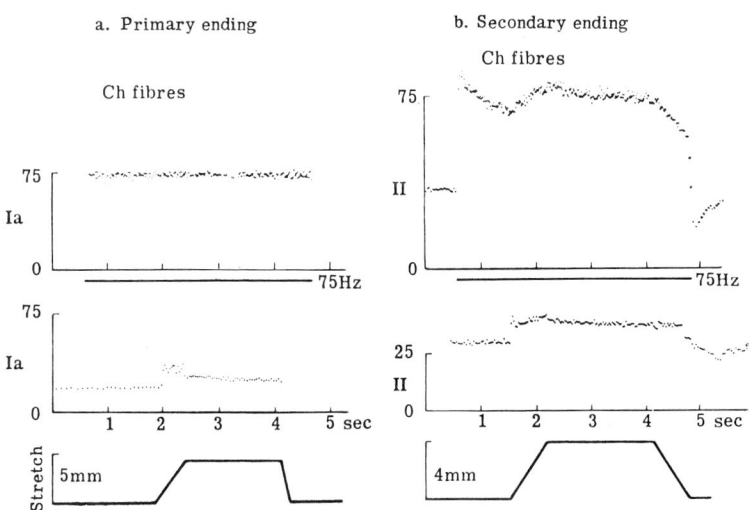

Fig. 7. Typical action of chain fibres on a primary and on a secondary ending during a ramp stretch with and without static γ axon stimulation. Details as in Fig. 3. (From Boyd, 1985a).

other muscles where the length signal is relatively large, however, the input from the dynamic bag$_1$ fibre 'breaks through the driving during muscle lengthening. During muscle shortening, even at velocities near the top of the physiological range, driving often maintains the Ia discharge either 1:1 as in Fig. 7a or 1:2. If two fusimotor axons to the two poles of the chain fibres are stimulated together, driving at the fusimotor frequency is often maintained during shortening of all muscles (Boyd, Murphy & Mann, 1985). It seems likely, then, that one of the principal functions of the chain fibres is to maintain the Ia input at a high value during muscle shortening and that this could be achieved by recruiting a pool of static γ motor neurones which innervated the chain fibres whether or not they also innervated the static bag$_2$ fibre.

Driving of secondary endings by chain fibres is rare, in the cat at least. If most of the chain fibres in a spindle pole are activated, a large bias is applied to the group II discharge (Fig. 7b) and the length sensitivity of the secondary ending is greatly increased both under dynamic and static conditions (Jami et al., 1980; Boyd, 1981; Boyd, Sutherland & Ward, 1985). This appears to be true whether or not the static bag$_2$ fibre is activated as well. If only a few chain fibres are active, however, such an increase in length sensitivity may not be seen and it may be necessary to activate more than one axon to the chain fibres to achieve it. It seems likely that recruitment of a pool of static γ motor neurones innervating chain fibres would have a large biassing action on secondary endings, at the same time greatly increasing their sensitivity both during changes in length and when moving from one static length to another. Under these conditions secondary endings would provide an excellent signal of muscle length.

RESPONSE OF INTRAFUSAL FIBRES TO SUCCINYLCHOLINE (SCh) OR ACETYLCHOLINE (ACh).

The transmitter at somatic fusimotor endings is ACh. It was of interest to determine, for each type of intrafusal fibre, whether ACh or SCh applied directly to the bath containing the spindle produced depolarisation block as in extrafusal muscle, or contraction as a result of action at the fusimotor end-plates, or a large contracture because the fibre was sensitive along much of its length. Six early experiments of this nature using SCh will now be described. The results were not published at the time because they could not be readily interpreted. Now that it is known that there are two types of nuclear bag fibre, the results fall neatly into place.

Concentrations of SCh up to $100 \mu g \ l^{-1}$ produced no contraction of the nuclear chain fibres in any of the six spindles. Contraction of chain fibres produced by repetitive stimulation of a γ axon selectively innervating them was abolished by 10 μg l^{-1} SCh in all of the five spindles in which this could be tested. In the experiment illustrated in Fig. 8(a-d) the slightly kinked chain fibres at the top of the intrafusal bundle moved 17 μm to the right along with their supplying axon when it was stimulated at 75Hz (b). Addition of 10μg l^{-1} SCh abolished this effect and did not produce any contracture of the chain fibres (c). After washing out of the SCh the response to fusimotor stimulation returned (d). In the experiment illustrated in Fig. 9 (squares) the degree of extension of the small primary sensory spirals round the chain fibres was recorded. Contraction of the chain fibres produced by γ stimulation was greatly reduced by 1μg l^{-1} SCh, intermittent neuromuscular block was observed with 5μg l^{-1}, and complete block with 10μg l^{-1}. Gladden (1976) likewise observed that ACh in concentrations up to 10μg l^{-1} produced no contraction in nuclear chain fibres, though she did not test the effect of fusimotor stimulation in the presence of ACh.

Maximal contraction (contracture) of the static bag$_2$ fibre was observed in three spindles about 15 min after the addition of 10μg l^{-1} of SCh. In three spindles the effect of stimulating a fusimotor axon selectively innervating the static bag$_2$ fibre in the presence of SCh was studied. The fusimotor induced contraction was reduced by 1μg l^{-1} SCh, and completely or almost completely abolished by 10μg l^{-1} SCh. The interaction of the effects of SCh and of fusimotor activity on the static bag$_2$ fibre are well illustrated in Fig. 8g-j and Fig. 9. Fig. 8g shows the Sb$_2$ primary sensory spiral at rest, the arrows indicating the dip between annulo-spirals. Each turn of the spiral is seen as a projection on the lower or upper edge of the intrafusal fibre since the spiral encircles the fibre superficially. Repetitive stimulation of a fusimotor axon to one pole of the fibre at 100Hz opened the spiral by 21% and the elevations of the spiral on the fibre were obviously flattened (Fig. 8h). Partial contracture was produced by 1μg l^{-1} SCh and the spiral opened 18% (Fig. 9, filled circles). Stimulation at 100Hz produced a further extension of 6% (Fig. 9, open circles). Almost complete contracture, causing 34% spiral extension, was produced by 5μg l^{-1} SCh and stimulation then produced an additional 2% extension only. Maximum contracture causing 36% extension was produced by 10μg l^{-1} (Fig. 8i) and then fusimotor activity caused no further extension (Fig. 8j). This degree of extension was precisely that produced by stimulation of axons to both ends of the static bag$_2$ fibre together in the absence of SCh (Fig. 9, half filled circle). It seems, then, that maximum contracture of this fibre produced by SCh is the same as the maximum contraction produced by repetitive stimulation of all the static γ axons supplying it. Thus, SCh must act at the fusimotor neuromuscular junctions by producing

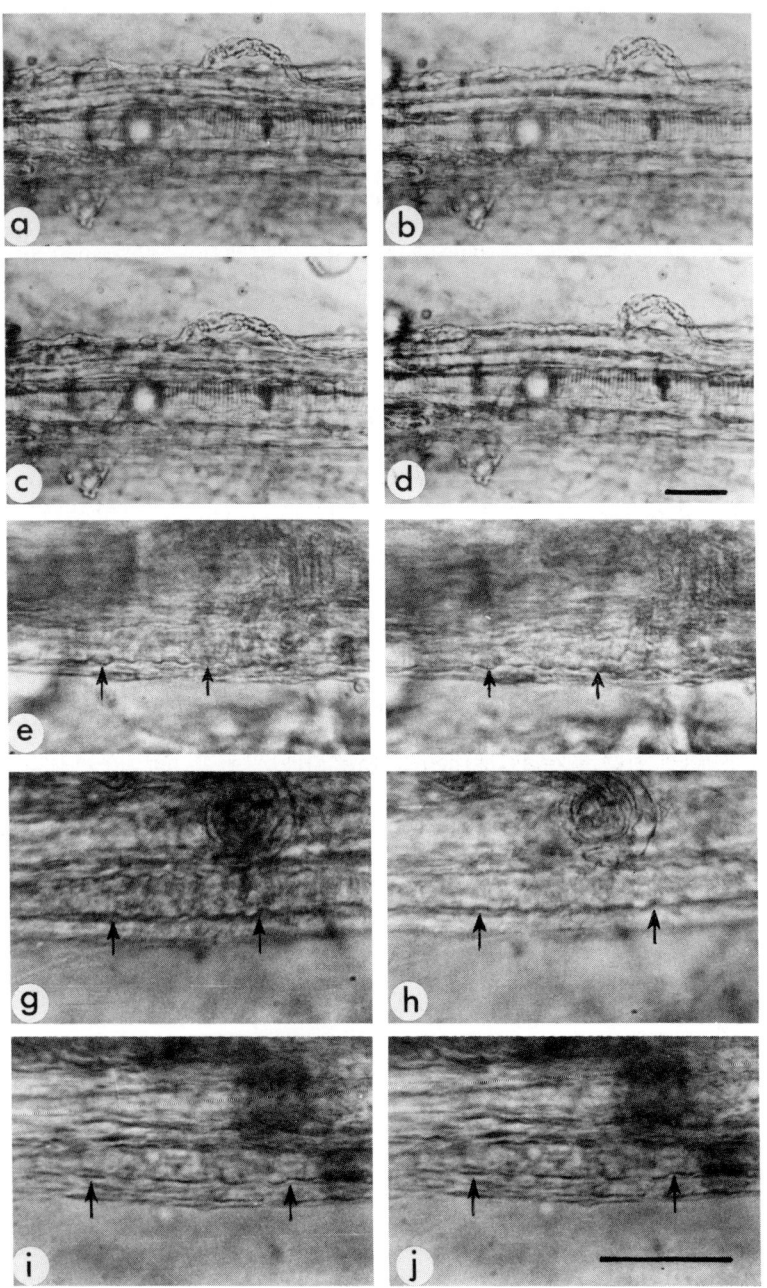

local depolarisation, and not along the length of the fibre. The fibre in question must have been the static bag_2 fibre not only because its contraction was both very marked and rapid, but because stimulation of a different axon caused only a small contraction of the other nuclear bag fibre, opening its primary spiral by 3.5%, which is typical of the dynamic bag_1 fibre. Gladden (1976), likewise, noted that $10\mu g\ l^{-1}$ ACh produced contracture of the static bag_2 fibre whereas $1\mu g\ l^{-1}$ ACh did not.

Contracture of the dynamic bag_1 fibre was observed in two spindles only, no doubt because it is small compared with that in the static bag_2 fibre, and because attention was not generally paid to both the nuclear bag fibres in the same spindle. In one case the dynamic bag_1 fibre contracture was produced by $10\mu g\ l^{-1}$ SCh, that is by the same concentration that produced static bag_2 fibre contracture in the same spindle; less concentrated solutions were not tested, however. In the other spindle (Fig. 8e,f; Fig. 9, triangles) partial contracture of the dynamic bag_1 fibre, opening its primary spiral by 6%, was produced by $1\mu g\ l^{-1}$ SCh and the effect was maximal at 8% with $5\mu g\ l^{-1}$ SCh (Fig. 8f). This limited data with SCh conforms with the finding of Gladden (1976) that the threshold of the dynamic bag_1 fibre to direct application of ACh is lower than that of the static bag_2 fibre.

The action of SCh on the primary sensory ending is well documented (Rack & Westbury, 1966). A very marked increase in dynamic sensitivity is produced (the Phase II response of Dutia, 1980), no doubt due to contracture of the dynamic bag_1 fibre. An increase in static sensitivity occurs some time later which Dutia attributed to contracture of the static bag_2 fibre, though it should be noted that this fibre biases the group Ia discharge rather than increasing the static sensitivity of the primary ending. SCh is used to

---

Fig. 8. The interaction of fusimotor stimulation and exposure to SCh on the nuclear chain fibres (a-d), the Db_1 primary sensory spiral (e,f) and the Sb_2 primary sensory spiral (g-j) in three isolated spindles. Scale bar, $50\mu m$.
  a. Rest, $800\mu m$ to right of primary sensory ending.
  b. Chain fibres and axon move $17\mu m$ during stimulation at 75Hz.
  c. Stimulation at 75Hz + $10\ \mu g\ l^{-1}$ SCh. No contracture. Fusimotor response abolished.
  d. SCh washed out. Response to 75Hz stimulation returned.
  e. Db_1 primary spiral at rest. Arrows five spiral turns apart.
  f. Db_1 spiral 8% extended by $5\ \mu g\ l^{-1}$ SCh.
  g. Sb_2 primary spiral at rest. Arrows seven spiral turns apart.
  h. Sb_2 spiral extended 21% by activation of one pole at 100Hz.
  i. Sb_2 spiral extended 36% by $10\ \mu g\ l^{-1}$ SCh.
  j. $10\ \mu g\ l^{-1}$ SCh + 100Hz stimulation. Extension still 36%.

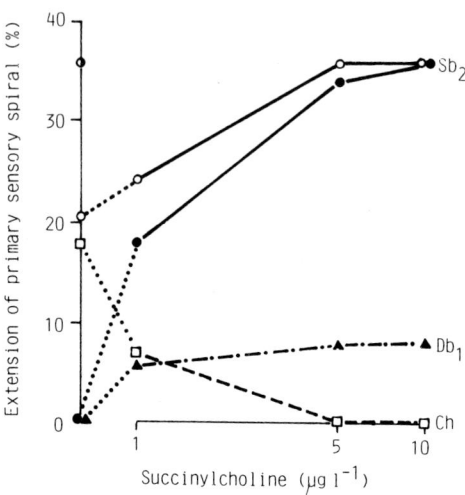

Fig. 9. Extension of the primary sensory spirals round the Db_1 fibre (triangles) and the Sb_2 fibre (circles) or the chain fibres (squares) in three different isolated spindles. Filled symbols, SCh contracture. Open symbols, maximum fusimotor activation. Fusimotor activation of Sb_2 fibre and chain fibres is progressively less effective as SCh concentration increases, while contracture in both nuclear bag fibres increases.

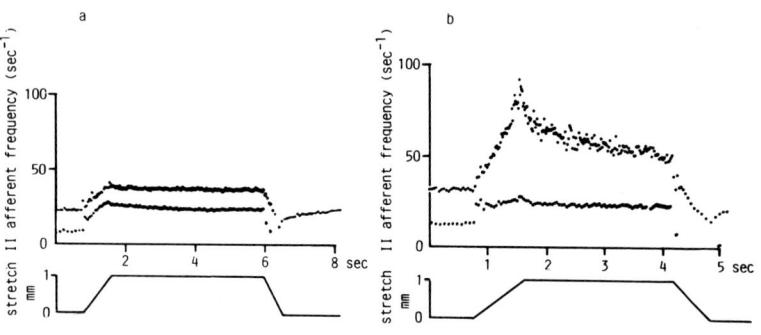

Fig. 10. Response of two secondary endings in isolated spindles to a ramp stretch in the presence and absence of SCh. a; 50 μg l^{-1} SCh, no increase in dynamic sensitivity. b; 50 μg l^{-1} SCh, marked increase in dynamic sensitivity.

discriminate between primary and secondary afferent responses in experiments in which the conduction velocity of afferent axons cannot readily be determined. It is usually assumed that secondary endings do not respond to SCh to any appreciable degree and Dutia (1980) confirmed that this is the case for a very large majority of secondary endings in the soleus muscle.

There are, however, exceptions to this rule resulting from the fact that some secondary endings have an appreciable number of terminals on the nuclear bag fibres (Banks et al., 1982). In the early experiments on isolated spindles now described the discharge in the group Ia axon from the primary ending tended to be phasic only. In four experiments useful data was obtained on the effect of SCh on secondary endings, in the absence of any extrafusal muscle which could modify spindle activity by itself responding to SCh. In one experiment, two previously silent secondary endings commenced firing steadily precisely when contracture of the static bag_2 fibre was induced by $10 \mu g\ l^{-1}$ SCh. Since it has been shown that the chain fibres do not contract in the presence of SCh this effect on the secondary endings must have been due to extension of secondary terminals on the static bag_2 fibre.

The results in the three other experiments are illustrated in Figs. 10 and 11. In all three cases an increase in the initial group II afferent discharge frequency

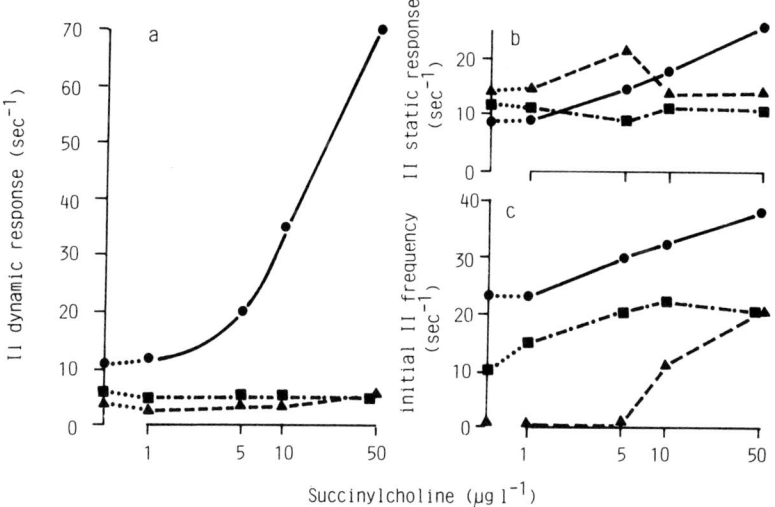

Fig. 11. Action of SCh on the initial frequency, dynamic response (peak to adapted value 2.5s later) and static response (difference between initial and final frequencies) of three secondary endings in isolated spindles.

was produced commencing at a concentration between $1\mu g\ l^{-1}$ and $10\mu g\ l^{-1}$ SCh (Fig. 11c). This increase could have been due to contracture in the static bag_2 fibre for which $1\mu g\ l^{-1}$ to $10\mu g\ l^{-1}$ is normally required, or possibly to the direct action of SCh on the secondary sensory terminals (the Phase I response of Dutia, 1980). SCh had no effect on the dynamic response of two of the secondary endings (Fig. 10a, and 11a, squares, triangles). The dynamic response of the third secondary ending increased markedly with increasing concentration of SCh, the increase being first noticeable with $5\mu g\ l^{-1}$ SCh and maximal with $50\mu g\ l^{-1}$ SCh (Fig. 10b and 11a, circles). The response of this ending resembles closely that of one secondary ending whose dynamic sensitivity was markedly increased by fusimotor induced contraction of the dynamic bag_1 fibre in the study of Boyd (1981). The incidence of such large dynamic effects is rare, however, though a modest increase in the dynamic sensitivity of about one secondary ending in ten can be produced by the dynamic bag_1 fibre even when the bias applied to the group II discharge is negligible (Boyd, Sutherland & Ward, 1985). It is likely that such effects are confined to secondary endings with conduction velocities towards the upper end of their range of velocities, and which lie adjacent to the primary ending.

FUSIMOTOR CONTROL OF INTRAFUSAL FIBRES

In early studies of the action of fusimotor axons on the afferent discharge from primary sensory endings it was established that a dynamic γ axon has a dynamic action on all the spindles it supplies whereas a static γ axon always has a static action (Crowe & Matthews, 1964). The present study has confirmed that a dynamic γ axon has a similar, dynamic action on up to three spindles. Further, dynamic bag_1 fibres, one in

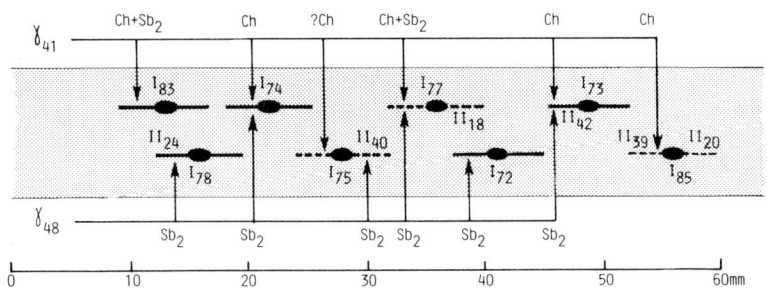

Fig. 12. Functional distribution of two static γ axons to eight spindles in the same 6cm length of tenuissimus muscle. Afferent and efferent axons designated by conduction velocity. (From Boyd, 1985b).

each of two spindles isolated from the same muscle, have been observed contracting by themselves in response to stimulation of a particular fusimotor axon.

A single static γ axon was shown by Barker et al. (1973) to innervate nuclear bag fibres, nuclear chain fibres, or both, in the different spindles it supplied. The nuclear bag fibre involved turned out subsequently to be the static bag$_2$ fibre. It was never conclusively demonstrated, however, that a single static γ axon supplied the static bag$_2$ fibre <u>only</u> in one spindle and the nuclear chain fibres <u>only</u> in a different spindle. Static γ motor neurones, in fact, fall into two populations. 'Static bag γ motor neurones', or γ_{sb}, supply the static bag$_2$ fibre in all the spindles they innervate, though they infrequently produce some contraction in chain fibres as well. 'Static chain γ motor neurones', or γ_{sc}, supply the chain fibres in all the spindles they innervate, and not infrequently produce some contraction in the static bag$_2$ fibre as well; it is the chain fibre activity which dominates the physiological picture, however. This is demonstrated in the experiment illustrated in Figs. 12-14. Eight group Ia afferent axons were isolated in dorsal root filaments, and all eight spindles were located in a 6cm length of tenuissimus muscle. Recordings were also made from six group II afferent axons from five of the eight spindles. Ramp stimulation and tonic stimulation tests were applied when the muscle was <u>in situ</u>, and ramp stretch during tonic stimulation was applied when the muscle was in the bath. All three types of tests were entirely consistent although only ramp stimulation tests are illustrated.

The action of two static γ axons was tested on all eight spindles (Fig. 12). One axon (γ_{41}) produced a nuclear chain action on six primary endings and five secondary endings and the chain fibres were seen to contract in one spindle which was isolated (I_{74}). The driving action of the nuclear chain fibres on three primary endings, and the typical powerful biassing action of the chain fibres on a secondary ending in a fourth spindle, are illustrated in Fig. 13. The static bag$_2$ fibre was involved in two spindles, shown by an irregular discharge at high frequency when the stimulation frequency exceeded 100Hz.

The second axon (γ_{48}) produced a static bag$_2$ fibre action on six primary endings and four secondary endings, with no evidence of chain fibre involvement in any of the six spindles. A biassing action on the Group Ia discharge without any driving action indicates activity in the static bag$_2$ fibre only in three spindles (Fig. 14a-c). This was confirmed by serial electron microscopy of one of the spindles (I_{74}). A relatively small biassing action on the secondary ending in a fourth spindle is consistent with static bag$_2$ fibre activity (Fig. 14d); the primary ending of this spindle gave a response similar to Fig. 14a-c.

In a second experiment two static γ axons had a chain fibre action in five spindles in the one case, and four in the other, which was confirmed by direct observation in one of the spindles; the static bag$_2$ fibre was involved in one spindle only in each case. A third static γ axon produced a static bag$_2$ fibre action in four spindles with no evidence of chain fibre involvement; one of these spindles was isolated and contraction was evident only in the static bag$_2$ fibre; the same axon had a static bag$_2$ fibre and chain fibre action in two further spindles.

The action of an additional ten static γ axons was tested on at least three spindles in nine tenuissimus muscles and the result was the same (Boyd, 1985b). Six static axons produced a chain fibre action in every spindle tested, three of them also activating the static bag$_2$ fibre in one or two of the spindles. Four static axons produced a static bag$_2$ fibre action in every spindle tested; two of them also activated the chain fibres in one spindle only.

These experiments clearly suggest that there are two distinct populations of static γ motor neurones. Static bag γ motor neurones bias the group Ia discharge in all the spindles they supply, and sometimes the discharge from secondary endings at any innervated pole; both sensory endings remain sensitive to length changes though this sensitivity may be reduced. Static chain γ motor neurones bias the group Ia

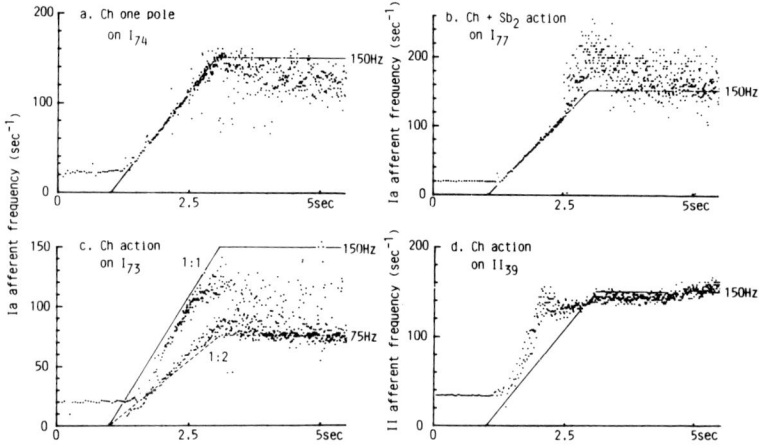

Fig. 13. A static γ axon (41m s^{-1}) producing chain fibre driving action on three primary endings and a large chain fibre biassing action on a secondary ending in a fourth spindle. Ramp stimulation 0-150Hz shown by continuous lines. Contraction of chain fibres observed in spindle a. (From Boyd, 1985b).

discharge of all the spindles they supply, and the discharge from secondary endings at any innervated pole. The length sensitivity of the primary ending is often abolished by driving so that the Ia input is maintained during muscle shortening; at the same time the secondary ending becomes very length sensitive both under dynamic and static conditions. The fact that γ_{sc} axons also innervate the static bag_2 fibre in some spindles does not modify this basic physiological picture provided that the γ output frequency does not exceed 100Hz.

PHYSIOLOGICAL ROLE OF THE THREE INTRAFUSAL SYSTEMS

It has been demonstrated that the dynamic intrafusal system, and the two static intrafusal systems, are to a large extent each subject to separate central control by three different populations of γ motor neurones. The action of dynamic β motor neurones within the spindles they supply is essentially similar to that of dynamic γ motor neurones; both make the primary sensory ending very sensitive to length changes but only during movement. Deviation from a planned trajectory such as obstruction of movement would thus generate a large signal within the central nervous system if the dynamic intrafusal system was active. The static bag_2 fibre

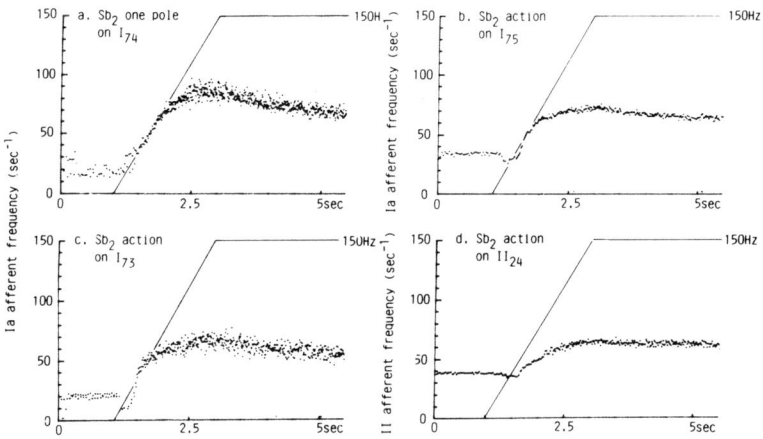

Fig. 14. A single static γ axon (48m s^{-1}) producing a static bag_2 fibre biassing action (no driving) on three primary endings and a small, static bag_2 fibre action on a secondary ending in a fourth spindle. Innervation of static bag_2 fibre alone confirmed in spindle a. (From Boyd, 1985b).

system biasses the primary sensory discharge, and to a lesser extent the secondary sensory discharge, at constant length which would keep the stretch reflex in operation and contribute to muscle tone; it could also maintain the spindle input during muscle shortening if the static bag γ motor neurone output frequency was increased to compensate for the shortening. The chain fibre system has a powerful biassing action on both primary and secondary endings, very powerful if both poles of the chain fibres are active. The group Ia discharge could be maintained during muscle shortening without the need to increase the output in static chain γ motor neurones; at the same time the secondary endings would provide an excellent length signal both when the limb was moving and when it moved from one position to another. This latter signal would be ideal for the conscious appreciation of position. Static β axons which innervate the long chain fibre in a minority of spindles may also produce driving of the primary ending the significance of which is unknown (Jami et al., 1985).

REFERENCES

Arbuthnott, E.R., Ballard, K.J., Boyd, I.A., Gladden, M.H. & Sutherland, F. (1982). The ultrastructure of cat fusimotor endings and their relationship to foci of sarcomere convergence in intrafusal fibres. J. Physiol., 331, 285-309.

Banks, R.W., Barker, D., Bessou, P., Pagès, B. & Stacey, M.J. (1978). Histological analysis of cat muscle spindles following direct observation of the effects of stimulating dynamic and static motor axons. J. Physiol., 283, 605-619.

Banks, R.W., Barker, D. & Stacey, M.J. (1982). Form and distribution of sensory terminals in cat hindlimb muscle spindles. Phil. Trans. R. Soc. B, 99, 329-364.

Barker, D. (1974). The morphology of muscle receptors. In Handbook of Sensory Physiology, Vol. 3, Pt. 2, Muscle Receptors. Ed. Hunt, C.C.. Springer-Verlag, New York, pp. 1-190.

Barker, D., Emonet-Dénand, F., Laporte, Y., Proske, U. & Stacey, M.J. (1973). Morphological identification and intrafusal distribution of the endings of static fusimotor axons in the cat. J. Physiol., 230, 405-427.

Boyd, I.A. (1971). Specific fusimotor control of nuclear bag and nuclear chain fibres in cat muscle spindles. J. Physiol., 214, 30-31P.

Boyd, I.A. (1976a). The response of fast and slow nuclear bag fibres in isolated cat muscle spindles to fusimotor stimulation, and the effect of intrafusal contraction on the sensory endings. Q. Jl exp. Physiol., 61, 203-254.

Boyd, I.A. (1976b). The mechanical properties of dynamic nuclear bag fibres, static nuclear bag fibres and nuclear chain fibres in isolated cat muscle spindles. Prog. Brain Res., 44, 33-50.

Boyd, I.A. (1981). The action of the three types of intrafusal fibre in isolated cat muscle spindles on the dynamic and length sensitivities of primary and secondary sensory endings. In Muscle Receptors and Movement Eds. Taylor, A. & Prochazka, A.. Macmillan, London, pp. 17-32.

Boyd, I.A. (1985a). Muscle spindles and stretch reflexes. In Scientific Basis of Clinical Neurology. Eds. Swash, M. & Kennard, C.. Churchill Livingstone, Edinburgh, pp. 74-97.

Boyd, I.A. (1985b). Static intrafusal systems in cat muscle spindles. Q. Jl exp. Physiol. (In press).

Boyd, I.A. & Gladden, M.H. (1985). The Muscle Spindle. Macmillan, London.

Boyd, I.A., Gladden, M.H., McWilliam, P.N. & Ward, J. (1977). Control of dynamic and static nuclear bag fibres and nuclear chain fibres by gamma and beta axons in isolated cat muscle spindles. J. Physiol., 265, 133-162.

Boyd, I.A., Gladden, M.H. & Ward, J. (1981). The contribution of mechanical events in the dynamic bag$_1$ intrafusal fibre in isolated cat muscle spindles to the form of the Ia afferent axon discharge. J. Physiol., 317, 80-81P.

Boyd, I.A., Murphy, P.R. & Mann, C. (1985). The effect of chain fibre 'driving' on the length sensitivity of primary sensory endings in the tenuissimus, peroneus tertius and soleus muscles. In The Muscle Spindle. Eds. Boyd, I.A. & Gladden, M.H.. Macmillan, London. (In press).

Boyd, I.A., Murphy, P.R. & Moss, V.A. (1985). Analysis of primary and secondary afferent responses to stretch during activation of the dynamic bag$_1$ fibre or the static bag$_2$ fibre separately in cat muscle spindles. In The Muscle Spindle. Eds. Boyd, I.A. & Gladden, M.H.. Macmillan, London. (In press).

Boyd, I.A. & Smith, R.S. (1984). The muscle spindle. In Peripheral Neuropathy. Eds. Dyck, P.J., Thomas, P.K., Lambert. E.H. & Bunge, R.. Saunders, Philadelphia, pp. 171-202.

Boyd, I.A., Sutherland, F. & Ward, J. (1985). The origin of the increase in the length sensitivity of secondary sensory endings produced by some fusimotor axons. In The Muscle Spindle. Eds. Boyd, I.A. & Gladden, M.H.. Macmillan, London. (In press).

Boyd, I.A. & Ward, J. (1975). Motor control of nuclear bag and nuclear chain intrafusal fibres in isolated living muscle spindles from the cat. J. Physiol., 244, 83-112.

Crowe, A. & Matthews, P.B.C. (1964). Further studies of static and dynamic fusimotor fibres. J. Physiol., 174, 132-151.

Dutia, M.B. (1980). Activation of cat muscle spindle primary, secondary and intermediate sensory endings by suxamethonium. J. Physiol., 304, 315-330.

Fukami, Y. (1985). Active force and sensory response of single isolated cat muscle spindles in vitro. In The Muscle Spindle. Eds. Boyd, I.A. & Gladden. M.H.. Macmillan, London. (In press).

Gladden, M.H. (1976). Structural features relative to the function of intrafusal muscle fibres in the cat. Prog. Brain Res., 44, 51-59.

Jami, L., Lan-Couton, D. & Petit, J. (1980). A study with the glycogen-depletion method of intrafusal distribution of axons that increase sensitivity of spindle secondary endings. J. Neurophysiol., 43, 16-26.

Jami, L., Petit, J. & Scott, J.J.A. (1985). "Driving" of spindle primary endings by static β axons. In The Muscle Spindle. Eds. Boyd, I.A. & Gladden, M.H.. Macmillan, London. (In press).

Kucera, J. (1984). Nonselective motor innervation of nuclear bag$_1$ intrafusal muscle fibers in the cat. Cell Tissue Res., 236, 383-391.

Kucera, J. (1985). Selective and non-selective motor innervation of intrafusal muscle fibers in the cat. In The Muscle Spindle. Eds. Boyd, I.A. & Gladden, M.H. Macmillan, London. (In press).

Laporte, Y., Emonet-Denand, F. & Hunt, C.C. (1985). Does stretch excite the bag$_1$ fibre? In The Muscle Spindle. Eds. Boyd, I.A. Gladden, M.H. Macmillan, London. (In press).

Rack, P.M.H. & Westbury, D.R. (1966). The effects of suxamethonium and acetylcholine on the behaviour of cat muscle spindles during dynamic stretching, and during fusimotor stimulation. J. Physiol., 186, 698-713.

Chapter Nine

HOW DO CRABS CONTROL THEIR MUSCLE RECEPTORS?

B.M.H. Bush and A.J. Cannone

Like crayfish, lobsters and other Decapod Crustacea, crabs are well endowed with proprioceptors and other mechanoreceptors in their walking legs and chelipeds (Bush & Laverack, 1982). These include joint receptors (chordotonal organs), muscle receptors and tendon (or apodeme) tension receptors, all to some extent functionally analogous with the corresponding sense organs in vertebrates. As in other animals, only the muscle receptors have an efferent innervation. This is comparable to the motor supply of the best known crustacean proprioceptors, the abdominal muscle receptor organs (MROs) of lobsters and crayfish (Fields, 1976). The limb muscle receptors, however, lack any peripheral inhibitory control upon the sensory neurones themselves, of the kind exerted by the 'accessory nerves' of the abdominal MROs.

Each of the five paired thoracic legs (including the 'first leg' or cheliped) possesses two muscle receptors. The 'thoraco-coxal muscle receptor organ' (TCMRO), the subject of this chapter, serves the first joint of the leg between thorax and coxopodite (the T-C joint), while the myochordotonal organ (MCO) is primarily concerned with control of the mero-carpopodite (M-C) joint in the middle of the leg. Both have their own efferent motor innervation, comprising two 'receptor motor neurones', one of which in the MCO (but not the TCMRO) is a peripheral inhibitor. These receptor motor neurones, like the motor neurones supplying each of the twelve power muscles of the leg, are subject to various reflex controls as well as tonic and phasic central drive.

TCMRO MORPHOLOGY AND SENSORY RESPONSES

For detailed accounts of the morphology, fine structure and physiology of the TCMRO in crabs, see Alexandrowicz and Whitear (1957), Whitear (1965), Bush and Roberts (1971), and Bush (1976, 1981). Only a brief outline is presented here (see Fig. 1).

How do crabs control their muscle receptors?

The receptor muscle (RM) of the crab TCMRO lies within and parallel to the muscle fibres of the leg 'promotor' muscle, which moves the leg forwards at its articulation with the thorax. (There is no muscle receptor for the antagonistic, 'remotor' muscle.) The TCMRO is thus stretched

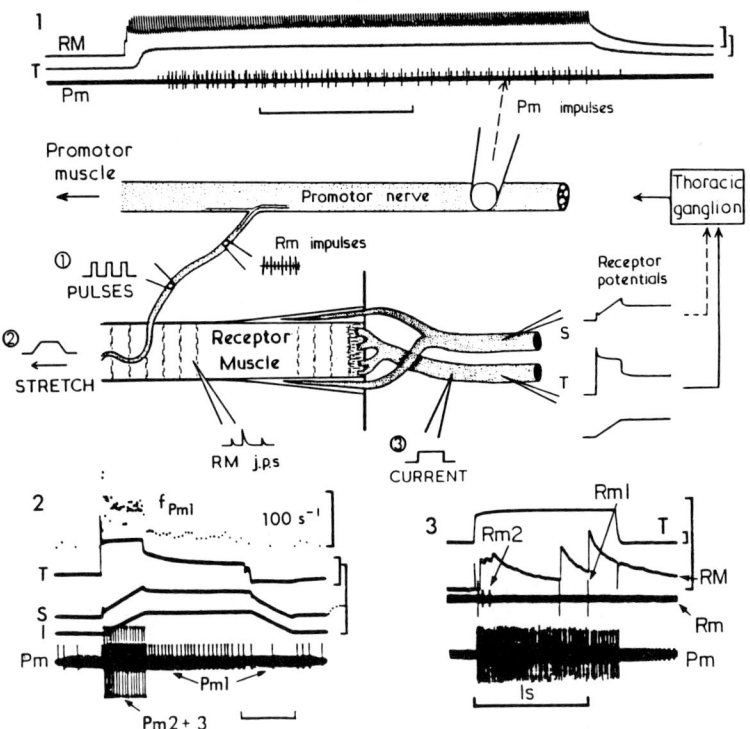

Fig. 1. Diagram of the crab thoraco-coxal muscle receptor organ (TCMRO) illustrating the sensory (S and T fibres) and receptor motor nerve (Rm) innervation at the proximal end of the receptor muscle (RM), and the main sites of stimulation and recording used to analyse the TCMRO-promotor reflex system. Accompanying records show typical afferent responses (T, S) and reflex responses of promotor nerve (Pm) and receptor motor nerve (Rm), with intracellular recording from the receptor muscle (RM) in 1 and 3, evoked by (1) repetitive pulse stimulation of the cut receptor motor nerve; (2) "trapezoidal" stretch of RM (dots represent instantaneous frequency of Pm1 impulses); (3) a depolarising current pulse injected into the T fibre. (After Bush, 1976; Cannone & Bush, 1980a, 1981a,b).

CALIBRATIONS (in all figures, except where otherwise indicated): vertical 20mV (intracellular: RM, S, T); 0.2mN (tension: t); time 1s; stretch amplitude: 1mm.

How do crabs control their muscle receptors?

by passive or active remotion (i.e. retraction) of the leg at the T-C joint, and it shortens with leg promotion (or protraction). In mammalian terms, then, the promotor muscle is the functional analogue of the 'extrafusal' muscle of a muscle spindle.

The sensory elements of the TCMRO comprise two afferent fibres of large diameter (40-70μm), the 'S' and 'T' fibres, whose cell bodies lie within the hemi-segmental thoracic ganglion. The numerous fine terminals of the T fibre insert into the tendinous proximal attachment of the receptor muscle to the endophragmal skeleton, while the S fibre innervates two elastic connective tissue strands which flank the RM proximally and merge with its enveloping sheath (Fig. 1). Morphologically, therefore, the S fibre's sensory terminals lie effectively in parallel with the muscular component of the TCMRO, whereas the T fibre is mechanically in series with it. Accordingly, the T fibre is sensitive to passive and active tension in the receptor muscle, whereas the S fibre is primarily sensitive to RM length (see below). To a rough approximation, the T fibre can be considered analogous to the primary (Group Ia) afferent of the mammalian muscle spindle, while the S fibre has some functional properties resembling the spindle secondary (Group II) afferent. Note that the T fibre is not analogous to the Group Ib afferents of mammalian Golgi tendon organs, since it terminates in series with a receptor muscle, not a main power producing muscle.

Physiology of the afferent responses

Intracellular microelectrode recordings at any point along the 5-10mm length of the S and T fibres reveal depolarising receptor potentials in response to RM stretch (Fig. 1(2); Bush, 1976, 1981). These are conducted to the thoracic ganglion of the CNS electrotonically, with little decrement, reflecting a high membrane resistance and consequently large length constant of up to 6cm (Mirolli, 1979). Small, graded, tetrodotoxin-sensitive transients sometimes occur on the rising phase of the depolarisation (the 'α' component: see Figs. 1(2), 2E, 3E), but all-or-none impulses are lacking. The S and T fibres are thus unusual among mechano-sensory neurones in being 'non-spiking', due at least partly to the presence of a fast outward K^+ current which shunts the normal, tetrodotoxin-sensitive, fast inward Na^+ current (Mirolli, 1983).

The characteristic waveforms of the S and T fibre receptor potentials can be attributed largely to the structural features of their mechanical linkage with the RM (Berger & Bush, 1979; Bush, 1976, 1981). The S fibre membrane potential follows length changes almost linearly, with little or no dynamic overshoot (Fig. 3B). In contrast, the membrane potential of the T fibre varies non-linearly with

RM length, and shows distinct dynamic peaks. These correspond to similar though smaller peaks in RM tension, as monitored at its distal end. With ramp stretches the amplitude of the T fibre's dynamic response increases approximately logarithmically with stretch velocity, as does the dynamic component of RM tension (Fig 3A). The tension waveform appears intermediate between the T and S fibre receptor potentials, reflecting the summed contributions of the muscular component and its surrounding elastic connective tissue sheath. When feedback from a tension transducer is used to produce servo-controlled changes in RM tension (as opposed to length), the T fibre response resembles the tension waveform more closely, while the S fibre membrane potential again follows the imposed length change.

EFFERENT (Rm) INNERVATION OF THE RECEPTOR MUSCLE

The receptor muscle (RM) of the crab's TCMRO receives multiterminal efferent innervation along its length from two fine motor axons, <u>receptor motor neurones</u> Rm1 and Rm2 (Bush,

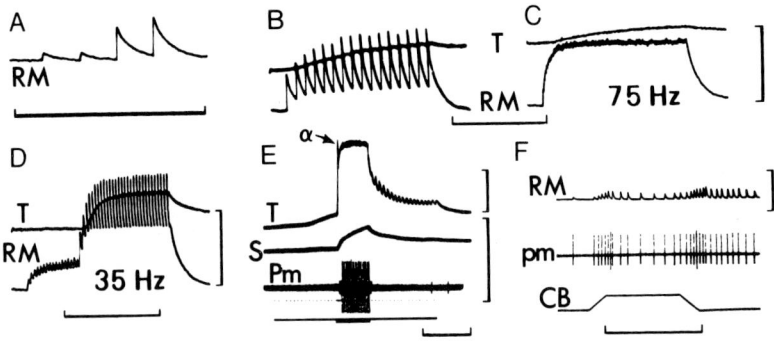

Fig. 2. Intracellular records from the receptor muscle (RM) and T fibre during repetitive pulse stimulation of the receptor motor nerve (except F). (A,D) Increasing intensity stimuli (at 5Hz and 35Hz) recruit first Rm2, giving small, unitary, excitatory junction potentials (e.j.p.s), then Rm1, producing large compound (Rm1 + Rm2) e.j.p.s. (B) 10Hz stimulation, above threshold for both Rm1 and Rm2; and (C) 75Hz, below threshold for Rm1, elicit small T fibre depolarisations. (E) 10Hz stimulation (thin bar), with superimposed 100Hz train (thick bar), evokes small and large T fibre depolarisations ('α' = active membrane response), a small S fibre response, and promotor reflex (Pm). (F) Stretch and release of CB chordotonal organ reflexly excites Rm2, monitored in RM and in a fine promotor nerve (pm). [After Bush, 1976; Cannone & Bush 1981a,b; and (F) S.I. Head, unpublished].

How do crabs control their muscle receptors?

1976; Cannone & Bush, 1981a). Rm1, the larger axon (diameter ca. 10μm), is specific to the receptor muscle, while Rm2 (ca. 4μm diameter) branches off from a motor axon which continues to the promotor muscle in which the RM lies (Figs. 1(3), 2F). Rm1 is thus analogous to the γ-efferents of mammalian muscle spindles, while Rm2 may be compared to the β-efferents which also innervate the extrafusal muscle.

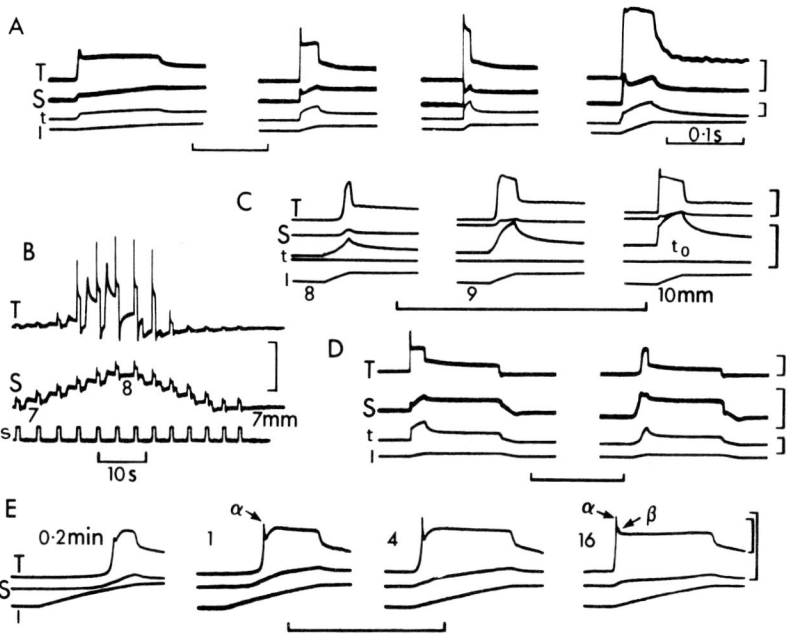

Fig. 3. Intracellularly recorded afferent (T and S) fibre responses and tension changes (t) elicited by trapezoidal stretches applied to the distal end of the receptor muscle (l = RM length change). (A) Four different stretch velocities at constant (mid) RM length. (B) Constant trapezoidal stimuli (bottom trace) at successively increasing and then decreasing initial RM lengths (not monitored, but reflected in S fibre membrane potential changes between stimuli). (C) Identical stretches at three initial RM lengths (8, 9 and 10mm; t_0 = zero tension). (D) Responses at 9mm RM length during increasing (left) and decreasing (right) length series (as in B but at 10s intervals). (E) Ramp stretches to slack RM (7mm length) after different intervals as indicated (min); 'α'= active membrane response, 'β'= initial response with taut RM. RM length range in situ: B, 7-10mm; others, 8-12mm). B from Potamon sp; other records (in all figures) from Carcinus maenas. [After Bush, 1976, 1981, and (A,D) unpublished; Cannone & Bush, 1980a].

Both Rml and Rm2 are excitatory; there is no evidence of inhibitory innervation of the TCMRO. Each receptor motor neurone elicits depolarising, excitatory junctional potentials (e.j.p.s) in the several electrically discrete muscle fibres making up the RM (Fig. 2). Rml e.j.p.s are always larger than those of Rm2, and evoke substantially greater summated levels of depolarisation at any given frequency. These RM e.j.p.s have a relatively slow time course, of around 100ms (cf. ca. 10ms for promotor muscle e.j.p.s), and therefore summate at relatively low frequencies. They also facilitate to a small extent, but never lead to spikes or even small active membrane responses. Thus Rml frequencies as low as 10Hz lead to significant RM muscle fibre depolarisation, while 75Hz trains even in Rm2 alone can evoke large, rapid depolarisations (Fig. 2B,C).

Tension development and afferent responses to Rm stimulation
In the absence of any background discharge, single Rml or Rm2 impulses have no overt effect on RM tension or afferent fibres. However, repetitive stimulation of Rml and/or Rm2, at frequencies sufficient to cause e.j.p. summation, evokes slow contraction of the receptor muscle (Fig. 4B). At a constant RM length, the rate and strength of tetanic contraction increases progressively with stimulation frequency.

Any active, isometric tension developed in the RM results in proportional depolarisation of the T sensory fibre, about 20mV/0.1mN (Fig. 4B,E; Bush & Godden, 1974; Bush, 1976). The S fibre usually shows little effect of isometric RM contraction, except at high Rm frequencies (e.g. 100Hz) when it may depolarise somewhat (Fig. 4B, 2E), though by only a fraction of the sometimes very large T fibre depolarisation (up to 25-30mV). During isotonic contraction the T fibre membrane potential changes little if at all, whereas the S fibre hyperpolarises, with a time course which closely parallels the recorded RM shortening.

Correlated with their different axon diameters and e.j.p. amplitudes, Rml has a much greater excitatory effect than Rm2. With moderate stimulation frequencies (e.g. 20-40Hz), Rml produces significant tensions and T fibre depolarisation, whereas Rm2 alone elicits little or no overt effect (Fig. 2D). At higher frequencies, however, Rm2 can also contribute significantly to tension development and consequently to T fibre depolarisation (Fig. 2C).

Influence of RM length on the responses to motor stimulation
The foregoing description of the effects of receptor motor activity is based upon observations with the RM held at a constant length around the middle of its normal range. Towards and below the minimum length in situ, the development of isometric tension and the consequent T fibre depolarisation in response to repetitive Rm stimulation is

commonly delayed in onset (Fig. 4E), by an amount inversely related to the initial RM length. Evidently, therefore, there is an inherent slackness in the RM at these shorter lengths, which must be taken up before any overt tension development, and hence T fibre depolarisation, is seen. Closely parallel

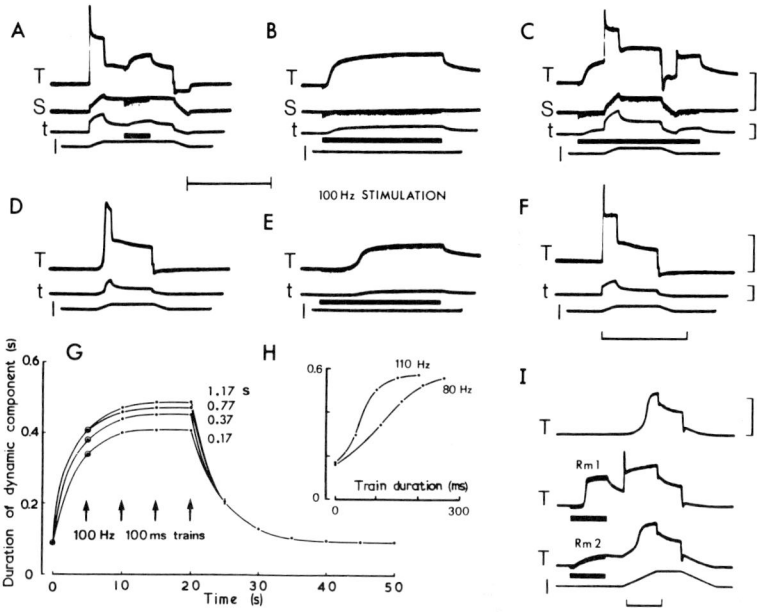

Fig. 4. Effect of trains of Rm stimuli upon RM tension (t) and afferent responses to trapezoidal RM length changes (l). (A-C) 100Hz Rm stimulation (bars), alone (B) and concurrent with RM stretch. (D-F) Tension and T fibre responses 10s before (D) and 10s after (F) Rm stimulation at 100Hz (E); same preparation as A-C. (G) Time-course of conditioning effect on T fibre dynamic response duration of a brief train of Rm stimuli preceding each of four successive, 0.6s duration ramp stretches (arrowed) at 5s intervals, and the ensuing decline in dynamic duration in the absence of conditioning stimuli; the four curves represent four different train-stretch intervals, as indicated. Similar curves are obtained when stimulus frequency or train duration are varied instead of train-stretch interval. (H) Variation in duration of fully conditioned T fibre dynamic component with duration of conditioning train (at two frequencies) 1s before each stretch, as in G. (I) Conditioning of T fibre dynamic response by stimulation of Rm1 and Rm2 individually at 100Hz (bars); top record shows the unconditioned response. RM length: A-C 10mm, D-I 8mm; in situ range 8-12mm. [After Bush & Godden, 1974; Cannone & Bush, 1981b, and (I) unpublished].

effects of RM length are seen in the passive tension changes and afferent responses to stretches applied to the RM.

TIME AND LENGTH-RELATED VARIATION IN RESPONSES TO RM STRETCH

When a constant stretch stimulus (e.g. a 200ms duration ramp of 0.5mm amplitude) is applied to the distal end of the RM in the absence of Rm efferent activity, the S and T fibre responses show systematic changes with time and pre-stretch (or 'initial') RM length. The first point to note is the variation in the early transient peak, seen in both the tension and the T fibre response at the onset of a ramp stretch (Fig. 3). This is commonly present in the response to the first pull after a period of rest at around mid in situ length, but in the absence of efferent activity it disappears progressively on repetition of the stretch at short intervals (e.g. 1-10s). Similar observations have been made on the isolated cat muscle spindle (Hunt & Ottoson, 1976).

More striking, and perhaps more functionally significant, is the variation in the latency of onset of the initial rapid depolarisation (and concomitant tension increment), and the consequent variation in the duration of the dynamic components in the tension and afferent response. At and above mid-length in situ, the passive tension increase, and the corresponding T (and S) fibre dynamic depolarisation, usually start sharply at the beginning of the stretch ramp (Fig. 3A). At shorter RM lengths, however, the tension and T fibre depolarisation at first increase slowly, before suddenly taking off more sharply some way up the ramp (Fig. 3C,E). This delay in onset of the dynamic response increases with decreasing RM length until, below physiological lengths, little or no tension rise or T fibre depolarisation may occur. The delay also increases in successive responses at any one short length (Fig. 5A). However, if the RM is left to rest for a long enough interval, e.g. one minute or more between successive stretches, the dynamic response progressively recovers its rapid onset; the longer the interval, at a given RM length, the earlier it starts in the ramp (Fig. 3E). As a corollary, the shorter the initial RM length, the longer it needs to be left before the full duration dynamic component is achieved; at sub-physiological lengths the dynamic response may never start immediately at the onset of the ramp.

An associated phenomenon appears when a constant, trapezoidal stretch stimulus is applied at each level in a stepwise sequence of increasing and then decreasing 'initial' RM lengths, over the physiological range. The passive tension change and T fibre dynamic response elicited at each of the shorter lengths are smaller, and more delayed in onset, during the decreasing length sequence than at the corresponding RM lengths during the increasing series (Fig.

3D). This length-dependent 'hysteresis' in the responses is more pronounced with shorter intervals between stretches.

These RM length- and time-dependent forms of variability in the afferent responses to a given stretch stimulus clearly constitute sources of ambiguity or 'error' in the sensory signals. They undoubtedly result from the concomitant variations in RM tension or 'tautness' at the onset of stretch, and accordingly are closely correlated with the initial resting tension and corresponding level of the T fibre membrane potential immediately prior to each stretch (see Figs. 3C, 5A). All these effects can therefore be expected to be subject to control by active tension development, mediated by receptor motor output to the RM.

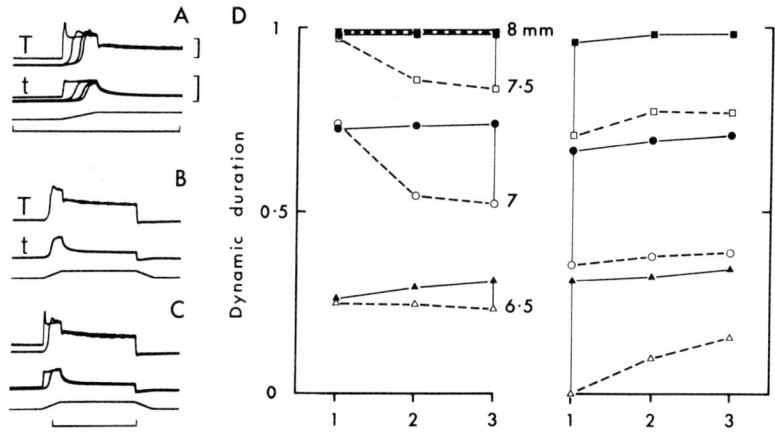

Fig. 5. Effect of receptor motor nerve stimulation on the dynamic responses of the receptor muscle (t, tension change) and T fibre to ramp stretches at short RM lengths. (A) Decline in the dynamic response, following a 20min rest, to four identical stretches at 10s intervals in a slack receptor (RM length 8mm; successive responses superimposed, left to right). (B) One 'habituated' response from a series of similar trapezoidal stretches at 10s intervals; and (C) the response following 4s of 50Hz stimulation, superimposed upon the previous response (shown in B). (D) Variation in relative duration of the dynamic component (maximum = "1") of T fibre response to three successive ramp stretches (at 10s intervals) at each of the RM lengths indicated, this being increased (left) and decreased (right) in a stepwise manner (as in Fig. 3B); broken lines, without Rm activity, solid lines, with continuous 7.5Hz Rm stimulation. In situ RM length range: A-C, 8-12mm; D, 6.5-10mm. (Bush & Cannone, unpublished).

How do crabs control their muscle receptors?

EFFECTS OF Rm ACTIVITY ON STRETCH-EVOKED RESPONSES

Receptor motor activity can have a pronounced effect upon the responses to RM stretch, even when, alone, it evokes no afferent response. This is particularly clear at short RM lengths where it is 'slack'. Rm activity usually causes the initial rapid tension rise and T fibre depolarisation to start sooner during the stretch stimulus, resulting in a longer duration dynamic response (Fig. 5C). Either continuous tonic stimulation (e.g. 1-10Hz), or brief phasic conditioning trains (e.g. 100Hz for 100ms, 1s before stretch), can produce such dynamic 'facilitation' of the stretch-evoked responses (Figs. 5D; 4D-I). At any given RM length below that required for a maximal, full duration dynamic response, the degree of this facilitation is directly related to the 'strength' (i.e. frequency and duration: e.g. Fig. 4H), and timing (e.g. Fig. 4G), of Rm activity. Comparable changes occur in the S fibre responses, though these are generally more evident at shorter RM lengths - i.e. the S fibre 'length threshold' is lower.

When Rm stimulus frequency is itself adequate to evoke a distinct tension increment and T fibre depolarisation, the responses to a concurrent stretch superimpose upon these (Fig. 4A,C). With small active tension increments the stretch-evoked responses sum nearly linearly, but larger contractions may result in the T fibre dynamic component amplitude being smaller than before, even though its duration is facilitated so as to occupy the full ramp duration (Figs. 4C, 8C3,4). This non-linear summation of the T fibre responses to RM stretch and active contraction may reflect an approach towards the reversal potential of the T fibre's transducer membrane for the stretch-evoked response, i.e. a reduction in the 'driving potential' for the stretch response due to the initial contraction-evoked depolarisation. That it is not purely due to some non-linear property of the tension-generating process is evident from the lack of any comparable reduction in the stretch-evoked dynamic tension increment when superimposed upon the active contractile response (Fig. 4C).

Owing to the slow rate of RM contraction, motor activity must start some time before the onset of stretch in order to produce maximal facilitation of the dynamic response. For instance, at around minimum in situ RM length, a 100ms 100Hz train must precede the stretch by at least 200-400ms to effect a full 'recovery' of the dynamic response (Figs. 4F,G). The shorter the RM length (and the lower the stimulus frequency), the longer the period of stimulation or pre-stretch interval needed. Moreover, when stretches alone are repeated at regular (e.g. 2-10s) intervals following a fully facilitated response, successive responses decline exponentially towards their 'habituated' (i.e. unfacilitated) level, at a rate proportional to the repetition frequency (Figs. 5A,D, 4G).

The general effect of these various actions of receptor

efferent activity is to reduce the variability in the stretch-evoked afferent responses which results from differences in 'initial length' and immediately preceding changes in length, including the hysteresis described in the previous section (see e.g. Fig. 5D). On the basis of such experiments, therefore, it seems plausible to suggest that a major role of the TCMRO's efferent innervation is to help maintain the sensitivity, or 'gain', of the receptor's dynamic response to RM length changes resulting, in vivo, from passive 'disturbances' or active 'intended' movements of the leg at the basal, thoraco-coxal joint.

PROMOTOR REFLEX RESPONSES TO RM ACTIVATION OR STRETCH

Depolarisation of the TCMRO's two afferent fibres, especially the T fibre, leads to reflex excitation of from 1 to 10 motor neurones supplying the promotor muscle of the T-C joint (Bush, 1976, 1981; Cannone & Bush, 1980a,b). The strength of this TCMRO-promotor 'stretch' reflex, as represented by the number of motor neurones recruited and their individual discharge frequencies, is proportional to the degree of depolarisation of the T and S fibres, however this is produced experimentally. Thus RM stretch, depolarising either afferent fibre by current injection, or isometric RM contraction, all elicit reflex promotor discharge (Fig. 1). However, since the afferent responses to RM contraction alone are generally quite small, the contraction-evoked reflex responses are seldom as strong as those resulting from RM stretch.

It seems unlikely, therefore, that indirect activation of the promotor muscle via a mechanism analogous to that postulated in the 'follow-up length servo' hypothesis for mammalian muscle spindles (see Matthews, 1972), constitutes a major role of the efferent innervation of the TCMRO. On the other hand, just as the T fibre dynamic response at short RM lengths is 'facilitated' in duration by even sub-threshold motor input to the RM, so too is the promotor stretch reflex: i.e. any delay in onset of the promotor response to a ramp stretch is reduced by prior Rm activity (Fig. 8C3). This lends support to the hypothesis that a major role for the TCMRO's efferent innervation is to maintain its dynamic sensitivity to length changes, both increasing and decreasing.

That RM shortening may also be a potent sensory stimulus is seen in the resulting suppression of promotor impulses, particularly when there is a significant background discharge. This 'negative dynamic response' in the promotor record is correlated with a hyperpolarising dynamic component in the T fibre and concomitant reduction in RM tension (Fig. 4A,C). All of these negative dynamic effects of RM shortening, sensory and reflex, are, like the positive dynamic responses to lengthening, accentuated by prior receptor motor activity (see Fig. 3 of Cannone & Bush, 1981b). Furthermore, motor neurones of the antagonistic remotor muscle, which commonly are reflexly excited by RM shortening, are correspondingly modulated by Rm activity, thus emphasising the potentially widespread importance of the TCMRO's efferent control.

REFLEX CONTROL OF RECEPTOR MOTOR NEURONES BY TCMRO INPUT

A trapezoidal stretch stimulus applied to the isolated distal end of the receptor muscle not only elicits a typical resistance or stretch reflex in promotor muscle motor neurones, but also reflexly activates both receptor motor neurones, Rm1 and Rm2 (Cannone & Bush, 1981a,c, 1982, 1983). In view of the established excitatory actions of both Rm1 and Rm2, this constitutes a dual efferent, positive feedback reflex. Like the promotor reflex, it is characterised by a pronounced dynamic (stretch phase) response, followed by a static (hold phase) response that shows lower impulse frequencies and may adapt rapidly (Fig. 6A,B). At the lower end of the RM length range there is a dynamic response alone; this may involve only Rm1, Rm2 usually having a higher recruitment threshold.

A comparison between the dynamic impulse frequencies of Rm1 and Rm2 in response to identical stretches applied at incrementally increasing receptor lengths reveals marked differences in their reflex firing characteristics (Fig. 6F). The mean dynamic frequency of Rm2 increases with RM length, whereas that of Rm1 remains remarkably constant for stretches over the entire receptor length range. Moreover, even during a single stretch at constant velocity, the instantaneous frequency of Rm1 is constant, whereas that of Rm2 increases with the ramp (Fig. 6A,B; E, lower record). These observations suggest that either the sensory drive to Rm1 differs from that onto Rm2, or that there is early saturation of transmission across the central synapse(s) for Rm1.

Current injection into the afferent fibres
Evidence for differential sensory drive comes from experiments where S and T fibres are selectively stimulated by current injection. Depolarising the T fibre excites both Rm1 and Rm2, the latter receptor motor neurone having the higher recruitment threshold (Fig. 7C,D). Hyperpolarising the T fibre abolishes the tonic output that characterises Rm1. This indicates that, as with the tonically active promotor motor neurones (Cannone & Bush, 1980b), the T fibre is responsible for a continuous synaptic drive, even at the most relaxed receptor lengths (cf. Blight & Llinás, 1980).

By contrast, while graded depolarisation of the S fibre also progressively excites Rm2, it inhibits any tonic firing in Rm1 (Fig. 7C,E). Conversely, hyperpolarising the S fibre

How do crabs control their muscle receptors?

at an extended receptor length, or concurrently with a stretch stimulus, increases the reflex firing of Rm1 (i.e. disinhibition), and reduces the frequency of the stretch mediated Rm2 response (Fig. 7F).

The pattern of reflex connectivity that emerges from these experiments is that Rm2 is reflexly activated by both S

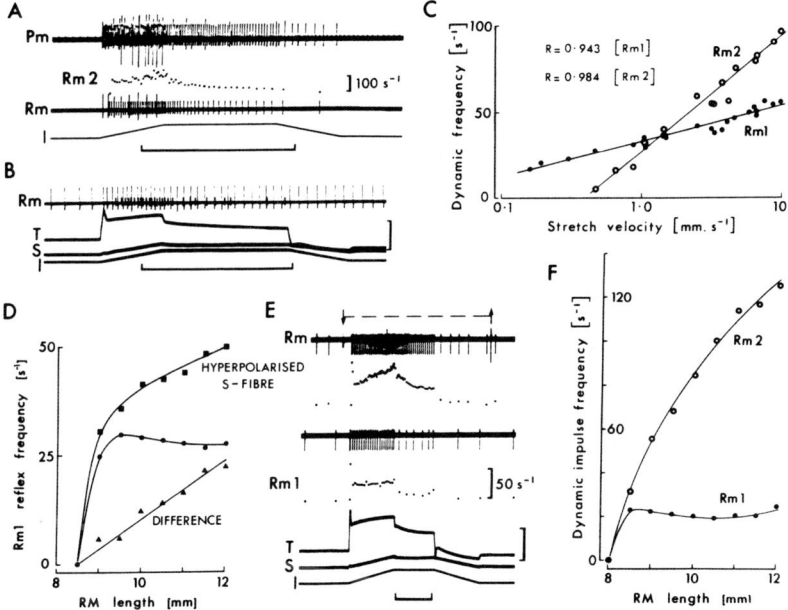

Fig. 6. Dynamic reflex responses of Rm1 and Rm2 to ramp-function RM stretch. (A) Strong and (B) weak Rm2 response (small spikes), together with promotor (Pm) reflex, or T and S fibre responses. (C) Dependence of mean dynamic response frequency upon stretch velocity. (D) Effect on Rm1 dynamic frequency, of S fibre hyperpolarization at different RM lengths (same protocol as E, but different preparation; stretch velocity 2.5mm s^{-1}): ●, stretch alone; ■, with hyperpolarizing pulses; ▲, difference = inhibitory contribution of S fibre; this reflects the linear dependence of S fibre membrane potential upon RM length. (E) Disinhibition of Rm1 by a strong hyperpolarizing pulse injected into S fibre (between arrows in upper trace) during a stretch stimulus, cf. stretch alone (lower Rm trace). (F) Constancy of Rm1 dynamic frequency at different RM lengths, contrasted with marked length sensitivity of Rm2 (same preparation and stimuli as A). Dots in A, E represent instantaneous frequencies of Rm2, Rm1 impulses. [After Cannone & Bush, 1981c, 1982, 1983, and (B) unpublished].

157

and T fibres, whereas Rm1 is excited by the T fibre but inhibited by the S fibre. In the case of Rm2 we have dual afferent positive feedback, whereas Rm1 manifests concurrent positive and negative feedback. The net result of this interaction between reflex excitation and inhibition is dependent upon the static and dynamic response characteristics of the S and T fibre receptor potentials, summarised in Fig. 3.

S fibre control of tonic Rm1 response to RM length
As noted previously, the S fibre is well adapted to signal receptor length, depolarising linearly with increasing length. In contrast, the T fibre shows a much smaller, non-linear variation in steady-state membrane potential with RM length. Accordingly, in the interaction between S fibre inhibition and T fibre excitation of Rm1, the S fibre influence tends to be dominant under purely tonic (static) conditions. This can be seen in preparations displaying ongoing tonic Rm1 activity, which probably reflect the in vivo condition more closely. Large amplitude stretches (4-6mm) encompassing the whole RM length range in such preparations often reveal an inhibitory input to Rm1, manifested as a reduction in its tonic frequency during the static phase of the stretch (Fig. 7A,B).

Dynamic equilibrium over RM length range
The almost constant Rm1 impulse frequency during the dynamic phase of constant velocity stretches over the entire RM length range (Fig. 6F) must be a consequence of precisely matched excitation and inhibition, irrespective of receptor length. That is, the rate of increase in S fibre inhibition with RM length matches step by step the increase in excitation exerted by the dynamically sensitive T fibre.

This relationship can be experimentally dissected by removing the inhibitory influence of the S fibre with a strong hyperpolarising current pulse during alternate identical stretch stimuli over the full length range (Fig. 6D,E). Thus the normal Rm1 dynamic reflex can now be compared with the reflex mediated by the T fibre alone, the difference being the inhibitory contribution of the S fibre. The result is self-explanatory. Although there are marked differences between the T and S fibres in their relative rates and levels of depolarisation with increasing receptor length (Fig. 3), such differences are evidently compensated for by differences in the input-output relationships across the central synapses, and/or in the length constants of the S and T fibres.

The foregoing balance between S and T fibre control of Rm1 does not hold for the reflex Rm responses to stretches of different velocities applied at an invariant RM length. This is because the S fibre is relatively insensitive to stretch velocity, whereas the T fibre is highly responsive to velocity (Fig. 3A; cf. Bush & Roberts, 1971). Thus, the

How do crabs control their muscle receptors?

reflex output of both Rm1 and Rm2 is strongly velocity dependent, the dominance of the T fibre being expressed in the form of logarithmic velocity relationships for both Rm1 and Rm2 (Fig. 6C), recalling that for the T fibre (op. cit.).

ROLES OF AUTOGENIC REFLEXES OF RM EFFERENTS

An important key to understanding the function of these 'autogenic' reflex actions may lie in the slow mechanical response of the receptor muscle to motor input, and the long

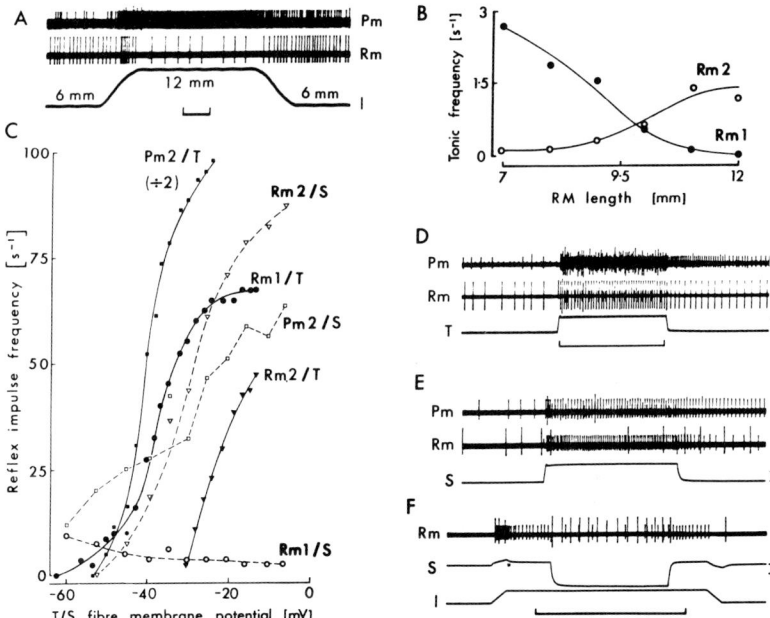

Fig. 7. Tonic influence of TCMRO afferent input upon promotor (Pm) and receptor motor (Rm) activity. (A) Responses to extending the RM from 6 to 12mm (physiological range 7-12mm) during sustained background discharge in both nerves. (B) Variation of tonic Rm1 and Rm2 discharge with resting RM length. (C) Dependence of tonic firing in Rm1, Rm2 and a representative promotor motor neurone (Pm2) upon S or T fibre membrane potential during intracellularly injected depolarising current pulses similar to those shown in D, E (S and T fibre data from different experiments); 'resting' membrane potentials at the pre-set RM length of 7mm were 66mV (T) and 60mV (S, cf. 19mV at maximum in situ length of 12mm). (F) A hyperpolarising pulse into the S fibre during a 1mm stretch at an initial RM length of 12mm disinhibits Rm1 (larger spikes) while greatly reducing Rm2 excitation (smaller spikes). [After Cannone & Bush, 1981c, 1983, and (D) unpublished].

159

lasting effects of such input. As noted above, Rm efferent activity has marked effects on the dynamic responses to imposed RM length changes. In particular, it counteracts the reduced sensitivity and the variability seen in the dynamic responses at shorter RM lengths, such as obtain towards the more 'promoted' (i.e. protracted) positions of the T-C joint. In vivo, Rm activation will result from central drive (see below), but the autogenic Rm reflexes could also play a part.

Positive feedback excitation of Rm1 and Rm2
By providing excitatory input to the RM each time it is stretched, the positive feedback reflex will contribute towards the maintenance of RM tension and hence to the sensitivity of the TCMRO afferents. Because the rate of tension development in response to the relatively low frequency (and brevity) of the reflex Rm discharge is likely to be slow, any phasic effect on the afferent responses to the same stretch eliciting the reflex is probably negligible. However, owing to the long-lasting influence of Rm activity on RM tension, each reflex Rm response to stretch would help to 'facilitate' the dynamic responses, afferent and reflex, to a subsequent change in RM length.

These predictions may be tested by comparing responses to RM length changes before and after opening the feedback loop by cutting the Rm nerve. Preliminary experiments indicate that the decline with successive stretch stimuli in the duration of the dynamic component of the T fibre response and resulting promotor reflex, seen at short RM lengths, is less pronounced before Rm nerve section. The intact Rm supply also reduces the hysteresis in the afferent and reflex responses to constant stretch stimuli applied at increasing and decreasing RM lengths. However, neither the decline in the dynamic responses, nor the hysteresis, was altogether prevented by the intact Rm innervation in these extensively dissected preparations. In the living animal the autogenic Rm reflexes might be expected to be stronger, probably more comparable to those seen in our occasional more active preparations. Moreover, a low frequency background discharge may normally be present in the Rm efferents, and this would supplement the above effects of the autogenic reflexes, since both phenomena are reduced by low frequency Rm stimulation (Fig. 5D).

Negative feedback reflex inhibition of Rm1
The positive feedback, excitatory reflexes are likely to be of special value in the more promoted positions of the T-C joint, when the RM tends to be relatively slack and the TCMRO afferent responsiveness therefore diminished. Near the other end of the T-C joint's arc in its more 'remoted' positions, the RM will be well stretched (i.e. taut), and accordingly the TCMRO sensitivity to imposed RM length changes will be high. Under these conditions the positive feedback excitation of the

How do crabs control their muscle receptors?

RM is less useful, and indeed could be maladaptive, leading to potential instability in the control loop. It is here that the S fibre mediated inhibition of Rm1 becomes quantitatively significant, and may be of functional importance.

Owing to the graded, length-related depolarisation of the S fibre, its reflex inhibitory effect on Rm1 increases progressively with RM length, as the T-C joint moves towards its fully remoted position. Since the steady-state membrane potential of the T fibre undergoes a much smaller variation with RM length, the overall result is the observed progressive decrease in tonic, TCMRO-mediated Rm1 frequency as RM length increases (Fig. 7A,B). For similar reasons, detailed above, the frequency of reflex Rm1 firing during the dynamic phases of constant velocity RM stretches fails to increase as the RM lengthens. Both these influences, tonic and phasic, of the S fibre mediated inhibition of Rm1 discharge, added to the slow response of the RM to its excitatory motor input, will reduce any tendency towards instability of the TCMRO reflex system. Further, in view of the low gain of the Rm2-TCMRO control pathway (compared to the Rm1-TCMRO pathway), the lack of a parallel reflex inhibition of Rm2 by either afferent need not be seen as undermining the foregoing hypothesis.

CENTRAL NERVOUS CONTROL OF RM EFFERENT ACTIVITY

Several lines of evidence bear out the expectation that the CNS plays a major role in regulating the discharge of Rm1 and Rm2, and hence TCMRO sensitivity to RM length changes. Many of the crab preparations used in our experiments showed periods of quasi-rhythmic motor bursts, either spontaneously or in response to various nerve or mechanosensory stimuli. In these preparations Rm1, and occasionally also Rm2, is commonly seen to discharge approximately in phase with the three (Pm1-Pm3) or more active promotor motor neurones, often starting almost simultaneously with the promotor burst (Fig. 8A,B). This behaviour is reminiscent of α-γ co-activation in mammals, and no doubt subserves a similar function.

In other experiments on lively preparations in which the RM motor supply was kept intact but Rm activity was not directly monitored, slow, small depolarisations of the T fibre occurred in association with, or closely following, bursts of activity in the promotor nerve. With the RM at constant length, these T fibre depolarisations must have been due to tonic, isometric RM contraction, so that Rm-promotor co-activation was again indicated. When, during such periods of fluctuating promotor activity, constant trapezoid stretches were applied at 5-10s intervals to the RM while held within the short part of its range, the positive and negative dynamic responses during and immediately following the promotor bursts and associated T-fibre depolarisations were 'facilitated', in the manner described previously (Fig. 8C). These observations are consistent with the hypothesis that co-activation of one or both receptor motor neurones together with central promotor drive, plays an important role in maintaining the TCMRO taut, and thus its readiness to respond effectively to imposed length changes or 'disturbances'. Such a role for Rm-promotor co-activation would be supplemented by any tonic background

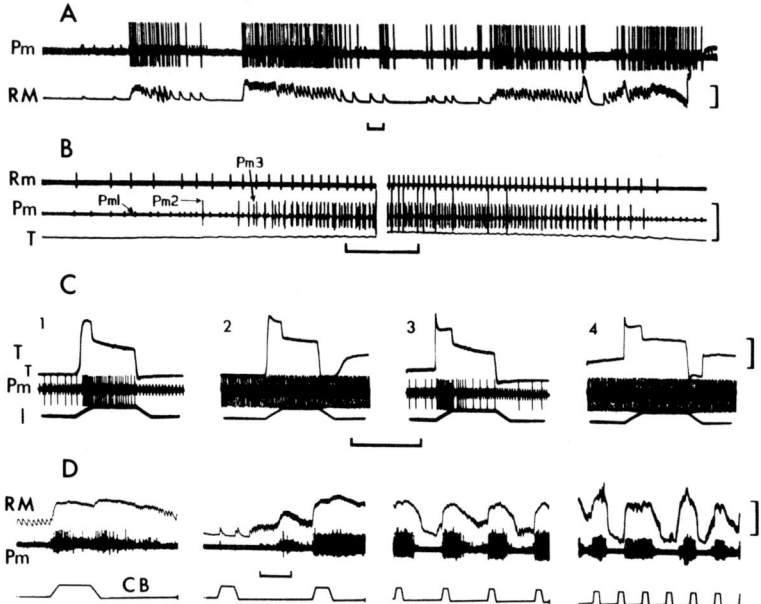

Fig. 8. Co-activation of receptor motor (Rm nerve, or RM e.j.p.s in the receptor muscle) with promotor motor neurones (Pm). (A,B) Spontaneous bursting activity in both nerves; B also shows T fibre depolarisation due to resulting isometric RM contraction (not recorded), with discrete bumps elicited by individual Rm1 impulses. (C) Rm activation (not recorded but inferred from changes in T fibre membrane potential) and responses to identical trapezoidal pulls applied to slack RM (8mm, cf. 11.5mm maximum in situ) at 10s intervals during spontaneous promotor activity; this ceased after record 4, and the next two responses resembled records 3 and 1, respectively. (D) Simultaneous reflex activation, by trapezoidal stretch (at 3s intervals) of the coxo-basal chordotonal organ (CB), of promotor and receptor motor neurones; in the 2nd and 4th records central co-activation partly overrides the reflex responses (records not sequential). [After Bush, 1976; Cannone & Bush 1981b; and (D) S.I. Head, unpublished].

How do crabs control their muscle receptors?

Rm activity, as well as by the positive feedback reflex excitation of the two Rm efferents, as proposed above.

Further strong evidence for Rm-promotor co-activation of central origin has recently been obtained by K.T. Sillar, working on the isolated thoracic CNS of the crayfish. Intracellular recordings of RM e.j.p. activity during rhythmically patterned motor bursts to the four basal muscles of a walking leg showed close synchrony between the onset of Rm and promotor discharge, alternating with activity in the antagonistic remotor nerve. Since the overall pattern of motor activity in these crayfish preparations resembles that presumed to occur in slow walking, this provides further support for the conclusion that Rm-promotor co-activation is likely to be an important mechanism in vivo.

If co-activation of the receptor motor innervation with that of its 'extrafusal', promotor muscle were the invariant mode of efferent control of the TCMRO, why should there be an independent receptor motor supply (Rm1)? One advantage is that it offers a greater potential for variable 'gain control' by the CNS of receptor sensitivity, to accommodate, for example, the wide repertoire of gaits and modes of locomotion - sideways, forward and backward walking, or swimming, in which the back legs (whose TCMRO has been the main subject of study here) play a major role. A second reason might be to allow independent setting of a postural 'baseline' or 'reference position' about which the T-C joint operates, so as to facilitate varying stance, locomotor control or adaptation to different substrata or media. A further advantage is that it permits the differential reflex control of Rm1 and Rm2 by the two afferent fibres, with the consequent potential for the 'automatic', sensory gain control postulated above.

REFLEX PATTERNS AND THEIR CENTRAL MODULATION

In addition to the autogenic Rm reflexes described above, the TCMRO efferents are also reflexly influenced by other proprioceptive inputs. The CB chordotonal organ, for example, a non-muscular elastic strand proprioceptor which monitors movement and position of the second, coxo-basal joint of the leg, modulates RM as well as promotor (and remotor) motor neurone activity (S.I. Head, unpublished observations). Both stretching and releasing CB, corresponding to leg depression and levation, respectively, commonly elicit facilitating bursts of e.j.p.s in the RM more or less synchronously with bursts of impulses in one or more promotor neurones (Fig. 8D). In one case these RM e.j.p.s were attributable to Rm2 impulses (Fig. 2F), but Rm1 is also implicated. The functional role of this inter-joint reflex pattern is not clear, but it may help to stabilise the basal (T-C) joint during sideways locomotion. During occasional periods of spontaneous rhythmic activity in these experiments, central excitatory drive interacted with and at times completely overrode the CB-evoked reflexes, again revealing close co-activation of Rm and promotor discharge (Fig. 8D, 2nd and 4th records).

Preparations exhibiting high levels of spontaneous motor discharge of central origin are generally characterised by greater variability in the recorded reflex patterns and responses to proprioceptor stimulation. During or immediately following bouts of rhythmic or sporadic bursting activity, the reflex responses to RM length changes (or T-C joint movement), for instance, may reverse, from the characteristic stretch or 'resistance' reflex to an 'assistance reflex': promotor firing is now inhibited by RM stretch and excited during shortening, while reciprocal changes occur in the antagonistic remotor nerve output (DiCaprio & Clarac, 1981). The receptor motor response pattern under such conditions, though quite often displaying co-activation with promotor activity, is less predictable than the promotor response. Furthermore, the Rm responses to current pulses injected into the S and T fibres during or between bouts of spontaneous motor output may vary widely, and are less clearly dependent on RM length. Evidently there is considerable functional plasticity in the central pathways of this whole sensori-motor system, which is undoubtedly more complex than the relatively well defined picture presented in this account.

Some insight into this complexity is provided by recent experiments by Sillar & Skorupski (1985; see also Sillar, 1985) on the isolated CNS preparation of the crayfish, with the TCMRO of one walking leg attached. During rhythmic motor output in the nerves to the four basal limb muscles, resembling that underlying walking, slow oscillations in membrane potential, phase-locked to the motor rhythm, are seen not only in individual impaled motor neurones, but also in the central processes of the S and T sensory neurones of the TCMRO. This central, presynaptic input to the afferents must be involved in the proprioceptive regulation of locomotory pattern generation, and may underly the 'sign reversal' observed in the TCMRO - promotor/remotor reflex during opposite phases of rhythmic motor output. It probably also has appropriate effects upon the receptor motor reflex, though this has yet to be established. Similar central modulation of T and S fibre input no doubt occurs in the crab system, and could explain some of the variability seen in our centrally active crab preparations.

Acknowledgements
We thank Stewart Head and Keith Sillar for permission to quote unpublished results, Robert Elson for constructive criticism of the manuscript, Sue Maskell for word processing it, and Jane Robbins for photographic assistance. The authors' work was supported by MRC and SRC research grants.

How do crabs control their muscle receptors?

REFERENCES

Alexandrowicz, J.S. & Whitear, M. (1957). Receptor elements in the coxal region of Decapoda Crustacea. J. mar. biol. Ass. U.K., 36, 603-628.

Berger, C.S. & Bush, B.M.H. (1979). A non-linear mechanical model of a non-spiking muscle receptor. J. exp. Biol., 83, 339-343.

Blight, A.R. & Llinás, R. (1980). The non-impulsive stretch-receptor complex of the crab: a study of depolarization-release coupling at a tonic sensorimotor synapse. Phil. Trans. R. Soc. B, 290, 219-276.

Bush, B.M.H. (1976). Non-impulsive thoracic-coxal receptors in crustaceans. In Structure and Function of Proprioceptors in the Invertebrates. Ed. Mill, P.J.. Chapman & Hall, London, pp. 115-151.

Bush, B.M.H. (1981). Non-impulsive stretch receptors in crustaceans. In Neurones Without Impulses. Eds. Roberts, A. & Bush, B.M.H. Cambridge University Press, London, pp. 147-176.

Bush, B.M.H. & Godden, D.H. (1974). Tension changes underlying receptor potentials in non-impulsive crab muscle receptors. J. Physiol., 242, 80-82P.

Bush, B.M.H. & Laverack, M.S. (1982). Mechanoreception. In The Biology of Crustacea, Vol. 3, Neurobiology: Structure and Function. Eds. Atwood, H.L. & Sandeman, D.C. Academic Press, New York, pp 399-468.

Bush, B.M.H. & Roberts, A. (1971). Coxal muscle receptors in the crab: the receptor potentials of S and T fibres in response to ramp stretches. J. exp. Biol. 55, 813-832.

Cannone, A.J. & Bush, B.M.H. (1980a). Reflexes mediated by non-impulsive afferent neurones of thoracic-coxal muscle receptor organs in the crab Carcinus maenas. I: Receptor potentials and promotor motoneurone responses. J. exp. Biol. 86, 275-303.

Cannone, A.J. & Bush, B.M.H. (1980b). Reflexes mediated by non-impulsive afferent neurones of thoracic-coxal muscle receptor organs in the crab Carcinus maenas. II. Reflex discharge evoked by current injection. J. exp. Biol., 86, 305-331.

Cannone, A.J. & Bush, B.M.H. (1981a). Reflexes mediated by non-impulsive afferent neurones of thoracic-coxal muscle receptor organs in the crab Carcinus maenas. III: Positive feedback to the receptor muscle. J. comp. Physiol., 142, 103-112.

Cannone, A.J. & Bush, B.M.H. (1981b). Reflexes mediated by non-impulsive afferent neurones of thoracic-coxal muscle receptor organs in the crab Carcinus maenas. IV. Motor activation of the receptor muscle. J. comp. Physiol., 142, 113-125.

Cannone, A.J. & Bush, B.M.H. (1981c). Positive feedback to a muscle receptor stabilized by concurrent self-inhibition. Brain Res. 229, 197-202.

Cannone, A.J. & Bush, B.M.H. (1982). Dual reflex motor control of non-spiking crab muscle receptor. I: Positive feedback tonically reduced and dynamically stabilized by concurrent inhibition of Rm1. J. comp. Physiol., 148, 365-377.

Cannone, A.J. & Bush, B.M.H. (1983). Dual reflex motor control of non-spiking crab muscle receptor. II. Reinforcement of Rm1 mediated positive feedback by dual afferent excitation of Rm2. J. comp. Physiol., 153, 309-320.

DiCaprio, R.A. & Clarac, F. (1981). Reversal of a walking leg reflex elicited by a muscle receptor. *J. exp. Biol.*, 90, 197-203.

Fields, H.L. (1976). Crustacean abdominal and thoracic muscle receptor organs. In *Structure and Function of Proprioceptors in the Invertebrates.* Ed. Mill, P.J. Chapman & Hall, London, pp. 65-114.

Hunt, C.C. & Ottoson, D. (1976). Initial burst of primary endings of isolated mammalian muscle spindles. *J. Neurophysiol.*, 39, 324-330.

Matthews, P.B.C. (1972). *Mammalian Muscles Receptors and their Central Actions.* Edward Arnold, London.

Mirolli, M. (1979). The electrical properties of a crustacean sensory dendrite. *J. exp. Biol.*, 78, 1-27.

Mirolli, M. (1983). Inward and outward currents in isolated dendrites of crustacea coxal receptors. *Cell. molec. Neurobiol.*, 3, 355-370.

Sillar, K.T. (1985). Comparative overview and perspectives. In *Coordination of Motor Behaviour.* Eds. Bush, B.M.H. & Clarac, F.. Cambridge University Press, London, pp. 303-316.

Sillar, K.T. & Skorupski, P. (1985). Central modulation of primary afferent neurons in the crayfish, *Pacifastacus leniusculus,* is correlated with rhythmic motor output of the thoracic ganglion. (Submitted).

Whitear, M. (1965). The fine structure of crustacean proprioceptors. II: The thoracico-coxal organs in *Carcinus, Pagurus* and *Astacus. Phil. Trans. R. Soc.*, B, 248, 437-456.

Section 3
Afferent Input during Normal Movements

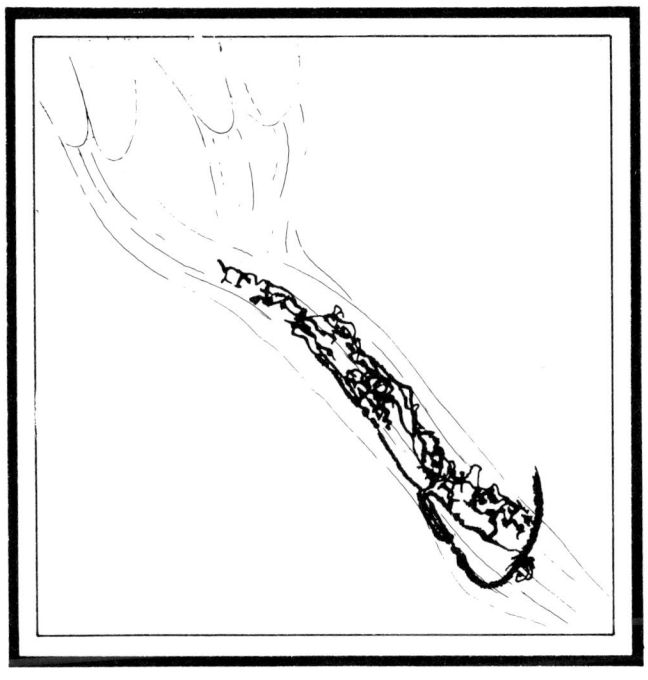

Chapter Ten

INTRODUCTION

A. Prochazka

Prior to 1967, the discharge properties of peripheral receptors, both vertebrate and invertebrate, had been examined and described in great detail. But one great weakness of all of this work was that the data had been obtained in immobile, usually anaesthetised animals. While this was not seen as a serious impediment to the understanding of the role of "passive" receptors, such as those in mammalian skin, it was clearly unsatisfactory for receptors whose sensitivities were under the control of efferents from the CNS (e.g. muscle spindles). Theories of motor control had tended to be built partly on what was known of segmental reflex organisation, and partly on the presumed behaviour of proprioceptors under natural conditions (e.g. Lundberg, 1969). Indeed to some extent the theories stood or fell, depending on the accuracy of these guesses regarding normal afferent activity.

The pressure for information grew and, in 1967, numerous groups tackled the technical obstacles involved with varying degrees of success (review: Prochazka, 1981). Three lines of approach have stood the test of time: human microneurography (Hagbarth & Vallbo, 1967), neurography in walking spinal or decerebrate cats (Severin et al., 1967), and recordings from first order afferents in the mid-brain of cats and monkeys (Davy & Taylor, 1967). In 1976, Loeb, the author of the next chapter, obtained some of the first recordings from limb muscle afferents in freely moving cats, and Zill, the author of the second chapter in this section, has now, remarkably, achieved the same in insects (see Chapter 12) and crustaceans (Clarac, Libersat & Zill, 1985).

The title of this section of the book (Afferent input during normal movements) afforded a good opportunity to assemble a composite picture of the normal behaviour of spindle and tendon organ afferents during stepping. Fig. 1 shows in schematic form the likely ensemble input to the CNS from spindle primaries, secondaries and tendon organs of the ankle extensors during a single step in a conscious cat. The patterns are based on the published data of Loeb and co-

Afferent input during normal movements

workers, as well as the author's own recordings to date. The summed firing rate scale on the ordinate is meant as a rough estimate only, and assumes a population of 100 spindle primary afferents, 100 spindle secondary afferents, and 60 tendon organ afferents (based on Matthews, 1972). Individual spindle secondary afferents and tendon organ afferents usually have time courses of firing similar to those shown (scaled down, and with variable offsets), whereas spindle primary afferents show a wider range of firing profiles (from "static fusimotor-dominated" to "dynamic fusimotor-dominated").

In the first chapter in this section, Loeb puts forward the interesting proposition that receptors may sense different things at different times, depending on the movement being

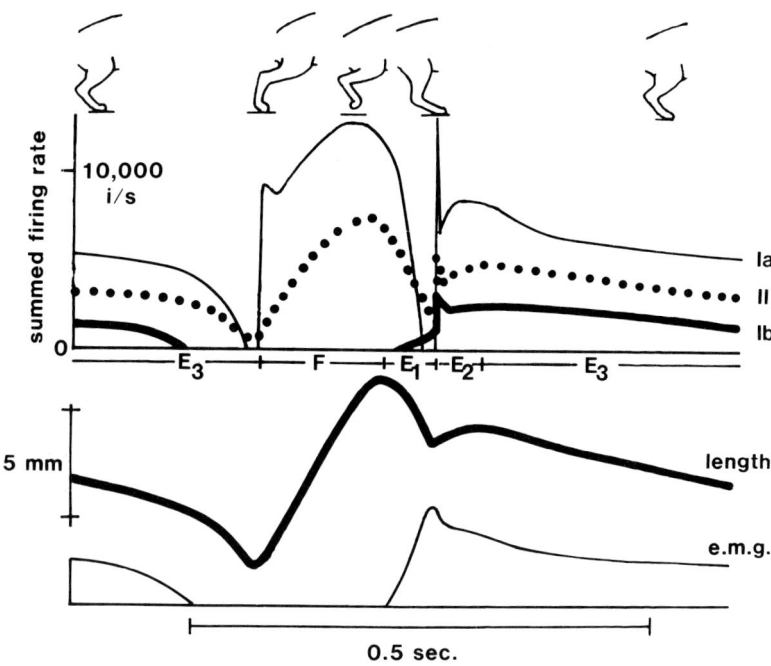

Fig. 1. Estimated time course of net firing in impulses per second (i/s) of three groups of muscle receptors (spindle primary, Ia; spindle secondary, II; tendon organ, Ib) of the triceps surae complex during a single locomotor step in the normal cat. Lower traces show gastrocnemius length variations and electromyogram (equals skeletomotor activation). The length and electromyogram of the soleus muscle, the second component of the triceps surae complex, are similar.

performed. Thus during stepping, a toe-pad skin receptor might signal foot contact, but under different circumstances it might sense foot slippage, or claw and joint position. A muscle spindle might sometimes sense acceleration, and sometimes position.

The reader may not accept this notion entirely. The CNS receives information from large numbers of mechanoreceptors, each of which signals, in isolation, certain limited features of the external world. A hair follicle afferent signals hair movement, regardless of whether this is caused by contact with clothing or by a puff of air. However, there can be little argument with the basic idea that the information from receptors is processed in different ways at different times (see also Chapter 5 by Prof Taylor and Dr Gottlieb). Light touch to a small patch of skin on the foot dorsum, ineffective during stance, triggers a rapid, stereotyped sequence of muscle contractions during swing. Loeb develops his theme by considering the different elements of a stimulus to which a given receptor might respond.

In the second chapter in this section, Zill discusses the first recordings obtained from afferents in intact, freely-moving insects. The discharge patterns of campaniform sensilla bear an uncanny resemblance to those of mammalian tendon organ afferents recorded during locomotion (Loeb, 1980; Appenteng & Prochazka, 1984). The very high rate of stepping observed in cockroaches results in large phase lags between the muscle activity generating a movement, and the afferent activity resulting from it. This prompts Zill to ask whether the afferent information can be of any use in the control of such movements. Every physically realisable control loop has a frequency beyond which the feedback signal will show progressively increasing lag and attenuation with respect to the command signal. This simply means that at such frequencies, control becomes progressively more "open-loop". However, it is conceivable that the information even at very high rates of stepping may be utilised by neural elements outside the immediate pattern generator to monitor the range of motion of the limb and its "bias" over a number of step cycles.

REFERENCES

Appenteng, K. & Prochazka, A. (1984). Tendon organ firing during active muscle lengthening in awake, normally behaving cats. J. Physiol., 353, 81-92.

Clarac, F., Libersat, F. & Zill, S. (1985). Single mechanoreceptor afferent units firing during locomotion in the shore crab (Carcinus maenas). J. Physiol. (In press).

Davey, M.R. & Taylor, A. (1967). Behaviour of jaw muscle stretch receptors during active and passive movements in the cat. Nature, Lond., 220, 301-

302.

Hagbarth, K.E. & Vallbo, A.B. (1967). Afferent responses to mechanical stimulation of muscle receptors in man. Acta Soc. Med. Upsal., 72, 102-104.

Loeb, G.E. (1980). Somatosensory unit input to the spinal cord during normal walking. Can. J. Physiol. & Pharmacol., 59, 627-635.

Lundberg, A. (1969). Reflex control of stepping. The Nansen Memorial Lecture V. Universitetsforlaget, Oslo, pp. 1-42.

Matthews, P.B.C. (1972). Mammalian Muscle Receptors and their Central Actions. Arnold, London, pp. 48-50.

Prochazka, A. (1981). Muscle spindle function during normal movement. In International Review of Physiology, Vol. 25, Neurophysiology, IV. Ed. Porter, R. University Park Press, Baltimore, pp. 47-90.

Severin, F.V., Orlovsky, G.N. & Shik, M.L. (1967). Work of the muscle receptors during controlled locomotion. Biophysics, 12, 575-586.

Chapter Eleven

WHAT THE CAT'S HINDLIMB TELLS THE CAT'S SPINAL CORD

G.E. Loeb

This is a paper about sensory information, rather than motor control. It is common for motor control physiologists to proceed by stating what information the somatosensory system should provide to service their particular theories of motor control. This often leads to attempts to "find" the right sensory information, to interpret what is found in narrow and even misleading ways, and even to criticise when what should be there is manifestly not. I wish only to summarise what sensory information is likely to be present and to speculate briefly about what might be done with it.

The amount of information is not given by the number of afferent fibres or even their activity rates. It is related to the number of significantly different states which such afferents can signal and the rate at which changes from one state to another can be detected. This is not the same thing as the rate of information utilisation. The utility of the information will depend on both the intrinsic information processing capacity of the receiver and the rate at which detectable new states arise. Obviously, an afferent which is inactive can contribute to neither information production nor utilisation. An afferent which is active but with a randomly "noisy" spike rate can signal relatively few distinguishable states. And an afferent which is steadily active at nearly the same firing rate for the entire range of conditions arising in a particular motor task is sending a high rate of virtually useless information; its inappropriately low sensitivity limits it to advising the CNS about that which it already knows.

The rate of extraction of information from physiological transducers about physical events is very difficult to estimate. One reason is that we do not really know what the transducers are measuring. When we describe the output of a sense organ in terms of spikes per gram or spikes per millimetre, we are making an implicit assumption that the organ is in the business of force or length transduction. Not only is there no reason to suppose that a particular receptor senses one and only one such simple physical parameter, there is no reason to suppose that it senses the same thing at all times. Such a problem is fairly obvious in the muscle spindles, with their well-known quantitative and qualitative changes in length and velocity sensitivity as a result of fusimotor input. It is also likely in other modalities such as stretch sensitive skin receptors of the toe pads, which can indicate vertical ground contact force during standing, impending slippage during tangential force application, and claw and joint position when the foot is in the air. It may even be a factor in Golgi tendon organs, whose output depends on motor unit recruitment order as well as actual force production (Stuart & Stephens, 1976). Whether and how the central nervous system can dynamically change its interpretation of input from a particular afferent remains unknown, although mechanisms such as presynaptic inhibition and interneurones carrying highly convergent, multimodal information are at least consistent with such a capability. Before proceeding to an enumeration of the known somatosensory modalities, I would like to offer what I hope will be a minimally presumptuous division of the physical transduction task.

1. Sense of effort. Without getting into the thorny question of efference copy and conscious sense of effort, it seems likely that a fairly wide range of specialised receptors contribute information to at least lower motor centres regarding the actual force output and general condition of the force-generating machinery in the muscles. In addition to the Golgi tendon organs, obvious candidates include slowly adapting skin receptors and some mechanically sensitive nociceptors influenced by contact with external objects. Several different chemical receptors have been suggested as contributing to the very important sense of fatigue, which together with pain has an important effect on the operating point of the motor neurone pools and actual output of the muscles.

2. Sense of position. Included in position are all the kinaesthetic senses of position in space, angular and translational velocity, and onset and acceleration of movement from a steady state, where the movement of the organism may be voluntary, or the result of applied forces, or both. Sense organs include rapidly and slowly adapting skin receptors, joint receptors, and muscle spindles.

3. Sense of contact. Many of the cutaneous modalities are suited to detecting the presence and nature of outside objects that do not influence directly the motion of the limb. These include thermal and nociceptors as well as various light touch receptors, many of which have strong, short latency effects on motor output.

In the following discussion, I will concentrate mostly on those modalities believed to be important to the control of locomotion in the cat hindlimb. It should be noted that this belief stems at least in part from the limited availability of data (generally biassed toward large diameter, rapidly conducting nerve fibres) and the usual notions about servocontrol and feedback signals. It must always be kept in mind that none of these sources of sensory information is indispensable to locomotion and that even heroic deafferentation procedures often produce only minimal behavioural deficits. This is yet another indication of a sophisticated, highly redundant, and dynamically reconfigurable system for information processing.

CUTANEOUS RECEPTORS

I have already alluded to the potential roles of slowly and rapidly adapting skin sensors in such traditional proprioceptive tasks as sense of position and effort. It is easy to record from such units in walking animals and to identify them using conventional physiological tests. However, it is not possible to interpret their activity beyond noting that it is often modulated in patterns that could be useful to the performance of such proprioceptive tasks (Loeb et al., 1977; Loeb, 1981).

A more productive line of inquiry has been the use of cutaneous information to trigger reflexive corrections in gait based on sense of contact. Obviously, such receptors are well suited to detecting the presence of objects obstructing the desired trajectory of the limb. Forssberg has studied extensively what he calls the "stumbling corrective reaction", in which such cutaneous input from the forward facing surfaces of the foot and shank during swing phase gives rise to an exaggerated flexion motion that is well suited to lifting the foot over and past the obstruction without breaking stride (Forssberg, 1979). This use of sensory feedback is certainly an improvement over more mindless length-servocontrol mechanisms which, left to their own devices, imply simply increasing motor output to knock the offending object out of the way. The responses of the locomotor pattern generator to similar inputs depend on their phase of occurrence in the step cycle, suggesting that transmission in the reflex pathways is modulated by the generator so as to cause different motor outputs at different points in the step cycle (Forssberg et al., 1976). Duysens and Loeb (1980) and Abraham and Loeb (1985) have pointed out that several different cutaneous and deep sites of electrical stimulation during walking produce rather similar patterns of short latency reflexes whether or not they include skin surfaces on the anterior surface of the leg, and that some of the longer latency reflexes may be the result of mechanical effects from shorter latency reflexes rather than the stimuli themselves. Thus, it is not certain

how much specific information about the nature of the
perturbation is signalled by these receptors and how many of
the details of the reflex represent a general response to any
perturbation which depends mostly on its timing with respect
to the step cycle. This is an important information processing
question, since it differentiates between the spinal cord as a
general perceptual machine and the spinal cord as a fast,
limited generator of preprogrammed movements selected on the
basis of motor context or functional state.

PROPRIOCEPTORS

Joint receptors
Presumably, all of the joints of the cat hindlimb contain
fairly numerous receptor structures scattered throughout the
capsule, internal ligaments, and adjoining connective tissue,
although only the cat knee joint has been studied in detail.
The receptors have the general form of biological strain
transducers, consisting of fine terminal arborisations
entwined in strands of connective tissue likely to be stressed
by motion of the joint and/or attached muscles (Boyd, 1954).
Their importance to the sense of kinaesthesia has been debated
as different studies produced different assessments of their
sensitivity to joint angle and off-axis motion, tested mostly
by applying passive motion to unloaded joints in anaesthetised
animals (Clark & Burgess, 1975). The recognition of the
importance of tension conveyed to the capsule by active
muscles (Grigg, 1976) seems to have convinced most workers
that whatever information they do convey, it is not simply
related to joint angle (Tracey, 1978). If they contribute to
the conscious or unconscious sense of limb position, they must
do so via a fairly sophisticated calculation capable of
cancelling the effects of off-axis loading and muscle tension.
We have published two records of knee joint afferents recorded
during normal walking from chronically implanted electrodes in
the L7 dorsal root ganglion, both showing activity only
partially related to joint angle and somewhat greater than the
activity occurring at similar joint angles during anaesthesia
(Loeb et al., 1977; Loeb, 1981).

Golgi tendon organs
The Golgi tendon organ (GTO) is emerging as perhaps the
closest thing to a selectively sensitive, linear transducer of
a single physical quantity in the somatosensory system. Each
of the several dozen per muscle is particularly sensitive to
the active tension of a small number of apparently randomly
combined muscle fibres, with little sensitivity to tension
resulting from the elastic properties of passively stretched
muscle. Because of the nature of their sampling, the output of
any single GTO has only a limited correlation with the total

active muscle tension (Stuart & Stephens, 1976). However, output does rise monotonically if not linearly as the tension of inserting muscle fibres increases either by increased rate of twitch or new fibre recruitment. Because of the generally orderly recruitment order of motor units during graded activation of a muscle, any nonlinearity could be cancelled by a relatively simple summing network for the activity of the GTO population as a whole (assuming that there might be some advantage in having a linear force feedback signal). The few published records of activity of GTOs during normal motor behaviour are consistent with their presumed sampling properties (Prochazka & Wand, 1980; Loeb, 1981); such records are few in number presumably because the afferents are difficult to activate by passive limb motion and are thus often not characterised.

Unfortunately, the simplicity of the GTO as a tension transducer contrasts with the complexity of its projections. In principle, local networks of positive length and negative force feedback (from spindles and GTOs, respectively) could be useful in regulating muscle output properties such as mechanical impedance or stiffness (Houk, 1979) or in preventing mechanically unstable or injurious states from arising during recruitment of the muscle. Localised effects of motor unit activation on both spindle and GTO output have been demonstrated (Cameron et al., 1981) and there appears to be some topographic selectivity in the monosynaptic excitation of spindle primaries onto homonymous motor neurones (Lucas & Binder, 1984; Lucas et al., 1984). However, the disynaptic feedback from GTOs (Lundberg et al., 1977) is mixed with other proprioceptive and cutaneous modalities (Czarkowska et al., 1981; Jankowska & McCrea, 1983) and projects nonspecifically to heteronymous and even antagonistic motor neurones (Brink et al., 1983; Harrison et al., 1983). In its differential effects on slow and fast twitch motor units, the connectivity of GTOs is more similar to that of cutaneous pathways (Powers & Binder, 1981). It is indeed paradoxical that the one biological transducer that appears most suited to a role in servocontrol shows no sign of being used in this manner.

Muscle spindles

The role of the muscle spindle afferents and their complex efferent control signals has become something of a holy grail in motor control. Teleological arguments abound regarding their presumed importance, invoking the large size and rapid conduction velocity of their primary afferents, the large monosynaptic effects onto homonymous and heteronymous motor neurones, the extraordinary elaboration of intrafusal contractile machinery in mammals, and the large numbers and complex arrangements of these transducers in precisely those muscles whose fine control would seem to be important (Richmond & Bakker, 1982). The properties of the isolated

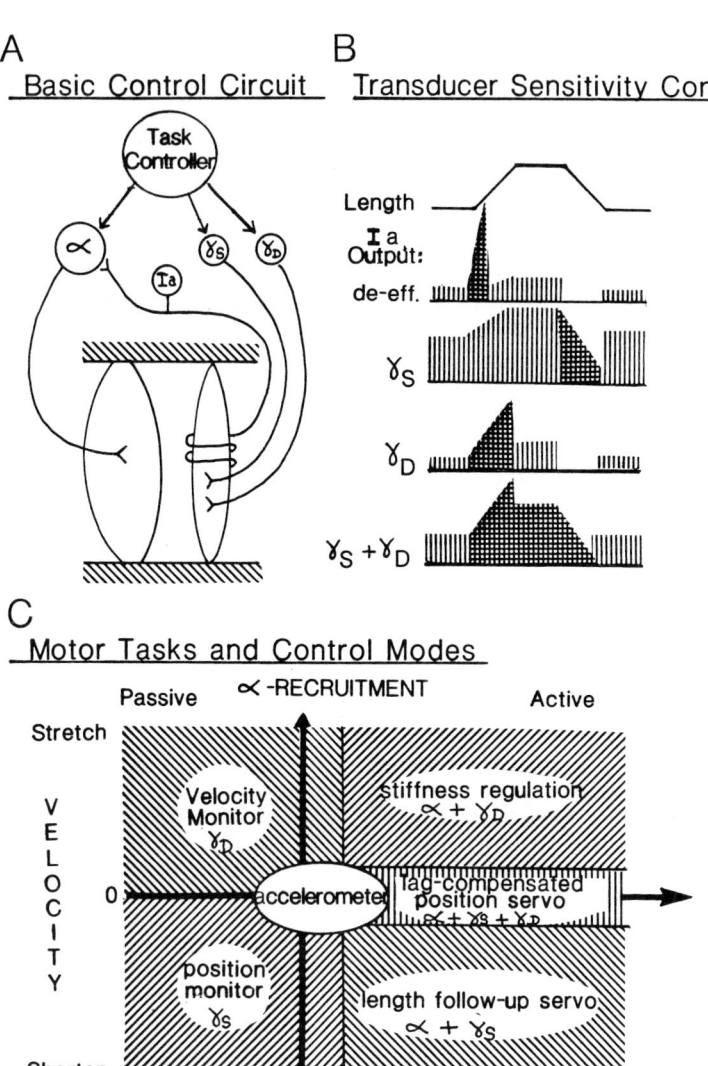

Fig. 1. A. A typical servocontrol circuit incorporating excitatory feedback from muscle spindle primary afferents (Ia) onto alpha motor neurones responsible for extrafusal muscle activity. The sensitivity of the spindle is controlled during the performance of a motor task by various patterns of activation of the static gamma (γ_S) and dynamic gamma (γ_D) motor neurones. B. Responses of an idealised Ia to a ramp stretch (top trace) under the influence of four different combinations of fusimotor activity. Cross-hatched areas indicate aspect of applied motion causing maximal modulation for each condition. C. Division of the work of the

spindle organ with and without fusimotor activity have been studied in great detail (e.g. Hulliger et al., 1977a,b; Boyd, 1981). There is an increasing wealth of data regarding the naturally occurring activity patterns of the primary and (to a lesser extent) secondary afferents in many cat hindlimb muscles as well as from other skeletomotor systems (particularly respiration, mastication, and human hand and ankle muscles; for review, see Loeb, 1984). The muscle spindles have been invoked as a critical sensory element in almost every theory of motor control ever formulated, but there remains no general consensus regarding either the information they convey or its actual use in motor control circuits.

The close anatomical relationship among the intra- and extrafusal motor neurones (gamma and alpha, respectively, plus dual function betas) and the terminal arborisations of the primary afferents in the motor nuclei of the ventral horn suggests a relatively tight control circuit such as shown in Fig. 1A. The intrafusal apparatus has been oversimplified here into just two distinctly different and separately controlled effects, shown under the control of static gamma and dynamic gamma motor neurones. A single monosynaptic excitatory connection is shown from a primary (Ia) afferent onto an extrafusal alpha motor neurone, although complex, strong oligosynaptic effects from both primary and secondary afferents (not shown) are well known, projecting to heteronymous motor neurones, gamma motor neurones, and inhibitory and excitatory interneurones influencing other segmental and remote motor nuclei (Murthy, 1978; Jankowska et al., 1981a,b). It is frequently postulated that the execution of a particular motor task involves some pattern of recruitment (or derecruitment) of the various alpha and gamma motor neurones, thereby creating a particular state of both muscle tone and transducer sensitivity.

The effects of the two principle types of fusimotor control on spindle primary afferent sensitivity are shown in Fig. 1B, again quite oversimplified. The afferent discharge in response to a ramp and hold stretch and shortening shown in the top trace is shown for the four qualitatively different states possible with two types of fusimotor input. In the absence of any gamma motor neurone activity, the response of a quiescent spindle to the stretch is characterised by a large transient at the very beginning of the stretch (Hassan & Houk,

---

muscles based on velocity of motion (ordinate, stretch positive, shortening negative) and level of extrafusal recruitment (abscissa, increasing to right). See text for explanation of type of servocontrol and motor neurone recruitment pattern suggested within each shaded region.

1975). This is followed by a much lower general sensitivity to increasing muscle length and a saturation in the off-state for applied shortening. In this and each of the other three activity traces, a region of crosshatching indicates the conditions producing the largest modulation of output, i.e. region of highest sensitivity. The hypersensitivity to short range stretches of the deefferented spindle thus provides an excellent early indicator of stretch, somewhat like an accelerometer.

Static gamma activity generally causes a biasing of the resting discharge of the spindle afferent up to high levels, where it may have little or no incremental sensitivity to actual length or stretch velocity (Boyd, 1981). However, by carefully adjusting and perhaps dynamically modulating the static gamma activity, the normal tendency of the primary afferent to be silenced by applied shortening can be eliminated (Appenteng et al., 1982). The cross-hatched region shown here indicates spindle activity so contrived to follow the applied shortening, or conversely, to signal any deviation from this trajectory by increases or decreases in its output.

Dynamic gamma activity has a somewhat opposite effect on the spindle afferent, causing it to be selectively sensitive to stretching movement (Boyd, 1981). Note that for both static gamma alone and dynamic gamma alone conditions, the spindle afferent is a relatively insensitive transducer of length; either the output is uniformly high or uniformly low for different steady state lengths.

Records from many spindle afferents and from some gamma motor neurones during motor behaviour suggest that concurrent use of both types of fusimotor influence may be part of many motor programmes, perhaps with complex temporal modulation of one or both of the intrafusal effects. This is further complicated by the nonlinear interaction of the two effects on afferent sensitivity, which include occlusions dependent on the applied length trajectories (Hulliger et al., 1977b). The bottom trace in Fig. 1B shows at least one state which can probably arise by appropriate combination of static gamma and dynamic gamma activity. In this state, the afferent output is well-modulated over a large range of both length and velocity conditions. In fact, the transducer provides a relatively symmetrical output for both lengthening and shortening which represents a sum of length and length-times-velocity terms.

KINEMATIC DIVISION OF MOTOR CONTROL TASKS

Fig. 1C shows a division of the work of muscles based on the motion experienced by the muscle (expressed as velocity on the ordinate) and the level of extrafusal recruitment (represented along the abscissa). The resulting two-dimensional space has been divided into six domains which might present significantly different types of motor control problems. I have indicated general patterns of motor neurone recruitment which would be appropriate for muscles operating in such states, along with a simple term evoking the utility of the spindle afferent signal under such conditions.

At the centre of the figure is an unshaded area in which alpha motor neurone levels are poised near threshold to respond to any perturbation from an isometric state. Detection of such a perturbation requires the sort of short-range hypersensitivity typical of the deefferented spindle, a function somewhat like that of an accelerometer. There is evidence to suggest that in relaxed human subjects, the spindles in some muscles are in such a state, with little or no fusimotor activity (for review, see Burke, 1981).

The left half of the plane indicates states in which the muscle is undergoing relatively large and/or rapid but passive length changes. For stretching motion (upper left), spindle activity would be preserved, although it may be useful to increase the normally low velocity-sensitivity of the deefferented spindle by dynamic gamma activity. For passive shortening (lower left), the spindle will continue to generate an output signal only if static gamma activity maintains tension on the transduction apparatus by actively contracting the chain fibres. Under such conditions, the velocity dependency of the output would tend to be relatively low, but with appropriate modulation of the static gamma activity, the length dependent sensitivity of the spindle output could be set at any desired level.

The right half of the plane includes regions in which the muscle is actively contributing significant tension to an ongoing motor behaviour. During active shortening (lower right), coactivation of alpha and static gamma motor neurones could be used to provide a length feedback signal which would tend to compensate for errors from the desired trajectory (Granit, 1975). In the active stretching condition (upper right), the spindle output could be used to linearise the stiffness of the muscle as suggested by Houk (1979). Under this condition, dynamic gamma activity could be modulated in conjunction with the alpha motor neurone output to adjust the spindle output as needed for the desired amount of muscle stiffness. During near-isometric activity of the muscle (around the x-axis), the low velocities and moderate excursions would cause relatively little modulation of spindle output unless static and dynamic gamma activity were carefully balanced to produce the high sensitivity state described by the bottom trace in Fig. 1B. Such a state is not unlike that inferred from spindle afferents recorded from human hand muscles during isometric contraction (for review, see Vallbo et al., 1979). This state might facilitate rapid motor responses to brief perturbations by using the fusimotor-enhanced velocity dependency of the spindle output to compensate for the activation delay inherent in the extrafusal muscle. Stein (1974) has commented on such a potential role for the muscle spindle primaries, which is analogous to the use of a differentiated feedback signal to compensate for the low frequency response of a control loop.

The division of the work of the muscles into the above-described six different motor control domains is certainly speculative, almost arbitrary. The particulars are intended only to suggest both the diversity of the work of the muscles and the range of transducer properties for which the muscle spindle can be programmed. If discrete domains exist at all within this large range of motion and force, it might be expected that there would be some corresponding heterogeneity in the circuitry for the recruitment of the various types of motor neurones and for the computation of feedback signals from the various sources of sensory information.

In fact, this notion of the division of motor control into discrete domains stems largely from recent data on the recruitment of alpha motor neurones and the activity of spindle afferents in the bifunctional cat hindlimb muscle, sartorius pars anterior. Fig. 2 summarises briefly the behaviour of the muscle as a whole, which has two periods of EMG activity during each locomotor step cycle, one during stance phase when it is rapidly lengthening, and one during swing phase when it is rapidly shortening. The activity patterns of 11 alpha motor neurones projecting to this muscle were recorded during unrestrained treadmill locomotion using floating microelectrodes in the ventral root. Four were recruited during stance phase and seven were recruited during swing phase; none was ever recruited during both phases of an unperturbed step cycle, even at faster speeds where the EMG activity in one or both phases increased greatly.

Fig. 2 also shows the activity of two muscle spindle primary afferents, recorded under similar conditions from the L5 dorsal root ganglion of other animals. While not nearly so clearly divided, the activity of these and other spindle afferents from this anatomically homogeneous part of the muscle was quite varied. Some afferents generated their greatest responses during the muscle stretch during stance phase, and were more strongly influenced by the velocity of the stretch than by its extent. Others were briskly active during the active muscle contraction in swing phase, despite its large negative (shortening) velocity. When low concentrations of lidocaine were infused around the femoral nerve via a chronically implanted nerve cuff and its attached percutaneous catheter, the activity in these and other spindle afferents could be made less modulated, more homogeneous, and generally consistent with the typical properties of deefferented spindles subjected to such length changes (Hoffer & Loeb, 1983). These concentrations of lidocaine preferentially block conduction in small diameter nerve fibres such as gamma motor neurones, leaving alpha motor neurone transmission intact, as indicated by unchanged EMG and length and force records from the quadriceps and sartorius muscles.

These recordings indicate that the normal locomotor pattern generator of the cat hindlimb has functionally divided the alpha motor neurone pool into two independent groups. They also suggest that these groups may include different patterns of gamma motor neurone recruitment which, in turn, cause spindle afferent sensitivity to be optimised for the very different kinematic conditions occurring in the two phases of muscle recruitment. It remains to be seen whether there is any

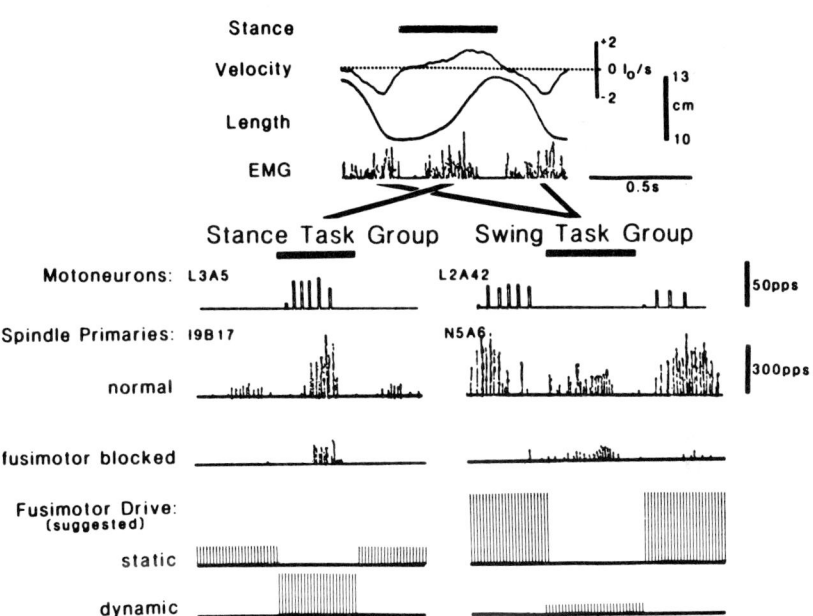

Fig. 2. Activity recorded from various single alpha motor neurones and spindle primary afferents from anterior sartorius muscle during unrestrained treadmill walking in the intact cat. Top centre group of traces shows kinematic division of the work of the muscle into active lengthening (during stance phase) and active shortening (during swing phase to either side). Activity of spindle afferents (different animals, similar conditions) is shown before and after infusion of lidocaine around the femoral nerve to block gamma motor axons. Two bottom traces indicate possible patterns of fusimotor effects that might account for the different patterns of spindle activity seen before the fusimotor block. (From Loeb, 1984, retouched).

segregation in the patterns of mono- and oligosynaptic connectivity among the various alpha and gamma motor neurones and the spindle afferents that are most active in each step cycle phase.

SUMMARY

The somatosensory apparatus of the cat hindlimb generates a great volume of information which is conveyed to the spinal cord by rapidly conducting nerve fibres making widespread connections to motor and interneurones. While the sensory endings of these myelinated afferents are highly detailed and diverse in their structure and sensitivities to various stimuli, most are clearly not designed to provide information about single, conventional physical variables such as pressure, joint angle, or muscle length. Furthermore, their responses to those forms of stimulation to which they respond best tend to be highly nonlinear, often with very high sensitivity to small inputs and greatly diminished sensitivity for inputs covering larger dynamic ranges. Some, such as the muscle spindle (and perhaps hair receptors in piloerectile tissue), may have their sensitivities adjusted by efferent signals from the spinal cord, both quantitatively and qualitatively. All of these sensory signals tend to converge within one or two synaptic links upon a great diversity of interneurones which appear also to convey both descending and segmental motor command signals to motor neurones. How these mixed signals are interpreted by the spinal cord and used as feedback to adjust motor output remains a mystery. It is this author's contention that we will need to know much more about the kinematic states and control problems faced by any particular muscle or synergistic group before we can interpret the patterns of convergence of sensory information onto a motor neurone pool in terms of any particular theory of motor control.

REFERENCES

Abraham, L.D. & Loeb, G.E. (1985). The distal hindlimb musculature of the cat: patterns of normal use. Expl. Brain Res. (In press).
Appenteng, K., Prochazka, A., Proske, U. & Wand, P. (1982). Effect of fusimotor stimulation on Ia discharge during shortening of cat soleus muscle at different speeds. J. Physiol., 329, 509-526.
Boyd, I.A. (1954). The histological structure of the receptors in the knee-joint of the cat correlated with their physiological response. J. Physiol., 124, 476-488.
Boyd, I.A. (1981). The action of the three types of intrafusal fibre in isolated cat muscle spindles on the dynamic and length sensitivities of primary and secondary sensory endings. In Muscle Receptors and Movement.

Eds. Taylor, A. & Prochazka, A.. MacMillan, London, pp. 17-32.
Brink, E., Harrison, P.J., Jankowska, E., McCrea, D.A. & Skoog, B. (1983). Post-synaptic potentials in a population of motoneurones following activity of single interneurones in the cat. J. Physiol., 343, 341-359.
Burke, D. (1981). The activity of human muscle spindle endings during normal motor behavior. In International Review of Physiology, Neurophysiology IV. Ed. Porter, R.. University Park Press, Baltimore, pp. 91-126.
Cameron, W., Binder, M., Botterman, B., Reinking, R. & Stuart, D. (1981). "Sensory partitioning" of cat medial gastrocnemius muscle by its muscle spindles and tendon organs. J. Neurophysiol., 46, 32-47.
Clark, F.J. & Burgess, P.R. (1975). Slowly adapting receptors in cat knee joint: can they signal joint angle? J. Neurophysiol., 38, 1448-1463.
Czarkowska, J., Jankowska, E. & Sybirska, E. (1981). Common interneurones in reflex pathways from group Ia and Ib afferents of knee flexors and extensors in the cat. J. Physiol., 310, 367-380.
Duysens, J. & Loeb, G.E. (1980). Modulation of ipsi- and contralateral reflex responses in unrestrained walking cats. J. Neurophysiol., 44, 1024-1037.
Forssberg, H. (1979). Stumbling corrective reaction: a phase-dependent compensatory reaction during locomotion. J. Neurophysiol., 42, 936-953.
Forssberg, H., Grillner, S., Rossignol, S. & Wallén, P. (1976). Phasic control of reflexes during locomotion in vertebrates. In Neural Control of Locomotion. Eds. Herman, R.M., Grillner, S., Stein, P.S.G. & Stuart, D.G.. Plenum, New York, pp. 647-674.
Granit, R. (1975). The functional role of the muscle spindles - facts and hypotheses. Brain, 98, 531-556.
Grigg, P. (1976). Response of joint afferent neurons in cat medial articular nerve to active and passive movements of the knee. Brain Res., 118, 482-485.
Harrison, P.J., Jankowska, E. & Johannisson, T. (1983). Shared reflex pathways of group I afferents of different cat hind-limb muscles. J. Physiol., 338, 113-127.
Hassan, Z. & Houk, J.C. (1975). Transition in sensitivity of spindle receptors that occurs when muscle is stretched more than a fraction of a millimeter. J. Physiol., 38, 673-689.
Hoffer, J.A. & Loeb, G.E. (1983). A technique for reversible fusimotor blockade during chronic recording from spindle afferents in walking cats. Expl. Brain Res. Suppl., 7, 272-279.
Houk, J.C. (1979). Regulation of stiffness by skeletomotor reflexes. A. Rev. Physiol., 41, 99-114.
Hulliger, M., Matthews, P.B.C. & Noth, J. (1977a). Static and dynamic fusimotor action on the response of Ia fibres to low frequency sinusoidal stretching of widely ranging amplitude. J. Physiol., 267, 811-838.
Hulliger, M., Matthews, P.B.C. & Noth, J. (1977b). Effects of combining static and dynamic fusimotor stimulation on the response of the muscle spindle primary ending to sinusoidal stretching. J. Physiol., 267, 839-856.
Jankowska, E. & McCrea, D.A. (1983). Shared reflex pathways from Ib tendon organ afferents and Ia muscle spindle afferents in the cat. J. Physiol., 338, 99-111.
Jankowska, E., McCrea, D. & Mackel, R. (1981a). Pattern of 'non-reciprocal'

inhibition of motoneurones by impulses in group Ia muscle spindle afferents in the cat. J. Physiol., 316, 393-409.
Jankowska, E., McCrea, D. & Mackel, R. (1981b). Oligosynaptic excitation of motoneurones by impulses in group Ia muscle spindle afferents in the cat. J. Physiol., 316, 411-425.
Loeb, G.E. (1981). Somatosensory unit input to the spinal cord during normal walking. Can. J. Physiol. & Pharmacol., 59, 627-635.
Loeb, G.E. (1984). The control and response of muscle spindles during normally executed motor tasks. Exercise & Sport Sci. Rev., 12, 157-204.
Loeb, G.E., Bak, M.J. & Duysens, J. (1977). Long-term unit recording from somatosensory neurons in the spinal ganglia of the freely walking cat. Science, N.Y., 197, 1192-1194.
Lucas, S.M. & Binder, M.D. (1984). Topographic factors in distribution of homonymous group Ia afferent input to cat medial gastrocnemius motoneurons. J. Neurophysiol., 51, 50-63.
Lucas, S.M., Cope, T.C. & Binder, M.D. (1984). Analysis of individual Ia-afferent EPSPs in a homonymous motoneuron pool with respect to muscle topography. J. Neurophysiol., 51, 64-74.
Lundberg, A., Malmgren, K. & Schomburg, E.D. (1977). Cutaneous facilitation of transmission in reflex pathways from Ib afferents to motoneurones. J. Physiol., 265, 763-780.
Murthy, K.S.K. (1978). Vertebrate fusimotor neurones and their influences on motor behavior. Prog. Neurobiol., 11, 249-307.
Powers, R.K. & Binder, M.D. (1981). Analysis of heteronymous group Ib synaptic input to the cat medial gastrocnemius motoneuron pool. Soc. Neurosci. Abstr., 7, 561.
Prochazka, A. & Wand, P. (1980). Tendon organ discharge during voluntary movement in cats. J. Physiol., 303, 385-390.
Richmond, F.J.R. & Bakker, D.A. (1982). Anatomical organization and sensory receptor content of soft tissues surrounding upper cervical vertebrae in the cat. J. Neurophysiol., 48, 49-61.
Stein, R.B. (1974). The peripheral control of movement. Physiol. Rev., 54, 215-243.
Stuart, D.G. & Stephens, J.A. (1976). The recruitment order of motor units and its significance for the behaviour of tendon organs during normal muscle activity. In The Motor System: Neurophysiology and Muscle Mechanisms. Ed. Shahani, M. Elsevier, Amsterdam, pp. 37-47.
Tracey, D. (1978). Joint receptors - changing ideas. Trends Neurosci., 1, 63-65.
Vallbo, A.B., Hagbarth, K.-E., Torebjörk, H.E. & Wallin, B.G. (1979). Somatosensory, proprioceptive, and sympathetic activity in human peripheral nerves. Physiol. Rev., 59, 919-957.

Chapter Twelve

PROPRIOCEPTIVE FEEDBACK AND THE CONTROL OF COCKROACH WALKING

S.N. Zill

Cockroach walking is a remarkably adaptable behaviour, for these animals are able to traverse walls, floors and ceilings at rates up to 24 steps per second (Delcomyn, 1971). In adapting locomotory patterns to a variety of terrains, cockroaches encounter problems that are common to all terrestrial animals. Walking movements must be monitored to be efficiently performed, and unexpected loads, due to variations in the environment, must be rapidly counterbalanced. To resolve these problems, the cockroach central nervous system must incorporate information provided by limb proprioceptors into the patterns of motor activity used in walking. While the economy of design of sensory and motor systems of these animals has permitted extensive investigations of the response properties and reflex effects of these receptors (Mill, 1976), the basic questions of how inputs from specific proprioceptors are used in the adaptation of walking and which parameters of locomotion are determined by these inputs have often remained unresolved. This has been due, in part, to the following difficulties that arise in the interpretation of data obtained from isolated preparations:

1. Determining the adequate stimulus of a receptor. Proprioceptors were defined by Sherrington (1906) as somatic sense organs that are stimulated by 'actions of the body itself'. However, as is apparent, all receptors of limb muscles and joints respond to a variety of external stimuli as well (Lissmann, 1950). Receptors that monitor forces resulting from muscle contractions, for example, also respond to external loads and simple predictions of their patterns of discharge in walking are not feasible.

2. Determining the effectiveness of reflexes in locomotion. A reflex implies a fixed relationship between a change in afferent input and changes in motor neurone activity (Sherrington, 1906). However, most reflexes studied have been found to be quite variable. Many reflexes show substantial changes in gain (intensity of motor neurone discharge) depending upon the behaviour (Forssberg et al., 1977) or state

Proprioceptive feedback and the control of cockroach walking

(Bässler, 1983) of the animal. Reflexes of both vertebrates and invertebrates can also be changed by training (Melvill Jones & Watt, 1971; Zill & Forman, 1983; Forman & Zill, 1984). In addition, since the time of the earliest account of proprioceptive reflexes, many reflexes have been demonstrated to show 'reflex reversals', that is complete changes in reflex sign (Graham Brown, 1911; Bässler, 1976; DiCaprio & Clarac, 1981; Vedel, 1982). In most cases, the functions of these reversals or the mechanisms underlying this reflex plasticity are unknown.

To resolve these problems it is essential to examine patterns of motor and sensory activity in animals that are walking freely. Recordings of the activity of proprioceptors in these preparations can permit direct evaluation of the relative contributions of internal and external stimuli to their actual responses in walking (Zill & Moran, 1981b). The effectiveness of reflexes can also be tested by examining correlations between changes in proprioceptive inputs and motor output implied by reflex actions (Barnes et al., 1972; Graham & Bässler, 1981). Further, knowledge of the exact patterns of activity in motor neurones to a number of leg muscles during walking can provide the information necessary to evaluate the relative contributions of reflex actions and central pattern generators in the control of locomotion (Delcomyn & Usherwood, 1973).

Cockroach walking has been one of the most intensively studied behaviours in which techniques of recording neuronal activity in freely moving animals have been utilised. In this chapter, I will first describe the basic methods used in these experiments; then examine the patterns of motor activity in leg muscles during walking and re-evaluate a model of their generation by the central nervous system; lastly, I will review those proprioceptive reflexes that can modify and adapt cockroach locomotion, and examine limitations of their effectiveness in rapid walking.

METHODS OF RECORDING ACTIVITY IN FREE MOVING PREPARATIONS

The methods used to record motor and sensory activities in legs of free-walking insects were first developed by Hoyle (1964). For myographic recordings, the animal is first restrained and small holes are made in the cuticle overlying leg muscles (Fig. 1). Long (60cm) lengths of fine (30-50μm) silver or copper wire are used as recording electrodes. These wires, insulated to their tips, are inserted through the holes in the cuticle into the appropriate muscles, tied to the legs and led to extracellular amplifiers. The animal is then released into a walking arena. Recordings may be obtained from these preparations for days to weeks following electrode implantation.

Subsequent studies have adapted these techniques for direct recordings from peripheral motor or sensory nerves (Runion & Usherwood, 1966; Burns & Usherwood, 1979). In these preparations, external cuticular landmarks are used to position the electrodes, and neural activity is evoked to insure placement close to the nerve. In nerves with many axons, only the activities of the largest fibres can be unequivocally identified. However, in some small nerves, activities of most axons can be individually discerned.

In the cockroach, most studies have recorded activities of muscles in proximal segments of the middle (mesothoracic) and hind (metathoracic) legs as illustrated in Fig. 1 (Delcomyn, 1969, 1971; Pearson, 1972). Those best studied are the paired antagonist extensor and flexor muscles of the trochanter (Etr and FlTr) and homologous antagonist muscles in the femur, the extensor and flexor tibiae (ETi and FlTi). The extensor muscles of each of these segments are innervated by single slow and fast excitatory motor neurones whose activity can be readily identified in myographic recordings, while the antagonist flexors are multiply innervated and recordings of single units are more difficult to obtain.

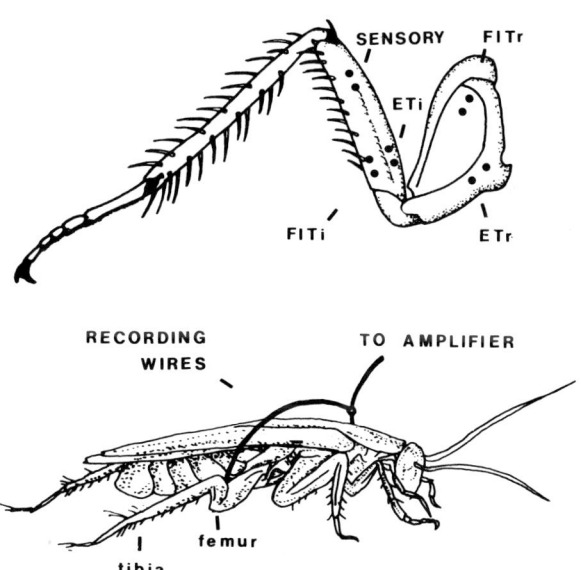

Fig. 1. Free walking preparation. Myographic activity in walking is recorded by pairs of fine wires inserted into small holes in the cuticle overlying the flexor (FlTr) and extensor (ETr) muscles of the trochanter and the homologous flexor (FlTi) and extensor (ETi) muscles of the tibia. Sensory activity of tibial campaniform sensilla is recorded by wires placed close to the dorsal nerve in the distal femur.

Proprioceptive feedback and the control of cockroach walking

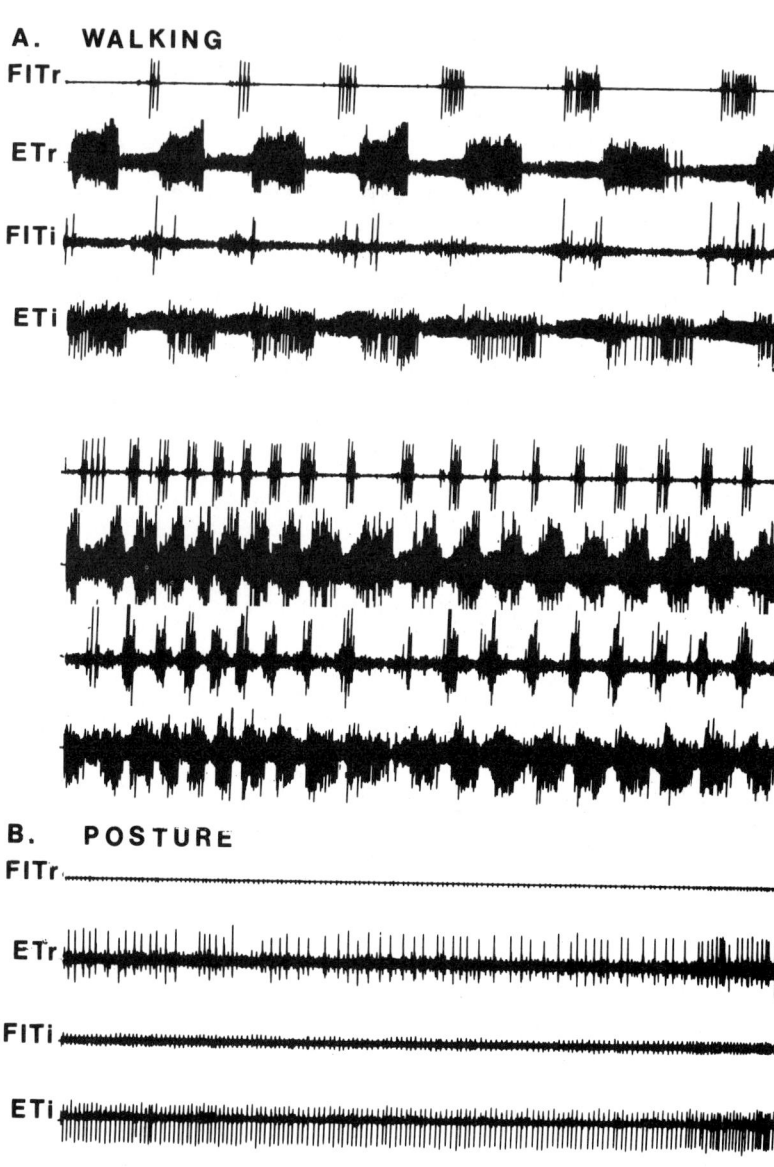

Fig. 2. Myographic activity during walking and standing.
A: Walking. In slow walking (upper four traces) the flexor (FlTr) and extensor (ETr) of the trochanter show reciprocal bursting, while firing of the tibial flexor (FlTi) continues into succeeding bursts of the

MOTOR ACTIVITY IN COCKROACH WALKING

Movements at joints of the cockroach leg were first determined by cinematographic studies (Hughes, 1952). In the middle and hindlegs, that are used predominantly as propulsive struts, the coxo-trochanteral and femoro-tibial joints, undergo simple alternating flexions (during swing) and extensions (during stance) in a single step cycle.

Activities of the slow motor neurones of the trochanteral extensor muscle of the middle and hindlegs were the first to be examined in freely moving animals. At moderate rates of stepping, these motor neurones are active during extension movements and are mostly silent during joint flexions (Delcomyn, 1969). These observations were subsequently extended by Pearson (1972) who recorded myographic activity of both the trochanteral extensor and flexor muscles and found mainly reciprocal bursting in antagonist motor neurones (Fig. 2). As is true of many animals, increasing rates of stepping are produced by shortening the stance phase and the duration of extensor bursts, although decreases in the duration of flexor bursts also occur.

The studies of Delcomyn (1969) and Delcomyn and Usherwood (1973) also showed that, at higher rates of stepping, the relationship between the timing of motor neurone activity and the resultant movements changed. During rapid walking, there is a substantial lag between the onset of extensor motor neurone activity and the corresponding extension movement (Fig. 3A). To achieve stepping rates over 5Hz, extensor bursts are initiated less than halfway through the preceding flexion movements. Delcomyn noted that while this phase shift would have little effect on the central generation of coordinated leg movement 'reflexes triggered by leg movements would occur at different stages of muscle activity' and the resultant effects might disrupt walking patterns.

Activities of muscles of the more distal leg segment, the tibial extensor and flexor, have also been studied and found to be more complex (Fig. 2). At slow rates of stepping (<2 Hz) strict reciprocity of antagonists does not occur, as flexor tibiae motor neurones continue to fire into the initial period of an extensor burst (Krauthamer & Fourtner, 1978). Preliminary studies using cinematography have shown that during this period the femoro-tibial joint is held immobilised by these co-contractions.

Recent studies by the author, illustrated in Fig. 2, have

---

tibial extensor (ETi). In rapid walking (lower four traces) activity of antagonist muscles at both joints is strictly reciprocal.
B. Posture. Both extensor muscles are active during standing as well as during the stance phase of walking.

Proprioceptive feedback and the control of cockroach walking

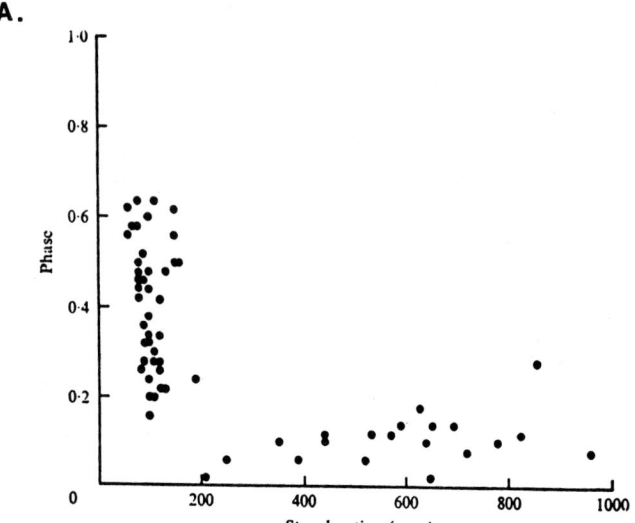

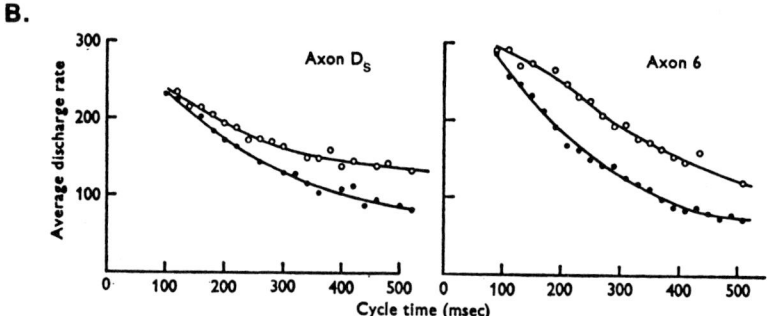

Fig. 3. A. Overlap of the beginning of extensor bursting on the preceding flexion movement as a function of the duration of the whole step. The overlap was determined by calculating the phase of initiation of bursts of the trochanteral extensor relative to flexion movement at the coxotrochanteral joint. The phase is zero if the extensor burst began as the flexion movement ended, 0.5 if the burst began halfway through the flexion movement. In rapid walking (over 10Hz), extensor bursts are initiated more than halfway through the preceding flexion movement. (From Delcomyn & Usherwood, 1973).

B. Changes in the effect of loading at different stepping rates. Discharge rates of the trochanteral extensor (D_s) and a trochanteral flexor (Axon 6) are plotted for different step durations of a cockroach that was freely walking (filled circles) or dragging a 1.5g weight (open circles). In slow walking, firing of both motor neurones is higher when loading is increased. In rapid walking (at 10Hz), the effect of loading is absent. (Data from Pearson, 1972).

examined the activities of all four of these muscles simultaneously in walking and permit the following observations. First, in slow walking, the trochanteral extensor leads the tibial extensor in onset of bursting, while in rapid walking bursts are usually synchronous. Second, there is often discrete patterning of activity in extensor bursts in slow walking; the trochanteral extensor often fires intensely at the onset of a burst and activity is either sustained or gradually declines in frequency during the burst; in contrast, tibial extensor activity is initiated at a slow rate and increases towards the end of the burst, which often outlasts that of the trochanteral extensor. This patterning does not occur in rapid walking which is characterised solely by intense, irregular activity in both motor neurones simultaneously. Third, extensors and flexors of both joints show strict reciprocity of activity at high rates of stepping so that the phase shifts shown by Delcomyn for the trochanteral extensor must also occur in the other leg muscles as well. Last, both extensors are used synergistically in postural support but there is no strict correlation of their spiking activity.

Several previous studies have shown, in addition, that increased loading of an animal or increased resistance to retraction in slow walking has a substantial effect upon patterns of activity in both flexor and extensor motor neurones that are related to sensory inputs. Pearson (1972) has shown that dragging a weight (1.5gm) results in large (50%) increases in discharge rates of slow flexor and extensor motor neurones in walking at 2Hz (Fig. 3B). In contrast, there is no effect of loading upon motor neurone frequency at stepping rates over 10 Hz. Similar effects have been noted for durations of flexor bursts as well.

In sum, these studies have shown that many parameters of walking activity are modifiable in slow to moderate rates of stepping, but in rapid walking modulation and adaptation of motor neurone activities are substantially reduced.

CENTRAL PATTERN GENERATION OF COCKROACH WALKING

Several models of the circuitry generating patterns of motor activity in insect walking have been proposed (see Delcomyn, 1981, for review). Most of these models agree upon the existence of oscillators within the central nervous system that generate the basic pattern of walking but they differ in respect of whether particular parameters are centrally generated or determined by proprioceptive feedback. I will reconsider one model, based upon activities generated by the isolated nerve cord of the cockroach (Pearson & Iles, 1970; Pearson, 1972, 1976), in the light of subsequent studies.

Reciprocal bursts of activity can be recorded in extensor

and flexor muscles of the cockroach leg if the animal is decapitated or if connectives anterior to the metathoracic ganglion are cut. In these preparations the animal is restrained on its back, and posterior connectives, linking the metathoracic ganglion to the abdomen, remain intact. Many of the parameters of bursting seen in these preparations are similar to walking. Durations of flexor bursts during prolonged sequences of activity are relatively constant compared to extensor bursts, and occur at frequencies seen in stepping. Further, flexor bursting persists after deafferentation of the leg. Based upon these observations, an endogenous flexor burst generator has been postulated to be present in the central nervous system that generates flexion movements of each leg. This model led to the important discovery of non-spiking interneurones as premotor elements in the insect central nervous system (Pearson & Fourtner, 1975); imposed changes in membrane potential of these interneurones could reset the bursting rhythm.

In these preparations, motor neurone bursting was inferred to represent attempts at walking based upon its similarity to patterns seen in locomotion and the absence of any other described behaviours with equivalent characteristics (Pearson, 1972). However, subsequent studies have shown that similar patterns of motor neurone activity can be seen in the same muscles during grooming of the cercus (an abdominal appendage), and in righting behaviour (Reingold & Camhi, 1977; Sherman et al., 1977), when an animal on its back attempts to turn upright (Fig. 4). Further, investigations of other insects, such as the stick insect (Bässler, personal communication) or locust (Hoyle, personal communication), have failed to demonstrate rhythmic bursting in headless preparations. Studies on the isolated nerve cord of the stick insect (Bässler & Wegner, 1983) have shown that only irregular bursting activity, closely resembling struggling or righting, can be evoked in these animals.

Recent studies, in which activity was recorded from free-moving, headless cockroaches, tend to support Bässler's findings (Bässler & Wegner, 1983). First, as noted by a number of previous authors, headless cockroaches show little spontaneous walking and evoked steps are often discoordinated and erratic (Reingold & Camhi, 1977). Second, upright headless cockroaches do not show regular bursting. However, regular bursting activity can be recorded in these preparations if they are turned on their backs and attempt to right themselves by struggling. During this activity in phase bursting of synergist muscles in different leg segments can occur, but often bursting in these muscles occurs out of phase as is characteristic of righting (Fig. 5). Lastly, bursting activity is inhibited if the animal is again turned upright (Zill, in preparation).

Thus, bursting activity shown by headless cockroaches

seems most likely to represent struggling during attempts at righting. However, this finding does not negate the idea of an endogenous flexor burst generator or imply that non-spiking interneurones, that have been shown to reset the rhythm of these bursts, are not also involved in the generation of walking. As several authors have noted (e.g. Reingold & Camhi,

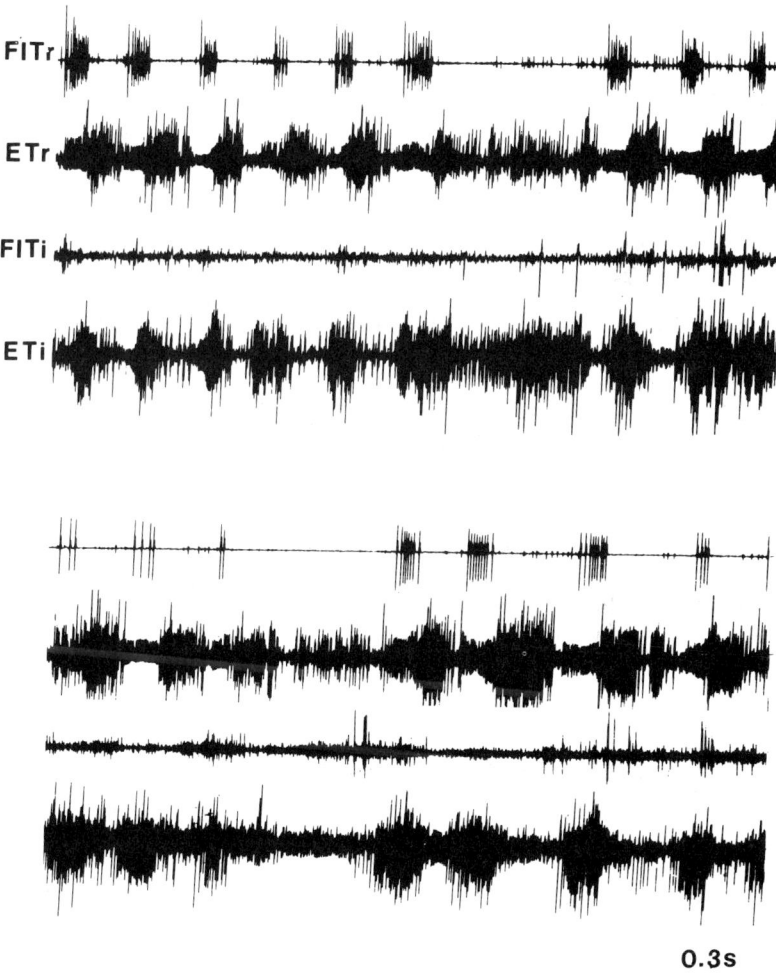

Fig. 4. Myographic activity during righting behaviour. A cockroach placed on its back shows rhythmic bursts of activity in motor neurones to leg muscles as it struggles to turn itself upright. Firing of antagonist flexor and extensor muscles is often reciprocal as in walking but trochanteral and tibial extensors often burst out of phase.

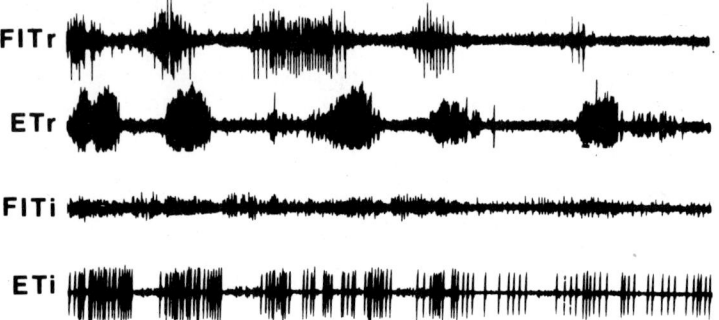

A. STRUGGLING ON BACK

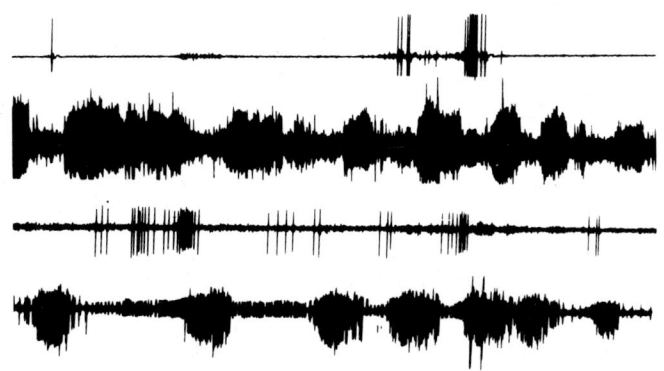

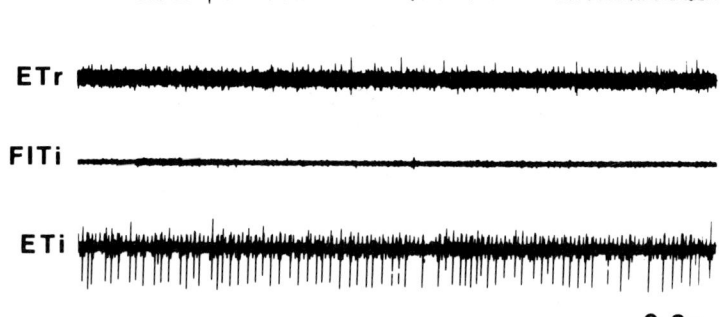

B. TURN UPRIGHT

0.3s

1977), the behaviours of righting and walking have many characteristics in common and many of the same premotor interneurones may be used in generating both behaviours. However, no insect preparation exhibits the exact pattern of walking activity in the absence of sensory inputs from the leg. Thus, the existence of central pattern generators for walking remains to be demonstrated (see Chapter 20 by Prof Pearson).

FUNCTIONS OF PROPRIOCEPTIVE REFLEXES IN WALKING

Proprioceptors of the legs of cockroaches have been demonstrated to elicit reflex effects that can, in concert, monitor and adapt all phases of walking. The potential functions of these reflexes are discussed below.

Sense organs that monitor joint angles
Three types of sense organs monitor joint angles in insect legs: hair plates, chordotonal organs and multipolar receptors (Nijenhuis & Dresden, 1952; Pringle, 1961). Hair plates consist of small groups of cuticular hairs located near joints that are bent during joint flexion (protraction). The trochanteral hair plate of the cockroach has been extensively studied (Wong & Pearson, 1976). It consists of phasic and tonic receptors that respond to joint flexion. Stimulation of these receptors causes reflex inhibition of trochanteral flexors. This reflex apparently functions to limit joint flexion during the swing phase of walking and also potentially aids in the initiation of extensor bursts during stance. Ablation of the hair plates produces exaggerated joint flexions in walking that result in overstepping and collision of operated legs with those anterior to them (Wendler, 1966). Hair plates could also function in load compensation by exciting the extensor motor neurone to counter any external disturbance of posture that produces joint flexion. In addition, inputs from the hair plate may be used in more complex functions, such as coordination of muscle contractions within a single leg, as these receptors have been shown to

---

Fig. 5. Myographic activity in a cockroach with cut thoracic connectives. Motor neurone bursting in these preparations has been studied as a model of walking.
A. With the animal placed on its back, rhythmic bursts of activity occur in motor neurones to leg muscles. However, bursting of the tibial and trochanteral extensors may be in or out of phase, as is characteristic of righting behaviour.
B. Bursting is inhibited when the animal is turned upright. Thus, bursting in these preparations is most likely the intermittent appearance of righting responses.

Proprioceptive feedback and the control of cockroach walking

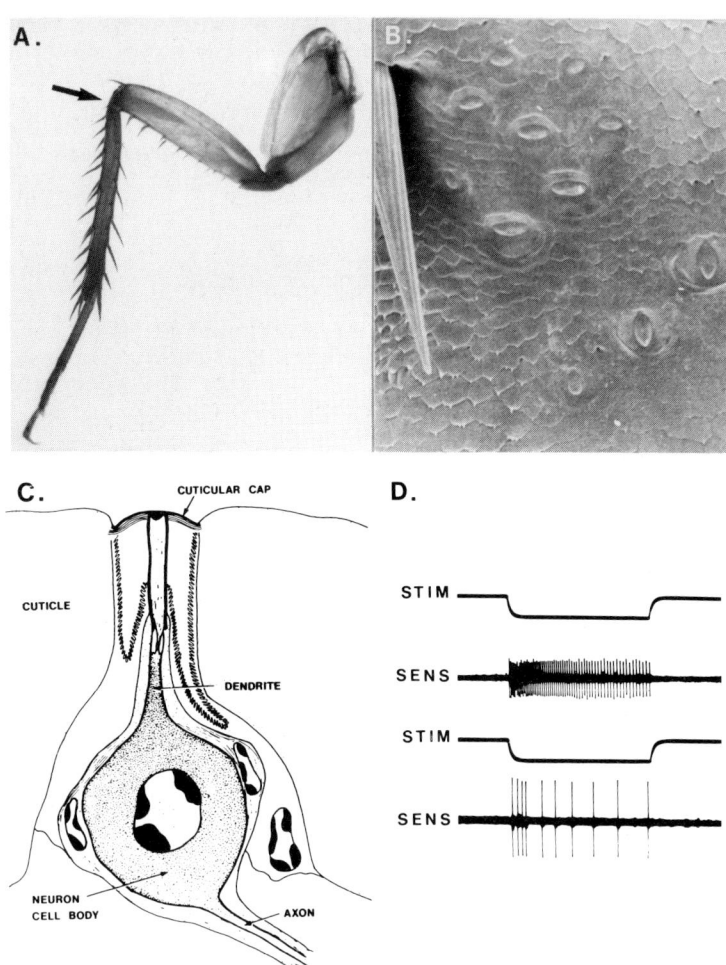

Fig. 6. Campaniform sensilla of the cockroach tibia. A group of campaniform sensilla (B) monitor cuticular strains in the tibia (A) that result from muscle contractions or external loads. Each sensillum consists of a single bipolar neurone whose dendrite inserts onto a thin cuticular cap in the exoskeleton (C). Tibial campaniform sensilla show a consistent orientation of their cuticular caps; proximal sensilla are oriented with the cap long axis perpendicular to the length of the tibia, while distal sensilla are oriented parallel to the tibial axis. Responses and reflex effects of these subgroups differ. Activity of these receptors can be recorded extracellularly in the distal femur and individual sensilla (SENS) identified by stimulation (STIM) of their cuticular caps with a fine wire (D).

Proprioceptive feedback and the control of cockroach walking

affect the activity of non-spiking interneurones (Pearson et al., 1976). Ablation of hair plates has been shown to produce disturbances of more complex behaviours in other insects (Liske & Mohren, 1984).

Chordotonal organs and multipolar receptors respond to more extensive ranges of joint angles (Guthrie, 1967; Young, 1970). Both types of receptors contain phasic and tonic units that signal joint movement and position (Becht, 1958). Chordotonal organs of the femur and coxa of the cockroach mediate resistance reflexes in both flexor and extensor motor neurones that oppose joint movement (Zill & Moran, 1982; Brodfuehrer & Fourtner, 1983). Similar reflexes are elicited by stimulation of multipolar receptors (Guthrie, 1967). These reflexes could readily function in load compensation if the animal were able to distinguish between changes in joint angle produced by external loads and those generated by its own movements. The ability to modulate resistance reflexes is clearly implied by the observation in freely moving animals that postures assumed in extreme extension or flexion do not generate activity in antagonist muscles (Fig. 2). Similar mechanisms have been postulated to underly the modulation of resistance reflexes during walking in other arthropods (Barnes et al., 1972). In addition to resistance reflexes, however, joint movements and chordotonal organ stimulation can produce reflexes of opposite sign that have been postulated to assist movements (Wilson, 1965; Bässler, 1976). The actual functions of these reflexes in walking have not been determined.

<u>Sense organs that monitor muscle tensions and cuticular strains</u>
Campaniform sensilla respond to strains in the cuticle that result from forces due to muscle contractions and loading of the leg by the animals weight (Pringle, 1938a,b; Zill & Moran, 1981a). Each sensillum consists of a single bipolar neurone

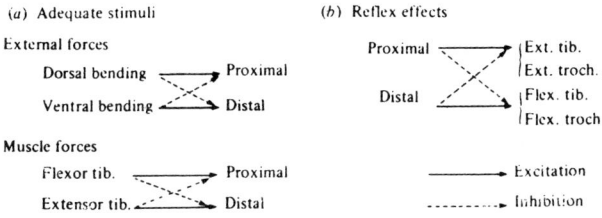

Fig. 7. Summary of adequate stimuli (a) and reflex effects (b) of tibial campaniform sensilla. (See text for discussion).

199

Proprioceptive feedback and the control of cockroach walking

A. WALKING

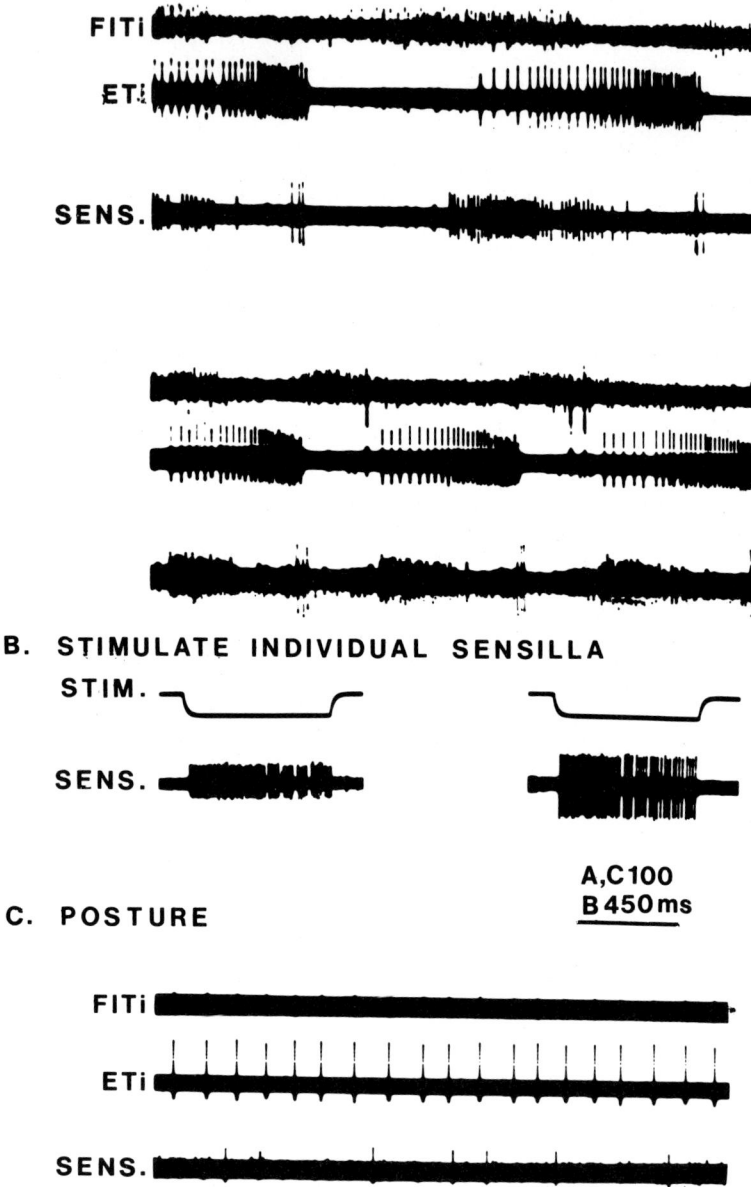

FITi
ETi

SENS.

B. STIMULATE INDIVIDUAL SENSILLA
STIM.
SENS.

A,C 100
B 450 ms

C. POSTURE

FITi
ETi
SENS.

whose dendrite inserts into an ovoid cap in the exoskeleton (Fig. 6). These receptors respond with directional sensitivity to forces exerted perpendicularly to the cap long axis (Zill & Moran, 1981a). The tibial group of campaniform sensilla consists of two subgroups with mutually perpendicular cap orientations: a proximal group that is oriented with the cap long axis perpendicular to the length of the tibia and a distal group whose cap orientation is parallel to the tibial axis. The proximal sensilla respond best to dorsal bending of the leg and to isometric contractions of the tibial flexor muscle; distal sensilla are activated by ventral bending and cuticular strains produced by contractions of the extensor muscle (Fig. 7a). Stimulation of individual receptors produces reflex effects in leg motor neurones that also depends upon cap orientation (Zill et al., 1981); proximal sensilla inhibit flexor motor neurones and excite extensors, while distal sensilla produce reflexes of opposite sign in that they excite flexors and inhibit extensors (Fig. 7b). These reflexes could function in two ways; first, they clearly form an inhibitory feedback pathway to limit forces in the cuticle that result from muscle contractions and thus prevent excessive muscle tensions. Second, with the leg placed upon a walking surface, these reflexes could function in load compensation by countering bending forces produced by the animal's weight upon the leg with bending resulting from reflex muscle contractions. In both cases the animal apparently attempts to balance the component of the force acting perpendicularly to the tibial long axis.

Thus, in sum, reflex effects of limb proprioceptors can effect both phases of walking. In stance, firing of extensor motor neurones can be adjusted for the animal's weight and unexpected loads by reflexes of campaniform sensilla, chordotonal organs and multipolar receptors. In swing, the amplitude and duration of flexor bursts are monitored and adjusted by hairplates, and potentially by inputs from other receptors monitoring joint angle as well.

Fig. 8. Activity of tibial campaniform sensilla in walking.
A. The firing patterns of two campaniform sensilla (SENS), one proximal (small spikes) and one distal (large spikes), were recorded in a free moving cockroach. Receptor firing is correlated with activity of the tibial extensor muscle. The proximal sensillum initiates bursting prior to extensor activity and ceases when extensor frequency exceeds 200Hz. The distal sensillum fires near the end of stance when extensor activity is highest (over 300Hz).
B. Individual sensilla are identified by stimulation of their cuticular caps.
C. The slow extensor is active in postural support and is accompanied by low frequency firing of proximal sensilla.

PROPRIOCEPTIVE MODULATION OF COCKROACH WALKING AND THE NEED FOR CENTRAL PATTERN GENERATION

While many limb proprioceptive sense organs of the cockroach have been shown to exhibit reflex effects upon motor neurones to leg muscles, it is necessary to ask whether these reflexes are functional during walking and whether there are limitations to the effectiveness of reflex actions during rapid walking (Hoyle, 1976). To provide some insight into these problems, I will review some recent experiments in which the activity of tibial campaniform sensilla have been recorded in freely moving animals, and consider these findings in relation to what is known about motor patterning in rapid walking.

Activities of tibial campaniform sensilla in walking

The activities of tibial campaniform sensilla in freely walking animals (Zill & Moran, 1981b) are correlated with the activities of the flexor and extensor tibiae muscles (Fig. 8). In slow walking, proximal sensilla fire in bursts that are initiated just prior to the onset of extensor activity in stance, decline as extensor activity increases within a burst and cease when motor neurone activity exceeds 200Hz. Distal sensillum activity typically occurs as a short intense burst near the end of stance, when slow extensor frequency is maximal (above 300Hz). These patterns of afferent and motor activity are consistent with the demonstrated adequate stimuli

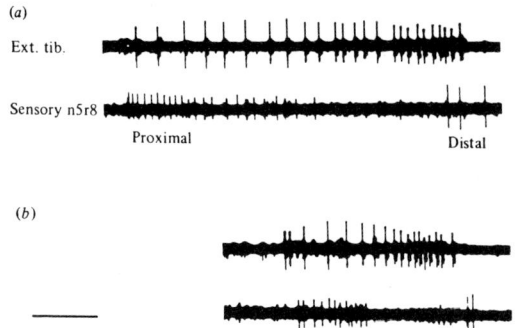

Fig. 9. Changes in phase of activity of campaniform sensilla at different walking speeds. In slow walking (a), activity of proximal sensilla is initiated prior to extensor bursting while distal sensilla fire before the end of an extensor burst. In rapid walking (b), sensory activity shifts in phase: proximal sensilla are active only after the initiation of extensor bursting while distal sensillum firing follows the termination of extensor activity.

and reflex effects of these receptors. At the initiation of stance, as the leg is placed upon the walking surface a dorsal bending of the tibia should occur from the animal's weight upon the leg and cause activity of proximal sensilla (Zill & Moran, 1981a). Firing of these receptors should reflexly excite the slow extensor motor neurone and afferent activity should be inhibited when bending forces are equalised by extensor tensions. Larger tensions in the extensor should induce ventral bending and firing of distal sensilla. Precisely this pattern of activity is seen in stance and indicates that reflexes of these receptors are active in walking and can adapt motor neurone activity to external loading while preventing excessive muscle tensions. Further, distal sensilla bursts could contribute to the timing of the onset of flexor activity during swing.

In rapid walking, however, activities of campaniform sensilla shift in phase relative to slow extensor firing. Proximal sensillum activity occurs only after the onset of extensor firing (Figs. 9, 11) while distal sensilla bursts

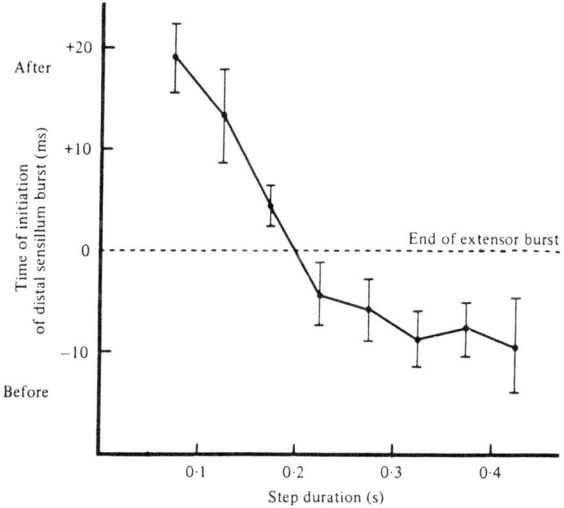

Fig. 10. Change in phase of distal sensillum firing. The time of initiation of distal sensillum firing relative to the termination of extensor bursting is plotted for whole steps of different durations. In slow walking, distal sensillum activity occurs before the termination of extensor bursting. In rapid walking, sensilla fire after the end of extensor activity. Reflex effects of these receptors, which inhibit the extensor, can therefore function in slow walking but are ineffective at high stepping rates.

follow the termination of slow extensor activity (Figs. 9, 10). The potential contribution of the reflex effects of these receptors in patterning of rapid walking is necessarily limited by these phase shifts: proximal sensilla cannot contribute to the initiation of extensor firing as they may at slower stepping rates, while distal sensilla cannot aid in the termination of extensor bursts. It is of interest to note that these phase shifts are accompanied by changes in the pattern of slow extensor tibiae activity within a burst. As discussed above, extensor firing frequencies increases during a burst in slow walking, thus providing maximal force in propulsion when the leg is retracted. In rapid walking, however, extensor firing is highly erratic within a burst. Thus at high rates of stepping, when reflexes are limited in efficacy, the patterning of motor neurone activity is less well adapted.

Limitations of proprioceptive feedback and the need for central pattern generation

In summary, both the shifts in phase of afferent firing and the lack of effective load compensation in rapid walking suggest that proprioceptive inputs are limited in their ability to modulate motor patterns at high rates of stepping (Pearson, 1972; Zill & Moran, 1981b). These conclusions are further supported by experiments examining the effects of limb amputation (Delcomyn, 1981). In slow walking, after amputation of the mesothoracic legs, animals change their gait in such a way that the centre of gravity is supported by the remaining legs. In rapid walking, however, animals may revert to patterns seen before amputation. These patterns can produce instability and slipping or falling often occurs. Thus, in very rapid walking, proprioceptive inputs may be completely ineffective in adapting motor patterns to loads.

A key element underlying this ineffectiveness may be the phase shifts between motor neurone firing and leg movement (Delcomyn, 1969; Delcomyn & Usherwood, 1973). Earlier contractions of the extensor muscle were assumed by Delcomyn to dampen or 'brake' the forward movement of the leg. However, it is also possible that initiation of slow motor neurone activity must occur far in advance of movement to allow for the development of effective muscle tensions. The activity of distal campaniform sensilla in walking tend to support this conclusion; slow extensor bursts at high frequencies (300-400Hz) do not produce sufficient tension to cause distal sensillum activity until 30ms after their initiation. In very rapid walking, however, extensor bursts do not exceed 20ms in duration. Thus, tensions that produce leg movements must lag substantially behind motor neurone activity. Delays inherent in the development of tension may therefore also require phase shifts in rapid movement.

Several observations suggest that these phase shifts effect the efficacy of reflexes of other proprioceptors of the

Proprioceptive feedback and the control of cockroach walking

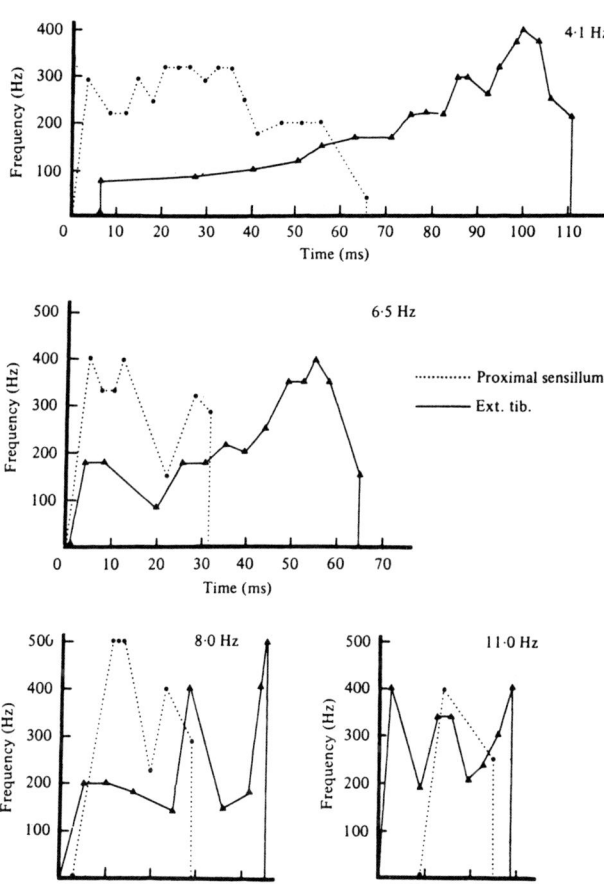

Fig. 11. Change in phase of proximal sensillum firing. The instantaneous firing frequency of a proximal sensillum and the slow extensor tibiae motor neurone are plotted for four steps at different walking speeds. In slow walking, proximal sensillum firing precedes the onset of extensor bursting and the frequency of sensory and motor activities are reciprocally related. In rapid walking, proximal sensillum bursting follows the onset of extensor bursting and motor activity is erratic. Thus, the reflex effects of these receptors, which cause excitation of the extensor, appear to be effective in slow walking but limited in rapid stepping.

leg. Reflex effects due to joint movement or chordotonal inputs have been shown to be consistently elicited at rates of up to 24Hz (Wilson, 1965; Brodfuehrer & Fourtner, 1983), the fastest a cockroach has been observed to run. However, at high rates of applied movement the extensor does not fire until the end of the flexion movement (Wilson, 1965; Zill, personal observation). Thus the effectiveness of these reflexes also seems limited and cannot account for extensor firing in the first half of swing (Fig. 3A).

Thus, central pattern generation seems to be required for very rapid walking in the cockroach. It should be noted that, under these conditions, it is not only flexor bursts in swing that must be patterned by the central nervous system (c.f. Pearson, 1976), but also extensor bursts providing support and propulsion in stance. Further studies of motor and sensory activity in walking and load compensation of freely moving animals may provide clues as to the nature of the central pattern generator in insect walking.

REFERENCES

Barnes, W.J.P., Spirito, C.P. & Evoy, W.H. (1972). Nervous control of walking in the crab, Cardisoma guanhumi. II. Role of resistance reflexes in walking. Z. vergl. Physiol., 76, 16-31.

Bässler, U. (1976). Reversal of a reflex to a single motoneuron in the stick insect Carausius morosus. Biol. Cybern., 24, 47-49.

Bässler, U. (1983). Neural Basis of Elementary Behavior in Stick Insects. Springer Verlag, Berlin.

Bässler, U. & Wegner, U. (1983). Motor output of the denervated thoracic ventral nerve cord in the stick insect Carausius morosus. J. exp. Biol., 105, 127-145.

Becht, G. (1958). Influence of DDT and lindane on chordotonal organs in the cockroach. Nature, Lond., 181, 777-779.

Brodfuehrer, P. & Fourtner, C.R. (1983). Reflexes evoked by the femoral and coxal chordotonal organs in the cockroach, Periplaneta americana. Comp. Biochem. Physiol., 74A, 169-174.

Burns, M.D. & Usherwood, P.N.R. (1979). The control of walking in Orthoptera. II. Motor neurone activity in normal free-walking animals. J. exp. Biol., 79, 69-98.

Delcomyn, F. (1969). Reflexes and locomotion in the American cockroach. Ph.D. Thesis, University of Oregon.

Delcomyn, F. (1971). The locomotion of the cockroach Periplaneta americana. J. exp. Biol., 54, 443-452.

Delcomyn, F. (1981). Insect locomotion on land. In Locomotion and Energetics in Arthropods. Eds. Herreid, C.F. & Fourtner, C.R.. Plenum Press, New York, pp. 103-125.

Delcomyn, F. & Usherwood, P.N.R. (1973). Motor activity during walking in the cockroach Periplaneta americana. I. Free walking. J. exp. Biol., 59, 629-642.

DiCaprio, R.A. & Clarac, F. (1981). Reversal of a walking leg reflex elicited

by a muscle receptor. J. exp. Biol., 90, 197-203.
Forman, R.R. & Zill, S.N. (1984). Leg position learning by an insect. 2. Motor strategies underlying learned leg extension. J. Neurobiol., 15, 221-237.
Forssberg, H., Grillner, S. & Rossignol, S. (1977). Phasic gain control of reflexes from the dorsum of the paw during spinal locomotion. Brain Res., 132, 121-129.
Graham, D. & Bässler, U. (1981). Effects of afference sign reversal on motor activity in walking stick insects (Carausius morosus). J. exp. Biol., 91, 179-183.
Graham Brown, T. (1911). Studies in the physiology of the nervous system. VIII. Neural balance and reflex reversal with a note on progression in the decerebrate guinea-pig. Q. Jl exp. Physiol., 4, 273-288.
Guthrie, D.M. (1967). Multipolar stretch receptors and the insect leg reflex. J. Insect Physiol., 13, 1637-1644.
Hoyle, G. (1964). Exploration of neuronal mechanisms underlying behavior in insects. In Neural Theory and Modeling. Ed. Reiss, R.. Stanford University Press, pp. 346-376.
Hoyle, G. (1976). Arthropod walking. In Neural Control of Locomotion. Eds. Herman, R.M., Grillner, S., Stein, P.S.G. & Stuart, D.G.. Plenum Press, New York, pp. 137-179.
Hughes, G.M. (1952). The co-ordination of insect movements. I. The walking movements of insects. J. exp. Biol., 29, 267-284.
Krauthamer, V. & Fourtner, C.R. (1978). Locomotory activity in the extensor and flexor tibiae of the cockroach Periplaneta americana. J. Insect Physiol., 24, 813-819.
Liske, E. & Mohren, W. (1984). Saccadic head movements of the praying mantis, with particular reference to visual and proprioceptive information. Physiol. Entomol., 9, 29-38.
Lissmann, H.W. (1950). Proprioceptors. Symp. Soc. exp. Biol., 4, 34-59.
Melvill Jones, G. & Watt, D.G.D. (1971). Observations on the control of stepping and hopping movements in man. J. Physiol., 219, 709-727.
Mill, P.J. (1976). Structure and Function of Proprioceptors in the Invertebrates. Chapman and Hall, London.
Nijenhuis, E.D. & Dresden, D. (1952). A micro-morphological study on the sensory supply of the mesothoracic leg of the American cockroach Periplaneta americana. Proc. Sect. Sci. K. ned. Akad. Wet. C, 55, 300-310.
Pearson, K.G. (1972). Central programming and reflex control of walking in the cockroach. J. exp. Biol., 56, 173-193.
Pearson, K.G. (1976). The control of walking. Scient. Am., 235, 72-86.
Pearson, K.G. & Fourtner, C.R. (1975). Nonspiking interneurons in walking system of the cockroach. J. Neurophysiol., 38, 33-52.
Pearson, K.G. & Iles, J.F. (1970). Discharge patterns of coxal levator and depressor motoneurones of the cockroach, Periplaneta americana. J. exp. Biol., 52, 139-165.
Pearson, K.G., Wong, R.K.S. & Fourtner, C.R. (1976). Connexions between hairplate afferents and motoneurones in the cockroach leg. J. exp. Biol., 64, 251-266.
Pringle, J.W.S. (1938a). Proprioception in insects. I. A new type of mechanical receptor from the palps of the cockroach. J. exp. Biol., 15,

101-113.

Pringle, J.W.S. (1938b). Proprioception in insects. II. The action of the campaniform sensilla on the legs. J. exp. Biol., 15, 114-131.

Pringle, J.W.S. (1961). Proprioception in arthropods. In The Cell and the Organism. Eds. Ramsay, J.A. & Wigglesworth, V.B. Cambridge University Press, London, pp. 256-282.

Reingold, S. & Camhi, J.M. (1977). A quantitative analysis of rhythmic leg movements during three different behaviors in the cockroach, Periplaneta americana. J. Insect Physiol., 23, 1407-1420.

Runion, H.I. & Usherwood, P.N.R. (1966). A new approach to neuromuscular analysis in the intact free-walking insect preparation. J. Insect Physiol., 12, 1255-1263.

Sherman, E., Novotny, M. & Camhi, J.M. (1977). A modified walking rhythm employed during righting behavior in the cockroach Gromphadorhina portentosa. J. comp. Physiol., 113, 303-316.

Sherrington, C. (1906). The Integrative Action of the Nervous System. Charles Scribner's Sons, New York.

Vedel, J.P. (1982). Reflex reversals resulting from active movements in the antenna of the rock lobster. J. exp. Biol., 101, 121-133.

Wendler, G. (1966). The co-ordination of walking movments in arthropods. Symp. Soc. exp. Biol., 20, 229-250.

Wilson, D.M. (1965). Proprioceptive leg reflexes in cockroaches. J. exp. Biol., 43, 397-409.

Wong, R.K.S. & Pearson, K.G. (1976). Properties of the trochanteral hair plate and its function in the control of walking in the cockroach. J. exp. Biol., 64, 233-249.

Young, D. (1970). The structure and function of a connective chordotonal organ in the cockroach leg. Phil. Trans. R. Soc. B, 256, 401-428.

Zill, S.N. & Forman, R.R. (1983). Proprioceptive reflexes change when an insect assumes an active, learned posture. J. exp. Biol., 107, 385-390.

Zill, S.N. & Moran, D.T. (1981a). The exoskeleton and insect proprioception. I. Responses of tibial campaniform sensilla to external and muscle-generated forces in the American cockroach, Periplaneta americana. J. exp. Biol., 91, 1-24.

Zill, S.N. & Moran, D.T. (1981b). The exoskeleton and insect proprioception. III. Activity of tibial campaniform sensilla during walking in the American cockroach, Periplaneta americana. J. exp. Biol., 94, 57-75.

Zill, S.N. & Moran, D.T. (1982). Suppression of reflex postural tonus: a role of peripheral inhibition in insects. Science, N.Y., 216, 751-752.

Zill, S.N., Moran, D.T. & Varela, F.G. (1981). The exoskeleton and insect proprioception. II. Reflex effects of tibial campaniform sensilla in the Americana cockroach, Periplaneta americana. J. exp. Biol., 94, 43-55.

Section 4
Reflexes

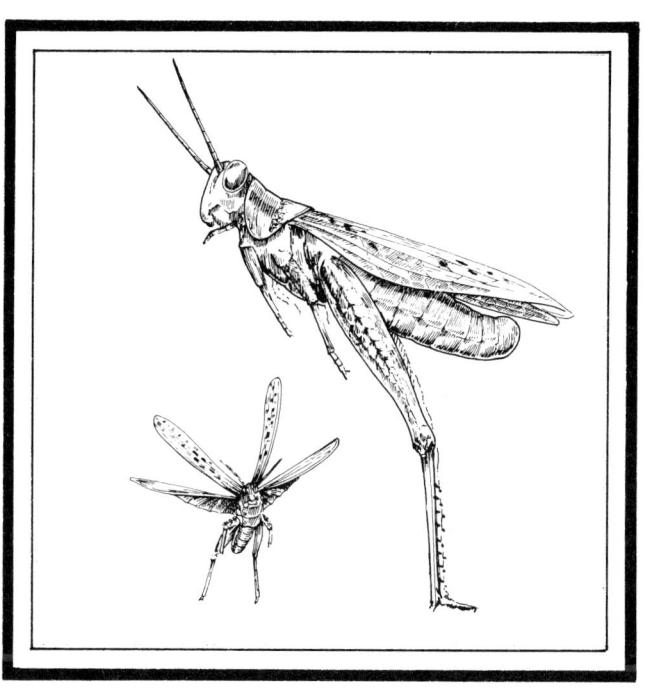

Chapter Thirteen

INTRODUCTION

W.J.P. Barnes

Reflexes are brief stereotyped movements carried out in automatic fashion in response to some sensory stimulus. Many avoid, escape from or minimise the effects of noxious stimuli. Thus, a boxer blinks and ducks as his opponent's fist approaches his face, while someone unlucky enough to step on a drawing pin quickly lifts his foot in response. But this book would not have been written if reflexes were not intimately involved in almost all aspects of motor control - in the maintenance of equilibrium, in the control of posture, and in locomotion and other movements. At the physiological level, the simplest reflex arcs involve only two neurones. But such arcs are only small components of most behavioural reflexes, which consist of many interacting arcs in parallel. Thus someone lifting his foot from a pin must not only coordinate activity in several muscles in the leg that is lifted, but adjust tension in many others to avoid falling over.

The reflex idea has a long history beginning in the seventeenth century with the French philosopher, René Descartes, who was the first to view all animal and much human behaviour in mechanistic terms. In the example shown in Fig. 1, the fire is said to have force enough to displace the skin. This pulls a tiny thread (c) which opens a pore (d) in the ventricle (F) of the brain, allowing animal spirit to flow out through hollow tubes in the nerves that go to the legs. The spirit inflates the muscles, causing the foot to withdraw. The details may be wrong, but the concept of an automatic reflex is clear enough.

Significant advances in our understanding of reflexes were made by one of the founders of modern physiology, Sir Charles Sherrington, around the beginning of this century. Rather than try to understand the movements of intact animals, a pretty daunting task, Sherrington attempted instead to reduce animal behaviour to its simplest elements by means of surgical operations, in which higher centres of the brain were removed under anaesthesia. Such animals cannot learn, but behave in an automatic way that conforms largely to Descartes' assumptions. With appropriate sensory stimulation they produce a wide variety of behaviours from stepping to grooming or scratching movements, from swallowing to righting behaviour. Sherrington suggested that such reflexes formed "the unit reaction in nervous integration" (1906). We now know there is also a central nervous component to most behaviours, though, as subsequent chapters of this book make clear, it is difficult to overestimate the importance of peripheral feedback on motor control.

As discussed in chapter 8, muscle spindles are extremely sophisticated proprioceptors that can be presumed to play an important role in motor control. One of the most influential ideas on what this role might be was proposed by Merton in 1953. Fig. 2 shows Merton's original drawing of the two possible motor pathways by which the central nervous system might be suggested to control movement. One pathway, involving the alpha motor neurones which innervate the main body of the

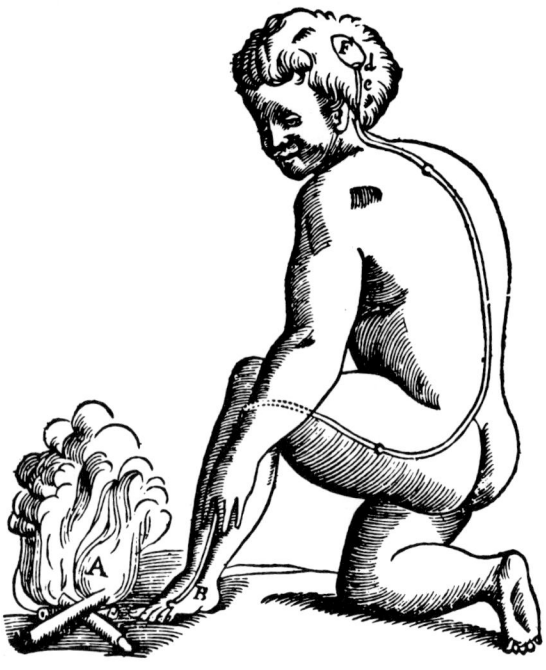

Fig. 1. An illustration from Descartes' book 'Traité de l'Homme' showing how reflex withdrawal of a limb from fire might be accomplished. The fire (A) causes a tiny thread (c) to be pulled which opens a pore (d) in the ventricle of the brain (F). Fluid flows out to the muscles of the foot (B), causing it to be withdrawn.

muscle, would be used for urgent movements, while the other roundabout route would be the normal pathway for normal movements. This second route involves the gamma motor neurones which innervate the muscle spindles. Contraction of the spindle intrafusal muscle fibres activates the spindle Ia afferents which in turn activate the alpha motor neurones. The main body of the muscle would then contract until it reached the length of the spindle intrafusal fibres. This degree of contraction would be maintained, since any increase in muscle contraction would lead to unloading of the spindles and thus a decrease in sensory activity and vice versa. This scheme of movement control is called a follow-up length servosystem.

To summarise 30 years of research in just a few words, it is probably fair to say that nobody now believes that movements are controlled in precisely this way. For instance, the stretch reflex component of this servosystem (spindle afferents synapsing with homonymous motor neurones) has long-latency components that are thought to involve either a trans-cortical loop (Marsden et al., 1976) or secondary spindle

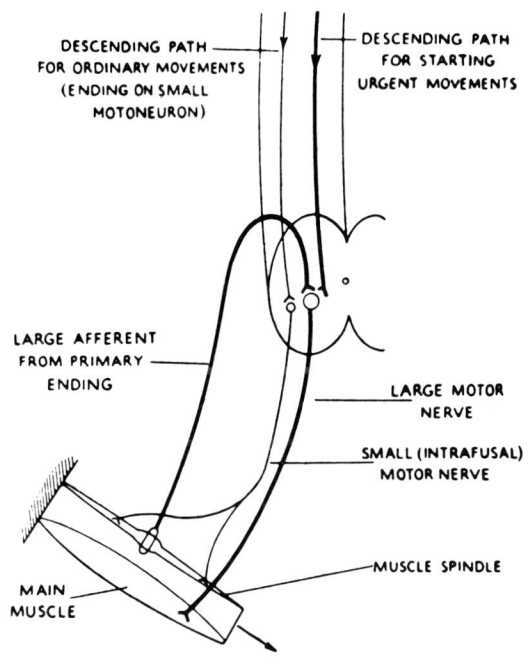

Fig. 2. Merton's original diagram of the two possible motor pathways by which movements might be controlled by the central nervous system. (From Merton, 1953).

afferents (Matthews, 1984a,b,c). It has also been shown that, during voluntary movements, alpha motor neurones are usually activated at the same time as gamma motor neurones (e.g. Sears, 1964; Severin et al., 1967), not subsequently as the servo-control hypothesis would predict. However, the suggestion that movements are servo-assisted rather than servo-controlled is still very much alive.

For spindles to play an effective role in adjusting for perturbations, it is important both that the gain of the reflex is high enough to influence the degree of muscle contraction to a reasonable extent, and that the spindles can indeed measure changes in joint angle produced by external disturbances. This latter question is addressed by Prof Rack in the chapter which follows. He shows that tendons are elastic and describes the implications of this for motor control. Such tendon elasticity also results in increased locomotor efficiency, for energy conservation by elastic storage in tendons occurs, not only in hopping animals such as kangaroos (Proske, 1980), but also in man (McMahan & Greene, 1978).

Turning to invertebrate animals on which I and the authors of chapter 15, Drs Watson and Burrows, do our research, let me ask why we choose to study lower animals. As zoologists, it is quite appropriate that part of the reason lies in our interest in the adaptations of animals. Taking an example from my own area of research, locomotion in decapod crustaceans, the reasons why, for instance, crabs usually walk sideways rather than forwards like most animals. But since such problems increasingly cut no ice with grant-giving bodies, it is fortunate that there is another compelling reason for our choice of experimental animal. This is that many invertebrates, notably gastropod molluscs, leeches, decapod crustaceans and insects, are in a variety of ways useful models of the more complex vertebrate system. This is not to say that their nervous systems are simple as the Scottish Electrophysiological Society suggested when it held its first international symposium in Glasgow in 1973 on 'Simple' Nervous Systems. Rather their nervous systems might be described as tractable, in that I think we stand more chance of understanding how they function than we do the nervous system of any higher vertebrate.

They possess at least two important advantages. First, they possess individually identifiable neurones that can be recognised in preparation after preparation. Secondly, there is relative economy in their nervous systems. This applies particularly to their motor systems, where, in insects and crustaceans, most muscles are innervatd by between 3 and 15 motor neurones (Wiersma & Ripley, 1952; Theophilidis & Burns, 1983).

Over the last fifteen years, techniques have been developed that allow intracellular recordings to be made from the neuropile of arthropod central nervous systems, while at the same time dye or heavy metal ion injection procedures have been evolved (Stretton & Kravitz, 1968; Pitman et al., 1972; Stewart, 1978) which permit correlated anatomical studies of the penetrated neurones. Dr Burrows and his colleagues have been in the forefront of this research. They have made significant advances in our understanding of the organisation of insect central nervous systems and have begun to unravel neural networks controlling a variety of simple behaviour patterns. One of their more surprising findings is that many of the local interneurones do not generate action potentials, but release transmitter in a graded fashion according to their level of depolarisation (Burrows & Siegler, 1976). Such non-spiking local interneurones are morphologically rather similar to another class of local interneurones which do spike. The former are involved in controlling and coordinating locomotory and postural activity at a local level (Burrows, 1980; Siegler, 1981a,b), while the latter seem to be involved in primary sensory integration and are the interneuronal component of local reflexes (Burrows & Siegler, 1982; Siegler & Burrows, 1983).

The tracer horseradish peroxidase (Muller & McMahan, 1976; Snow et al., 1976) can be visualised under the electron microscope following its injection into a neurone. This permits correlated ultrastructural and physiological studies to be undertaken, which greatly increase our understanding of neuronal networks by providing information on the distribution of synapses. Such a study of local reflexes in the locust forms the basis of Watson and Burrows's chapter. It illustrates very forcibly the complexity of even the simplest of reflexes and the many similarities that exist between the reflexes of invertebrates and vertebrates.

REFERENCES

Burrows, M. (1980). The control of sets of motoneurones by local interneurones in the locust. J. Physiol., 298, 213-233.

Burrows, M. & Siegler, M.V.S. (1976). Transmission without spikes between locust interneurones and motoneurones. Nature, Lond., 262, 222-224.

Burrows, M. & Siegler, M. (1982). Spiking local interneurons mediate local reflexes. Science, N.Y., 217, 650-652.

Descartes, R. (1680). Traité de l'Homme.

Marsden, C.D., Merton, P.A. & Morton, H.B. (1976). Servo action in the human thumb. J. Physiol., 257, 1-44.

Matthews, P.B.C. (1984a). Evidence from the use of vibration that the human long-latency stretch reflex depends upon spindle secondary afferents. J. Physiol., 348, 383-415.

Matthews, P.B.C. (1984b). The contrasting stretch reflex responses of the long and short flexor muscles of the human thumb. J. Physiol., 348, 545-558.

Matthews, P.B.C. (1984c). Observations on the time course of the electromyographic response reflexly elicited by muscle vibration in man. J. Physiol., 353, 447-461.

McMahan, T.A. & Greene, P.R. (1978). Fast running tracks. Scient. Am., 112, No.6, 112-121.

Merton, P.A. (1953). Speculations on the servo-control of movement. In The Spinal Cord. Ed. Wolstenholme G.E.W., Churchill, London, pp.247-255.

Muller, K.J. & McMahan, U.J. (1976). The shapes of sensory and motor neurones and the distribution of their synapses in ganglia of the leech: a study using intracellular injection of horseradish peroxidase. Proc. R. Soc. B, 194, 481-499.

Pitman, R.M., Tweedle, C.D. & Cohen, M.J. (1972). Branching of central neurons: intracellular cobalt injection for light and electron microscopy. Science, N.Y., 176, 412-414.

Proske, U. (1980). Energy conservation by elastic storage in kangaroos. Endeavour, N.S., 4, 148-153.

Sears, T.A. (1964). Efferent discharges in alpha and fusimotor fibres of intercostal nerves in the cat. J. Physiol., 174, 295-315.

Severin, F.V., Shik, M.L. & Orlovsky, G.N. (1967). Work of the muscles and single motor neurones during controlled locomotion. Biophysics., 12, 762-772.

Sherrington, C. (1961). The Integrative Action of the Nervous System. Yale University Press, New Haven. (Reprint of the original 1906 edition).

Siegler, M.V.S. (1981a). Posture and history of movement determine membrane potential and synaptic events in nonspiking interneurons and motor neurons of the locust. J. Neurophysiol., 46, 296-309.

Siegler, M.V.S. (1981b). Postural changes alter synaptic interactions between nonspiking interneurons and motor neurons of the locust. J. Neurophysiol., 46, 310-323.

Siegler, M.V.S. & Burrows, M. (1983). Spiking local interneurons as primary integrators of mechano-sensory information in the locust. J. Neurophysiol., 50, 1281-1295.

Snow, P.J., Rose, P.K. & Brown, A.G. (1976). Tracing axons and axon collaterals of spinal neurons using intracellular injection of horseradish peroxidase. Science, N.Y., 191, 312-313.

Stewart, W.W. (1978). Functional connections between cells as revealed by dye-coupling with a highly fluorescent naphthalimide tracer. Cell, 14, 741-759.

Stretton, A.O.W. & Kravitz, E.A. (1968). Neuronal geometry: determination with a technique of intracellular dye injection. Science, N.Y., 162, 132-134.

Theophilidis, G. & Burns, M.D. (1983). The innervation of the mesothoracic flexor tibiae muscle of the locust. J. exp. Biol., 105, 373-388.

Wiersma, C.A.G. & Ripley, S.H. (1952). Innervation patterns of crustacean limbs. Physiologia comp. Oecol., 2, 391-405.

Chapter Fourteen

STRETCH REFLEXES IN MAN: THE SIGNIFICANCE OF TENDON COMPLIANCE

P.M.H.Rack

There can be no doubt that incoming sensory information plays a major part in shaping motor activity. There remains, however, uncertainty and disagreement about the level of the nervous system at which these reflex adjustments may occur, and about the roles of the different sensory receptors.

During the last 40 years, dramatic advances in servo-engineering have directed attention toward the simpler 'low-level' reflexes. In particular, it has sometimes been thought that stretch reflexes might behave as elementary position controlling servomechanisms. Physiologists have often thought in terms of some model such as Fig. 1A; afferent activity from muscle spindles was considered to convey information about limb position, which was fed back to motor neurones in such a way that the resulting muscular activity corrected for unintended displacements. In this way it was supposed that the muscle spindles might play the dominant role in reflex control and adjustment of position or movement.

Position controlling servomechanisms of this type are well known to engineers, and at first sight such a system would appear to have many advantages for the neural control of limbs. A spinal reflex which could automatically maintain position, and could translate relatively simple neural 'instructions' into the appropriate movements would save the

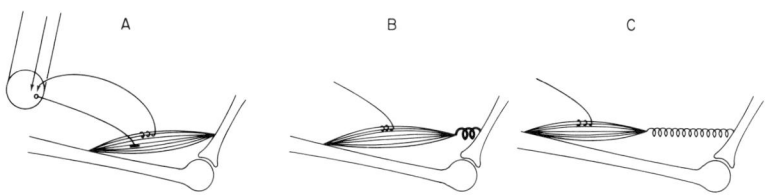

Fig. 1. For explanation see text.

higher centres of the nervous system from much computational work. The wide distribution of the spindles, and the richness of their sensory innervation supported the idea that they might have this important function, and the fusimotor fibres could act either as an input to the reflex pathway, or as a gain setting mechanism.

There are, however, features of the real biological system which do not tally with this simple model, and we are now obliged to re-think some aspects of this servocontrol hypothesis. A system such as that shown in Fig. 1A could only provide a reliable control of position if the sensors (the muscle spindles) were to 'see' changes in position accurately, and if their signals were transmitted rapidly through a 'high gain' pathway to achieve the necessary corrections. In fact, these conditions are not fulfilled: when compared with the movements to be controlled, the neural pathways are relatively slow (Rack, 1981), the gain around the reflex pathway is usually rather small (Matthews, 1972; Vallbo, 1974; Brown et al., 1982), and, as I shall show in the present paper, the muscle spindles cannot always 'see' a reliable representation of position or movement.

Muscles are coupled to bone by elastic tendons
Unlike Fig. 1A, most mammalian muscles are in fact attached to bone by a tendon at one or both ends, and these tendons have elastic properties. In Figs. 1B and 1C the model has been modified to include this spring-like attachment. In Fig. 1B the tendon is represented as a short and rather stiff spring, and this model would have properties rather similar to Fig. 1A; an imposed movement would still be transmitted to the muscle spindles with only slight attenuation, and any shortening of the muscle fibres would still lead to an almost equivalent movement of the bones. Models such as Figs. 1A or 1B are quite compatible with the idea that the muscle spindles may act as the sensors in a position-controlling servomechanism.

However, the situation looks very different in Fig. 1C, where the tendon is represented as a long and relatively compliant spring. If an external movement were applied to the bony attachment, much of it might be lost in this springy tendon, so that only a relatively small movement would reach the muscle spindles. When the muscle fibres of Fig. 1C contracted, they could do so at the expense of the springy tendon, and could thus shorten with little or no movement of the bones. If Fig. 1C were a realistic representation of the actual situation then the muscle spindles could hardly be expected to provide an adequate representation of the joint position, though they could of course signal the lengths of the muscle fibres.

It will be clear from Fig. 1 that we have a problem that is essentially quantitative; if the tendon is always very stiff compared with the muscle fibres, then the stretch reflex physiologist can probably ignore it, but if tendons have a compliance that is comparable to or greater than the muscle fibres, then some re-thinking will be necessary.

In larger animals (such as man), muscle fibres are often considerably shorter than their tendons of attachment, and the sum of the muscle fibres has an effective cross sectional area that is much larger than that of the tendon. These anatomical features are obvious in some of the forearm muscles whose tendons extend down into the digits (Fig. 2A). Many other limb muscles also have surprisingly long tendinous attachments, though they do not always appear as well-defined tendons at one or other end of the muscle. Fig. 2B is a diagram to show the pennate arrangement of the fibres in human triceps surae; when knee and ankle are both at 90°, each fibre is only .30-55mm long, but it is coupled to bone by a tendon of insertion

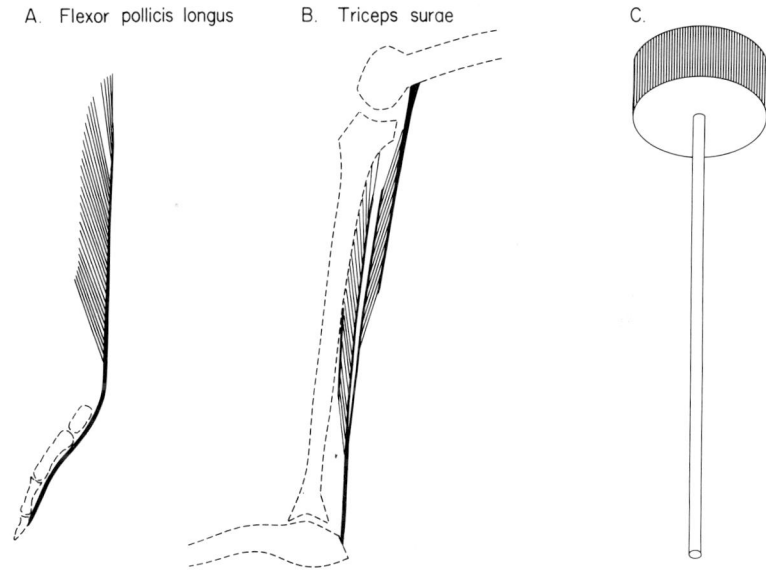

Fig. 2. A: The relation of muscle fibres to tendon in human flexor pollicis longus. B: diagram of the human triceps surae complex, consisting of gastrocnemius and soleus muscles connected to heel by means of the Achilles tendon. C: the muscle fibres and tendons of gastrocnemius and soleus have been 're-arranged' to show their relative lengths and cross-section areas. Data from Alexander & Vernon (1975), and Rack et al. (1983).

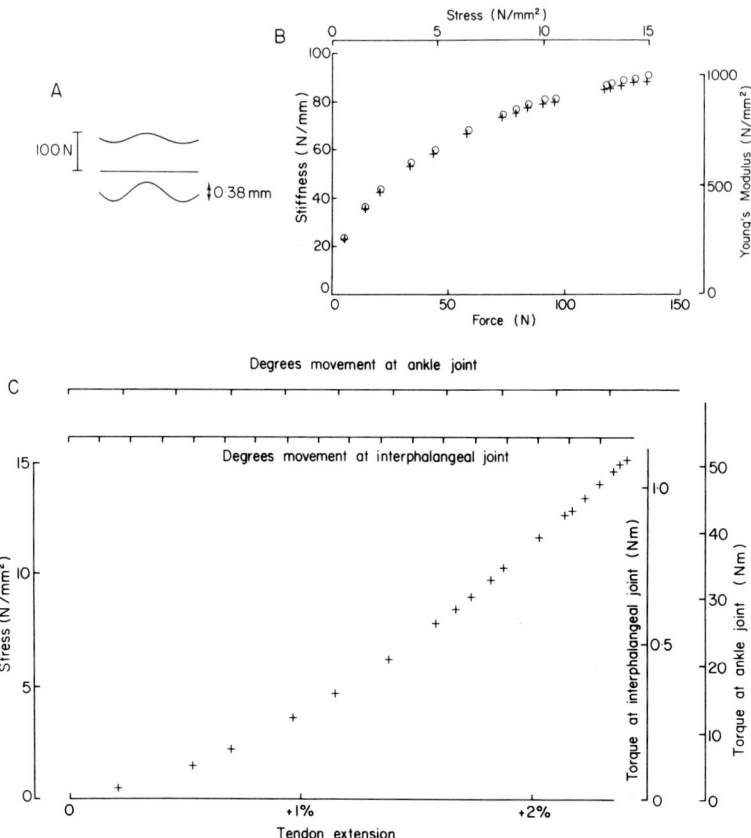

Fig. 3. Mechanical properties of a human flexor pollicis longus tendon.
A: Force and length records obtained during sinusoidal stretching of 100mm of tendon (cross section area 9.1 mm^2).
B: The stiffness of the tendon (change in force/change in length) has been plotted as a function of the mean force. At each force o indicates the average stiffness for all the cycles of frequencies 10-14 Hz; + is from cycles at 2-6 Hz. At the top of the figure the mean force is expressed as a stress; on the right the stiffness has been expressed as a Young's modulus. (Reproduced from Rack & Ross, 1984).
C: The same data as in B has been re-plotted as a length-tension diagram (by numerical integration of compliance with respect to force). The lower scale shows extension, and the left-hand scale shows stress. Above and to the right the inner scales relate to the flexor pollicis longus tendon; the horizontal scale shows the angular joint movement which would correspond to the tendon extensions shown below, while the vertical scale shows the torque at the joint which would correspond to

and a tendon of origin which together add up to a length of 300-400mm.

In Fig. 2C the muscle fibres and tendons of Fig. 2B have been "re-arranged" to show their relative proportions. The muscle fibres are less than one sixth of the length of the tendon, but the cross-section area of the combined muscle fibres is about 140 times larger than that of the tendon. (Data from Alexander & Vernon, 1975; and Rack et al. 1983). Diagrams such as Figs. 2A & C suggest that we have to consider the possibility that tendons may yield significantly under load.

The mechanical properties of tendons
Although there have been numerous studies of the mechanical properties of tendons (eg. Rigby et al, 1959; Diamant et al, 1972; Woo et al., 1980, Ker, 1981), most of the measurements have been made on tendons that were subjected to relatively large stretching forces. Therefore the information which is available in the literature cannot be used to predict the behaviour of a tendon during many everyday activities. In succeeding paragraphs I shall describe some of the properties of tendons when they are subjected to more moderate forces. An account of the experimental methods and the detailed results has already appeared (Rack & Ross, 1984).

1) Flexor pollicis longus tendons were obtained from human subjects during post mortem examinations carried out shortly after death (with the relatives consent), and the mechanical properties of these tendons were investigated. Lengths of tendon were securely clamped at each end, and mounted in a machine which imposed controlled stretching movements while recording the resulting force changes. Repeated sinusoidal stretches were used (see Fig. 3A) at a range of different frequencies, and with a range of different initial loads.

2) Measurements of formalin-fixed cadavers established the relationship between the movement of this tendon and the angular movements of the interphalangeal joint of the thumb.

3) Normal conscious human subjects undertook an experiment in which the interphalangeal joint of the thumb was

---

each tendon stress. (Each millimetre of tendon movement was taken to correspond to 7.45° of joint movement). The uppermost and right-most scales relate to the ankle joint. These have been calculated by assuming that the material of the Achilles tendon has a similar stress/strain relation to the tendon of flexor pollicis longus. The following data were required to construct these scales: the Achilles tendon lies 45-50mm from the joint axis, has a cross-section area of about 70 mm^2 (see Alexander and Bennet-Clark 1977), and each muscle fibre acts through 300-400mm of tendinous fibres.

driven through sinusoidal flexion-extension movements, while records were made of the resisting forces (see Fig. 4). By using mechanical and anatomical information obtained from the post mortem specimens, it was then possible to predict how much of each imposed movement would have reached the muscle fibres.

Elastic properties of the flexor pollicis longus tendon
Measurements of fresh specimens confirmed that the tendons are essentially elastic. During each cycle of a sinusoidal stretching movement, the force increased approximately in phase with length (Fig. 3A), and alterations in the frequency of stretching between 2 and 16Hz made little difference to this force. These are the results that one would expect of a spring-like structure. However, unlike a linear spring, tendons become stiffer (less compliant) as the load upon them is increased. A given extension was thus met by a larger increment in force when the initial tension on the tendon was already large. This result is illustrated in Fig. 3B, where the resistance to extension is plotted as a function of the mean force. The stress and the Young's modulus of the tendon material may be read from scales above and to the right of the figure. By integration of the data (or by direct measurements) one can obtain a stress/strain curve (Fig. 3C), which is similar to ones obtained for other tendons (Ker, 1981; Woo et al, 1980).

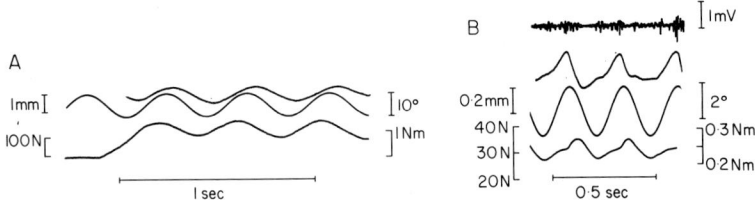

Fig. 4. Mechanical properties of the interphalangeal joint of the thumb.
A: The joint was driven through sinusoidal flexion-extension movements at 2.8Hz (middle trace). Percutaneous tetanic stimulation of the flexor pollicis longus was followed by a rise in flexing force (lower record) to about 1Nm (130N at the tendon). The upper record shows the movement of the flexor pollicis longus muscle after subtraction of the movement 'lost' in the tendon. Tendon stiffness taken to be 90N/mm (see Fig. 3B).
B: The joint was moved at 4Hz, while the subject exerted a 1/5 maximal mean flexing torque (0.22Nm which is 29N at the tendon). Force and length records are as in A; tendon stiffness (from Fig. 3B) taken to be 54N/mm. The uppermost record is an electromyogram from flexor pollicis longus.

Examination of the interphalangeal joints of a number of thumbs showed that each millimetre movement of the tendon was equivalent to an angular movement of $7.2°$ - $7.7°$. This measurement enables one to express extension of the tendon in terms of angular movement at the joint, and this has been done in Fig. 3C (above the figure); the force on the tendon can also be converted to a torque at the joint (scale on the right). Those two scales then indicate the torque/angle relationship that one would see if the tendon were securely anchored at the point where the most distal muscle fibres are inserted, so that movement was only possible by extension of this free distal part of the tendon.

Figure 3C gives some insight into the effects that the tendon compliance may have on physiological control of joint position. The torques in this figure are within the working range of the thumb; a normal subject can exert 1.0Nm for a short time, and sustain 0.25Nm for a number of minutes. Fig. 3C shows that an increase in torque from zero to 0.25Nm would be accompanied by almost a millimeter stretch in the tendon, which is equivalent to more than $7°$ of movement at the joint. Measurements of the tendon thus indicate that even if a stretch reflex from the muscle spindles were to have sufficient gain to give a very good control of the muscle fibre length, the joint would still yield considerably in response to this 0.25Nm load. The free distal part of the tendon would account for more than $7°$ of movement, so that the total movement attributable to the tendon alone would be appreciably larger.

Stretch reflexes in thumb and finger muscles

Other long flexor and extensor muscles of the fingers and thumb have fibre arrangements that are generally similar to the flexor pollicis longus, and their mechanical properties are presumably similar to those shown in Fig. 3. Thus, apart from the small intrinsic muscles of the hand, our fingers appear to be controlled by muscles which act through relatively compliant couplings. In considering the behaviour of these muscles, we can therefore quite reasonably refer to such diagrams as Fig. 1C.

Since the spindles in these muscles will only see that part of the movement which reaches the muscle fibres, a stretch reflex that arises from the spindles alone could only control the lengths of these muscle fibres, and even if the gain of this reflex were very high indeed, it could not (on its own) control joint position in a way that was proof against displacing forces. When these muscles are active, some of the movement of the muscle fibres may be 'lost' in stretching the tendons, and would thus fail to appear as movements of fingers or thumb; a compliant tendon would in this way reduce the precision with which the nervous system can control the positions of the digits.

At first sight this compliance of the coupling between muscle and joint would seem to be a shortcoming, but in other ways it may be an advantage: by absorbing part of the movement, the tendon provides an immediate buffer which reduces the changes in force that might accompany either a sudden muscle contraction or an unexpected external movement. A compliant tendon thus makes it 'easier' to maintain a steady and controlled force, in which minor fluctuations are ironed out, though it appears to add to the problems of maintaining an accurate position. We use our fingers and thumb as the two components of a pincer for gripping and handling, and on many occasions it is probably more important to maintain an accurately controlled force than to have precise control of actual position.

Tendons are compliant when they are under low tension, but stiffer when forces are high (Fig. 3); hence, when the gripping pressure is small the fingers and thumb will meet displacing movements with only a small increase in force, but a more forcible grip will be correspondingly more resistant to any disturbance. It is tempting to think that this relationship between force and stiffness contributes to the delicacy with which we can handle fragile objects, but still permits a secure power grip.

The tendons of other limb muscles

The part of the Achilles tendon which extends beyond the muscle fibres is accessible for direct measurements, but one must use indirect methods of estimating the mechanical properties of the more proximal parts which lie on the surfaces of gastrocnemius and soleus. Each of the muscle fibres is attached through approximately the same length of tendon; the lower fibres have long tendons of origin, whereas the upper fibres act through long tendons of insertion, (Fig. 2B). In order to make definite calculations, I shall make simplifying assumptions that loads are distributed uniformly through the different parts of the gastrocnemius-soleus complex, and that the tendon has the same effective cross-section area irrespective of whether the fibres are gathered together at the heel or spread over the surfaces of the muscles (Rack et al. 1983); this second assumption is supported by results obtained from cat soleus muscles in which it was possible to make more complete measurements (Rack & Westbury, 1984).

The information which is available suggests that different tendons are made of essentially similar material in which strain is related to stress in the same way. The main differences in their mechanical properties can thus be presumed to arise from differences in their lengths and their cross section areas. By assuming that the Achilles tendon is made of the same material as the tendon of flexor pollicis longus, it is possible to re-calculate the scales above and to the right of Fig. 3C in a way that indicates how far the tendinous attachments of the human triceps surae would yield with particular loads on the ankle joint. This calculation led to the uppermost and right-most scales on Fig. 3C; by reading the data points in relation to these scales, one can see how torque at the ankle joint would change with each extension of the Achilles tendon, and how angular movement at the joint would relate to tendon extension. (For numerical details see the legend to Fig. 3).

A number of assumptions were involved in constructing the ankle scales of Fig. 3C, so they must be regarded as approximate. However, it is notable that a force sufficient to raise the body weight on the ball of one foot (perhaps 70Nm) takes us right off the figure, and would involve much more than 2% elongation of the tendon. In order to take up this 'give' in the tendon, the muscle fibres would need to shorten by more than 8mm (15-20% of their initial length) before they began to lift the body off the ground. Once again, it will be clear that the muscle spindle lengths bear no simple relation to the joint position, and a reflex mechanism that was based only on muscle spindle afferents could hardly be expected to provide adequate control of the position of the ankle joint.

Tendon properties complicate the investigation of reflex responses

In recent years, measurements of the responses to imposed joint movements have often been used to investigate the stretch reflexes of conscious human subjects (eg. Hammond, 1960; Marsden et al, 1972; Lee & Tatton, 1982; Matthews 1984). In many experiments the perturbation was applied while the subject was already exerting some voluntary torque; in that situation the muscle fibres would have an appreciable stiffness, so that the amount of the movement which reached the muscle spindles would have depended on the relative stiffness of muscle fibres and tendon.

The movement of muscle fibres and spindles would not, however, be merely an attenuated version of the joint movement. Muscle fibres have quite different mechanical properties from tendons; whereas tendons are essentially elastic, the force/velocity relation of active muscle fibres impedes any rapid change in their length, and they behave as a visco-elastic material. An extension of the muscle-tendon combination would therefore be distributed between muscle fibres and tendon in a way that reflects the difference in their mechanical properties. The tendon would absorb most of the faster movements, whereas slower movements would be more completely transmitted to the muscle spindles.

The waveform of the movement that reaches the spindles may thus be quite different from the movement imposed at the joint. A 'step' extension of the joint will lead to a much less abrupt movement of the muscle spindles, and a sinusoidal movement of the joint will be transmitted to the spindles with both attenuation and phase change. From a knowledge of the tendon properties, it is possible to compute the muscle movements which would accompany given movements of the thumb; this has been done in Fig. 4.

The relationship between joint movement and muscle length was relatively straightforward when the muscle was continuously stimulated so that there could be no reflex modulation of its activity; this was the situation in Fig. 4A. During sinusoidal movement of the joint (middle trace) the flexor pollicis longus muscle was stimulated percutaneously to give a tetanic contraction, and a high flexing force (lower trace). The force rose to approximately 1Nm (130N at the tendon) and fluctuated sinusoidally with the movement, but the maximum force preceded maximum extension by about $40°$. Along with these force changes there would have been changes in the tendon length: reference to Fig. 3B indicates that with this level of mean force the tendon would have yielded by about 0.011mm/N (90N/mm). By subtracting this stretching of the tendon from the total, we obtained the calculated movement of the muscle itself (the uppermost record of Fig. 4A). It will be noted that the movement of the muscle was about 70% of the movement of the muscle-tendon combination at the joint, but it was delayed by $40°$ behind the joint movement, and approximately $80°$ behind the force fluctuations. Similar results were obtained when the muscle was continuously activated in a maximum voluntary contraction.

When a vigorous stretch reflex modulated muscle activity within each cycle of movement, this added a further complication. In Fig. 4B when the subject was exerting a smaller voluntary force, and the flexor pollicis longus electromyogram showed a burst of activity in response to each lengthening movement. This e.m.g. activity was followed by shortening of the muscle at a time when the extension movement at the joint was barely completed, so that the muscle movement now preceded the imposed joint movement, which in turn preceded the force fluctuations. Thus, in Fig. 4B the situation was quite different from Fig. 4A, and the presence of an active stretch reflex has altered the distribution of movement between muscle and tendon.

Fig. 4B illustrates the response to only one frequency of movement; when the frequency was increased the reflex delay occupied more of the cycle of movement and the timing of the activation was correspondingly different. The relationship between the muscle movement and joint movement thus changed with frequency in ways that complicated any analysis of reflex performance (Rack et al. 1983; Rack & Ross, 1984).

What does the spindle stretch reflex control?
It will be clear that stretch reflexes arising from the spindles of active muscles cannot provide the direct low-level

servocontrol of limb position that has sometimes been attributed to them. What then can they do?

The location of the spindles enables them to 'see' changes in the length of the muscle fibres, so that monosynaptic stretch reflexes arising from the spindles would tend to prevent or reduce sudden changes in the lengths of these fibres; estimates of the gain of the thumb stretch reflex (force/length) show that for frequencies of movement up to about 5Hz, the stretch reflex is indeed much more effective in controlling muscle fibre length than in controlling joint position (Rack & Ross, 1984). Thus, in considering the human stretch reflex it is probably useful to think of some arrangement such as Fig. 5, in which the monosynaptic reflex from muscle spindles has a rapid effect on muscle fibre activity, and thereby plays a part in controlling the proximal end of the relatively compliant tendon. The more precise control of joint position presumably depends on signals from other receptors acting through longer neural pathways.

In the larger terrestial animals (such as ourselves) compliant tendons are used to conserve energy during locomotion, and to achieve an optimal coupling between muscles and load in jumping and throwing movements (Cavagna, 1970).

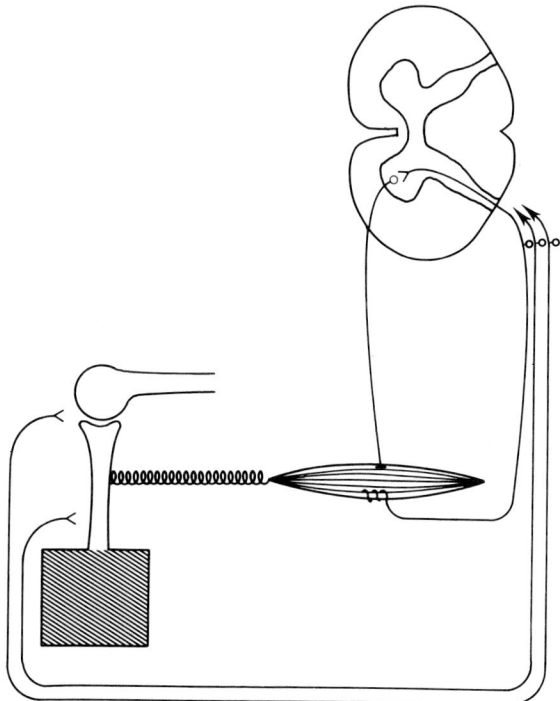

Fig. 5. For explanation see text.

The precise control of the proximal ends of the tendons must therefore be a matter of extreme importance, and it should be no surprise that the muscle spindles are particularly highly developed and richly innervated.

REFERENCES

Alexander, R.McN. & Bennet-Clark, H.C. (1977). Storage of elastic strain energy in muscles and other tissues. Nature, Lond., 265, 114-117.

Alexander, R.McN. & Vernon, A. (1975). The dimensions of knee and ankle muscles and the forces they exert. J. Hum. Mov. Stud., 1, 115-123.

Brown, T.I.H., Rack, P.M.H. & Ross, H.F. (1982). A range of different stretch reflex responses in the human thumb. J. Physiol., 332, 101-112.

Cavagna, G.A. (1970). Elastic bounce of the body. J. appl. Physiol., 29, 279-282.

Diamant, J., Keller, A., Baer, E., Litt, M. & Arridge, R.G.C. (1972). Collagen: ultrastructure and its relation to mechanical properties as a function of ageing. Proc. R. Soc. B, 180, 293-315.

Hammond, P.H. (1960). An experimental study of servo-action in human muscular control. In Proceedings of 3rd conference on medical electronics. Institution of Electrical Engineers, London, pp. 190-199.

Ker, R.F. (1981). Dynamic tensile properties of the plantaris tendon of sheep (Ovis aries). J. exp. Biol., 93, 283-302.

Lee, R.G., & Tatton, W.G. (1982). Long latency reflexes to imposed displacements of the human wrist: dependence on duration of movement. Expl. Brain Res., 45, 207-216.

Marsden, C.D., Merton, P.A. & Morton, H.B. (1972). Servo-action in the human thumb. J. Physiol., 257, 1-44.

Matthews, P.B.C. (1972). Mammalian muscle receptors and their central action. Edward Arnold, London.

Matthews, P.B.C. (1984). Evidence from the use of vibration that the human long-latency stretch reflex depends upon spindle secondary afferents. J. Physiol., 348, 545-558.

Rack, P.M.H. (1981). Limitations of somatosensory feedback in control of posture and movement. In Handbook of Physiology, Section 1, The Nervous System, Vol. 2, Motor Control. Ed. Brooks, V.B., American Physiological Society, Bethesda, pp. 229-256.

Rack, P.M.H. & Ross, H.F. (1984). The tendon of flexor pollicis longus; its effects on the muscular control of force and position at the human thumb. J. Physiol., 351, 99-110.

Rack, P.M.H., Ross, H.F., Thilmann, A.F. & Walters, D.K.W. (1983). Reflex responses at the human ankle: the importance of tendon compliance. J. Physiol., 344, 503-524.

Rack, P.M.H., & Westbury, D.R. (1984). Elastic properties of the cat soleus tendon and their functional importance. J. Physiol., 347, 479-495.

Rigby, B.J., Hirai, N., Spikes, J.D. & Eyring, H. (1959). The mechanical properties of rat tail tendon. J. gen. Physiol., 43, 265-283.

Vallbo, A.B. (1974). Human muscle spindle discharge during isometric voluntary contraction. Amplitude relations between spindle frequency and torque. Acta physiol. Scand., 90, 319-336.

Stretch reflexes in man: the significance of tendon compliance

Woo, S.L.-Y, Ritter, M.A., Amiel, D., Sanders, T.M., Gomez, M.A., Kuei, S.C., Garfin, S.S. & Akeson, W.H. (1980). The biomechanical and biochemical properties of swine tendons - long term effects of exercise on digital extensors. Connect. Tissue Res., 7, 177-183.

Chapter Fifteen

THE SYNAPTIC BASIS FOR INTEGRATION OF LOCAL REFLEXES IN THE LOCUST

A.H.D. Watson and M. Burrows

The object of this paper is to show how electrophysiological and ultrastructural studies of identified neurones can be combined to investigate the organisation of the neural pathways that underlie local reflexes of the legs in the locust. We will discuss our results in the light of what is known of vertebrate reflexes and consider whether there are common features to the organization of reflex pathways in these two different nervous systems.

In highly organised regions of the brain such as the cerebellum or retina of a vertebrate, or the optic lobes of an insect, it is possible to recognise classes of neurone in the electron microscope without recourse to selective staining, and thus correlate structure with function. Many regions of the brain do not, however, have such a regular structure, and a correlation cannot therefore be made in this way. The advent of staining methods in which compounds such as Horseradish Peroxidase (HRP) are injected intracellularly into neurones through the recording electrode, has opened these regions of the nervous system for studies correlating the physiology of neurones with the distribution and structure of their synapses. This powerful approach is particularly valuable in the nervous systems of certain invertebrates where many neurones can be characterised anatomically and physiologically as individuals. Such an identified neurone may play a unique role in the behaviour of the animal. A knowledge of its fine structure will contribute not only to an understanding of this role, but will also test preconceptions of the meaning of synaptic structure and distribution. Our study of reflexes in the locust demonstrates the potential complexity of even the simplest behaviour and emphasises that an evaluation of this complexity requires careful interpretation of both structure and function.

Reflexes in invertebrates and vertebrates
  The simplest reflexes in both vertebrates and invertebrates are those mediated by only two neurones; a

Integration of local reflexes in the locust

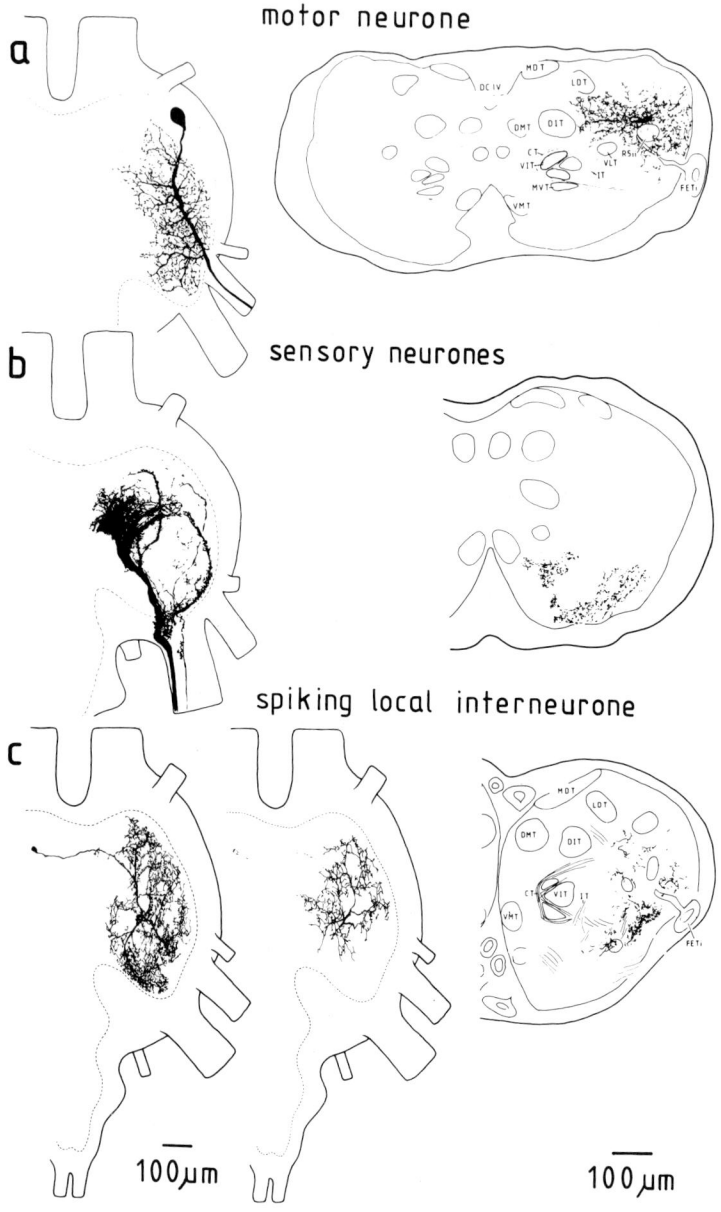

sensory afferent and a motor neurone. In the vertebrate stretch reflex, which has probably received more attention than any other, axon terminals of Ia afferent sensory neurones synapse directly onto homonymous alpha motor neurones. In a study which elegantly combined anatomical and physiological methods, Redman and Walmsley (1983a,b) labelled these sensory neurones and the motor neurones with which they were shown to connect physiologically. Each Ia afferent makes only a small number (sometimes as few as three) synaptic contacts on a given motor neurone, and the actual position of these on the dendrites corresponds to the positions predicted from electrotonic analysis. At the base of the locust wing hinge there is a stretch receptor that is excited during elevation of the wing in flight. This single sensory neurone makes synapses, which on physiological criteria appear to be monosynaptic, onto wing depressor motor neurones (Burrows, 1975). In the cockroach leg, sensory neurones from a hair plate which are excited when the femur is flexed, similarly make direct synaptic connections with a motor neurone that causes femoral extension (Pearson, et al., 1976). All three examples are of negative feedback loops, which result in the reduction of the stimulus that evoked the initial sensory discharge.

Most reflex pathways in both vertebrates and invertebrates are, however, more complex and involve three or more neurones. In vertebrates, Ia afferents also reflexly inhibit motor neurones to antagonistic muscles via an interposed inhibitory interneurone (Eccles et al., 1956). A propriospinal neurone receiving direct input from Ia afferents (Hultborn et al., 1971) and whose spikes lead to short latency inhibitory post synaptic potentials (IPSPs) in alpha motor neurones (Jankowska & Roberts, 1972) was subsequently revealed

---

Fig. 1. The three components of a local reflex pathway in the locust Schistocerca gregaria (Forskal). a) Motor neurone; the slow extensor tibiae. b) Sensory neurones; the central terminals of some hair afferents in the tibia. c) Spiking local interneurone; one that evokes IPSPs in the slow extensor motor neurone when certain tibial hairs are moved. The motor and interneurone were stained by the intracellular injection of cobalt, the sensory neurones by backfilling from the cut ends of their axons. All stains were intensified with silver. The drawings on the left are of one half of the metathoracic ganglion viewed from the dorsal surface, with the outline of the neuropile dashed. The branches of the interneurone are shown in two drawings, the ventral branches on the left, the dorsal ones on the right. The right hand column of drawings are of three adjacent and superimposed 10μm transverse sections taken at approximately the same level for all three types of neurone. The tracts and commissures are labelled according to the scheme of Tyrer and Gregory (1982).

Integration of local reflexes in the locust

by intracellular dye injection (Jankowska & Lindstrom, 1972). The pathway for recurrent inhibition of spinal motor neurones also appears to be a three neurone arc (Eccles et al., 1962) in which the propriospinal Renshaw neurone is the central element. In the locust three neurone reflex pathways mediating local adjustments of single legs have recently been described (Burrows & Siegler, 1982; Siegler & Burrows, 1983). The three neurones of these reflexes are; 1) sensory afferent,

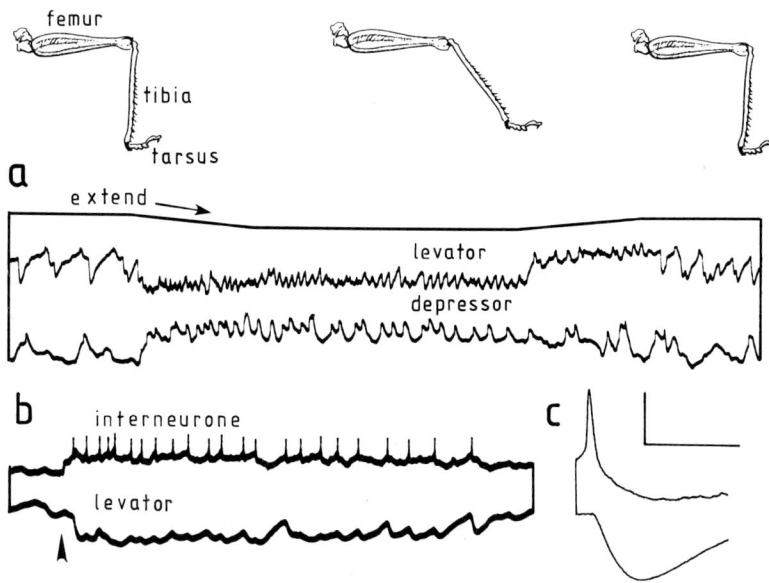

Fig. 2. A local reflex of a hind leg of the locust. If the tibia is forcibly extended, the tarsus compensates by depressing. a) Extension of the tibia from a femoro-tibial angle of 90° to 130°, as indicated in the diagrams and monitored in the first trace, evokes an inhibition in the levator tarsi motor neurone (second trace) and excitation with spikes in a depressor tarsi motor neurone (third trace). The recordings, from the same animal but not made simultaneously, are from cell bodies. b) A spiking local interneurone and the levator tarsi motor neurone recorded simultaneously. Touching hairs on the distal tibia (as here, arrow) or extending the tibia, evokes spikes in the interneurone and IPSPs in the motor neurone. c) Signal averaging (64 sweeps) shows that an interneurone spike evokes an IPSP in the motor neurone with a latency which suggests a direct connection. Calibration: vertical (a,b) levator 8mV, depressor 16mV, interneurone 4mV; horizontal (a,b) 400ms, (c) 10ms.

2) spiking local interneurone, 3) motor neurone (Fig. 1). It is the cellular nature of these reflexes as revealed by combined physiological and ultrastructural methods that will now be considered.

Three neurone reflex pathways in the locust
Segmental ganglia in the thorax of the locust contain the somata of a few thousand interneurones and motor neurones in a cortex that surrounds the neuropile where synaptic interactions take place. The somata of most sensory neurones are in the peripheral nervous system, close to the sensory structure they innervate. Within the ganglionic neuropile, several discrete regions and many tracts and commissures can be reliably recognised (Tyrer & Gregory, 1982). The labelled branches of identified neurones can therefore be mapped accurately and described in relation to other structures in the ganglion.
  Each ganglion acts with considerable autonomy in the control of the legs of its segment. For example, a forcible extension of the tibia of a hind leg leads to a depression of the tarsus of the same leg (Burrows & Horridge, 1974). In this reflex the depressor tarsi motor neurones are excited, while the single levator tarsi motor neurone is inhibited (Fig. 2a). The reflex persists even when the metathoracic ganglion is isolated from the rest of the nervous system (Burrows & Horridge, 1974). Numerous other specific local reflexes are activated by stimulation of particular groups of sensory hairs and campaniform sensilla on a hind leg.
  The sensory neurones mediating these reflexes do not synapse directly upon the motor neurones, but instead a spiking local interneurone is interposed. These small interneurones have central processes restricted to one ganglion which overlap the sensory terminals of leg afferents and the branches of leg motor neurones (Fig. 1). They are excited when a single hair on a particular region of the leg is moved (Fig. 3a). Each afferent spike from a hair evokes an excitatory postsynaptic potential (EPSP) in a particular interneurone (Fig. 3b,c) (Siegler & Burrows, 1983). In this way the entire surface of the leg is mapped onto a set of these interneurones, each of which has a different and restricted receptive field. In addition, an interneurone may also synapse directly onto a restricted number of motor neurones (Burrows & Siegler, 1982). Thus in the tarsal reflex, extension of the tibia, or mechanical stimulation of a group of hairs on the distal end of the tibia, depolarizes a particular spiking local interneurone and causes it to spike (Fig. 2b). Each of the spikes then evokes an IPSP in the levator tarsi motor neurone (Fig. 2b,c). To date, the number of motor neurones found to be connected to any one spiking local interneurone is small. Where a reflex requires the coordinated activity of several motor neurones, at least one

Integration of local reflexes in the locust

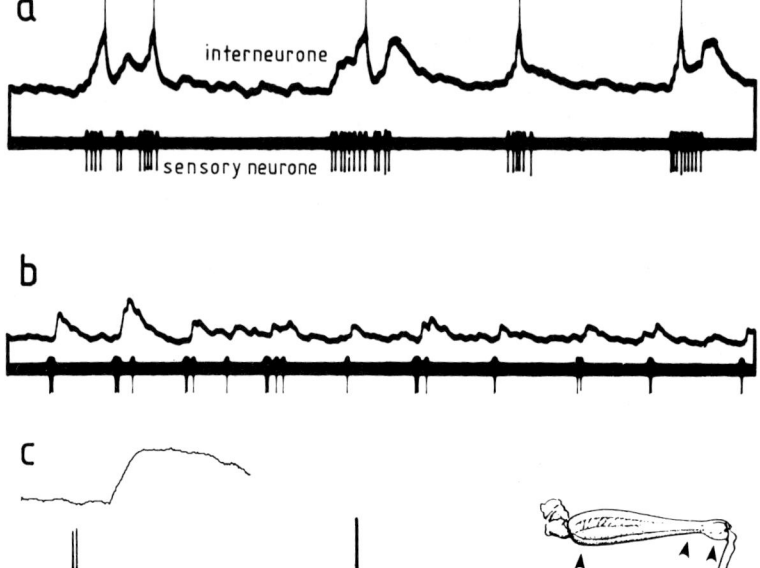

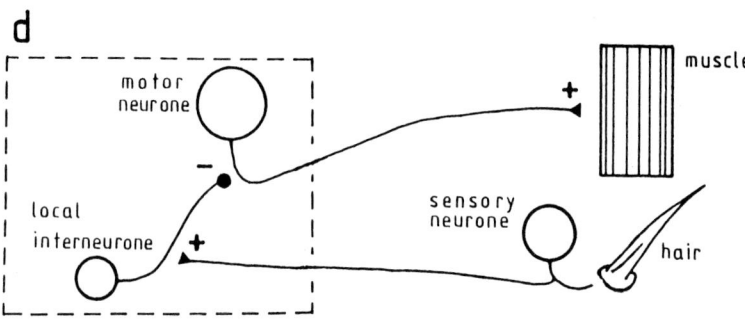

Fig. 3. Connections between sensory neurones and a local interneurone that underlie a local reflex in the locust. Intracellular recordings are from the cell body of an interneurone, and extracellular recordings are from the axon of a hair afferent in the tibia. The interneurone is held hyperpolarised with a steady current injected through the recording electrode. a) Moving a hair evokes a burst of spikes in its afferent, and a depolarisation and spikes in the interneurone. b) More hyperpolarising current is injected into the interneurone to enhance the

additional link will be required. This may be one of the non-spiking local interneurones, which control sets of motor neurones in the locust thoracic nervous system (Burrows, 1980). We will now consider the synaptic relationships of each of the three elements involved in these reflexes.

Sensory afferents
The central projections of the sensory neurones from hairs on the leg are difficult to stain by backfilling, due to the length of their axons, and the axons themselves are too small to penetrate with intracellular electrodes near to the ganglion. The terminals of afferents from similar hairs on the ventral surface of the locust thorax have, however, been successfully stained with HRP and examined in the electron microscope (Watson & Pflüger, 1984). The axon terminals of these neurones in the ventral neuropile of the prothoracic ganglion (Pflüger & Tautz, 1982) are in a similar position to those of sensory neurones from leg hairs that synapse upon spiking local interneurones in the metathoracic ganglion (Siegler & Burrows, 1983).

The most striking feature of the sensory hair afferents is that input as well as output synapses are present, often no more than $1\mu m$ apart (Fig. 4a,b). Some of the contacts are reciprocal (Fig. 4). The output synapses are characterised by the presence of a row of presynaptic densities approximately 85nm apart and by spherical electron lucent vesicles approximately 37nm in diameter (37.2±0.4 S.E., n=266). Two classes of processes can be identified in the surrounding neuropile. The first contains spherical, clear vesicles which are statistically identical in size to those in the labelled terminals, and probably therefore represent other afferent terminals. The second contains significantly larger spherical vesicles, and is the only type to make synapses onto the afferent terminals. This implies that direct interactions between primary afferents do not take place.

The presence of input synapses onto primary afferent terminals is a widespread phenomenon; indeed there are few ultrastructural descriptions of afferents where inputs are not recorded. For example, in invertebrates they are described on afferents from fly tarsal sensilla (Geisert & Altner, 1974), locust wing stretch receptors (Altman et al., 1980), and for leech sensory neurones (Muller, 1979; Muller & McMahan, 1976). In the vertebrate spinal cord, inputs from both axons and

---

amplitude of EPSPs evoked by the afferent spikes. c) Signal averaging (64 sweeps) indicates that the connection is probably direct. d) Diagram of the pathways in the reflex inhibition of the levator tarsi motor neurone. Calibration: vertical 4mV; horizontal (a,b) 200ms, (c) 22ms.

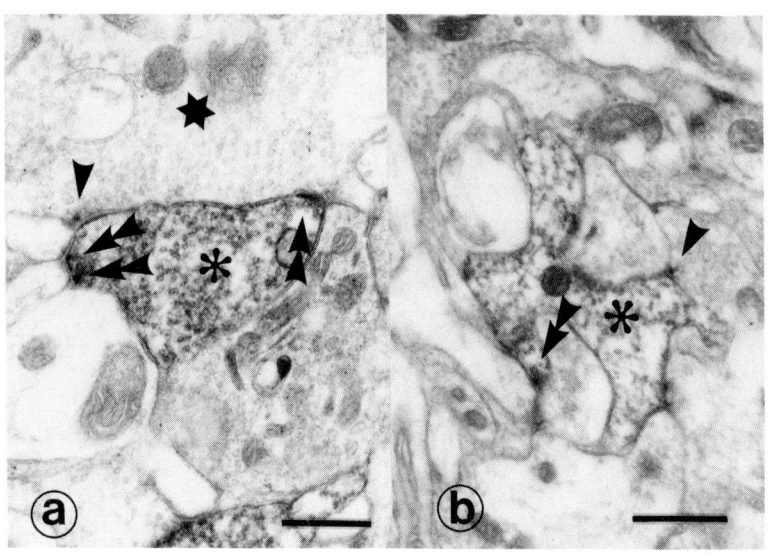
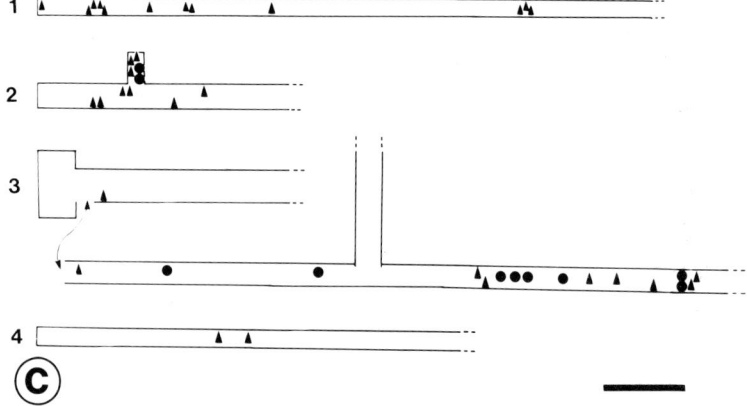

Fig. 4. Sensory neurones innervating hairs on the prosternum of the locust, labelled with HRP. a) A spine (asterisk) containing synaptic vesicles which makes three output synapses (double arrows). One of the postsynaptic processes (star) makes a reciprocal synapse (single arrow) back onto the sensory terminal. Scale bar = 0.5μm. b) A terminal (asterisk) that receives one input (arrow) and makes one output synapse (double arrow). Scale bar = 1μm. c) A diagram of a reconstruction from serial sections, of terminal branches (numbered 1-4) showing the distribution of input (circles) and output (triangles) synapses. Scale bar = 5μm.

dendrites occur on the terminals of Ia afferents (Fyffe & Light, 1984), and on a variety of sensory neurones innervating the skin (Rethelyi et al., 1982; Maxwell et al., 1982; Semba et al., 1983; Ralston et al., 1984). These ultrastructural observations imply that there is the means for the primary sensory information to be modified before it reaches the first synapse within the central nervous system, and that these effects may be important in reflex behaviour. The most widely observed physiological manifestation of this is presynaptic inhibition (Eccles et al., 1962; Takeuchi & Takeuchi, 1966; Krasne & Bryan, 1973; Kennedy et al., 1974). It can be brought about either by an increase in sodium conductance causing a depolarisation, or by an increase in chloride conductance which leads to a depolarisation or hyperpolarisation depending on the relative values of the chloride equilibrium potential and the resting potential in the terminals (Eccles et al., 1962; Takeuchi & Takeuchi, 1966; Rudomin, 1980).

The terminals presynaptic to vertebrate spinal afferents predominately contain flattened vesicles which have elsewhere been associated with inhibitory synapses (Uchizono, 1967; Atwood & Morin, 1970). In the crayfish, inhibitory neurones can act postsynaptically on the muscle fibres and presynaptically on the terminals of the excitatory axons. At each type of synapse flattened synaptic vesicles are present and GABA is released, resulting in an increase in the chloride permeability of the postsynaptic membrane.

If presynaptic inhibition is caused by a depolarisation resulting from changes in sodium permeability there would be no reason to suppose that the vesicles should be flattened. The observation that vesicles presynaptic to the sensory hair afferents in the locust are round does not, therefore, exclude them from a role in presynaptic inhibition. However, synapses on sensory terminals can also mediate presynaptic facilitation or 'dishabituation' (Kandel et al., 1976; Bailey et al., 1981). Facilitation of Ia afferent output occurs in the cat, but it is not certain if this is brought about by synapses directly onto the afferent terminals (Rudomin et al., 1974).

Spiking local interneurones
The spiking local interneurones found so far to process sensory inputs from a hind leg of a locust have two distinct fields of branches (Fig. 1c). The first lies close to the ventral surface of the ganglion in the region of the neuropile to which the terminals of the leg afferents project (Siegler & Burrows, 1983). It is composed of a small number of large diameter (2-5μm) branches from which a dense feltwork of fine (0.5μm or less) processes arise. It is linked to the second field by a single, glial wrapped process that runs, along with those from many other interneurones of the same type, in a particular tract. The second field consists of main branches

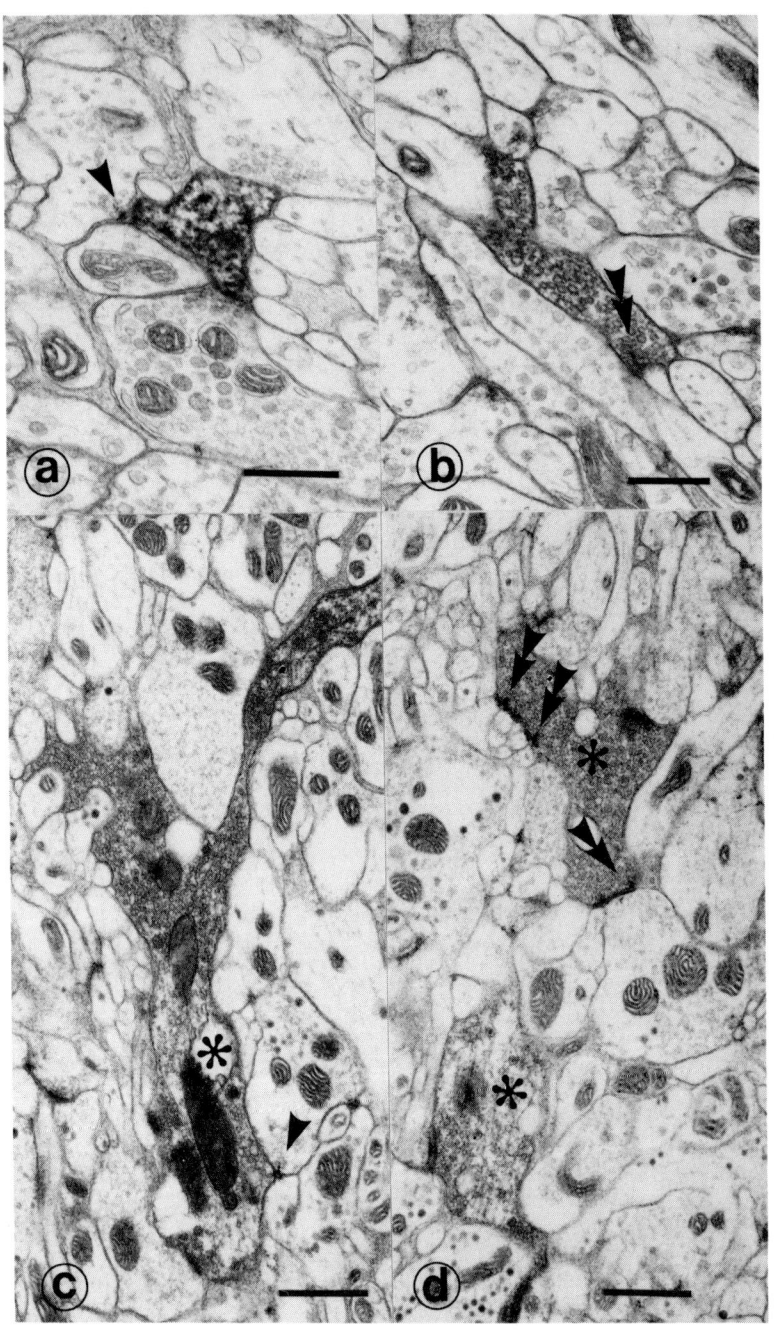

Integration of local reflexes in the locust

1-3μm in diameter which divide less profusely, into side branches of uneven diameter, and which have a varicose appearance (Siegler & Burrows, 1984).

The electron microscope reveals that the ventral field receives mostly input synapses that are found mainly on the smallest diameter branches (Fig. 5a). Not all of the inputs are from sensory neurons, as some come from profiles containing large dense-core vesicles of a type not seen in afferent terminals. Output synapses are also made from the ventral field though they are much less frequent (Fig. 5b). In the fine branches these synapses often occur at regions of increased diameter containing synaptic vesicles. In the larger branches the vesicles are clustered close to the presynaptic membrane, but are not associated with an increase the diameter of the branch.

The dorsal field makes mostly output synapses (Fig. 5d). The changes in branch diameter which are visible in the light microscope can be followed through serial sections in the electron microscope. Branches which at one point are only 0.2μm in diameter may, within a few microns, increase to 1-2μm. These varicosities are generally filled with synaptic vesicles and may be the site of as many as twenty output synapses. Input synapses are also made onto these varicosities but in much smaller numbers (Fig. 5c). For example, a varicosity that made eighteen output synapses received only two inputs.

It therefore appears that the ventral branches are predominantly, but not exclusively the sites of synaptic input and that the dorsal branches are predominantly, but not exclusively, the sites of synaptic output. The inputs onto the dorsal branches, though few in number, are strategically placed to control the much greater number of outputs. The physiological correlate of this could be the context dependence of the leg reflexes in the locust. This is clearly of behavioural importance because reflexes must be tailored appropriately to the posture of the animal or the phase of walking. Inputs onto the terminals of the spiking local interneurone would be one way of achieving this goal.

There is little ultrastructural information available for equivalent interneurones in vertebrates. The dendrites of neurones thought to be Renshaw cells have been examined in the electron microscope and appear to be postsynaptic only

---

Fig. 5. A spiking local interneurone labelled with HRP. a) A ventral branch receives an input synapse (arrow). Scale bar = 0.5μm. b) An output synapse (double arrow) from a small diameter ventral branch. Scale bar = 0.5μm. c,d) Sections from the same dorsal branch (asterisk), which in (c) receives a single input (arrow) and in (d) makes several output synapses (double arrows). Scale bars = 1μm.

(Lagerbäck & Ronnevi, 1982). It has been proposed that spinocervical tract cells might, in addition to transmitting information along the spinal cord, mediate spinal reflexes through their axon collaterals (Rastad et al., 1977), on which output synapses only were observed (Rastad, 1981).

Motor neurones

Invertebrate muscle fibres can be supplied by three different types of neurone. 1) Conventional motor neurones which produce EPSPs in a muscle fibre. Many muscle fibres do not support action potentials so that the force they produce depends upon the size and temporal pattern of EPSPs, and upon the activation of neurones with different properties. Individual motor neurones are classed as 'fast' or 'slow' in arthropods depending on the response they evoke in a muscle, although the types form a continuum. 2) Inhibitory motor neurones which produce IPSPs in a muscle fibre or inhibit the excitatory motor neurones presynaptically. 3) Modulatory

TABLE 1

A Comparison of Motor Neurone Pool Sizes in Mammals and Arthropods

Muscle	Motor Neurones		
	Excitatory	Inhibitory	
Man			
External Rectus.	2,970	–	Feinstein et al., 1955
Platysma.	1,096	–	
Medial Gastrocnemius.	579	–	
Lumbricales.	96	–	
Stapedius	1,160	–	Fullerton et al., 1983
Crustacea: Brachyura.			
Abductor of dactylopodite.	1	2	Wiersma & Ripley, 1952
Adductor of dactylopodite.	2	1	
Flexor of carpopodite.	4	1	
Insecta: Locust			
Extensor tibiae.	2	1	Hoyle & Burrows, 1973
Flexor tibiae.	9	2	Phillips, 1981

neurones which have little direct effect on the membrane potential of a muscle fibre, but which potentiate the force produced by the excitatory motor neurones. They also have potentiating effects on the terminals of the excitatory motor neurones (O'Shea & Evans, 1979). In general, few motor neurones innervate a particular muscle (see Table 1) so that most are identifiable according to their action, and their morphology within the central nervous system. By contrast, a vertebrate muscle is innervated only by excitatory motor neurones, but there are large numbers of them (Table 1). Gradation of muscular force is brought about by recruiting more motor neurones which stimulate additional sets of muscle fibres.

The differences in motor innervation between invertebrates and vertebrates have consequences for reflex organisation. For example, in invertebrates the production of force by a muscle may require the simultaneous activation of excitatory motor neurones and modulatory neurones, and inhibition of inhibitory neurones. Some motor neurones may be set aside for particular actions. For example, the fast motor neurone to the extensor tibiae muscle (FETi) of a locust hind leg is used only in jumping and kicking. In the cockroach, some motor neurones may be reserved for particular reflex functions, and may not contribute to the normal generation of locomotory movements (Zill & Moran, 1982).

Ultrastructural examination of locust motor neurones reveals that while the branches of some are exclusively postsynaptic (Fig. 6a,b), those of others are also presynaptic (Fig. 6b-d) (Watson & Burrows, 1982). Only one leg motor neurone, the FETi, has so far been found to possess output synapses (Fig. 6a,b). Its branches are involved in extremely complex arrays of serial and reciprocal contacts with surrounding processes (Fig. 6e). Physiological evidence indicates the FETi makes a direct excitatory chemical synapse with some motor neurones of the antagonistic flexor muscle (Hoyle & Burrows, 1973). This connection is important during defensive kicks or jumps in which the extensor tibiae muscle provides most of the power. To achieve this it must co-contract with the flexor muscle, an outcome that is ensured by the synaptic coupling (Heitler & Burrows, 1977). By contrast, the slow excitatory and the inhibitory neurone to the same muscle make no ganglionic outputs, and the modulatory neurone makes very few (Watson, 1984).

The dendrites of vertebrate motor neurones do not make chemical output synapses. Close membrane appositions and occasional gap junctions have, however, been observed (Sotelo & Taxi, 1970; Matthews et al., 1971), that may form the substrate for short latency (0.2ms) connections between motor neurones (Gogan et al., 1977). In the frog, these electrical coupling potentials apparently exist only between motor neurones innervating the same muscle, or between those with

Integration of local reflexes in the locust

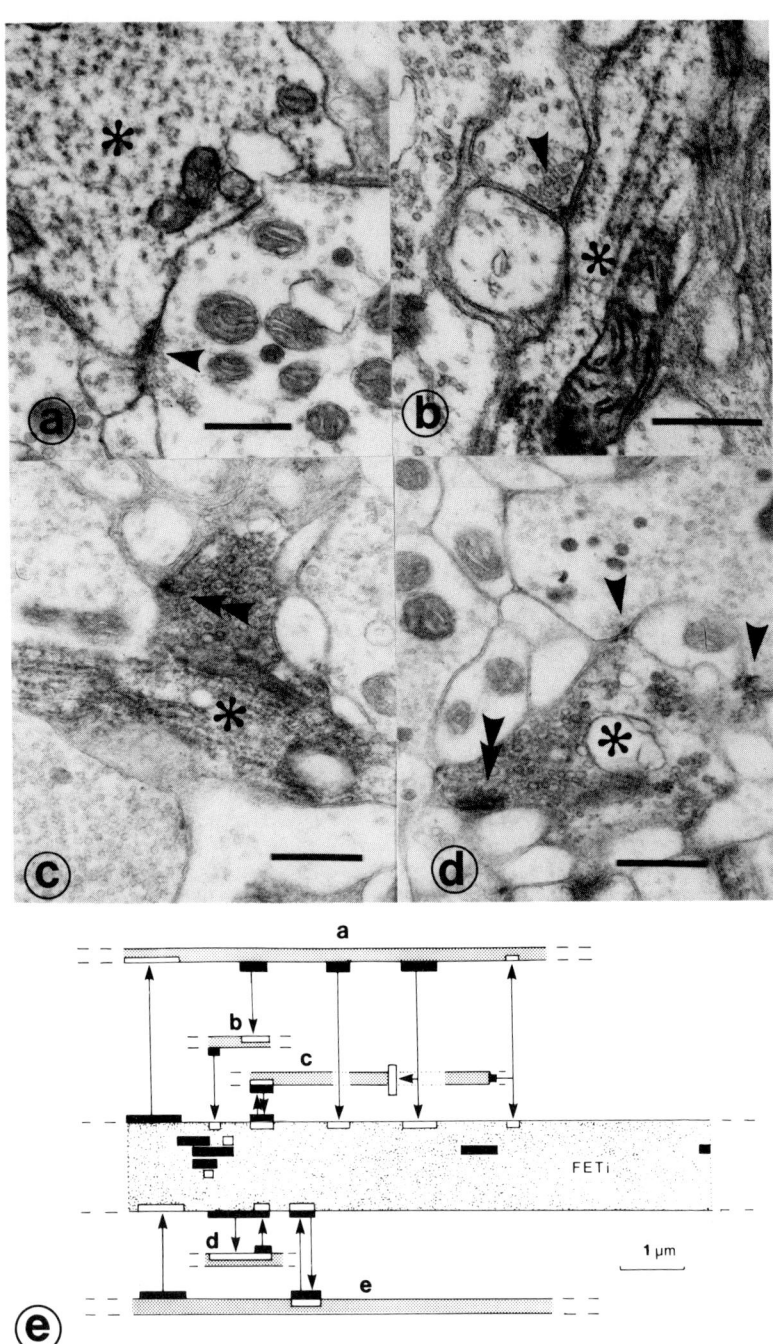

related functions (Westerfield & Frank, 1982; Collins, 1983). They are therefore seen as a means of coordinating the activity of functionally homologous motor neurones. In insects, electrical coupling between motor neurones is rare. The only extant example is between a supernumerary levator tarsi motor neurone of the locust and its normal homologue (Siegler, 1982). Even where large numbers of motor neurones innervate the same muscle, as in the wing depressor muscles of the dragonfly (Simmons, 1977), electrotonic contacts are not present. In other invertebrate phyla electrical coupling is more common. In the buccal ganglion of the mollusc Navanax, electrotonic coupling between pharyngeal motor neurones is explained by the necessity for vigorous and well coordinated contractions of the pharyngeal musculature in this active predator (Spira et al., 1980). In the leech, bilateral pairs of motor neurones innervating the longitudinal and the annulus erector muscles are electrically coupled to ensure that the muscles on each side of the midline contract together (see Muller et al., 1982). In the lobster stomatogastric ganglion, the electrical junctions between some of the motor neurones are an integral part of central pattern generating mechanisms (Eisen & Marder, 1982).

Many vertebrate motor neurones synapse chemically onto other neurones within the spinal cord by way of axon collaterals. Some of these collaterals are assumed to synapse onto Renshaw cells mediating recurrent inhibition (Lagerbäck et al., 1981b). In addition, labelling pairs of motor neurones for electron microscopy shows that the collaterals of one may synapse, perhaps in an excitatory fashion, onto the dendrites of the other (Lagerbäck et al., 1981a).

Though there are similarities between vertebrate and invertebrate motor neurones in that they may make chemical output synapses within the central nervous system, there are qualitative differences in the likely action of these synapses. In vertebrate motor neurones, the terminals of the axon collaterals will be invaded by a propagated orthodromic spike, whereas the output synapses of a locust motor neurone will be invaded by an antidromic spike that spreads decrementally. Moreover, because of the juxtaposition of

---

Fig. 6. Motor neurones labelled with HRP (asterisks). a,b) Input synapses (arrows) onto the slow extensor tibiae motor neurone. Scale bars = 0.5μm. c,d) Input (single arrows) and output synapses (double arrows) on central processes of a tergosternal motor neurone. Scale bars = 0.5μm. e) A diagrammatic reconstruction from serial sections of a branch of the fast extensor tibiae motor neurone (FETi) and five adjacent processes (a-d) which synapse with it. The black bars show the lengths of the presynaptic densities at each synapse, and the white bars represent input synapses.

input and output synapses in a locust motor neurone two further factors must be considered. First, the spike may directly alter synaptic inputs by way of reciprocal synapses and local circuit interactions, and hence influence the probability of subsequent motor spikes. Second, the output synapses may function in a graded fashion independent of spike activity. In this type of action each output will be more strongly influenced by neighbouring input synapses than by distant ones. The appearance of complex circuits like the one in Fig. 6e must be tempered by the observation that a physiological connection between two neurones may be composed of many tens or hundreds of anatomical synaptic sites (King, 1976, Watson & Burrows, 1983). It cannot be assumed that the anatomical complexity of Figure 6e represents an equally complex physiological relationship without some knowledge of the distribution of synapses from a particular source over the dendritic field of the motor neurone. Furthermore, the synaptic relationships of the motor neurones (and spiking interneurones) must reconcile the exigences of two modes of operation; simultaneous activation of output synapses after an action potential and independent activation in local circuit interactions.

Conclusions
The greatest similarities between the neurones of reflex arcs in invertebrates and vertebrates are found in the sensory afferents. Examination of their ultrastructure firmly establishes the principle that gating of sensory input at the first synapse within the central nervous system may be both widespread and of considerable importance. Comparisons of the interneurones mediating reflexes are made difficult by the problems of identifying and characterising them in vertebrates. In the locust it is clear that presynaptic 'dendrites' in both local spiking interneurones and motor neurones could play a significant role in the control of reflex motor output. Local circuit interactions between such 'dendrites' would, in general, not be mediated by action potentials. In more complex reflex pathways involving more than three neural elements, non-spiking interneurones coordinate groups of motor neurones to different muscles (Burrows, 1980). Non-spiking interactions therefore play an important part in motor control in insects. They allow a greater flexibility of response than would otherwise be possible from a small group of motor neurones innervating each muscle. Further flexibility is achieved by integrating at the sarcolemma itself the actions of several different types of neurone. By contrast, the activity of vertebrate muscles with their large motor pools can be finely graded by regulating motor neurone recruitment entirely within the central nervous system.

Despite the great potential for flexibility of response

in the insect thoracic nervous system, reflexes are still predictable if the behavioural context remains the same. The flexibility allows a range of discrete motor responses, with some scope for modulation, to originate from the same elements. A behavioural reflex should be seen not as a single arc but as a series of parallel interacting arcs that control several muscles and the different types of neurone that innervate each participating muscle.

Acknowledgements
This work was supported by a grant from the SERC (U.K.) and by NIH grant NS16058 to M.B.

REFERENCES

Altman, J.S., Shaw, M.K. & Tyrer, N.M. (1980). Input synapses onto a sensory neurone revealed by cobalt-electron microscopy. Brain Res., 189, 245-250.
Atwood, H.L. & Morin, W.A. (1970). Neuromuscular and axo-axonal synapses of the crayfish opener muscle. J. Ultrastruct. Res., 32, 351-369.
Bailey, C.H., Hawkins, R.D., Chen, M.C. & Kandel, E.R. (1981). Interneurons involved in mediation and modulation of gill-withdrawal reflex in Aplysia. IV. Morphological basis of presynaptic facilitation. J. Neurophysiol., 45, 340-360.
Burrows, M. (1975). Monosynaptic connexions between wing stretch receptors and flight motoneurones of the locust. J. exp. Biol., 62, 189-219.
Burrows, M. (1980). The control of sets of motoneurones by local interneurones in the locust. J. Physiol., 298, 213-233.
Burrows, M. & Horridge, G.A. (1974). The organization of inputs to motoneurones of the locust metathoracic leg. Phil. Trans. R. Soc. B, 269, 49-94.
Burrows, M. & Siegler, M.V.S. (1982). Spiking local interneurons mediate local reflexes. Science, N.Y., 217, 650-652.
Collins, W.F. (1983). Organisation of electrical coupling between frog lumbar neurones. J. Neurophysiol., 49, 730-744.
Eccles, J.C., Fatt, P. & Landgren, S. (1956). Central pathway for direct inhibitory action of impulses in largest afferent fibres to muscle. J. Neurophysiol., 19, 75-95.
Eccles, J.C., Schmidt, R.F. & Willis, W.D. (1962). Presynaptic inhibition of the spinal monosynaptic pathway. J. Physiol., 161, 282-297.
Eisen, J.S. & Marder, E. (1982). Mechanisms underlying pattern generation in lobster stomatogastric ganglion as determined by selective inactivation of identified neurons. III. Synaptic connections of electrically coupled pyloric neurons. J. Neurophysiol., 48, 1392-1415.
Feinstein, B., Lindegard, B., Nyman, E. & Wohlfart, G. (1955). Morphologic studies of motor units in normal human muscles. Acta anat., 23, 127-142.
Fullerton, B.C., Joseph, M.P., Guinan, J.J. & Norris, B.E. (1983). Stapedius motoneurons in the cat. Soc. Neurosci. Abstr., 9, 1085.
Fyffe, R.E.W. & Light, A.R. (1984). The ultrastructure of group 1a afferent

fibre synapses in the lumbosacral spinal cord. Brain Res., 300, 200-210.
Geisert, B. & Altner, H. (1974). Analysis of the sensory projection from the tarsal sensilla of the blowfly (Phormia teranovae Rob.-Desv., Diptera). Cell Tissue Res., 150, 249-259.
Gogan, P., Gueritaud, J.P., Horcholle-Bossavit, G. & Tyc-Dumont, S. (1977). Direct excitatory interactions between spinal motoneurones of the cat. J. Physiol., 272, 755-767.
Heitler, W.J. & Burrows, M (1977). The locust jump. I. The motor programme. J. exp. Biol., 66, 203-219.
Hoyle, G. & Burrows, M. (1973). Neural mechanisms underlying behavior in the locust Schistocerca gregaria. I. Physiology of identified motorneurons in the metathoracic ganglion. J. Neurobiol., 4, 3-41.
Hultborn, H., Jankowska, E. & Lindstrom, S. (1971). Recurrent inhibition of interneurones monosynaptically activated by 1a afferents. J. Physiol., 215, 613-636.
Jankowska, E. & Lindstrom, S. (1972). Morphology of interneurones mediating 1a reciprocal inhibition of motoneurones in the spinal cord of the cat. J. Physiol., 222, 805-823.
Jankowska, E. & Roberts, W.J. (1972). An electrophysiological demonstration of the axonal projections of single spinal interneurones in the cat. J. Physiol., 222, 597-622.
Kandel, E.R., Brunelli, M., Byrne, J. & Castellucci, V. (1976). A common presynaptic locus for the synaptic changes underlying short term habituation and sensitization of the gill withdrawal reflex of Aplysia. Cold Spring Harb. Symp. quant. Biol., 60, 465-482.
Kennedy, D., Calabrese, R.L. & Wine, J.J. (1974). Presynaptic inhibition: Primary afferent depolarization in crayfish neurons. Science, N.Y., 186, 451-454.
King, D.G. (1976). Organisation of crustacean neuropile: I. Patterns of synaptic connections in lobster stomatogastric ganglion. J. Neurocytol. 5, 207-237.
Krasne, F.B., & Bryan, J.S. (1973). Habituation: regulation through presynaptic inhibition. Science, N.Y., 182, 590-592.
Lagerbäck, P-A. & Ronnevi, L-O. (1982). An ultrastructural study of serially sectioned Renshaw cells. II. Synaptic types. Brain Res., 246, 181-192.
Lagerbäck, P-A., Ronnevi, L-O., Cullheim, S. & Kellerth, J-O. (1981a). An ultrastructural study of the synaptic contacts of alpha motoneurone axon collaterals. I. Contacts in lamina IX and with identified alpha motoneurone dendrites in lamina VII. Brain Res., 207, 247-266.
Lagerbäck, P-A., Ronnevi, L-O., Cullheim, S. & Kellerth, J-O. (1981b). An ultrastructural study of the synaptic contacts of alpha motoneurone axon collaterals. II. Contacts in lamina VII. Brain Res., 222, 29-41.
Matthews, M.A., Willis, W.D. & Williams, V. (1971). Dendritic bundles in lamina IX of cat spinal cord: a possible source for electrical interaction between motoneurons? Anat. Rec., 171, 313-328.
Maxwell, D.J., Bannatyne, B.A., Fyffe, R.E.W. & Brown, A.G. (1982). The ultrastructure of hair follicle afferent fibre terminations in the spinal cord of the cat. J. Neurocytol., 11, 571-582.
Muller, K.J. (1979). Synapses between neurons in the central nervous system of the leech. Biol. Rev., 54, 99-134.

Muller, K.J. & McMahan, U.J. (1976). The shapes of sensory and motor neurons and the distribution of their synapses in the ganglia of the leech: a study using intracellular injection of HRP. Proc. R. Soc. B, 194, 481-499.

Muller, K.J., Nicholls, J.G. & Stent, G.S. (Eds.) (1982). Neurobiology of the leech. Cold Spring Harbor Labs., Cold Spring Harbor, New York.

O'Shea, M. & Evans, P.D. (1979). Potentiation of neuromuscular transmission by an octopaminergic neurone in the locust. J. exp. Biol., 79, 169-190.

Pearson, K.G., Wong, R.K.S. & Fourtner, C.R. (1976). Connexions between hair-plate afferents and motoneurones in the cockroach leg. J. exp. Biol., 64, 251-266.

Pflüger, H-J. & Tautz, J. (1982). Air movement sensitive hairs and interneurones in Locusta migratoria. J. comp. Physiol., 145, 369-380.

Phillips, C.E. (1981). Organization of motor neurons to a multiply innervated insect muscle. J. Neurobiol., 12, 269-280.

Ralston, H.J., Light, A.R., Ralston, D.D. & Perl, E.R. (1984). Morphology and synaptic relationships of physiologically identified low-threshold dorsal root axons stained with intra-axonal horseradish peroxidase in the cat and monkey. J. Neurophysiol., 51, 777-792.

Rastad, J. (1981). Quantitative analysis of axodendritic and axosomatic collateral terminals of two feline spinocervical tract cells. J. Neurocytol., 10, 475-496.

Rastad, J., Jankowska, E. & Westman, J. (1977). Arborisation of initial axon collaterals of spinocervical tract cells stained intracellularly with horseradish peroxidase. Brain Res., 135, 1-10.

Redman, S. & Walmsley, B. (1983a). The time course of synaptic potentials evoked in cat spinal motoneurones at identified group 1a synapses. J. Physiol., 343, 117-133.

Redman, S. & Walmsley, B. (1983b). Amplitude fluctuations in synaptic potentials evoked in cat spinal motoneurones at identified group 1a synapses. J. Physiol., 343, 135-145.

Rethelyi, M., Light, A.R. & Perl, E.R. (1982). Synaptic complexes formed by functionally defined primary afferent units with fine myelinated fibers. J. comp. Neurol., 207, 381-393.

Rudomin, P. (1980). Information processing at synapses in the vertebrate spinal cord: presynaptic control of information transfer in monosynaptic pathways. In Information processing in the nervous system. Eds. Pinsker, H.M. & Willis, W.D., Raven Press, New York, pp. 125-155.

Rudomin, P., Nunez, R., Madrid, J. & Burke, R.E. (1974). Primary afferent hyperpolarization and presynaptic facilitation of 1a afferent terminals induced by large cutaneous fibers. J. Neurophysiol., 37, 413-429.

Semba, K., Masarachia, P., Malamed, S., Jacquin, M., Harris, S., Yang, G. & Egger, M.D. (1983). An electron microscope study of primary afferent terminals from slowly adapting type 1 receptors in the cat. J. comp. Neurol., 221, 466-481.

Siegler, M.V.S. (1982). Electrical coupling between supernumerary motor neurones in the locust. J. exp. Biol., 101, 105-119.

Siegler, M.V.S. & Burrows, M. (1983). Spiking local interneurons as primary integrators of mechanosensory information in the locust. J.

Neurophysiol., 50, 1281-1295.

Siegler, M.V.S. & Burrows, M. (1984). The morphology of two groups of spiking local interneurones in the metathoracic ganglion of the locust. J. comp. Neurol., 224, 463-482.

Simmons, P.J. (1977). The neuronal control of dragonfly flight. II. Physiology. J. exp. Biol., 71, 141-155.

Sotelo, C. & Taxi, J. (1970) Ultrastructural aspects of electrotonic junctions in the spinal cord of the frog. Brain Res., 17, 137-141.

Spira, M.E., Spray, D.C. & Bennett, M.V.L. (1980). Synaptic organisation of expansion motoneurons of Navanax inermis. Brain Res., 195, 241-269.

Takeuchi, A. & Takeuchi, N. (1966). On the permeability of the crayfish neuromuscular junction during synaptic inhibition and the action of GABA. J. Physiol., 183, 433-449.

Tyrer, N.M. & Gregory, G.E. (1982). A guide to the neuroanatomy of locust suboesophageal and thoracic ganglia. Phil. Trans. R. Soc. B, 297, 91-123.

Uchizono, K. (1967). Inhibitory synapses on the stretch receptor neurones of the crayfish. Nature, Lond., 214, 833-834.

Watson, A.H.D. (1984). The dorsal unpaired median neurons of the locust metathoracic ganglion: neuronal structure and diversity, and synapse distribution. J. Neurocytol., 13, 303-327.

Watson, A.H.D. & Burrows, M. (1982). The ultrastructure of identified locust motor neurones and their synaptic relationships. J. comp. Neurol., 205, 383-397.

Watson, A.H.D. & Burrows, M. (1983). The morphology, ultrastructure and distribution of synapses on an intersegmental interneurone of the locust. J. comp. Neurol., 214, 154-169.

Watson, A.H.D. & Pflüger, H-J. (1984). The ultrastructure of prosternal sensory hair afferents within the locust central nervous system. Neuroscience, 11, 269-279.

Westerfield, M. & Frank, E. (1982). Specificity of electrical coupling among neurons innervating forelimb muscles of the adult bullfrog. J. Neurophysiol., 48, 904-913.

Wiersma, C.A.G. & Ripley, S.H. (1952). Innervation patterns of crustacean limbs. Physiologia comp. Oecol., 2, 391-405.

Zill, S.N. & Moran, D.T. (1982). Suppression of reflex postural tonus: a role of peripheral inhibition in insects. Science, N.Y., 216, 751-753.

Section 5
The Control of Equilibrium

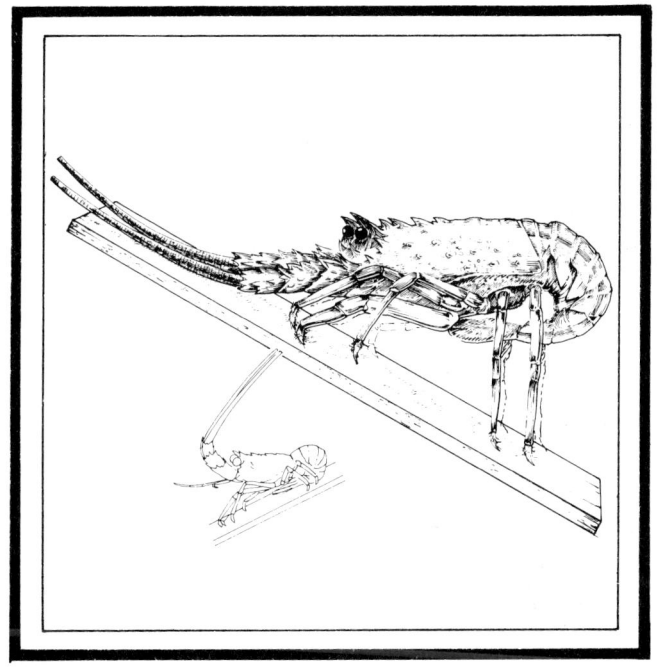

Chapter Sixteen

INTRODUCTION

W.J.P. Barnes

Animals continually need to make adjustments in order to maintain normal posture and balance. These movements consist of a series of <u>righting reflexes</u> aimed at restoring the body's equilibrium, brought into play whenever the body deviates from its desired orientation. Such reflexes will obviously be finely developed in a tight-rope walker traversing a high wire at a circus, but are also involved in maintaining the body in its normal upright stance and maintaining balance while walking or running.

In addition to righting responses, there is a second class of equilibrium responses, <u>compensatory responses</u>, which do not in themselves correct deviations from equilibrium. Instead they compensate for such deviations. Thus, as Dr Neil describes in Chapter 18, a lobster compensates for imposed body tilt by moving its stalked eyes so that they retain their normal orientation in space. Since effective vision depends upon the stabilisation of the external visual scene on the retina of the eye, such compensatory responses also operate in the horizontal plane and compensate for body turns or rotation of the visual field.

Equilibrium reactions are evoked by a variety of sense organs, notably eyes, organs of balance and proprioceptors. Laboratory experiments can be designed so that only single sensory modalities are stimulated, but most naturally-occurring equilibrium responses involve the interaction of information from a number of different sources. Since the stimulus can usually be controlled by the experimenter and responses can be easily measured, equilibrium reactions have proved particularly amenable to study. Besides their intrinsic interest, their study has obvious medical importance for the understanding of oculomotor and vestibular disorders, and significance in space research for the study of, for example, space motion sickness and vestibular adaptation to weightlessness.

That there are close similarities between the equilibrium reactions of vertebrates and invertebrates first became apparent to me when, as a graduate student, I studied optokinetic responses in crabs (Barnes & Horridge, 1969a,b). These responses consist of a compensatory slow phase of nystagmus in which the stalked eyes of the crab follow the movements of a vertically striped drum rotated around it, and a saccadic fast phase of nystagmus in which the eyes quickly flick back to their starting position. I also read the papers of Ter Braak on similar reflexes studied in rabbits (Ter Braak, 1936; Rademaker & Ter Braak, 1948) and was struck by the extraordinary similarity of the reflex in these two very different animals. In both, the control system incorporates a negative feedback loop since the movement of the eyes in following the stripes reduces the apparent velocity of the stimulus. Secondly, in both rabbits and crabs, the reflex is controlled by a system of high but not infinite gain. This means that, under normal conditions with the feedback loop closed (both eyes seeing and free to move), the eyes move almost as fast as the drum, while under open loop conditions (one eye fixed and seeing, the other blind and free to move), eye velocities may exceed drum velocities by factors in excess of 10. Furthermore, both species exhibit this response over a very wide range of stimulus velocities extending down to $0.003° \text{ s}^{-1}$, less than the speed of movement of the sun across the sky.

Crabs also exhibit the equivalent of vestibular nystagmus in that they show compensatory eye responses to acceleration and deceleration. Indeed, the sense organs involved in this reflex, the statocysts, each possess two circular canals, one horizontal and one vertical (Sandeman & Okajima, 1977). Since the vertical canals of each statocyst are orientated at right angles to each other, the system, like the vertebrate labyrinth, provides information about body movement in all three planes.

More recently, it has been shown that, both in crabs (Varju & Sandeman, 1982) and man (Barlow & Freedman, 1980), eye movements in the horizontal plane can be elicited in response to proprioceptive inputs. In crabs, proprioceptors at the base of the legs are implicated in the response, while in man, where the reflex is very weak, neck proprioceptors are thought to be involved.

At the time I did my work, little attempt had been made to investigate multisensory convergence in any equilibrium reaction; nor had the interaction of equilibrium responses with behavioural acts such as locomotion been analysed to any degree. Recent work by a number of scientists, including Drs Berthoz and Neil the authors of Chapters 17 and 18, has done much to remedy this deficit.

In his chapter, Dr Berthoz discusses the interaction of compensatory eye reflexes with saccadic orienting movements of the eyes. When something of interest appears in our visual world, we usually turn our eyes towards it to have a look.

These saccadic eye movements are frequently followed by turning movements of the head and trunk in the same direction. At the time such orienting movements occur, compensatory eye movements are inappropriate since they would antagonise the orienting movements. Indeed, as Dr Berthoz describes, they are centrally inhibited by gaze signals. In the intervals between such saccadic orienting movements, however, vestibulo-ocular and optokinetic reflexes combine to fixate the eyes in space. This has been particularly clearly demonstrated by Collewijn (1977), who recorded eye and head movements in unrestrained rabbits. As Fig. 1 shows, a hand moved in front of the animal was tracked with great interest. The head followed the hand movements smoothly, while eye movements consisted of a series of saccadic orienting movements in the same direction that the hand was moved, separated by periods when compensatory movements of the eye in its orbit (middle trace) fixated the eye in space (lower trace).

Recent experiments of my own on the eye movements of freely moving land crabs present a comparable picture. Here, all large amplitude eye movements appear to be compensatory except for protective eye retractions and, as described below,

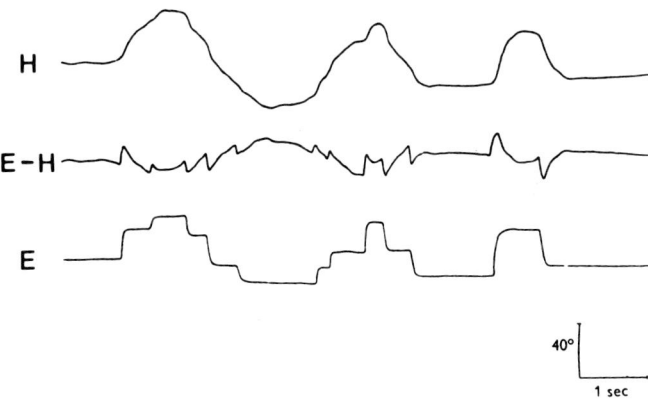

Fig. 1. Eye and head movements of an unrestrained rabbit. The hand of the observer was moved backwards and forwards in front of the animal and was tracked by means of both eye and head movements. The movements of the head in space (H) followed the hand smoothly, while movements of the eye in its orbit (E-H) were a combination of orienting saccadic movements and slower compensatory movements. Since compensation for head movements was complete, the movements of the eye in space (E) consisted solely of saccades. Movement to the right is indicated by a downward movement of the traces. (From Collewijn, 1977).

fast phases of nystagmus. Orienting movements of the eyes towards objects of interest appear to be absent, but as the land crab, Cardisoma, lacks a fovea this is not too surprising. Fig. 2 shows that, during turns to the right, the eyes move to the left thus tending to fixate the eyes in space, and vice versa. Fast phases of nystagmus, in which the eye moves quickly in the same direction as the turn, occur whenever the eye approaches the end of its traverse. These responses are not driven solely by the eyes for, when the crab is blinded (by painting the eyes over with opaque paint), the movements persist though the gain is reduced. That at least some of the eye movement is produced by proprioceptive input is shown by experiments in which a blinded crab is firmly held above a floating ball which is free to turn under its legs. During sideways locomotion, the crab's normal mode of

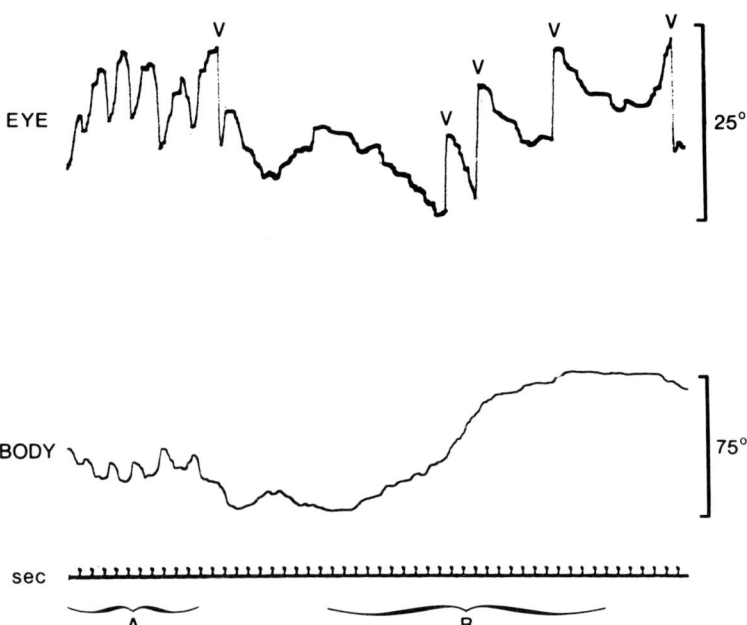

Fig. 2. Eye and body movements of a freely moving land crab, Cardisoma guanhumi. During forwards walking (A), small amplitude oscillations of the body were compensated by movements of the eyes in the opposite direction. Subsequently a 75° right turn (B) was accompanied by compensatory eye movements to the left with interposed fast phases of nystagmus (V) in the same direction as the turn. Movement to the right is indicated by an upward movement of the traces. Methods used to record eye and body movements are described in Barnes (1982).

progression, no large amplitude eye movements occur, but whenever the crab rotates the ball (i.e. turns), the eyes move in the same direction as the ball is rotated (i.e. in the opposite direction to the turn). Although turning would obviously stimulate the crabs statocysts, it doesn't appear that these sensory structures contribute in any important way to the responses, since there is no sign of any response to the deceleration that occurs when turning stops, even in blinded animals.

In Chapter 18, Dr Neil provides a wide-ranging review of recent work on the control of equilibrium in a different group of decapod crustaceans, lobsters and crayfish. Whereas both my own work and that of Dr Berthoz has been concerned with movements in the horizontal plane, Dr Neil's chapter mainly describes experiments relating to righting reactions and compensatory responses to movements in the vertical plane. It thus provides interesting comparisons with similar experiments carried out by Sandeman and others on crabs, reviewed in a previous symposium of the Scottish Electrophysiological Society (Sandeman, 1981). Much of the work has been carried out by his own research group or in collaboration with others including Schöne, Cattaert and myself. It covers compensatory eyestalk reflexes, righting responses of swimmerets (swimming appendages of the abdomen), uropods (tail fan) and antennae, and escape swimming during which the lobster moves rapidly backwards by means of a number of fast abdominal flexion movements.

A first major theme of Dr Neil's chapter is multisensory interaction. By careful control of stimulus variables so that visual, statocyst and proprioceptive inputs are utilised both singly and in different combinations, he shows how different sensory modalities interact to produce good compensation over a wide range of stimulus velocities. Whereas previous workers have assumed that the different inputs simply summate algebraically (e.g. Olivo & Jazak, 1980) though may also exhibit range fractionation (Sandeman, 1977), Dr Neil shows how the operation of open loop statocyst and proprioceptive control systems adjust the visual input so that the range of operation of the closed loop optokinetic response is effectively increased.

A second major theme of this chapter is the interaction of equilibrium responses with the neural control mechanisms governing locomotion. Dr Neil argues that this integration is so complete that any attempt to separate equilibrium and locomotor behaviours into different categories (as we have in the Symposium) is largely artificial. Let me defend my position by saying that such a separation remains one of convenience, and does represent a distinctive approach to the study of motor control.

Thus the two chapters that follow provide contrasting approaches to the same topic. Chapter 17 covers vertebrates, is concerned with movements in the horizontal plane and is an electrophysiological analysis of the neuronal components of equilibrium responses. In contrast, Chapter 18 is concerned with invertebrates, deals with movements in the vertical plane and, in the main, is an analysis at the behavioural level.

REFERENCES

Barlow, D. & Freedman, W. (1980). Cervico-ocular reflex in the normal adult. Acta oto-lar., 89, 487-496.
Barnes, W.J.P. (1982). Recording of eye and body movements in freely moving crabs. J. Physiol., 329, 19-20P.
Barnes, W.J.P. & Horridge, G.A. (1969a). Interaction of the movements of the two eyecups in the crab Carcinus. J. exp. Biol., 50, 651-671.
Barnes, W.J.P. & Horridge, G.A. (1969b). Two-dimensional records of the eyecup movements of the crab Carcinus. J. exp. Biol., 50, 673-682.
Collewijn, H. (1977). Eye- and head movements in freely moving rabbits. J. Physiol., 266, 471-498.
Olivo, R.F. & Jazak, M.M. (1980). Proprioception provides a major input to the horizontal oculomotor system of crayfish. Vision Res., 20, 349-253.
Rademaker, G.C.J. & Ter Braak, J.W.G. (1948). On the central mechanism of some optic reactions. Brain, 71, 48-76.
Sandeman, D.C. (1977). Compensatory eye movements in crabs. In Identified Neurons and Behavior of Arthropods. Ed. Hoyle, G. Plenum, New York, pp. 131-147.
Sandeman, D.C. (1981). Equilibrium and proprioceptive systems, and the central nervous system of arthropods. In Sense Organs. Eds. Laverack, M.S. and Cosens, D.J. Blackie, Glasgow, pp. 276-294.
Sandeman, D.C. & Okajima, A. (1972). Statocyst-induced eye movements in the crab Scylla serrata. I. The sensory input from the statocyst. J. exp. Biol., 57, 187-204.
Ter Braak, J.W.G. (1936). Untersuchungen über optokinetischen Nystagmus. Archs neérl. Physiol., 21, 309-376.
Varju, D. & Sandeman, D.C. (1982). Eye movements of the crab Leptograpsus variegatus elicited by imposed leg movements. J. exp. Biol., 98, 151-173.

Chapter Seventeen

CONTROL OF EYE-HEAD COORDINATION BY BRAIN STEM NEURONES

A. Berthoz

The maintenance of equilibrium is dependent upon several basic mechanisms. A first class consists of compensatory stabilising reflexes which keep the body, head or gaze (sum of eye, head and body positions) fixed in space. These reflexes are regulated by sensory feedback from proprioceptors which signal relative body segment motion, and from vestibular, visual and tactile receptors which signal head or body motion with respect to space. The interaction of these mechanoreceptors in motion perception and in the control of posture has recently been studied extensively and the role of vision in this multisensory integration process has been particularly stressed. But equilibrium control also implies other mechanisms which do not use the feedback mode, an example of which is the important preparatory postural activity which precedes movement. I shall deal in this paper with the interaction between stabilising reflexes and orienting movements which are made when an animal orients to a prey or visual target. In this situation the compensatory reflexes may not be appropriate and have therefore to be modulated or suppressed if they antagonise the orienting behaviour. In this short review I will give a few examples of the mechanisms which underly the interaction between vestibular reflexes and orienting reactions and, because orienting involves eye and head movements, I will focus on the vestibulo-ocular and vestibulo-collic reflexes (which stabilise gaze in space) and the mechanisms underlying active changes of gaze which are made during head and eye turning to a visual target. For the sake of simplicity these examples will only concern movements in the horizontal plane (that is in the plane of the horizontal canals). The main point of this paper will be to show that eye movement signals deeply modify the activity of all neuronal structures which are involved in the control of equilibrium. Feedback sensory regulation of these neural stations forms only a small part of the control mechanism. A very powerful signal of eye position or velocity gives a feedforward or corollary influence which adapts the reflexes

to the needs of planned orientation.
The physiological study of sensori-motor systems underlying equilibrium has often been done in either anaesthetised or decerebrate preparations. The limit of this approach is that such preparations do not display all oculomotor or orienting activity. Under these conditions most neuronal structures exhibit activity which seems to be dominated by vestibular influence. We shall see that in the alert animal the so-called vestibular relays are themselves modified by gaze signals. Therefore this kind of study has to be made on fully alert animals.
I will first review current hypotheses and facts concerning the saccadic eye movement network and the way in which gaze position and velocity are processed by brain stem neurones. I will then point out some properties of the basic organisation of the vestibulo-ocular and vestibulo-collic reflexes and show how the activity of these stabilising reflexes is modulated by the saccadic orienting system. Lastly I will describe a newly discovered reticulo-spinal pathway which mediates the oculomotor influence on neck (and possibly body) muscle activity.

METHODS

All experiments are made on alert cats. The originality of our approach is that it allows a complete description of neuronal activity from function to morphology. This is achieved by intracellular (intra-axonal) recording of neuronal activity and injection of horseradish peroxidase in neurones whose physiological properties are first studied by their response to natural stimuli (vestibular or optic) or during spontaneous movements. We believe that this approach is necessary to allow for a complete identification of neurones and for adequate knowledge of their role in the interaction between automatic and voluntary control of movement and equilibrium. Detailed methods have been described in Yoshida et al. (1982) and Vidal et al. (1983).

Surgical Procedures
Experiments are performed on adult cats. The animals undergo the following surgical procedures which are carried out under pentobarbital anaesthesia (40mg/kg). The implantations which are actually performed on each cat are dependent upon the type of problem under study and the recorded neurones. They will however be summarised and will be specified in each case.
   a) A coil of teflon-coated, stainless steel wire is implanted on the eyeball to measure eye movements with the search coil method (Fuchs & Robinson, 1966).
   b) Bipolar electrodes made of insulated $300\mu m$ silver wire are implanted bilaterally to stimulate the abducens nerves at

their exit from the brainstem for antidromic activation of abducens motor neurones. Correct implantation is verified by observing the eye movement due to the twitch contractions of the lateral rectus muscle elicited by single shock stimulation with pulses of 100 microsecond duration.

c) Fine silver wire electrodes are placed bilaterally near the oval and round windows to stimulate the vestibular nerves. The mastoids on both sides are exposed via a lateral approach and small holes are made with a fine dental drill close to the ampulla of the horizontal semicircular canal. Electrodes are placed so as to evoke the eye movements without any sign of current spread to the facial nerve. The threshold for eye movement tested by a train of impulses (100-150 impulses/s and 50-100 microsecond duration) is generally about 1.5-2 V.

d) Stimulating silver wire electrodes are placed chronically on the superior colliculus for orthodromic activation of tecto-reticulo-spinal neurones and in the spinal cord for antidromic activation of vestibulo-spinal, reticulo-spinal, and tectal efferent neurones.

e) Electromyographic bipolar electrodes made of teflon-coated stainless steel wires (100μm diameter) are implanted chronically in several neck muscles : splenius, longissimus capitis, obliquus capitis cranialis and caudalis.

f) A small portion of the occipital and temporal bone is removed to allow access of recording microelectrodes to the brainstem. An opening of about 5mm in diameter is made and a funnel-shaped chamber is formed with dental cement. The dura is removed and the cerebellar surface is covered with pieces of silicone film (Silastic). The chamber is closed with semi-solid bone wax between experiments.

g) A metal platform with three stereotaxically positioned bolts is cemented onto the skull to restrain the animal's head during recording sessions. At the same time, the tip of a hypodermic needle is fixed to the chamber close to the opening as a reference point for stereotaxic fixation of the head.

Recording conditions
Following recovery from surgical procedures, the animals are placed in the stereotaxic apparatus with their head fixed in a 25 degree nose-down position in order to orient the horizontal semicircular canals in the horizontal plane. The body of the animal is gently restrained with a cloth bag and an elastic bandage.

Eye movement recording is made by a SKALAR magnetic search coil system with a band-width of DC to 200 Hz (3 dB) providing both horizontal and vertical components of eye movements in one eye. The sensitivity is 0.1 V per degree with a resolution of about 5' of arc. Calibration of eye movements is made by rotating the field coils around the animal by steps of 5, 10 and 20 degrees. Extracellular unit activity is recorded with glass micropipettes filled with 2 M NaCl (DC resistance of 3-5 MΩ). The animal, together with the stereotaxic frame, micromanipulator and solenoids for the generation of magnetic field, is firmly fixed on a turntable for vestibular stimulation.

Identification of Reticular Neurones
Reticular neurones can be identified by their location with respect to the abducens nucleus according to the results of Peterson (1977, 1980). Only recently have we succeeded in implanting chronically a spinal stimulating electrode. In other cases an electrophysiological method can be used. The microelectrode tracts penetrated the brainstem at an angle of about 30° with respect to the stereotaxic vertical. The antidromic field potentials due to VIth nerve stimulation were recorded after recording each reticular neurone in order to assess its position with respect to the abducens nucleus.

Data processing
Neural activity, horizontal and vertical eye position signals and the velocity signal from the turntable are recorded on separate channels of FM magnetic tape for off-line analysis. Unit activity is discriminated, and instantaneous firing rate is calculated by an HP 5451 computer with a specially constructed system giving a time resolution of 10 microseconds on spike detection. Eye movements and table signals are sampled and processed to provide horizontal and vertical eye velocity.

Intra-axonal injection of horseradish peroxidase (HRP)
After proper identification of each axon, HRP can be, in some cases, injected electrophoretically by passing positive current through the micro-electrode for several minutes (total charge, 4-40nA min). After a period of 12-24h, animals are anaesthetised and perfused through the ascending aorta with a normal saline solution followed by 3 litres of fixative (0.8% paraformaldehyde, 1.0% gluteraldehyde, pH 7.4). The brain is stored at 4°C in a solution of phosphate buffer (pH 7.4) and 30% sucrose and subsequently cut in 100μm thick coronal or parasagittal sections on a freezing microtome. Following incubation in a solution of $CoCl_2$, the sections were reacted with diaminobenzidine and H_2O_2 in a modification of the technique described by Bishop et al. (1976). The sections are then mounted on gelatin-coated slides, dehydrated, cleared with xylene, and cover-slipped. The drawing and reconstructions shown are made with the aid of a drawing tube.

RESULTS

The saccadic system : an orienting system
The saccade of the eye requires fast (pulse-like) changes in

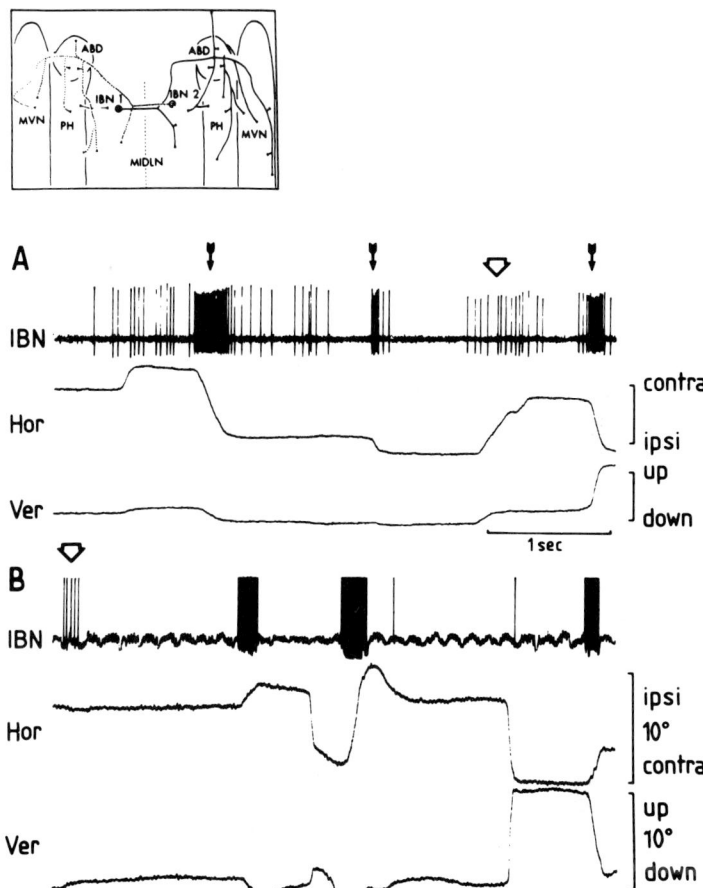

Fig. 1. Activity of two inhibitory burst neurones whose axons were injected with HRP during spontaneous saccades in darkness. A: activity of neurone IBN1 (see insert drawing). From top to bottom: extracellular recording of the discharge of the IBN, horizontal (hor) and vertical (ver) components of eye angular displacement (calibration given for $10°$). Filled arrows: ipsilateral (left) burst. Open arrow: contralateral discharge. B: activity of neurone IBN2. Same notations as in A.
Inset: Drawing that reconstructs the axonal arborisation of the two inhibitory burst neurones (somata indicated by IBN1 and IBN2) whose axons were injected with HRP. Both neurones were injected in the same cat. ABD - abducens nucleus. MVN - medial vestibular nucleus. PH - prepositus hypoglossi nucleus. MIDLN - midline.
From Yoshida et al., 1982.

eye position plus maintained tonic (step-like) activity. The current view is that this "pulse-step" sequence is accomplished by a saccadic generator whose neuronal network is located in the ponto-medullary reticular formation in the vicinity of the abducens nucleus for horizontal saccades. Van Gisbergen and Robinson (1977) have proposed a stimulating, although not yet proven, model to represent the mechanisms of saccade generation. A set of burst neurones would provide the pulse. They would be part of a local feedback high gain circuit. They would normally be under the tonic inhibitory action of neurones which would pause during saccades. The saccades would be triggered when "desired eye position" is different from "actual eye position". An important feature of this hypothetical network is that eye position is derived from burst neurones which are supposed to code eye velocity, by a neural mathematical "integration" performed through a yet unidentified circuit. This internal construction of eye position would be possible because the mechanical impedance of the eye globe and the load under which it moves do not change as is the case for limbs. A given motor command, in the adult animal, always gives rise to an identical eye movement. It is therefore most probable that the saccadic generator can produce an internal signal related to eye position. A few neurones predicted by this hypothetical model have been identified to date (review in Robinson, 1981). Our group has used the technique described above to study inhibitory burst interneurones whose function is to block the antagonist abducens motor neurones during a saccade (Hikosaka & Kawakami, 1977; Yoshida et al., 1982). The remarkable feature of these inhibitory premotor interneurones is that they code the three essential parameters of saccades: <u>saccadic amplitude</u> which is coded by the number of spikes in the burst, <u>saccade duration</u> which is equal to burst duration and <u>saccadic velocity</u> which is coded by the instantaneous frequency after peak burst frequency. The fact that the number of spikes in the burst relates so nicely to eye displacement means that, as predicted by the model, an integrator neurone or a circuit placed at the output of this burst neurone could provide an eye position signal independent of proprioceptive information from extraocular muscles. Fig. 1 shows recordings of the activity of two such inhibitory burst neurones (IBNs) in the same cat. They have been injected with HRP and their pattern of axonal projections has been reconstructed. The interesting observation here is that these neurones do not only project to contralateral abducens motor neurones but to several other targets such as the prepositus nucleus, the vestibular nucleus and the ponto-medullary reticular formation on the contralateral side. Therefore the signals carried by the firing frequency of these neurones are addressed in a corollary fashion to structures in the brain stem which are involved in gaze or equilibrium control. The prepositus nucleus is in fact a candidate for contributing to the generation of eye position signals (Baker & Berthoz, 1975; Lopez-Barneo et al., 1982).

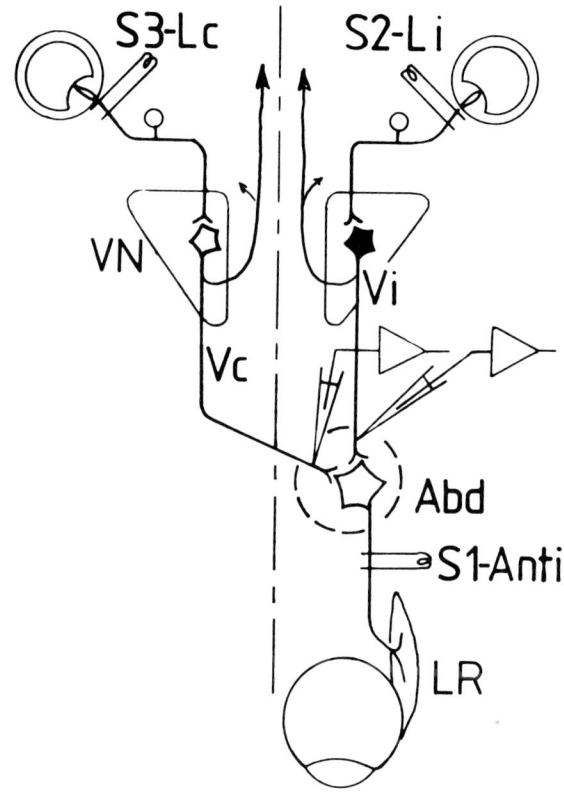

Fig. 2. Experimental paradigm for the study of second order vestibular neurones in the alert cat. Stimulating electrodes are implanted on the abducens (Abd) nerve for antidromic identification (S1-Anti), and on the vestibular nerves in the labyrinth (S2-Li and S3-Lc) for orthodromic identification of second order vestibular neurones. Intra-axonal penetration of either ipsilateral (Vi) or contralateral (Vc) second order neurones is performed with glass micropipettes through the intact cerebellum. In some cases injections of HRP is made after physiological study of the properties of the neurones. LR - lateral rectus muscle.

Modulation of vestibular neurone firing rate by eye movement signals

The horizontal vestibulo-ocular reflex (VOR) stabilises eye position in space during angular rotation in the plane of the semi-circular canals (Fig. 2). It is organised centrally as a disynaptic arc whose central interneurones are the so-called "second order vestibular neurones" whose somata are located in the medial vestibular nucleus. The organisation of this reflex is very complex. It was thought until recently that second order vestibular neurones were only sensory relays receiving excitation from primary afferents of vestibular receptors. However, Baker and Berthoz (1974) and Maeda et al. (1971) demonstrated that their neuronal activity was modulated during vestibular nystagmus by signals related to eye position. More recently, Yoshida et al. (1981), McCrea et al. (1981) and Berthoz et al. (1981) have shown that their activity is modulated both by horizontal eye position and eye velocity signals. Indeed, the eye position sensitivity of these neurones can be as high as 8 spikes per second per degree of horizontal angular eye displacement. Figs. 3 and 4 show the eye position sensitivity of two such neurones, together with the response of one of them to a combination of vestibular and optokinetic nystagmus. The discharge rate of second order vestibular neurones is in fact a combination of several signals; a) head horizontal angular velocity, b) horizontal eye position and c) horizontal eye velocity during rapid eye movements. The activity of these neurones can be suppressed

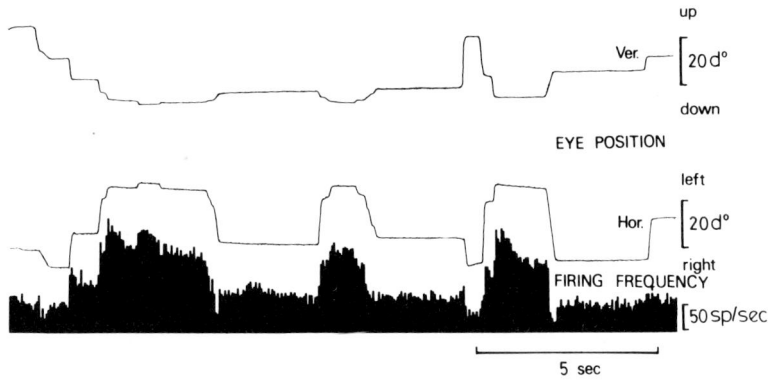

Fig. 3. Horizontal eye position sensitivity of a second order vestibular neurone whose soma was located in the right vestibular nucleus. From top to bottom: vertical (Ver) and horizontal (Hor) components of eye position, firing rate (50msec bins). (From Yoshida, Berthoz & Vidal, unpublished observations).

during ipsilateral saccades which means that the VOR can be cancelled during orienting eye movements. Modulation of their discharge rate by optokinetic stimuli as well as by cervical (neck) proprioception has also been shown. The latter underlies the cervico-ocular reflex which in most mammals

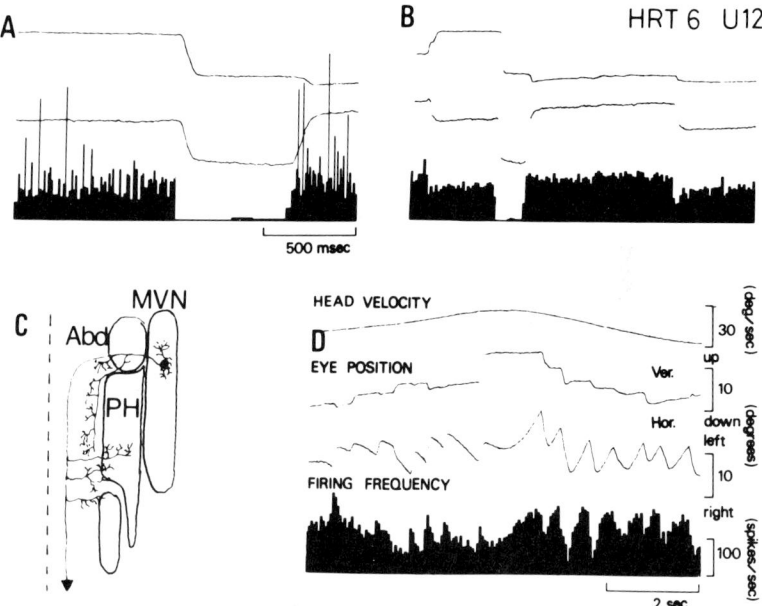

Fig. 4. Morphological and physiological characteristics of a second-order ipsilateral vestibular neurone.
A. Detail of the discharge characteristics, at fast sweep speed, of the neurone during spontaneous saccades in the dark. From top to bottom: vertical and horizontal eye movements, instantaneous firing rate. Amplitude calibrations as in D.
B. Discharge of the same neurone in darkness with a slower sweep and an averaged firing rate (50msec bins). Time and amplitude calibrations as indicated in D.
C. Morphology of this neurone revealed by reconstruction after intracellular injection of HRP. The soma lies in the medial vestibular nucleus (MVN); the axon descends ipsilaterally to the spinal cord, giving off collaterals around the abducens nucleus (Abd), caudal prepositus hypoglossi (PH), medullary reticular formation, and nucleus of Roller.
D. Modulation of the discharge rate of the same neurone during sinusoidal rotation in the illuminated laboratory. Combination of optokinetic and vestibular nystagmus.
From Berthoz et al., 1981.

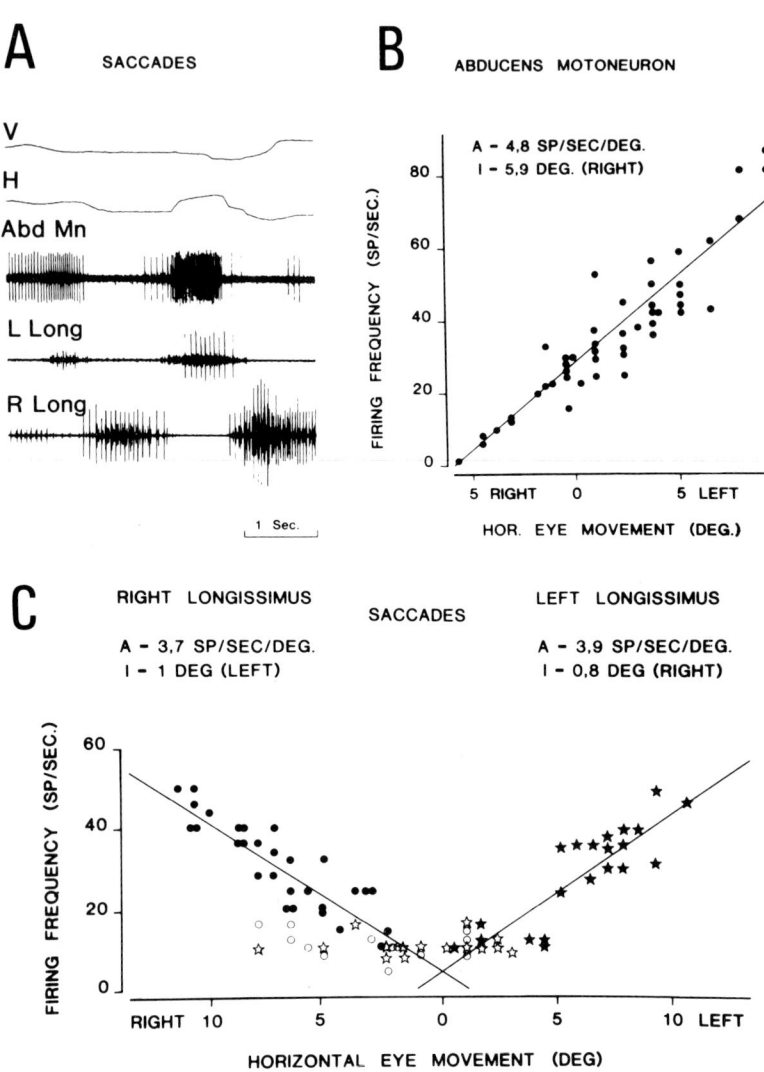

Fig. 5. Firing characteristics of an abducens motor neurone (which innervates the lateral rectus muscle of the orbit) and two motor units of left and right longissimus capitis muscles of the neck recorded simultaneously in an alert cat.

A. Simultaneous recording of eye movement, dorsal neck muscle activity and firing rate of an abducens motor neurone in the alert cat. From top

including primates and man has a rather low gain and can be considered as a back-up system (see review in Berthoz, 1985). Intracellular injection of HRP in these neurones shows that these highly elaborate signals are not only sent to the ipsi- or contralateral abducens nucleus for operation of the VOR. Axon collaterals project to structures of the brain stem involved in eye and head coordination (prepositus nucleus, nucleus of Roller, medial reticular formation, etc.) and also to the spinal cord (McCrea et al., 1980; McCrea et al., 1981) (Fig. 4).

Gaze signals in neck muscles
During head rotation, a compensatory stabilising vestibulo-collic reflex (VCR) is produced in the horizontal plane by the action of the semi-circular canals on neck motor neurones. This reflex has a disynaptic component which relays through the vestibular nuclei, and has been described by Wilson, Peterson and their colleagues (see reviews in Wilson & Peterson, 1978 and Peterson & Goldberg, 1982). This reflex, whose dynamic stabilising properties have been studied in the decerebrate cat by Berthoz and Anderson (1975), is operating in a closed loop fashion (contrary to the VOR) and is interacting with the cervico-collic (neck-neck) reflex. These reflexes are not the only ones acting on neck muscles. For instance, visual world motion can also modulate tonic neck muscle activity. Recently, Vidal et al. (1982) have discovered that, in the fully alert cat, these reflexes are in fact greatly modified by orienting behaviour. As shown in Fig. 5, the main finding is that, during spontaneous saccades in the alert head fixed-cat, there is a close correlation between the activity of dorsal neck muscles (obliquus capitis, longissimus capitis, splenius capitis, which are known to contribute to horizontal rotation of the head), and the horizontal component of eye position in the orbit. If the eye globe moves to the left, there is an increase of left neck motor neurone firing rate proportional to eye position and to abducens motor neurone discharge. Neck muscles are organised in a push-pull

---

to bottom: vertical (V) and horizontal (H) components of eye angular displacement, extracellular recording from an abducens motor neurone, EMG of left and right longissimus capitis muscles.
B. Rate position curve for the abducens motor neurone. A is the slope in spikes per second per degree and I, the intercept of the regression line with the abscissa.
C. Rate position curve for two isolated motor units belonging respectively to the left (stars) and right (dots) muscles. Unfilled stars and dots indicate firing frequencies measured at the threshold when the motor unit stops firing during the same fixation.
From Berthoz et al., 1982.

fashion. Their discharge threshold corresponds to an eye position which is 0° to 5° contralateral to mid-position of the eye in the orbit. When the cat is looking straight ahead, a tonic co-contraction may occur. When gaze is directed to one side the ipsilateral neck muscle is activated, while contralateral neck motor neurones seem to be actively inhibited. The remarkable fact is that, under these conditions when the animal is fully alert, the vestibulo-collic reflex induced by horizontal head rotation in a given muscle can be completely suppressed or facilitated depending upon whether the beating field of vestibular nystagmus (or the mean gaze position) is directed to, respectively, the contralateral or ipsilateral side of the orbit (in other words depending upon whether the cat is "looking" in the opposite or in the same direction as head motion). This coupling is also present in the head-free animal (Crommelinck et al. 1982).

The coupling between abducens and neck motor neurones which is so strong in the cat is also present in the monkey (Lestienne et al., 1984), although the primate is more able to decouple eye and head (see review in Berthoz, 1985). The fact that a common signal is sent to eye and neck muscles is compatible with the hypothesis of Bizzi et al. (1971). Darlot et al. (1985) have studied the interaction between vestibulo-collic reflexes and oculo-collic (eye-neck) coupling in the frontal plane of the alert cat. They found that there is a quasi-linear addition of these two motor effects.

Neuronal subsystems carrying gaze signals to neck motor neurones
The reconstruction of axonal branching of second order vestibular neurones described above has revealed that an important number of neurones projecting to the abducens nucleus also project to the spinal cord. Therefore a common neuronal network subserves the vestibulo-ocular and vestibulo-collic reflexes in the horizontal plane. However, the gaze signals carried by second order vestibular neurones analysed so far cannot account for the activity recorded in neck muscles (see discussion in Berthoz et al., 1982; Vidal et al., 1983). Following these observations we have suggested that another system could carry these gaze signals to neck motor neurones and have demonstrated that ponto-medullary reticular neurones in the vicinity of the abducens nucleus have discharge patterns which are closely related to dorsal neck muscle activity (Figs. 6 and 7). We have therefore suggested that reticulo-spinal neurones are responsible for the gaze signals recorded in neck muscles. A final proof supporting this hypothesis has recently been obtained by Berthoz and Grantyn (1985) who have applied to these neurones the methods of intracellular injection of HRP in the alert cat together with electrophysiological identification.

We have demonstrated that some of these neurones receive

monosynaptic excitation from the contralateral superior colliculus and are antidromically activated from the spinal cord. Their discharge pattern is similar to ipsilateral neck muscle activity during active orienting reactions. Two of these neurones have been injected with intracellular HRP. Their evaluation is in progress but it is clear that these findings support the hypothesis of Vidal et al. (1983), also proposed by Peterson (1980), Peterson and Fukushima (1982), Grantyn et al. (1980) and Grantyn and Grantyn (1980), concerning the existence of a reticulo-spinal neuronal activity underlying eye-head synergy. It is most probable that the superior colliculus, whose connections to cat neck motor neurones was studied by Anderson et al. (1971), provides a phasic input to these reticular neurones during head turning triggered by a visual target (Grantyn et al., 1979; Grantyn & Berthoz, 1985). In addition, a tonic gaze signal is probably added in the brain stem by the output from the integrator placed after the saccadic generator (see first section of Results).

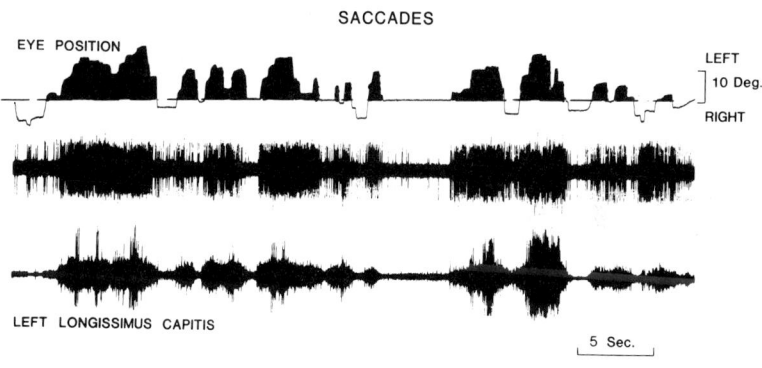

Fig. 6. Recording from a group of reticular neurones located 1mm below the anterior border of the left abducens nucleus in an alert cat. The horizontal component of eye position has been separated into two zones. The horizontal line represents the approximate mid-position of the eye in the orbit. The dark and light areas indicate when the eye globe is to the left and right, respectively. Note that the left longissimus capitis muscle (bottom trace) is discharging together with the ipsilateral eye position as shown by Vidal et al., 1982. The group of reticular units (middle trace) are discharging together with the ipsilateral neck muscle. (From Vidal, Corvisier & Berthoz, unpublished observations).

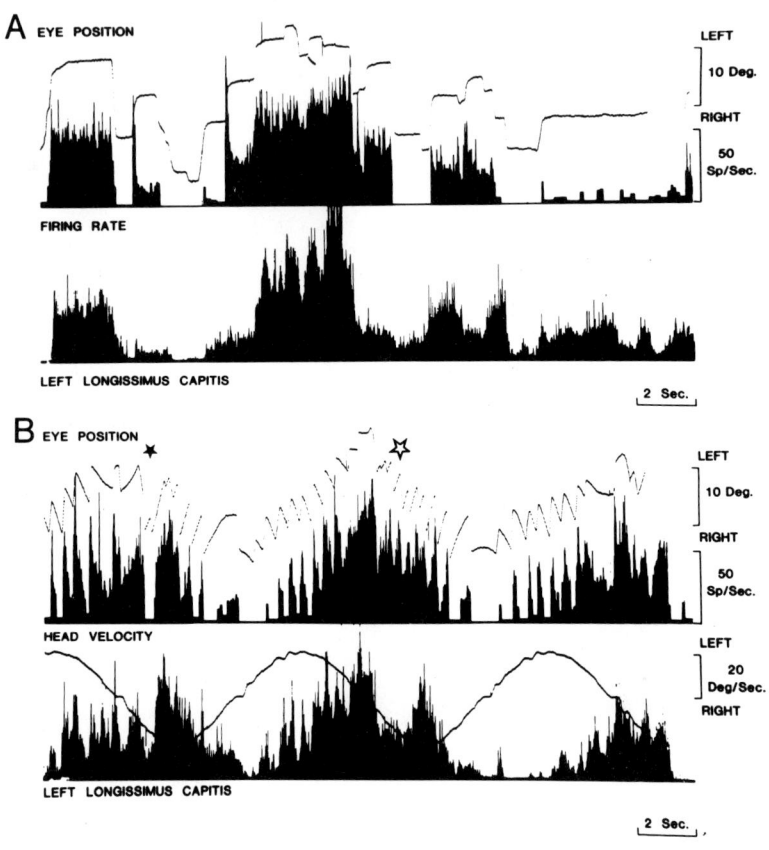

Fig. 7. Instantaneous firing rate of a "burst-tonic" reticular neurone recorded extracellularly in the immediate vicinity of the abducens nucleus (from Vidal et al., 1983).
A. Firing pattern during spontaneous fixations. From top to bottom: horizontal component of eye position, firing rate of the neurone, rectified integrated EMG activity of the left longissimus capitis muscle. Arrows show the corresponding burst of EMG activity during the burst of discharge of the neurone.
B. Firing pattern during sinusoidal head rotation in the light. From top to bottom: horizontal component of eye position, firing rate of the neurone, head velocity trace partially overlapping the integrated rectified EMG activity of the left longissimus capitis muscle. Stars indicate times at which head velocity is the same but eye position in the orbit is different: note that EMG follows eye position.

CONCLUSIONS

Several conclusions can be drawn from this brief review. They will be expressed as statements or as hypotheses in keeping with the spirit of this Symposium devoted to the general role of feedback in motor control.

a) Vestibular second order neurones are not only sensory relay neurones. They perform very specific computations related to determination of head motion in space and the vestibular input is only one of several afferent inputs. These computations in the alert animal are made in relation to intended or actual gaze. In fact this property is rather general and it is now obvious that there is no such thing as a pure sensory relay neurone in the central nervous system. Results emerging from the study of all sensory modalities indicate that sensory information is either under the influence of efferent control or that it is immediately integrated at the first central synapse.

b) The signals constructed by the saccadic generator, by vestibular second order neurones or by reticular neurones are not only sent to motor nuclei where they exert their direct motor influence. They are also addressed, by their axon collateral network which may appear diffuse, to extremely specific targets belonging either to synergistic motor centres which are involved in activating or inhibiting a given set of muscles, or to premotor neurones which use the particular signal for elaborating other motor responses or even for perceptual activity. These ideas would support a hypothesis of highly specific axonal targets determined by the nature of the signal generated by any given neurone. Profuse axonal branching of reticular neurones may therefore not be, as was thought early in this century, a sign of general unspecific activation or inhibition, but be the reflection of the highly useful nature of a given signal elaborated by a given neurone for many sensory or motor neuronal operations.

c) In all cases, stabilising reflexes such as VOR, VCR, etc. are modified by a gaze signal which has probably nothing to do with peripheral feedback, but which adapts the performance of stabilisation to the intended motor behaviour. This observation indicates that oculomotor or rather gaze orientation is an important component of equilibrium and that there is a need for future research bridging the gap between oculomotor and spinal neurophysiology.

REFERENCES

Anderson, M.E., Yoshida, M. & Wilson, V.J. (1971). Influence of superior colliculus on cat neck motoneurons. J. Neurophysiol., 34, 898-907.
Baker, R. & Berthoz, A. (1974). Organization of vestibular nystagmus in oblique oculomotor system. J. Neurophysiol., 37, 195-217.

Baker, R. & Berthoz, A. (1975). Is the prepositus hypoglossi nucleus the source of another vestibulo-ocular pathway? Brain Res., 86, 121-127.

Berthoz, A. (1985). Adaptive mechanisms in eye-head coordination. In Adaptive Mechanisms in Gaze Control. Eds. Berthoz, A. & Melvill Jones, G.. Elsevier-North Holland, Amsterdam. (In press).

Berthoz, A. & Anderson, J.H. (1971). Frequency analysis of vestibular influence on extensor motoneurons. II. Relationship between neck and forelimb extensors. Brain Res., 34, 376-380.

Berthoz, A. & Grantyn, A. (1985). Neuronal mechanisms underlying eye-head coordination. Prog. Brain Res., 64. (In press).

Berthoz, A., Vidal, P.P. & Corvisier, J. (1982). Brain stem neurons mediating horizontal eye position signals to dorsal neck muscles of the alert cat. In Physiological and Pathological Aspects of Eye Movements. Eds. Roucoux, A. & Crommelinck, M. Dr W. Junk Publ., The Hague, pp.385-398.

Berthoz, A., Yoshida, K. & Vidal, P.P. (1981). Horizontal eye movement signals in second-order vestibular nuclei neurons in the alert cat. Ann. N.Y. Acad. Sci., 374, 144-156.

Bishop, G.A., McCrea, R.A. & Kitai, S.T. (1976). Afferent projection to the nucleus interpositus anterior (NIa) and lateral nucleus (LN) of the cat cerebellum. Anat. Rec., 184, 360.

Bizzi, E., Kalil, R.E. & Tagliasco, V. (1971). Eye-head coordination in monkeys: evidence for centrally patterned organization. Science, N.Y., 173, 452-454.

Crommelinck, M., Roucoux, A. & Verhaart, C. (1982). The relation of neck muscle activity of horizontal eye position in the alert cat. II. Head free. In Physiological and Pathological Aspects of Eye Movements. Eds. Roucoux, A. & Crommelinck, M. Dr W. Junk Publ., The Hague, pp. 379-384.

Darlot, C., Denise, P. & Droulez, J. (1985). Modulation by horizontal eye position of the vestibulo-collic reflex induced by tilting in the frontal plane in the alert cat. Eur. Neurosci. Assoc. Abstr. (In press).

Fuchs, A.F. & Robinson, D.A. (1966). A method for measuring horizontal and vertical eye movement chronically in the monkey. J. appl. Physiol., 21, 1068-1070.

Grantyn, A. & Berthoz, A. (1985). Burst activity of identified tecto-reticulo-spinal neurons in the alert cat. Expl. Brain Res., 57, 417-421.

Grantyn, A. & Grantyn, R. (1980). Reticular substrates for coordination of horizontal eye movements and their relationship to tectal efferent pathways. In The Reticular Formation Revisited. Eds. Hobson, J.A. & Brazier, M.A.B.. Raven Press, New York, pp. 211-225.

Grantyn, A., Grantyn, R., Robine, K.-P. & Berthoz, A. (1979). Electroanatomy of tectal efferent connections related to eye movements in the horizontal plane. Expl. Brain Res., 37, 149-172.

Grantyn, R., Baker, R. & Grantyn, A. (1980). Morphological and physiological identification of excitatory pontine reticular neurons projecting to the cat abducens nucleus and spinal cord. Brain Res., 198, 221-228.

Hikosaka, O. & Kawakami, T. (1977). Inhibitory reticular neurons related to the quick phase of vestibular nystagmus - their location and projection. Expl. Brain Res., 27, 377-396.

Lestienne, F., Vidal, P.P. & Berthoz, A. (1984). Gaze changing behaviour in head restrained monkeys. Expl. Brain Res., 53, 349-356.

Lopez-Barneo, J., Darlot, C., Berthoz, A. & Baker, R. (1982). Neuronal activity in prepositus nucleus correlated with eye movement in the alert cat. J. Neurophysiol., 47, 329-352.

Maeda, M., Shimazu, H. & Shinoda, Y. (1971). Rhythmic activities of secondary vestibular efferent fibers recorded within the abducens nucleus during vestibular nystagmus. Brain Res., 34, 361-365.

McCrea, R.A., Yoshida, K., Berthoz, A. & Baker, R. (1980). Eye movement related activity and morphology of second order vestibular neurons terminating in the cat abducens nucleus. Expl. Brain Res., 40, 468-473.

McCrea, R.A., Yoshida, K., Evinger, C. & Berthoz, A. (1981). The location, axonal arborization and termination sites of eye movement related secondary vestibular neurons demonstrated by intra-axonal HRP injection in the alert cat. In Progress in Oculomotor Research, Vol.12. Eds. Fuchs, A.F. & Becker, W. Elsevier-North Holland, Amsterdam, pp. 379-386.

Peterson, B.W. (1977). Identification of reticulo-spinal projections that may participate in gaze control. In Control of Gaze by Brainstem Neurons. Eds. Baker, R. & Berthoz, A. Elsevier-North Holland, Amsterdam, pp. 143-152.

Peterson, B.W. (1980). Participation of pontomedullary reticular neurons in specific motor activity. In The Reticular Formation Revisited. Eds. Hobson, J.A. & Brazier, M.A.B. Raven Press, New York, pp. 171-192.

Peterson, B.W. & Fukushima, K. (1982). The reticulo-spinal system and its role in generating vestibular and visuo-motor reflexes. In Brainstem Control of Spinal Mechanisms. Eds. Sjölund, B. & Björklund, A. Elsevier-North Holland, Amsterdam, pp. 225-251.

Peterson, B.W. & Goldberg, J. (1982). Role of vestibular and neck reflexes in controlling eye and head position. In Physiological and Pathological Aspects of Eye Movements. Eds. Roucoux, A. & Crommelinck, M. Dr W. Junk Publ., The Hague, pp. 351-364.

Robinson, D.A. (1981). The use of control systems analysis in the neurophysiology of eye movements. Annu. Rev. Neurosci., 4, 463-503.

Van Gisbergen, J.A.M. & Robinson, D.A. (1977). Generators of micro- and macrosaccades by burst neurons in the monkey. In Control of Gaze by Brainstem Neurons, Eds. Baker, R. & Berthoz, A. Elsevier-North Holland, Amsterdam, pp. 301-308.

Vidal, P.P., Corvisier, J. & Berthoz, A. (1983). Eye and neck motor signals in periabducens reticular neurons of the alert cat. Expl. Brain Res., 53, 16-28.

Vidal, P.P., Roucoux, A. & Berthoz, A. (1982). Horizontal eye position related activity in neck muscles of the alert cat. Expl. Brain Res., 46, 448-453.

Wilson, V.J. & Peterson, B.W. (1978). Peripheral and central substrates of vestibulospinal reflexes. Physiol. Rev., 58, 80-105.

Yoshida, K., Berthoz, A. Vidal, P.P. & McCrea, R. (1981). Eye-movement-related activity of identified second order vestibular neurons in the cat. In Progress in Oculomotor Research, Vol. 12. Eds. Fuchs, A. & Becker, W. Elsevier-North Holland, Amsterdam, pp. 371-378.

Yoshida, K., McCrea, R., Berthoz, A. & Vidal, P.P. (1982). Morphological and physiological characteristics of inhibitory burst neurons controlling horizontal rapid eye movements in the alert cat. J. Neurophysiol., 48, 761-784.

Chapter Eighteen

MULTISENSORY INTERACTIONS IN THE CRUSTACEAN EQUILIBRIUM SYSTEM

D.M. Neil

Equilibrium reactions have long been distinguished as a separate category of motor response since they are so clearly and consistently elicited, and involve stereotyped movements of the majority of body appendages. They can be conveniently divided into two categories: compensatory movements, in which an appendage maintains itself in a constant position with reference to space, and righting responses, in which the appendages move actively to restore the whole body to its normal position. Both these types of reaction may be elicited by a number of different sensory systems, and both may involve some form of feedback control. Therefore they provide convenient systems in which to study such aspects of motor behaviour as the organisation of reflex pathways, multisensory convergence, bilateral and intersegmental co-ordination of outputs and the role of feedback. We are only beginning to exploit their potential for these various lines of investigation.

In this article I will review current work on multisensory interactions in motor control within the equilibrium systems of the long-bodied macruran decapod crustaceans, i.e. the lobsters and crayfishes. Equilibrium control in crabs is organised in a similar way, but shows particular adaptations to the problems posed by their shortened bodies, and sideways gait (Sandeman, 1983).

Although it is customary to consider motor behaviour in crustaceans in terms of such distinct categories as equilibrium, posture and locomotion (Sandeman, 1981; Page, 1982; Evoy & Ayers, 1982), in fact these are artificial subdivisions which have no correspondence to the organisation of the nervous system. A complex behavioural act such as swimming or walking involves a great variety of neuronal processes: central pattern generation, feedback from proprioceptors and exteroceptors, modulation and gating of central pathways, many of which utilise the same neuronal elements and pathways. On this view it is naive to expect to identify distinct equilibrium and locomotory behaviours.

Multisensory interactions in the crustacean equilibrium system

Rather, we must attempt to understand how inputs from sensory systems which detect body orientation feed into the circuits controlling both the disposition of the body and legs and the generation of propulsion. A first attempt to analyse these relationships is presented in the final section of this review.

MULTISENSORY CONTROL OF EQUILIBRIUM

Eyestalk Movements

The stalked eyes of decapod crustaceans are mobile in three dimensions (Fig. 1), and show optokinetic nystagmus (OKN) in the horizontal plane. This reaction has many features in common with vertebrate OKN, and has been subjected to intensive investigation, especially in crabs (reviewed by Neil, 1982). Image movement over the eye elicits a corrective eye response which reduces, but never completely abolishes retinal slip. The control system incorporates a variable-gain amplifier, and a negative feedback loop which acts to minimise the error between input and output. It is particularly sensitive at low stimulus velocities, when open-loop gain reaches large values.

Similar visual responses occur in vertical planes, where they form components of equilibrium behaviour. They show many

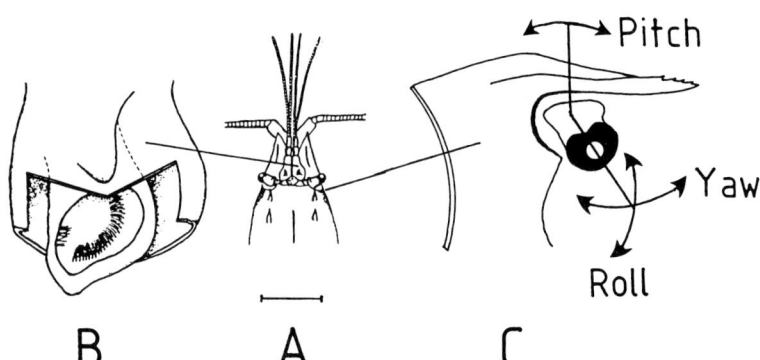

Fig. 1. Sensory and motor structures involved in eye movement control in the crayfish. A. Dorsal view of the animal (with rostrum removed) showing the statocysts in the bases of the antennules, and the stalked compound eyes. B. Enlarged drawing of the right statocyst showing the invaginated sac and the crescent of sensory hairs (with statolith removed). C. Lateral view of the right eye indicating the range of its compensatory movement about different axes in space. Scale bar = 20mm in A, 1mm in B, 4mm in C. (Modified from Stein, 1975).

Multisensory interactions in the crustacean equilibrium system

of the characteristics of the horizontal OKN, but lack the nystagmus fast phase. These optomotor reactions have been examined in various species (Hisada et al., 1969; Neil, 1975b), and their response characteristics over a wide range of stimulus frequencies have recently been examined in the spiny lobster, Palinurus elephas (Neil & Schöne, 1979; Neil et al., 1983). With increasing stimulus velocity the gain of the response falls steadily, until a washout is reached at a velocity of $2°\ s^{-1}$ (Fig 2A). Stimuli encountered by lobsters

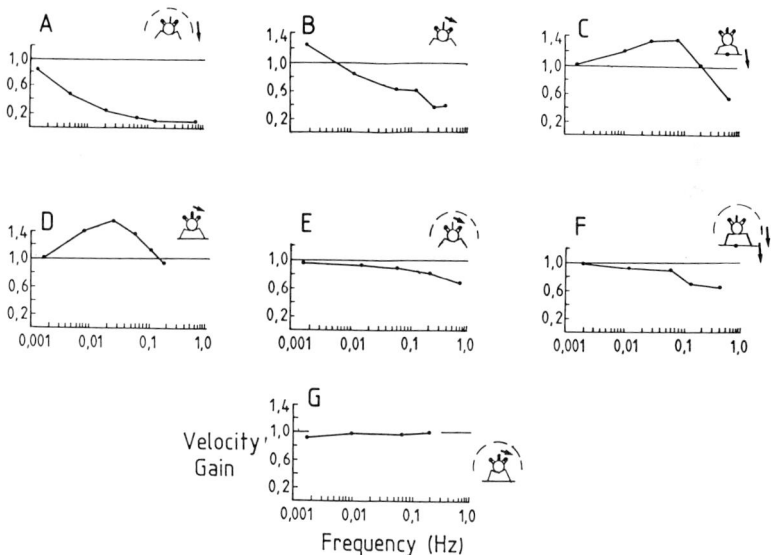

Fig. 2. Comparison of compensatory eye responses of the spiny lobster, Palinurus elephas, in the roll plane under different stimulus conditions. Data obtained from ±20° sinusoidal oscillation of the body, the platform under the legs or the high-contrast visual surround at a number of different frequencies. Results expressed as velocity gain i.e. response velocity/stimulus velocity. A. Optokinetic stimulation alone. B. Statocyst stimulation alone (body oscillation in darkness). C. Leg proprioceptor stimulation alone (platform oscillation in darkness). D. Combined stimulation of statocysts and leg proprioceptors (body oscillation against fixed platform in darkness). E. Combined stimulation of statocysts and optokinetic system (body oscillation in light). F. Combined stimulation of leg proprioceptors and optokinetic system (platform oscillation in light). G. Combined stimulation of all three sensory modalities (body oscillation against fixed platform in light). (Redrawn from data in Neil & Schöne, 1979, Schöne et al., 1983 and Neil et al., 1983).

when exposed to buffeting by surge or tidal currents often exceed these values (Herrnkind, 1983), but in fact optomotor responses can continue to contribute to compensatory eye reflexes under these circumstances. In order to understand how this can occur, it is necessary to consider the inputs from other sensory systems.

Responses to body tilt are mediated by the statocyst organs, which are in many ways analogous to the vestibular system of vertebrates. They are formed as cuticular invaginations at the bases of the first cephalic appendages, the antennules. In the statocyst sac a cushion of sensory hairs is formed into a crescent, and bears a lith formed from a concretion of sand grains (Fig 1). These hairs are primarily responsive to the shearing force imposed upon the overlying lith, which in turn is proportional to the angle of body tilt. Response direction is expressed in the pattern of excitation of the polarised hairs around the sensory crescent (Schöne, 1975; Takahata & Hisada, 1979). In a series of behavioural experiments on crayfish, lobsters and prawns, Schöne (1951, 1954) showed that there is a linear relationship between the tilt stimulus and the resulting compensatory eye movement. We do not know if linearity is preserved at all levels in the nervous system between the sensory and motor neurones, or is imposed upon a series of essentially non-linear transformations by some process of calibration, involving perhaps the visual system. The latter process seems the more likely since a formal integration step may be incorporated in the mechanics of the final eye movement (Fay, 1975; Silvey & Sandeman, 1976; Mellon & Lorton, 1977) and since recalibration of the eye reflex indeed occurs when one statocyst is destroyed, or its lith removed (Schöne, 1954; Neil, 1975c).

In the spiny lobster <u>Palinurus</u> the eye response to body tilt produces only a partial compensation (Fig. 2B) (Schöne et al., 1983), unlike the virtually complete compensation displayed by other species (Schöne, 1954; Hisada et al., 1969; Neil, 1975a). This may be due to the rather primitive nature of its statocyst, which has an ill-defined crescent of sensory hairs and separate statoconial grains rather than a consolidated lith. As might be expected, the sensory responses of the hairs in this statocyst to body tilt are inconsistent (Cohen et al., 1953).

Perhaps for this reason, spiny lobsters show a very strong responsiveness to the changing position of the body relative to the proximal joints of the walking legs, which accurately reflects imposed roll disturbances to a standing animal. The changes in leg joint angles are detected by proprioceptors, notably the chordotonal organ at the coxo-basal joint (CB CO) for movements in vertical planes, which initiate strong equilibrium reactions, including compensatory eye movements (Schöne et al., 1976).

Multisensory interactions in the crustacean equilibrium system

Experimental investigation of the response characteristics of the leg-driven eye response can be conveniently performed by suspending the animal over a pivoted platform, and tilting the platform to impose movements of the legs relative to the body. If such experiments are performed in darkness, the eye response to leg movements normally exceeds the equivalent roll stimulus in terms of both velocity and amplitude (Fig. 2C)(Neil & Schöne, 1979). This high-gain response occurs over a wide range of stimulus velocities and has an upper cut-off well beyond the wash-out of the optokinetic response (Neil et al., 1983). By reacting in this way the eye overcompensates in its movement, and fails to maintain space constancy. In this sense the leg-driven response deviates as much from true compensation as does the statocyst driven response, and it is this common discrepancy which the visual reflexes are required to correct.

It is important to note that the movement of the eye under the influence of both leg proprioceptor and statocyst stimulation is essentially open-loop, since there appears to be no proprioceptive feedback of eye position onto these reflex pathways (Sandeman, 1964). What has not previously been established is how such strong ballistic responses are influenced by the action of optokinetic feedback, especially at those velocities where this visual response is weak.

The manner in which the three sensory modalities dependent upon visual, gravitational and proprioceptive cues interact in the control of compensatory eye movements has been investigated by simultaneously presenting two or more of these stimuli in either in-phase or antiphase combinations (Neil et al., 1983; Schöne et al., 1983). As might be expected from their open-loop nature, the combination of statocyst and leg proprioceptor stimulation which simulates a roll disturbance to a standing animal produces a response which exceeds that obtained when either system acts alone (Fig. 2D), although there are indications that the response may saturate.

Combining an optokinetic stimulus with the gravitational stimulus in the configuration which occurs naturally during body roll improves the eye response towards full compensation (Fig 2E). The equivalent relationship between the visual stimulus and that delivered to the legs through the tilting platform actually reduces the extent of eye movement, although in doing so it again makes it a more precise compensation (Fig. 2F). When all three stimuli are combined in this manner, eye movements closely match those of the stimulus over a wide range of velocities, which might be expected to cover the normal operating range of this compensatory response (Fig. 2G).

The optomotor system extends its closed-loop effect into the upper velocity range of the multimodal stimulus because the true visual stimulus derives from the difference between the velocity of the eye, as driven by other modalities, and the velocity of the visual surround. This value is equivalent to the "slip speed" of images over the retina in a purely optokinetic response. Calculated values for the experiments on Palinurus are shown in Fig. 3, and indicate that at all stimulus velocities employed the true visual stimulus remains within the range of high sensitivity in the optomotor system.

In order to study the behaviour of the multisensory system under conditions of poor optomotor performance, it is necessary to provide a high velocity visual stimulus in combination with a mechanosensory stimulus (eg. platform tilt) which is either opposite in its direction of drive, or has a lower velocity. Under both these conditions the resulting eye movement does indeed follow the mechanosensory stimulus, even if the visual stimulus dictates otherwise (Neil et al., 1983). The effect of the optomotor system can therefore be overridden when it operates outside its sensitive range. These results

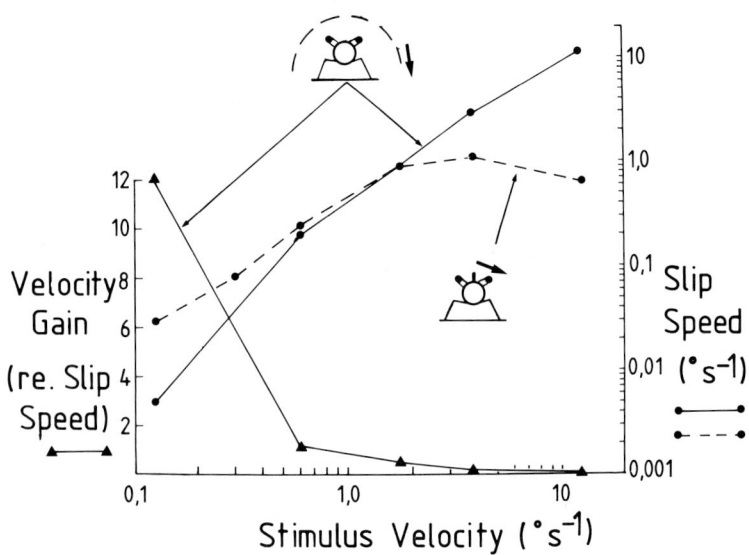

Fig. 3. Slip speed (stimulus velocity - response velocity) and velocity gain re. slip speed (response velocity/slip speed) for compensatory eye movements of Palinurus. The closed-loop optokinetic response has a high slip speed and consequent low velocity gain at high stimulus velocities. Under open-loop conditions (animal tilted against platform in darkness) the slip speed never exceeds $1°\ s^{-1}$. This, in turn, will represent the highest stimulus velocity encountered by the optokinetic system under these multimodal conditions. (Redrawn from data in Schöne et al., 1983).

suggest that under multisensory conditions the control systems for eye compensation may not involve any strong inhibitory interactions, but behave in a manner predictable from their individual characteristics.

We are therefore able to define with a certain precision the manner in which different sensory modalities interact to control an equilibrium reaction. The open-loop, high-gain mechanosensory responses initiate eye movement in the appropriate direction, maintaining the visual stimulus to the optomotor system within its sensitive range. This closed-loop visual response, by dint of its negative feedback loop, adds the fine control to the response so that the eye movement becomes an accurately matched compensation. Thus the visual apparatus is maintained in a constant positional relationship to the environment in the face of external disturbances. This precise control appears to be necessary to enable the eye to detect image movements in the field of view which are not self-induced, but are due to presence of predators, prey or mates.

This interpretation of multimodal interaction is somewhat at variance with that derived from previous studies on compensatory eye movements in crustaceans. It suggests that there is much more than an algebraic summation of fixed inputs from each modality, as concluded by Fay (1973), Mellon and Lorton (1977) and Olivo and Jazak (1980), since the response characteristics of each sensory system change with stimulus velocity in different ways. Moreover, the concept of range-fractionation in the velocity-sensitivity of the multimodal response is less straightforward than descibed by Sandeman (1977), and cannot be represented simply by the response curves for the individual modalities. Rather, as I have described, they reflect the way in which an open-loop mechanosensory system with high-velocity sensitivity permits a closed-loop visual system with low-velocity sensitivity to contribute feedback control over a wide velocity range of multisensory stimulation.

Righting Reactions
In the context of feedback and motor control, righting reactions form part of a formal feedback loop since they actively restore the animal's primary orientation, which represents the controlled set-point. They have proved suitable systems in which to study the neuronal mechanisms underlying equilibrium control since many are distant from the sensory systems involved, and offer many convenient sites for monitoring neuronal traffic.

The role of vision in releasing righting behaviour is not clear: optomotor swimming behaviour can easily be demonstrated (Fraenkel & Gunn, 1961), but its specific effect on righting reactions under experimental conditions appears to be minimal (Neil et al., 1983; Neil & Schöne, unpublished observations).

Multisensory interactions in the crustacean equilibrium system

The clearest effects originate from statocyst stimulation, which have thus received most attention (Davis, 1968; Yoshino et al.,1980; Miyan, 1982). In certain species, notably the spiny lobsters and the crayfish, inputs from leg proprioceptors onto cephalic, thoracic and abdominal appendages involved in righting behaviour are also significant (Schöne et al., 1976; Suzuki & Hisada, 1979; Neil, 1982).

The majority of righting reactions are dynamic, and involve the generation of forces to oppose external disturbances, or those which result from the animal's own locomotion. Large appendages such as the claws of Homarus may be redisposed to change the moment of inertia in an appropriate way (Fig. 4A)(Kennedy & Davis, 1977). Appendages which act as paddles for swimming can change attitude to redirect their powerstroke thrusts, eg. swimming legs (Fraser, 1975) and swimmerets (Fig. 4B)(Davis, 1968; Miyan, 1982). Phasic movements of the giant antennae of the spiny lobster Palinurus can act in a similar way either against the water

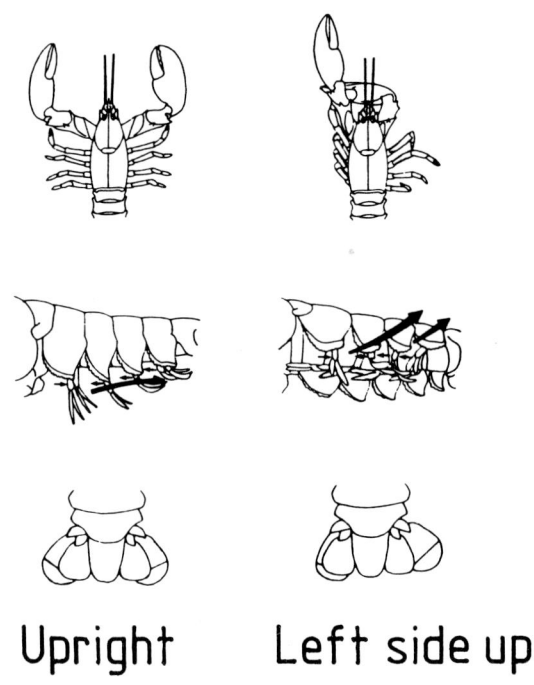

Upright   Left side up

Fig. 4. Righting reactions of claws and legs (above), swimmerets (middle) and uropods (below) in the lobster Homarus americanus in response to a body roll which takes the left side up. (From Kennedy & Davis, 1977).

itself, or more effectively against fixed objects in the environment which offer purchase. The walking legs are most important in maintaining equilibrium, and produce stabilising forces through the action of resistance reflexes at the various joints (Bush, 1965). These reflexes are driven by proprioceptors which detect changes in joint angle, and activate muscles at the stimulated joint which oppose the effect of the disruption (reviewed by Page, 1982).

Other righting forces arise because certain appendages act as steering elements. When a lobster is tilted in roll, the flattened appendages of the terminal segment, the uropods, adopt asymmetric positions with the upper one closed and the lower one opened (Fig. 4C). This asymmetric attitude of the uropods cannot by itself cause righting, but must redirect other forces produced either by the animal's locomotor systems, or by external factors. We know much about the neuronal mechanisms which control such steering reactions in a

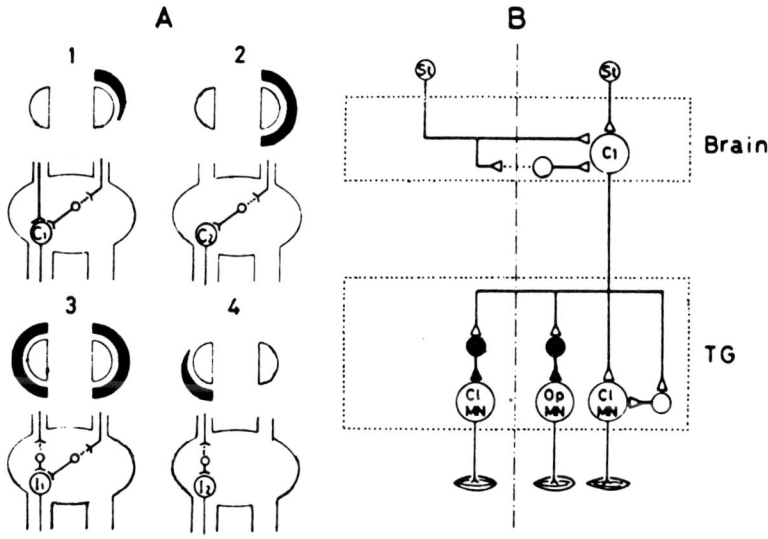

Fig. 5. Functional connections of four statocyst interneurones (C1, C2, I1 & I2) in the crayfish, Procambarus clarkii. A Polysynaptic inputs onto the interneurones from statocyst receptors. Shading around the statocyst sensory crescent indicates connections from receptors in this region. B Output connections of one interneurone (C1) with the uropod motoneurones in the terminal abdominal ganglion (TG). Open triangles - excitatory synapses; filled triangles - inhibitory synapses. St - statocyst; Cl MN - closer motor neurone; Op MN - opener motor neurone. (A from Takahata & Hisada, 1982; B from Takahata et al., 1985).

number of different macruran species, but surprisingly little about their effectiveness in producing righting torque, or even the contexts in which such reactions occur. Recent studies described here attempt to clarify their role in equilibrium behaviour.

In the crayfish, statocyst information is transmitted from the brain to the posterior ganglia via four pairs of interneurones, each of which responds maximally to particular directions of body tilt (Fig. 5A)(Takahata & Hisada, 1982). They are connected selectively to afferents coding restricted angular ranges, and their tonic firing levels exactly reflect the position of the animal relative to the upright. These interneurones pass by either ipsilateral or contralateral routes to supply the equilibrium motor systems of the thorax and abdomen. Their output connections onto the motor system controlling uropod steering are made through parallel monosynaptic and polysynaptic connections in the terminal abdominal ganglion (Fig. 5B). Interestingly, no single interneurone makes complete reciprocal connections onto the opening and closing muscles of the uropods on the two sides, and it therefore seems necessary for several to be activated in unison to produce the complete steering reaction (Takahata et al., 1985).

How do asymmetric uropod movements cause righting? Any passive braking effect during roll would be increased by opening both uropods, to maximize the surface area presented for reaction against the water. Davis (1968) suggests that the opened lower uropod intercepts the flow of water set up by the ipsilateral swimmerets, which maintain a rearward beat. This is unconvincing, since the magnitude of any righting torque produced will be very small, and it is often observed that the lower swimmerets cease beating during body tilts (Miyan, 1982).

A more likely explanation suggests itself from the behavioural conditions under which uropod steering occurs. This response is variable in occurrence, and is often not initiated when the body is tilted. However, it appears reliably when the abdomen is actively extending (Takahata et al., 1981), a behaviour which can be elicited by removing the substrate from beneath the legs (Larimer & Eggleston, 1971;, Page, 1975, 1981). This behavioural state, which Cattaert (1984) designates the "startle reaction", is characterised by a defined sequence of movements (Bowerman & Larimer, 1974), and may be regarded as a normal prelude to rapid tail flexion (Wine & Krasne, 1982).

It seems appropriate therefore to consider the uropod steering response in relation to the tail flick itself, especially since Webb (1979) has demonstrated that these terminal appendages make a major contribution to force generation. Experiments on the spiny lobster, Jasus lalandii and the Norway lobster, Nephrops norvegicus (Newland, unpublished observations) demonstrate that when tail flicks are induced with the animal held in a tilted position, phasic asymmetrical movements of the uropods reliably accompany abdominal flexion, even if they are absent during the preparatory phase. Furthermore, it has been directly demonstrated by force measurements that a righting torque is produced by these tilted tail flicks (Fig. 6). Therefore, it seems appropriate to regard the association between uropod steering and the tail flick as functionally significant, and to consider escape swimming as an important behavioural context in which righting reactions are brought into play. This exemplifies a motor behaviour in which equilibrium and locomotion are inseparable entities. Even so, such an association may not be universally present: some crayfish

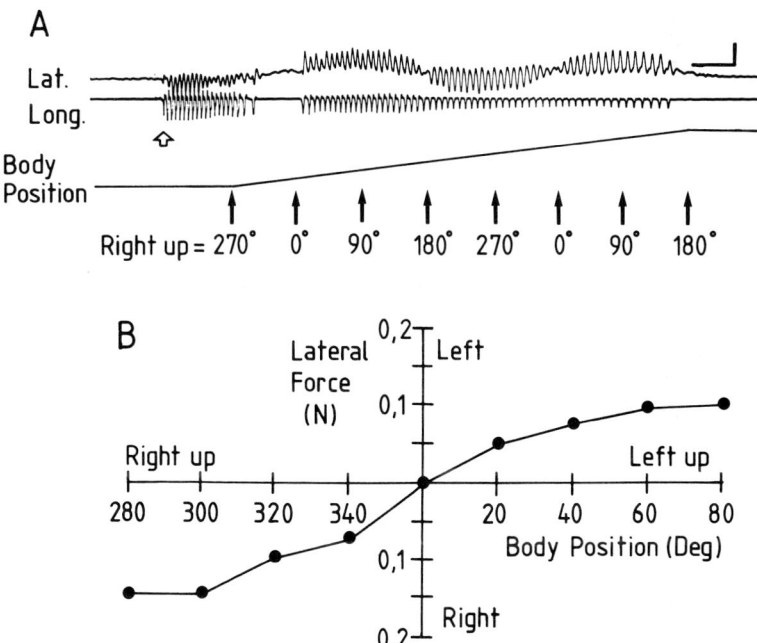

Fig. 6. Forces produced by tilted tail flicks in the lobster Nephrops norvegicus. A Recordings of longitudinal (Long.) and lateral (Lat.) forces in a sequence of escape swimming executed whilst the animal was rolled about its longitudinal axis. Scale bars 5s, 0.1N (Lat.) & 10N (Long.). B Plot of lateral forces produced by tail flicks evoked with the animal held at different roll angles. (P. Newland, unpublished results).

species show reversed uropod responses to body tilt, which suggests that their role in righting behaviour depends not on locomotory thrusts, but on other forces such as those produced by water currents passing over the body (Yoshino et al., 1982). Such variation in homologous behaviour poses interesting problems regarding both its function and its neuronal control (Hisada & Neil, 1985).

The uropod response to roll exemplifies one type of righting reaction to tilt about a particular horizontal axis as detected by a single sensory modality. This is a first stage in building up a picture of how different righting reactions driven by the various sensory systems are integrated to control equilibrium in all directions.

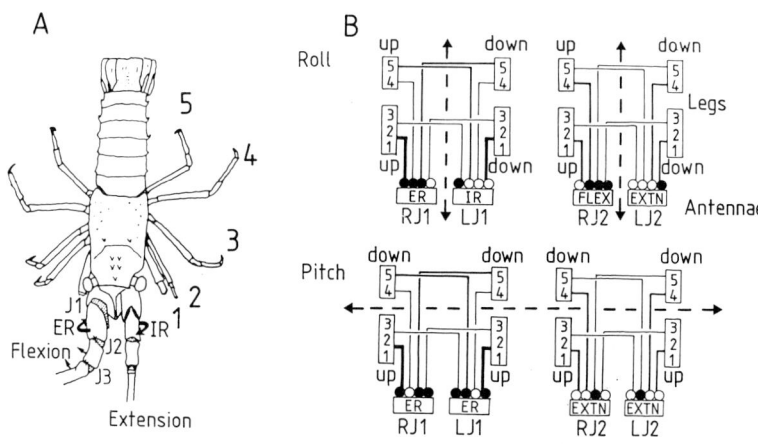

Fig. 7 A. Dorsal view of the spiny lobster Palinurus in its normal stance, showing the disposition of the legs. The antennae are depicted with the posture adopted when the right legs are levated and the left legs depressed on a pivoted platform. In the left antenna, internal rotation (IR) at the joint J1 is accompanied by extension at the outer joints J2 and J3. In the right antenna, external rotation (ER) at J1 occurs with flexion at J2 and J3. B. Diagrams showing the functional connections between the proprioceptive systems of the legs and the joints J1 and J2 of the two antennae. The diagrams have the same orientation, and represent the same polarities of antennal response, as in A. Upper figures show the response to roll, lower figures the response to pitch (dashed lines indicate axis of tilt). The strength and directionality of the inputs are represented by the thickness of the lines and the symbols respectively. Closed symbols represent ER & flexion, open symbols IR & extension. Pathways controlling the J1 joints and J2 joints are shown separately, and the observed movement at each of the four joints is indicated. (From Neil et al., 1984).

Multisensory interactions in the crustacean equilibrium system

A Righting Reaction Driven by Leg Proprioceptors
Recent studies on Palinurus have demonstrated that the reflex drive from leg proprioceptors onto the righting reactions of the antennae can be as strong as it is onto compensatory eye movements (Clarac et al., 1976; Neil et al., 1982). Spiny lobster giant antennae provide convenient systems in which to study vector coding by the leg system, since they are moved in three dimensions by the action of the adjacent uniplanar joints J1 and J2 which operate in mutually orthogonal planes (Fig. 7A). In response to platform tilt, antennal movements are not compensatory, but occur in a direction appropriate to produce righting forces. This remains true when the animal is turned around a vertical axis, so that its body axis changes relative to the tilt axis of the platform, simulating the action of water currents impinging from different directions. Changes in the contributions of joints J1 and J2 to the movement of an antenna, and changes in the co-ordination between the two antennae account for this persistent correlation between the input and output vectors of this righting reaction (Neil et al., 1984).

The reflex organisation underlying these reactions has been elucidated by measuring the effect of separately stimulating small groups of legs. Each leg acts as a single element in the detecting system, but the strength and directionality of its reflex effect depend upon its segmental position. These differences are not simply bilateral, but also exist between the front and rear legs on one side. Thus, homologous receptors in legs of the two ipsilateral groups may produce opposite reflex effects in the antennal motor system (Fig. 7B). These effects are reinforced by the stance of the animal, which maintains a clear separation between front and rear legs, and imposes a morphological polarization upon the detection system. Interleg reflexes at the most proximal joints (T-C) may be implicated in regulating this posture (Clarac, 1981).

Taking account of the wiring arrangement shown in Fig. 7B, the reflex system from legs to antennae behaves in a predictable manner, suggesting that the weighted inputs are handled by the nervous system in a relatively simple way. The extensive vector decomposition and recomposition performed by the reflex pathways seems to be accomplished without the involvement of complex pattern generators, and is therefore economic of neuronal machinery.

Many parallel interneurones are involved in coding the movements of the legs at the C-B joint in the Norway lobster, Nephrops norvegicus (Priest, 1983). Those which respond to single legs carry the necessary information for driving the antennal reflexes, although their outputs would have to converge on motor or premotor neurones in different combinations to produce effects such as those we observe at the two antennal joints of Palinurus (Neil et al., 1984). We know nothing at present about this level of organisation. Other interneurones have been shown to pool the sensory inputs from a number of legs, either ipsilaterally, or, as described by Wiersma (1958), bilaterally in a "push-pull" fashion. Such integrated signals indicate that the leg detection system is, to some extent, "hard-wired" to respond to roll stimuli, and these interneurones might be expected to project to those motor outputs, e.g. eyes and uropods, which preferentially respond in this plane (Hisada & Neil, 1985).

EQUILIBRIUM REACTIONS AND POSTURAL REACTIONS

Having considered in detail the control circuitry of equilibrium behaviour, as conventionally defined, it is instructive to point out the relationship between these activities and other motor behaviours, in so far as we know them. A few examples suffice to illustrate that the distinction between equilibrium reactions and other postural reactions is an artificial one.

Substrate Contact and Equilibrium
It has been known since the work of Alverdes (1926) that leg contact with the substrate causes a significant depression of equilibrium reactions. Thus, tilting an animal together with the platform upon which it stands induces virtually no compensatory eye responses (Stein & Schöne, 1972; Schöne & Neil, 1977), or antennal responses (Priest, 1983). This inhibitory effect is apparently mediated by tactile receptors on the terminal segments of the legs, since supporting the legs beneath their middle segments does not cause response depression (Schöne et.al., 1978). These results suggest that different reflexes are expressed when the animal is standing, rather than swimming in the mid-water.

Resistance Reactions and Equilibrium
In the spiny lobster antenna it is possible to activate intrinsic resistance reflexes, through proprioceptors at the antennal joints, in association with reflex drive from leg proprioceptors. The equilibrium reaction is here being used primarily as a controllable motor output, more convenient in such a study than the voluntary movements often made by these appendages (Vedel & Clarac, 1975; Vedel & Monnier, 1983). There is no evidence that antennal movements induced by leg stimulation in turn elicit resistance reactions in the antennal muscles, although the system remains responsive to additional, imposed movements of the antenna (Barnes & Neil, 1982). These and other data suggest that the antennal motor system incorporates both negative feedback and efference copy elements.
    Simultaneously imposed movements of the antenna and the

legs in different timing relationships demonstrate that both reflex pathways retain access to the antennal muscles, with very little mutual inhibition (Fig. 8)(Neil et al., 1982). Movement and final position are resolved by mechanical integration of the effects of the activated muscles. This process is economic of neuronal machinery, and may be more widespread in the crustacean nervous system than has previously been realized (Burrows & Horridge, 1968; Olivo & Mellon, 1980).

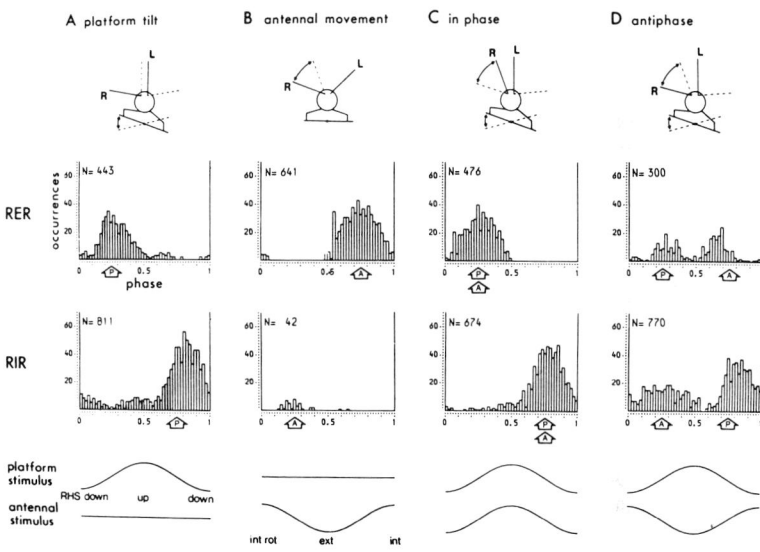

Fig. 8. Activity in the external rotator (RER) and internal rotator (RIR) muscles acting at the joint J1 of the right antenna of Palinurus, in response to different reflex inputs. Stimuli were imposed oscillations of the platform under the legs (25° excursion), or of the antenna itself (18° excursion) at 0.48 Hz. Responses are presented as phase histograms of spike activity summed over 25 cycles. Inset diagrams show the positions of the platform and antennae at phase position 0.5 (complete lines), and phase positions 0 and 1 (dashed lines). Stimulus waveforms are shown below. Open arrowheads beneath the histograms indicate the expected peaks of responses to platform oscillation (P) and antennal movement (A). A Platform tilt. B Antennal movement. C In phase movement of platform and antenna, producing compatible reflex drives. D Antiphase movements of platform and antenna, producing incompatible reflex drives. (From Barnes & Neil, 1982).

Avoidance Reactions and Equilibrium

In the crayfish, a strong mechanical stimulus to one side of the abdomen induces a sequence of escape swimming directed away from the point of stimulation (Reichert & Wine, 1983). A similar response in the spiny lobster Jasus is seen in individual tail flicks induced by brisk mechanosensory stimulation of the legs on one side (Newland, unpublished). This is accomplished by asymmetric uropod positioning, in association with rotation of the abdomen, to redirect the tail flick (Fig. 9). These steering reflexes are identical in form to those induced by body tilt (see Fig. 4), but are now directed to move the animal away from the point of stimulation, rather than to bring it back to an upright position. Thus we see that one motor pattern can subserve different behaviours, if appropriate wiring exists from the respective sensory systems involved.

Does the leg stimululation used to induce the avoidance reaction (Fig. 9) simply activate the proprioceptors

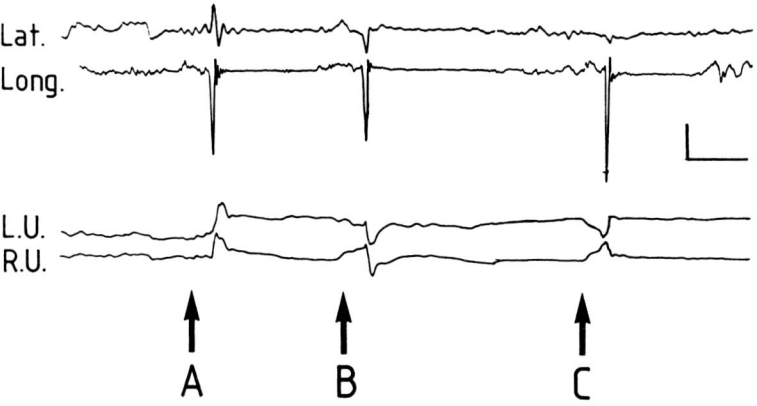

Fig. 9. Rapid tail flicks in the spiny lobster, Jasus lalandii, induced by stimulation of the legs. Correlated records of the lateral (Lat.) and longitudinal (Long.) forces produced by abdominal flexion (downward deflections illustrate forces directed to left and forwards, respectively), together with transverse movements of the uropods (downward deflections illustrate movement to left, i.e. left uropod (L.U.) opens, right uropod (R.U.) closes). A. Stimulate right legs. Uropods move to the right (an asymmetric response) and a lateral force is directed to the right. B. Stimulate left legs. Uropods move to the left (an asymmetric response) and a lateral force is directed to the left. C. Stimulate all legs. Uropods open symmetrically, and no lateral force is produced. Scale bars - 2s, 3N (Lat.), 30N (Long.) & 60°. (P. Newland, unpublished results).

responsible for the long range equilibrium response? Recent experiments on spiny lobster swimmerets suggest otherwise (Fig. 10)(Neil, Cattaert & Clarac, in preparation). In response to brisk levation of the legs on one side the ipsilateral swimmerets open, while the contralateral swimmerets remain closed. However, a more gentle levation of these legs, which simulates the stimulus imparted by a tilting platform, induces the ipsilateral swimmerets to close. Because opposite movements of the swimmerets are involved, it can be clearly seen that two distinct reflex effects are present. They display differences in response threshold which presumably reflect a selective connection to different mechanosensory systems in the leg. This finding offers an explanation for the results of Page (1981) and Page and Jones (1982) that in the crayfish a rapidly imposed depression of the leg induces short-latency reactions in motor systems throughout the body, but no eye movement. This apparent discrepancy with the clear effect of leg movements on compensatory eye responses is almost certainly due to the fact that the intense stimulus used by Page (1981) induced startle responses, rather than proprio-ocular equilibrium reactions.

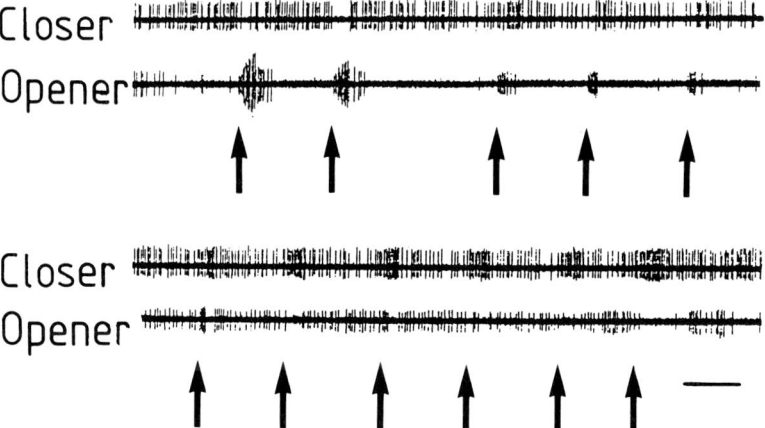

Fig. 10. Transition in the reflex effect with intensity of leg stimulation in the swimmeret of Jasus. Myographic recordings from the opener and closer muscles of a right swimmeret during a sequence of discrete levation movements of the right legs (arrows). Stimuli were imposed with diminishing velocity, and the reflex responses show a gradation from clear opening (first stimulus, upper panel), through co-activation, to clear closing (last stimulus, lower panel). Time scale - 3s. (From Neil, Cattaert & Clarac, in preparation).

CONCLUSIONS

The maintainance of equilibrium is a basic requirement for all animals, and it is perhaps not surprising to find that many sensory systems contribute to this process. Studies of the multisensory control of equilibrium reactions in decapod crustaceans provide important insights into the detailed nature and effect of these converging influences.

In the compensatory eye response we see clearly the necessity to employ both open-loop mechanosensory detectors and closed-loop optokinetic feedback. The properties of each system are not merely complementary, but interact to confer a precision of response not achievable with either system alone.

It is now clearly established that, in addition to a true vestibular system, the standing legs form a detection system which is equally capable of encoding the stimulus vectors of an equilibrium disturbance, and of initiating appropriate compensatory and righting reactions.

Some equilibrium reactions depend for their hydrodynamic effect upon the action of other, propulsive, motor systems. As an extension of this concept, many examples are now known which demonstrate that equilibrium reactions do not form a uniquely different category of motor behaviour, but grade into, or share a common identity with other motor behaviours, both postural and locomotory. A more realistic interpretation of neural integration in motor control may emerge once we abandon such a strict categorisation of observed behaviours.

Acknowledgements

I owe an outstanding debt to Professor H. Schöne for introducing me to this area of research, and collaborating over many years. The Max-Planck Gesellschaft, British Council and Carnegie Trust funded this work. Joint studies with Dr F. Clarac were funded by the European Science Foundation and N.A.T.O. Other work reported here was supported by research grant GRB27104, and two research studentships from the British S.E.R.C.

REFERENCES

Alverdes, F. (1926). Stato-, Photo- und Tangoreactionen bei zwie Garneelenarten. Z. vergl. Physiol., 4, 699-765.

Barnes, W.J.P. & Neil, D.M. (1982). Reflex antennal movements in the spiny lobster, Palinurus elephas. II. Feedback and motor control. J. comp. Physiol., 147, 269-280.

Bowerman, R.F. & Larimer, J.L. (1974). Command fibers in the circumoesophageal connectives of the crayfish. I. Tonic fibres. J. exp. Biol., 60, 95-117.

Burrows, M. & Horridge, G.A. (1968). The action of the eyecup muscles of the crab, Carcinus, during optokinetic movements. J. exp. Biol., 49, 223-

250.
Bush, B.M.H. (1965). Leg reflexes from chordotonal organs in the crab, Carcinus maenas. Comp. Biochem. Physiol., 15, 567-587.
Cattaert, D. (1984). Polymorphisme d'expression d'un systeme locomoteur vestigial chez le homard Homarus gammarus. These de 3° Cycle, Université de Bordeaux I.
Clarac, F. (1981). Postural reflexes coordinating walking legs in a rock lobster. J. exp. Biol., 90, 333-337.
Clarac, F., Neil, D.M. & Vedel, J.-P. (1976). The control of antennal movements by leg proprioceptors in the rock lobster, Palinurus vulgaris. J. comp. Physiol., 107, 275-292.
Cohen, M.J., Katsuki, Y. & Bullock, T.H. (1953). Oscillographic analysis of equilibrium receptors in Crustacea. Experientia, 11, 434-435.
Davis, W.J. (1968). Lobster righting responses and their neural control. Proc. R. Soc. B, 144, 480-495.
Evoy, W.H., & Ayers, J. (1982). Locomotion and control of limb movements. In The Biology of Crustacea, Vol. 4. Eds. Atwood, H.L. & Sandeman, D.C. Academic Press, New York, pp. 61-105.
Fay, R.R. (1973). Multisensory interaction in the control of eyestalk rotation in the crayfish (Procambarus clarkii). J. comp. physiol. Psychol., 84, 527-533.
Fay, R.R. (1975). Dynamic properties of the compensatory eyestalk rotation response of the crayfish (Procambarus clarkii). Comp. Biochem. Physiol., 51A, 101-103.
Fraenkel, G. & Gunn, D.L. (1961). The Orientation of Animals: Kineses, Taxes and Compass Reactions. Dover, New York.
Fraser, P.J. (1975). Three classes of inputs to a semicircular canal interneurone in the crab, Scylla serrata, and a possible output. J. comp. Physiol., 104, 261-271.
Herrnkind, W.F. (1983). Movement patterns and orientation. In The Biology of Crustacea, Vol. 7. Eds. Vernberg, F.J. & Vernberg, W.B.. Academic Press, New York, pp. 41-105.
Hisada, M. & Neil, D.M. (1985). The neuronal basis of equilibrium behaviour in decapod crustaceans. In Co-ordination of Motor Behaviour. Eds. Bush, B.M.H. & Clarac, F. Society for Experimental Biology Seminar Series 24. Cambridge University Press, Cambridge, pp. 229-248.
Hisada, M., Sugawara, K. & Higuchi, T. (1969). Visual and geotactic control of compensatory eyecup movement in the crayfish, Procambarus clarki. J. Fac. Sci. Hokkaido Univ. Ser. 6, 17, 224-239.
Kennedy, D. & Davis, W.J. (1977). Organization of invertebrate motor systems. In Handbook of Physiology, Section 1, The Nervous System, Vol. 1, Cellular Biology of Neurons. Ed. Kandel, E.R. American Physiological Society, Bethesda, pp. 1023-1087.
Larimer, J.L. & Eggleston, A.C. (1971). Motor programs for abdominal positioning in the crayfish. Z. vergl. Physiol., 74, 388-402.
Mellon, DeF. & Lorton, E.D. (1977). Reflex actions of the functional divisions in the crayfish oculomotor system. J. comp. Physiol., 121, 367-380.
Miyan, J.A. (1982). The neuronal basis of the swimmeret equilibrium reaction in the lobster Nephrops norvegicus (L). Ph.D. Thesis, University of

Glasgow.
Neil, D.M. (1975a). The control of eyestalk movements in the mysid shrimp, Praunus flexuosus. J. exp. Biol., 62, 487-504.
Neil, D.M. (1975b). The optokinetic responses of the mysid shrimp, Praunus flexuosus. J. exp. Biol., 62, 505-518.
Neil, D.M. (1975c). The mechanism of statocyst operation in the mysid shrimp, Praunus flexuosus. J. exp. Biol., 62, 685-700.
Neil, D.M. (1982). Compensatory eye movements. In The Biology of Crustacea, Vol. 4. Eds. Atwood, H.L. & Sandeman, D.C.. Academic Press, New York, pp. 133-163.
Neil, D.M., Barnes, W.J.P. & Burns, M.D. (1982). Reflex antennal movements in the spiny lobster, Palinurus elephas. I. Properties of reflexes and their interaction. J. comp. Physiol., 147, 259-268.
Neil, D.M., Priest, T.D., Miyan, J.A., Wotherspoon, R.M. & Schöne, H. (1984). Co-ordinated equilibrium responses at two joints in the spiny lobster antenna in relation to the pattern of movements imposed upon the legs. J. comp. Physiol. A, 155, 351-363.
Neil, D.M. & Schöne, H. (1979). Reactions of the spiny lobster, Palinurus vulgaris to substrate tilt. II Input-output analysis of eyestalk responses. J. exp. Biol., 79, 59-67.
Neil, D.M., Schöne, H., Scapini, F., & Miyan, J.A. (1983). Optokinetic responses, visual adaptation and multisensory control of eye movements in the spiny lobster, Palinurus vulgaris. J. exp. Biol., 107, 349-366.
Olivo, R.F. & Jazak, M.M. (1980). Proprioception provides a major input to the horizontal oculomotor system of crayfish. Vision Res., 20, 349-353.
Olivo, R.F. & Mellon, DeF. (1980). Oculomotor activity during combined optokinetic and proprioceptive stimulation in the crayfish. Soc. Neurosci. Abstr., 6, 101.
Page, C.H. (1975). Command fiber control of crayfish abdominal movement. II. Generic differences in the extension reflexes of Orconectes and Procambarus. J. comp. Physiol., 102, 77-84.
Page, C.H. (1981). Thoracic leg control of abdominal extension in the crayfish, Procambarus clarkii. J. exp. Biol., 90, 85-100.
Page, C.H. (1982). Control of posture. In The Biology of Crustacea, Vol 4. Eds. Atwood, H.L. & Sandeman, D.C.. Academic Press, New York, pp. 33-59.
Page, C.H. & Jones K.A. (1982). Abdominal motoneurone responses elicited by flexion of a crayfish leg. J. exp. Biol., 99, 339-347.
Priest, T.D. (1983). An equilibrium reflex in decapod Crustacea mediated by basal leg proprioceptors. Ph.D Thesis, University of Glasgow.
Reichert, H. & Wine, J.J. (1983). Coordination of lateral giant and non-giant systems in crayfish escape behavior. J. comp. Physiol., 153, 3-15.
Sandeman, D.C. (1964). Functional distinction between oculomotor and optic nerves in Carcinus (Crustacea). Nature, Lond., 201, 302-303.
Sandeman, D.C. (1977). Compensatory eye movements in crabs. In Identified Neurons and Behavior of Arthropods. Ed. Hoyle, G. Plenum, New York, pp 131-147.
Sandeman, D.C. (1981). Equilibrium and proprioceptive systems and the central nervous system of arthropods. In Sense Organs. Eds. Laverack, M.S.& Cosens, D.J.. Blackie, Glasgow, pp. 276-294.
Sandeman, D.C. (1983). The balance and visual systems of the swimming crab:

their morphology and interaction. In Multimodal Convergences in Sensory Systems. Ed. Horn, E. Fortschr. Zool., 28, 213-229.

Schöne, H. (1951). Die statische Gleichgewichts-orientierung bei dekapoden Crustaceen. Verh. dt. zool. Ges., 16, 157-162.

Schöne, H. (1954). Statocystenfunktion und statische Lageorientierung bei dekapoden Krebsen. Z. vergl. Physiol., 36, 241-260.

Schöne, H. (1975). On the transformation of the gravity input into reactions by statolith organs of the "fan" type. Fortschr. Zool., 23, 120-126.

Schöne, H. & Neil, D.M. (1977). The integration of leg position receptors and their interaction with statocyst inputs in spiny lobsters (Reactions of Palinurus vulgaris to substrate tilt). Mar. Behav. Physiol., 5, 45-59.

Schöne, H., Neil, D.M. & Scapini, F. (1978). The influence of substrate contact on gravity orientation. Substrate orientation in spiny lobsters V. J. comp. Physiol., 126, 293-295.

Schöne, H., Neil, D.M., Scapini, F. & Dreissman, G. (1983). Interaction of substrate, gravity and visual cues in the control of compensatory eye responses in the spiny lobster, Palinurus vulgaris. J. comp. Physiol., 150, 23-30.

Schöne, H., Neil, D.M., Stein, A. & Carlstead, M.K. (1976). Reactions of the spiny lobster, Palinurus vulgaris to substrate tilt. I. J. comp. Physiol., 107, 113-128.

Silvey, G.E. & Sandeman, D.C. (1976). Integration between statocyst sensory neurons and oculomotor neurons in the crab Scylla serrata. I. Horizontal compensatory eye movements. J. comp. Physiol., 108, 35-43.

Stein, A. (1975). Attainment of positional information in the crayfish statocyst. Fortschr. Zool., 23, 109-119.

Stein, A. & Schöne, H. (1972). Uber das Zusammenspiel von Schwereorientierung und Orientierung zur Unterlage beim Flusskrebs. Verh. dt. zool. Ges., 65, 225-229.

Suzuki, Y. & Hisada, M. (1979). Abdominal adductor muscle in crayfish: physiological properties and neural control. II. Abdominal movements. Comp. Biochem. Physiol., 62A, 841-846.

Takahata, M. & Hisada, M. (1979). Functional polarization of statocyst receptors in the crayfish Procambarus clarkii Girard. J. comp. Physiol., 130, 201-207.

Takahata, M. & Hisada, M. (1982). Statocyst interneurons in the crayfish Procambarus clarkii Girard. I. Identification and response characteristics. J. comp. Physiol., 149, 287-300.

Takahata, M., Yoshino, M. & Hisada, M. (1981). The association of uropod steering with postural movement of the abdomen in crayfish. J. exp. Biol., 91, 341-345.

Takahata, M., Yoshino, M. & Hisada, M. (1985). Neuronal mechanisms underlying the crayfish steering behaviour as an equilibrium response. J. exp. Biol., 114, 599-617.

Vedel, J-P. & Clarac, F. (1975). Neurophysiological study of the antennal motor patterns in the rock lobster Palinurus vulgaris. II. Motoneuronal discharge patterns during passive and active flagellum movements. J. comp. Physiol., 102, 223-225.

Vedel, J-P. & Monnier, S. (1983). A new muscle receptor in the antenna of the rock lobster Palinurus vulgaris : mechanical, muscular and

proprioceptive organisation of the two proximal joints J0 and J1. Proc. R. Soc. B, 218, 95-110.
Webb, P.W. (1979). Mechanics of escape responses in crayfish (Orconectes virilis). J. exp. Biol., 79, 245-263.
Wiersma, C.A.G. (1958). On the functional connections of single units in the central nervous system of the crayfish, Procambarus clarkii (Girard). J. comp. Neurol., 110, 421-471.
Wine, J.J. & Krasne, F.B. (1982). The cellular organization of crayfish escape behavior. In The Biology of Crustacea, Vol 4. Eds. Atwood, H.L. & Sandeman, D.C. Academic Press, New York, pp. 241-292.
Yoshino M., Takahata, M. & Hisada, M. (1980). Statocyst control of the uropod movement in response to body rolling in crayfish. J. comp. Physiol., 139, 243-250.
Yoshino M., Takahata, M. & Hisada, M. (1982). Interspecific differences in crustacean homologous behaviour: neural mechanisms underlying the reversal of uropod steering movement. J. comp. Physiol., 145, 471-476.

Section 6
The Control of Movement

Chapter Nineteen

INTRODUCTION

E. Jankowska

The five contributors to this section on feedback in the control of movement obviously could not discuss all aspects of this problem and could not compare observations made on different movements and in different species. I have, therefore, primarily attempted to relate the reported observations to each other and to the organisation of the reflex activity in general.

Each contributor to this section deals with a particular animal species and a particular form of motor behaviour. It appears, however, that the same forms of feedback control may operate in various species and in a number of quite different movements. For instance, it appears that the frequency of rhythmic movements can be increased, or decreased, by imposed sensory signals from the movement apparatus. The phenomenon is referred to as 'entrainment' and has been found in the case of locust flight (see Chapter 20 by Prof Pearson), fish swimming (see Chapter 21 by Drs Wallén and Williams), cat walking (see e.g. Andersson et al., 1981) and lobster walking (see Chapter 24 by Dr Clarac), being discussed in this volume in most detail by Drs Williams and Wallén. Another common phenomenon is that the kind of reaction evoked by a stimulus acting during an ongoing movement greatly depends on the phase of the movement, so that the same stimulus may evoke a flexion when it coincides with the flexion phase of a movement, and an extension during the extension phase. This phenomenon is most often referred to as 'phase dependent reflex reversal' and may serve to gate out any inappropriate reactions. Again, it has been observed in several animals: in the cat (see Chapter 23 by Drs Rossignol and Drew), in fish (see Chapter 21 by Drs Wallén and Williams) and in decapod crustaceans (see Chapter 24 by Dr Clarac); it will be discussed in particular by Drs Rossignol and Drew. Since the motor apparatus and the nervous system of these various animals are clearly organised in a different way, it would be of great interest to understand how they achieve the same results and which common neuronal mechanisms are behind them.

Previous chapters have discussed the problem of how feedback information is processed. They make it very clear that many forms of motor control use multimodal feedback and feedforward information from tactile, proprioceptive, visual, acoustic or any other receptors, which is integrated in different combinations at a premotor neurone level. We can thus no longer think in terms of feedback systems based on only one kind of information (or on changes in only one parameter of stimuli) in other than artificial experimental situations. Neuronal systems receiving different types of feedback information have been found in various forms of motor behaviour, represented in this volume by the escape reaction of the cockroach (see Chapter 4 by Prof Camhi), as well as cat and lobster walking (see Chapter 23 by Drs Rossignol and Drew and Chapter 24 by Dr Clarac), fish swimming (see Chapter 21 by Drs Wallén and Williams) and locust flight (see Chapter 20 by Prof Pearson and Chapter 22 by Prof Rowell and Drs Reichert and Bacon). Rowell, Reichert and Bacon pay most attention to this problem, discussing how such information is processed and forwarded to effectors in the locust. In this context, I would like to mention the first neuronal pathways for which multimodal sensory feedback and feedforward has been demonstrated in the cat, the neuronal pathways of flexion-extension movements, since these movements are one of the main elements of many complex movements. Also, knowledge of their neuronal organisation is quite advanced (see e.g. Lundberg, 1979a). When the convergence of various muscle, skin and joint afferents (which were called flexor reflex afferents, or FRA, by R.M. Eccles & Lundberg, 1959) was first described in the pathways of the flexion-extension reflexes, the meaning of this convergence was not quite apparent. However, both in view of what we have learned in the meantime about other multimodally steered movements, and in view of the fact that we find multimodal input in all investigated spinal neuronal pathways in the cat (see e.g. Lundberg, 1979a,b), it should now be easier to understand. There is now also fairly detailed knowledge on how the reflex neuronal networks dealing with the multimodal information operate (see e.g. Baldissera et al., 1981; Jankowska & Lundberg, 1981; Jankowska, 1983).

With respect to the question of how the multimodal feedback information is used, the chapters in this Section show how the ongoing motor activity is adjusted to the requirements of the situation, how the errors in its performance are corrected and, generally, how various reflex mechanisms are incorporated into larger entities of movements. It will be noted, however, that all the contributors to this Section have chosen to discuss the problems of feedback control in rhythmic movements, although in general there is no reason why the discussion of feedback control of movement should be restricted to the control of rhythmic movements. The recent interest in rhythmic movements seems to have its source in the hope that, in such movements, it is easier to identify the involved neurones and to investigate their organisation. Selverston (1980 see p. 353) states for instance that "it is only rhythmic behaviour which has the possibility of being explainable in terms of all the neurones involved". The analysis of non-rhythmic movements, with the most advanced examples in the studies of locust jump (see Heitler and Burrows, 1977; Pearson, 1982) and withdrawal movements (see Chapter 15 by Drs Watson and Burrows) does not appear to give us any less chance of understanding the mechanisms of these movements. There are also possibilities for such studies in vertebrates (see e.g. Alstermark et al., 1981). Furthermore, the heuristic value of rhythmic movements is not as high as expected, since their networks appear to be much more complex than anticipated (see e.g. Chapter 2 by Prof Davis). In the particular case of cat locomotion, the involved neuronal networks have in fact been "tagged" (c.f. Lundberg, 1979a) by properties other than their rhythmic activity.

The basic organisation of the neuronal machinery of movements has been discussed in Section 1. However, it might be useful to stress here one of the features of this organisation which has not been considered, although it is particularly relevant for the feedback control of movement. It could be labelled the "Principle of Economy" to match the four principles of Prof Davis. It appears that the same basic neuronal networks subserve both simple reflex movements and other movements (see e.g. Baldissera et al., 1981; Jankowska & Lundberg, 1981). At least, we do not find in the cat any neurones, or neuronal networks, which are exclusively involved in either local reflexes or in centrally initiated movements. Thus one of the cornerstones of spinal cord physiology becomes that not only the same muscles and motor neurones but also the premotor neuronal networks are shared in different movements. This is particularly true for networks of basic motor synergies such as the coordination of antagonist muscles and interlimb coordination. The same kind of organisation (i.e. use of the same neuronal networks for different purposes) appears to exist in at least some of the invertebrate systems. The closest to the vertebrate organisation is perhaps the case when the locust uses the same interneurones in such movements as a jump and a defensive kick (Heitler & Burrows, 1977; Steeves & Pearson, 1982). We shall also learn from Prof Rowell and his colleagues that the same system or systems of neurones are used for correcting deviations from a straight flight course signalled by different receptors (ocelli, eyes, wind hairs), and for which various effectors are apparently involved. Rowell, Reichert and Bacon's conclusion that the insect nervous system operates by allowing the minimal number of neurones to be used for forwarding the maximal information is certainly true both for insects and the cat, although with a different order of magnitude for the number of involved neurones.

Another important feature of the neuronal organisation of movement is that the basic neuronal networks may include a variety of networks which subserve different variants of movements, the most appropriate ones being selected at any moment. Since the input to these networks is multimodal and originates from both the periphery and central nervous structures, they may be selected by peripheral reflex mechanisms as well as by the use of central commands. However, how the basic neuronal networks, the functional units of movements, are incorporated into larger entities of movements is still far from being solved and should be analysed for each type of movement. With respect to cat locomotion Prof Grillner (1981; see p. 1216) has recently described how reflex mechanisms change the speed of locomotion, its direction, the locomotory posture, positioning of limbs and adaptation of the step, and concluded in a very convenient way that "the spinal cord contains a number of small networks that can somehow be connected in alternative ways.... The number of networks activated and the type of coupling chosen decide whether a fish will swim forward or backward... or in which direction a cat will walk, or whether it will walk or gallop". Although Prof Grillner only considered neuronal networks of rhythmic movements, his description would apply to any movement (and may thus be generalised to include non-rhythmic movements) when the relevant basic networks are taken into account.

Which neuronal networks are considered to be parts of "rhythmic pattern generators" by those who use this term to describe endogenous mechanisms of rhythmic movements is usually undefined, and the meaning of this term is sometimes stretched to the extent that it is applied to any part of the nervous system involved in the investigated movements. For this reason most "pattern generators" may be considered as a modern version of a "black box" with a still-unknown content. One of the few cases in which its content has been investigated is presented by Prof Pearson, with the conclusion that the central pattern generators do not form the basis of insect flying and walking, and that reflex mechanisms play a dominant role in the patterning of movements in these animals. The relative contribution of endogenous versus reflex mechanisms may clearly differ in various movements and observations made on insects should caution us against attributing too much to endogenous factors. Both these and other observations reported in this Section should thus encourage more systematic comparative studies of the neuronal mechanisms of feedback and feedforward control, and of their role in motor control.

REFERENCES

Alstermark, B., Lundberg, A., Norrsell, U. & Sybirska, E. (1981). Integration in descending motor pathways controlling the forelimb in the cat. 9. Differential behavioural defects after spinal cord lesions interrupting defined pathways from higher centres to motoneurones. Expl. Brain Res., 42, 299-318.

Andersson, O., Forssberg, H., Grillner, S. & Wallén, P. (1981). Peripheral feedback mechanisms acting on the central pattern generators for locomotion in fish and cat. Can. J. Physiol. & Pharmacol., 59, 713-726.

Baldissera, F., Hultborn, H. & Illert, M. (1981). Integration in spinal neuronal systems. In Handbook of Physiology, Sect. I. The Nervous System, Vol. II. Motor Control. Ed. Brooks, V.B., American Physiological Society, Bethesda, pp. 509-595.

Eccles, R.M. & Lundberg, A. (1959). Synaptic action in motoneurones by afferents which may evoke the flexion reflex. Archs ital. Biol., 97, 199.

Grillner, S. (1981). Control of locomotion in bipeds, tetrapods, and fish. In Handbook of Physiology, Sect. I. The Nervous System, Vol. II. Motor Control. Ed. Brooks, V.B., American Physiological Society, Bethesda, pp. 1179-1236.

Heitler, W.J. & Burrows, M. (1977). The locust jump. I. The motor programme. J. exp. Biol., 66, 203-219.

Jankowska, E. (1983). Interneuronal organization in reflex pathways from proprioceptors. Proc. int. Union Physiol. Sci., 15, 61.

Jankowska, E. & Lundberg, A. (1981). Interneurones in the spinal cord. Trends Neurosci., 4, 230-233.

Lundberg, A. (1979a). Multisensory control of spinal reflex pathways. In Reflex Control of Posture and Movement, Eds. Granit, R. & Pompeiano, O., Prog. Brain Res., 50, 11-28.

Lundberg, A. (1979b). Integration in a propriospinal motor centre controlling the forelimb in the cat. In Integration in the Nervous System, Eds. Asanuma, H. & Wilson, V.J., Igaku-Shoin, Tokyo & New York, pp. 47-65.

Pearson, K.G. (1982). Neural circuits for jumping in the locust. J. Physiol., Paris, 78, 765-771.

Selverston, A.I. (1980). Are central pattern generators understandable? Behav. & Brain Sci., 3, 535-571.

Steeves, J.D. & Pearson, K.G. (1982). Proprioceptive gating of inhibitory pathways to hindleg flexor motoneurons in the locust. J. comp. Physiol., 146, 507-515.

Chapter Twenty

ARE THERE CENTRAL PATTERN GENERATORS FOR WALKING AND FLIGHT IN INSECTS?

K.G. Pearson

One of the most influential concepts in the field of motor control is that central pattern generators are primarily responsible for the generation and patterning of rhythmic motor activity (Delcomyn, 1980; Selverston, 1980). Although the definition of a central pattern generator is not precise (Selverston, 1980), it is generally taken to mean a centrally located system capable of generating, in the absence of input from peripheral receptors, a rhythmic motor pattern similar to that occurring in the normal animal. Sensory input in the intact system is considered to modulate the activity of the central pattern generator, i.e. increase the overall repetition rate or modify the intensity of activity, but not function in any important way to establish the basic pattern of the motor output. The purpose of this chapter is to examine whether this view is reasonable for the flight and walking systems of insects.

First, however, it is worthwhile to consider the criteria used to establish the existence of a central pattern generator and illustrate some of the difficulties in the use of the central pattern generator concept. Basically, the only criterion for the existence of a central pattern generator is that, following deafferentation, a rhythmic motor pattern can be generated which resembles the normal motor pattern in the intact animal. Deafferentation usually leads to changes in the pattern of the motor output which can vary from slight to large depending on the system. For those systems in which there is little change in motor pattern the central pattern generator concept is obviously appropriate. The difficulty comes when we consider those systems in which there are large changes in motor output pattern following deafferentation. Clearly a central circuit must exist in each of these systems for generating rhythmic motor activity in the absence of sensory input. What are we to call this central circuit? To call it a 'central pattern generator' is not appropriate because it generates a motor pattern quite unlike that in the intact animal. Better terms are 'central rhythm generator' or 'central oscillator' for these make no inference regarding the motor pattern. However, this semantic difficulty is not the major problem with the central pattern generator concept when applied to systems in which there are large changes in the motor pattern following deafferentation. A much more serious difficulty is whether the 'central pattern generator' or 'central oscillator' (or whatever we call it) retains its identity as a functional unit in the intact system. Implicit in the central pattern generator concept is that the final motor pattern in the intact system can be explained by sensory input <u>modulating</u> the activity in a central pattern generator. However, it is quite conceivable that sensory influences on the central pattern generator may be so complex that it becomes meaningless to regard the central pattern generator as a functional unit. This idea can be easily illustrated by an analogy with known characteristics of the neuronal networks in the pyloric system of the lobster stomatogastric ganglion. In this system, a rhythmic motor pattern can be generated by the reciprocal inhibitory interaction of two of the fourteen neurones, but this pattern is quite unlike the normal pyloric pattern (Miller & Selverston, 1982). As more neurones are included in the network containing the elementary oscillator, this oscillator progressively loses its identity until eventually it cannot be regarded as a functional unit in the generation of the final motor pattern. Thus the final motor pattern cannot be explained simply by the influences other neurones have on this elementary oscillator. In a similar manner, therefore, it is conceivable that inclusion of sensory inputs into circuits generating central oscillatory activity may modify these circuits to such an extent that the central oscillator, or central pattern generator, loses its identity.

With these ideas in mind let us now examine the evidence for central pattern generators in the flight and walking systems of insects, and consider whether this evidence is sufficient for us to accept the central pattern generator concept for these systems.

<u>Flight in insects</u>
An early observation which contributed greatly to the acceptance of the concept of central patterning of motor activity was that rhythmic motor activity could be generated in the flight system of the locust following removal of all sensory input from wing receptors (Wilson, 1961). This observation has been confirmed repeatedly (Kutch, 1974; Altman, 1975; Robertson & Pearson, 1982) so there is now no doubt that rhythmic activity can be generated in the absence of sensory feedback from wing receptors. This ability of deafferented preparations to generate rhythmic activity has allowed the identification of interneurones in the flight system and the determination of many of their interconnections (Robertson & Pearson, 1982, 1983, 1984). It should be noted,

however, that the motor pattern generated in a deafferented preparation has features which differ from those of the normal flight pattern. Some of the differences are as follows:

1. The overall frequency of the rhythm in deafferented preparations is about half that in intact animals (Wilson, 1961).

2. In deafferented preparations significantly more spikes are generated per cycle in most flight motor neurones, and the usual order of recruitment of different depressor motor neurones is lost (Hedwig & Pearson, 1984).

3. Deafferentation causes a change from a phase-constant pattern (Waldron, 1967) to a latency-constant pattern (Hedwig & Pearson, 1984). In the former the phase of the depressor activity in the elevator cycle remains constant for variation in cycle time, whereas in the latter depressor bursts follow elevator bursts with a constant latency regardless of cycle time.

4. In deafferented preparations there is no rigid phase shift of forewing activity relative to hindwing activity. In normal animals hindwing activity leads forewing activity by 5 to 7ms.

It is clear, therefore, that the motor pattern generated in deafferented preparations is not a slowed down version of the normal flight pattern. Thus the production of the flight motor pattern cannot be simply explained by a tonic sensory influence acting on a central rhythm generator, as was proposed some time ago by Wilson and his colleagues (Wilson, 1961; Wilson & Gettrup, 1963; Wilson & Wyman, 1965). It is now apparent that sensory input from wing receptors has a role in the _patterning_ of motor activity. This role has been demonstrated directly in a number of recent experiments. First, the frequency of the flight rhythm can be increased significantly by imposed phasic sensory signals from wing stretch receptors (Pearson et al., 1983). Second, stimulation of stretch receptors, campaniform sensilla, tegula and head hairs alters the timing of motor activity (Neumann et al., 1982; Horsmann et al., 1983; Möhl & Neumann, 1983; Wendler, 1983). Third, the motor rhythm can be entrained by imposed movements of the wings (Wendler, 1974), rhythmic activation of head hairs (Horsmann et al., 1983), and by rhythmic stimulation of wing stretch receptors (Pearson et al., 1983).

Even though sensory input is clearly involved in patterning motor activity, this does not necessarily mean that the production of the flight pattern cannot be explained by sensory influences on a central rhythm generator. Obviously a system for centrally generating a motor rhythm exists and it may eventually be possible to account for the flight pattern by sensory modulation of the central rhythm generator. However, the critical question is: _does this central rhythm generator continue to function as a unit in the intact animal?_ Currently this question cannot be answered conclusively.

Patterning of rhythmic motor activity

Nevertheless there are reasons for thinking that sensory input might reorganise the circuits involved in the central generation of rhythmic activity. First, it is known that afferents from wing stretch receptors make extensive connections to many interneurones in the flight system, including some of those involved in generating the central rhythmicity (Pearson et al., 1983; Reye, unpublished). Presumably other sensory afferents also make widespread connections to flight interneurones (Kien & Altman, 1982). Second, some sensory receptors are clearly elements in the system generating the rhythmicity in an intact animal. The wing stretch receptors, for example, fulfil every criterion currently used for identifying elements in rhythm generating systems (Pearson et al., 1983). Third, it can easily be imagined that, in the presence of sensory input, additional central neuronal elements may be introduced into the rhythm generating system and that some interneurones functioning to generate rhythmicity in a deafferented preparation may not be functioning in the presence of sensory input. In other words, the central rhythm generator may be physically altered (additional or fewer elements) by sensory input. It is noteworthy that some interneurones only weakly rhythmically active in deafferented preparations receive very strong input from the wing stretch receptors (Reye, unpublished) and would therefore be expected to show strong rhythmicity in the intact animal. Although none of these observations force us to abandon the central pattern generator concept for the locust flight system, they do alert us to the possibility that sensory input may profoundly alter the characteristics of the central rhythm generating network and that the central rhythm generator may not exist as a functional unit in the intact animal.

Walking in insects

The conclusion that central pattern generators form the basis for rhythmic leg movements during insect walking comes primarily from observations on the cockroach (Pearson & Iles, 1970, 1973; Pearson, 1972). In deafferented preparations recordings from the nerves supplying extensor and flexor muscles of either the hindleg or the middle leg show periods of rhythmic reciprocal activity. During regular sequences of rhythmic activity the durations of the flexor bursts remain relatively constant and reasonably close to the duration of leg protraction during walking. This was the main observation leading to the conclusion that the rhythmic motor activity in the deafferented preparation was associated with the system generating motor activity for walking. Before considering whether central pattern generators form the basis of the motor programme for walking, let us note some further observations on the walking system of the cockroach:

1. In deafferented preparations there is no strong

correlation between the duration of a flexor burst and the following extensor burst (Pearson & Iles, 1970). In other words a short duration flexor burst can be followed by a long duration extensor burst and vice versa. This is quite unlike the situation in the walking animal where the durations of these bursts are strongly correlated (Pearson, 1972).

2. There is no strict correlation between the intraburst discharge rates of sequential flexor and extensor bursts in deafferented preparations. Sometimes the extensors can discharge at high rates and at other times they can be completely inactive for similar flexor bursts (Pearson & Iles, 1970). Again this is unlike the walking animal where there is a strong correlation between extensor and flexor discharge rates (Pearson, 1972).

3. The periods of regular rhythmic reciprocal activity in deafferented preparations rarely occur spontaneously and are difficult to evoke by sensory stimulation. For the majority of the time there are infrequent, irregular flexor bursts of varying duration (from 100ms to many seconds). This type of pattern does not give the impression of an endogenous rhythm generator for walking.

4. Recordings from corresponding nerves on opposite sides of deafferented preparations have not shown reciprocal patterns of activity typical of those occurring in walking animals (Pearson & Iles, 1973). When one side is generating rhythmic activity the other usually is not. Similarly, burst activity in adjacent ipsilateral segments is not strongly reciprocal. Usually bursts are generated in one or the other but not both segments. When both segments are rhythmically active there is a tendency for flexor bursts not to occur simultaneously. However, this pattern differs from the pattern seen in walking animals where there is a strict reciprocal relationship between flexor bursts in adjacent segments (Pearson & Iles, 1973).

From all this we see that the pattern of motor activity generated in deafferented preparations is, in general, quite unlike that occurring in the walking animal. It is certainly possible for reciprocal activity to be generated in a limited set of motor neurones in a single segment. This pattern can be similar to that in the same motor neurones in a walking animal. However, given their relatively rare occurrence and the absence of data on the patterns in other motor neurones this is not a sound basis for concluding that a central pattern generator is responsible for patterning motor activity during walking. This conclusion is further supported by the lack of evidence for strong intersegmental coupling of oscillators in deafferented preparations.

The notion that central pattern generators do not form the basis of motor programming in insect walking is also supported by recent observations in the stick insect (Bässler & Wegner, 1983). In this animal recordings from a variety of peripheral nerves in deafferented preparations have completely failed to reveal patterned activity similar to that occurring during walking. Nevertheless they do show patterns resembling those for other rhythmic leg movements such as searching and rocking.

If central pattern generators do not form the basis of insect walking then what is the basis for the generation of the rhythmic motor output? A possible clue comes from observations in the cockroach in which relatively constant duration flexor bursts are often generated independently of cycle time or the intensity of antagonistic extensor activity. This suggests that a central mechanism exists for the generation of flexor bursts resembling to a reasonable extent flexor bursts in a walking animal. Similarly in the stick insect relatively constant duration flexor bursts are often generated in the deafferented nerve cord (Bässler & Wegner, 1983). Given that there is a flexor burst generating system, the generation of rhythmicity during walking could be dependent on sensory input timing of the onset of activity in this generator. For example, one can imagine that the flexor burst generator is turned on by a sensory signal near the end of leg retraction. In the cockroach it has been proposed that this signal is a decrease in activity of campaniform sensilla (Pearson, 1972). Also in the walking system of the cat there is now considerable evidence that the transition from stance to swing depends on proprioceptive input from the hip (Andersson et al., 1981).

Currently the simple idea that sensory input modulates a central pattern generator to produce the motor pattern for insect walking does not appear valid. The evidence for central pattern generators is very weak or, in the case of the stick insect, completely absent. Even if there is some rudimentary central oscillating system it seems that sensory input plays such a dominant role in insect walking that this central oscillator could not be regarded as a functional unit in the intact system.

Conclusions

My main conclusion is that the idea of central pattern generators being primarily responsible for the patterning of motor activity in the flight and walking systems of insects cannot be accepted at this time. In both systems major changes in the motor pattern are produced by deafferentation and sensory input has been demonstrated to play a role in the patterning of motor activity. Even though rhythmic motor activity can be generated in both systems following deafferentation it is not obvious that the normal pattern can be explained simply by a sensory modulation of the central rhythm generator (or central pattern generator). It appears more likely that sensory input forms an integral part of the pattern generating system. The necessity of intimately incorporating sensory input into the pattern generating system seems obvious, namely to ensure that the motor pattern is appropriate for the biomechanical state of the peripheral structures. In other words, for effective movements the motor output must be timed exactly to the position, movements, forces etc. of peripheral structures. One way this is done is to use sensory signals in a feedforward manner to initiate specific parts of the motor pattern. An example is found in the walking systems of insects and cats where sensory signals trigger the onset of flexor activity (Pearson, 1972; Andersson et al., 1981). It may also be the case in the locust flight system that stretch receptor input is responsible for activating depressor motor neurones (Möhl & Neumann, 1983). The same conclusion has come from the analysis of other motor systems. Examples are the jumping system of the locust where sensory input signalling the status of the leg ensure the jump is not triggered until a well developed co-contraction of the hindleg flexor and extensor muscles is achieved (Pearson et al., 1980), and in human speech movements where sensory input functions in a feedforward open-loop manner to pattern motor output appropriately for the exact position of the lips (Abbs & Gracco, 1984).

Summary
Re-examination of old data on cockroach walking, together with recent observations on the walking system of the stick insect and the flight system of the locust, has raised the question of whether the concept of central pattern generators is applicable to the walking and flight systems of insects. On balance, current data indicate that it is not. Instead it appears likely that the motor patterns for walking and flight are generated by neuronal systems in which sensory and central elements cannot be clearly distinguished functionally. The integration of sensory input into the pattern generating system ensures that the timing of the motor pattern is appropriate for the biomechanical state of peripheral structures.

Acknowledgements
I would like to thank G. Boyan, I. Gynther and D. Reye for their valuable comments on initial drafts of this manuscript.

REFERENCES

Abbs, J.H. & Gracco, V.L. (1984). Control of complex motor gestures: orofacial muscle responses to load perturbations of lip during speech. J. Neurophysiol., 51, 705-723.
Altman, J.S. (1975). Changes in the flight motor pattern during the development of the Australian plague locust, Chortoicetes terminifera. J. comp. Physiol., 97, 127-142.

Andersson, O., Forssberg, H., Grillner, S. & Wallén, P. (1981). Peripheral feedback mechanisms acting on the central pattern generators for locomotion in fish and cat. Can. J. Physiol. & Pharmacol., 59, 713-726.

Bässler, U. & Wegner, U. (1983). Motor output of the denervated ventral nerve cord in the stick insect Carausius morosus. J. exp. Biol., 105, 127-145.

Delcomyn, F. (1980). Neural basis of rhythmic behavior in animals. Science, N.Y., 210, 492-498.

Hedwig, B. & Pearson, K.G. (1984). Patterns of synaptic input to identified flight motoneurons in the locust. J. comp. Physiol. 154, 745-760.

Horsmann, U., Heinzel, H.G. & Wendler, G. (1983). The phasic influence of self-generated air current modulations on the locust flight motor. J. comp. Physiol., 150, 427-438.

Kien, J. & Altman, J.S. (1979). Connections of the locust wing tegula with metathoracic flight motoneurons. J. comp. Physiol., 133, 299-310.

Kutsch, W. (1974). The influence of wing sensory organs on the flight motor pattern in maturing adult locusts. J. comp. Physiol., 88, 413-424.

Miller, J.P. & Selverston, A.I. (1982). Mechanisms underlying pattern generation in lobster stomatogastric ganglion as determined by selective inactivation of identified neurons. IV. Network properties of the pyloric system. J. Neurophysiol., 48, 1416-1432.

Möhl, B. & Neumann, L. (1983). Peripheral feedback-mechanisms in the locust flight system. In Biona Report 2. Ed. Nachtigall, W. Gustav Fischer, Stuttgart, pp. 81-87.

Neumann, L., Möhl, B. & Nachtigall, W. (1982). Quick phase-specific influence of the tegula on the locust flight motor. Naturwissenschaften, 69, 393-394.

Pearson, K.G. (1972). Central programming and reflex control of walking in the cockroach. J. exp. Biol., 56, 173-193.

Pearson, K.G., Heitler, W.J. & Steeves, J.D. (1980). Triggering of the locust jump by multimodal inhibitory interneurons. J. Neurophysiol., 43, 257-278.

Pearson, K.G. & Iles, J.F. (1970). Discharge patterns of coxal levator and depressor motoneurones in the cockroach, Periplaneta americana. J. exp. Biol., 52, 139-165.

Pearson, K.G. & Iles, J.F. (1973). Nervous mechanisms underlying intersegmental coordination of leg movements during walking in the cockroach. J. exp. Biol., 58, 725-744.

Pearson, K.G., Reye, D.N. & Robertson, R.M. (1983). Phase-dependent influences of wing stretch receptors on flight rhythm in the locust. J. Neurophysiol., 49, 1168-1181.

Robertson, R.M. & Pearson, K.G. (1982). A preparation for the intracellular analysis of neuronal activity during flight in the locust. J. comp. Physiol., 146, 311-320.

Robertson, R.M. & Pearson, K.G. (1983). Interneurons in the flight system of the locust: distribution, connections and resetting properties. J. comp. Neurol., 215, 33-50.

Robertson, R.M. & Pearson, K.G. (1984). Neural circuits in the flight system of the locust. J. Neurophysiol. (Submitted for publication).

Selverston, A.I. (1980). Are central pattern generators understandable? Behav. & Brain Sci., 3, 535-571.

Waldron, I. (1967). Mechanisms for the production of the motor output pattern in flying locusts. *J. exp. Biol.*, 47, 201-212.

Wendler, G. (1974). The influence of proprioceptive feedback on locust flight coordination. *J. comp. Physiol.*, 88, 173-200.

Wendler, G. (1983). The locust flight system: functional aspects of sensory input and methods of investigation. In *Biona Report* 2. Ed. Nachtigall, W. Gustav Fischer, Stuttgart, pp. 113-125.

Wilson, D.M. (1961). The central nervous control of flight in a locust. *J. exp. Biol.*, 38, 471-490.

Wilson, D.M. & Gettrup, E. (1963). A stretch reflex controlling wingbeat frequency in grasshoppers. *J. exp. Biol.*, 40, 171-185.

Wilson, D.M. & Wyman, R.J. (1965). Motor output patterns during random and rhythmic stimulation of locust thoracic ganglia. *Biophys. J.*, 5, 121-143.

Chapter Twenty-one

THE ROLE OF MOVEMENT-RELATED FEEDBACK IN THE CONTROL OF
LOCOMOTION IN FISH AND LAMPREY

P. Wallén & T.L. Williams

In its natural habitat, a fish must continuously adapt its swimming movements to the existing environmental conditions as it pursues its various aims. A coral reef-dwelling fish, for example, will manoeuvre its body into a position suitable for nibbling vegetation from a coral, while simultaneously coping with often powerful water currents. To capture its victim a predatory fish must be able not only to accelerate suddenly, but also to avoid any obstacles such as rocks or seaweed. A swimming fish will benefit from a quick withdrawal reflex in response to a noxious stimulus, but it must be able to integrate this response with the swimming movement.

It is thus evident that the fish when swimming must deal with an immense amount of sensory information in controlling its movements. Proprioceptive input from such structures as the vestibular apparatus or movement-detecting mechanoreceptors along the body will signal the position and movement of the body itself. Exteroceptive information regarding the motion of the surrounding water, as well as the position and movements of other animals and objects, will be provided by the lateral line system and cutaneous receptors. Of course visual and auditory signals, not to mention olfactory input, will also play a significant role in shaping the animal's movements. Bearing in mind the complex task of information processing and motor command generation that must be undertaken by the central nervous system, we have limited this review to a discussion of some aspects of the role played by movement-related feedback in controlling the undulatory body movements of the swimming fish or lamprey.

The body musculature is arranged in myotomes corresponding to spinal cord segments, of which there are about 100 in the dogfish or lamprey. The swimming movements are characterised by alternating activation of the two sides of each muscle segment, which are sequentially activated in a rostral to caudal direction along the body. This produces an undulation which travels towards the tail, thus propelling the fish forward through the water. The electromyographic pattern of intra- and intersegmental coordination during swimming was first described in the spinal dogfish (Grillner, 1974), a preparation which for nearly a century has been known to produce spontaneous swimming movements (e.g. Steiner, 1886; Bethe, 1899; Gray & Sand, 1936; Le Mare, 1936). At any one segmental level, the bursts of activity occupy a constant fraction of the cycle, irrespective of the velocity of swimming. In addition, the rostro-caudal delay in activation between successive segments is also a constant fraction of the cycle, so that there is a constant phase-coupling between segments (Grillner, 1974). This pattern of coordination has also been demonstrated for intact fish (eel, trout, dace; Grillner & Kashin, 1976) and for the intact and spinal lamprey (Wallén & Williams, 1984).

Looking back through the literature on fish locomotion, one finds the question repeatedly raised as to whether this sequential, rhythmic activation of muscle segments along the body is generated centrally or is dependent upon a chain of reflexes between the periphery and the spinal cord. In 1935 Erich von Holst reported a series of experiments on swimming in the tench. Using visual inspection and filming, he found rhythmic swimming movements after transection of all dorsal roots except those innervating the pectoral fins. He therefore concluded that sensory feedback was not necessary for generating the swimming rhythm in fish (von Holst, 1935b). It should be noted, however, that these conclusions presuppose an absence of ventral root afferents or intraspinal mechanoreceptors (see below). To investigate further the mechanism of the progression of the undulation along the body, von Holst cut all dorsal and ventral roots in the middle third of the body (using mostly eels), immobilised this part mechanically, and prevented movement transmission through the water with a dividing wall. He found that the undulations set up in the rostral part reappeared in the caudal part after a delay corresponding to the normal speed of the travelling wave, and he therefore concluded that its propagation is not controlled by a chain of reflex arcs (von Holst, 1935a).

These conclusions of von Holst's were supported at the time by Gray (1936) who made similar observations, also on eels. Later, however, Gray (1950) came to support the opposite view, following experiments in his laboratory by Lissmann (1946a,b) on the deafferented, spinal dogfish. Lissmann found that after extensive bilateral dorsal root transection the preparation did not display rhythmic activity, neither spontaneously nor in response to sensory stimulation of the tail fin. He concluded that a persistent locomotory rhythm must depend on afferent excitation and thus found no support for the 'central hypothesis'. Today we may presume that Lissmann's findings were due to a low general level of excitability in the spinal cord resulting from lengthy and extensive surgery.

About two decades later, Roberts utilised 'pharmacological deafferentation' by paralysing the spinal dogfish with curare (Roberts, 1969a). In contrast to dorsal root transection, curare paralysis rules out any contribution from possible ventral root afferents (see Grillner et al., 1976) or intraspinal mechanoreceptors (discussed below). Although rhythmic activity was recorded in abdominal spinal nerves after paralysis, the rhythm was slow and irregular and did not persist, and the data were accordingly taken as evidence against the 'central hypothesis'. It should be noted, however, that Roberts also reported the important observation that sensory signals are able to influence the efferent rhythm recorded during paralysis (Roberts, 1969a).

A few years later, Grillner, Perret and Zangger (1976) were able to record spontaneous, well-coordinated rhythmic activity in the spinal dogfish, both after dorsal root transection and after curare paralysis. The rhythm (recorded in ventral roots) could persist for several hours, with a range of burst rates overlapping that of the actively swimming animal. Furthermore, the pattern of segmental coordination was similar to that which had been described using electromyographical techniques in the swimming spinal dogfish (Grillner, 1974). With these data at hand, the 'reflex chain' hypothesis had to be abandoned, and this first demonstration of 'fictive' swimming finally settled the matter in favour of a central, rhythm-generating mechanism in the spinal cord.

Further advances toward understanding the mechanisms of central pattern generation and feedback modulation (see Grillner et al., 1983) have followed upon the development of an in vitro preparation of the isolated spinal cord of the lamprey (Rovainen, 1967) which generates fictive locomotor activity (Cohen & Wallén, 1980). The spinal cord is removed from the animal along with the notochord and placed in a lamprey Ringer solution. Bath addition of those excitatory amino acids which activate either N-methyl-D-aspartate receptors (Watkins, 1981; Grillner, McClellan, Sigvardt, Wallén & Wilén, 1981) or kainate receptors (Brodin & Grillner, 1984; Brodin et al., 1985) induces rhythmic activity in the ventral roots (cf. Poon, 1980). This activity has been shown quantitatively to have the same phase coordination, both segmental and intersegmental, as the electromyographic activity recorded during swimming in the intact or spinal lamprey, thus providing a firm identification of fictive swimming with the active locomotor rhythm (Wallén & Williams, 1984).

Thus throughout the years the general view regarding the neural mechanisms of coordinated locomotor movements has shifted between an often extreme 'central' hypothesis and a likewise extreme 'reflex-chain' one. Today it is clear that we are dealing with a central mechanism of rhythm generation, but we also know from studies in dogfish and lamprey as well as other vertebrates that the centrally generated rhythm is subject to a very profound influence from sensory signals (Grillner, 1975; Grillner et al., 1976; Andersson et al., 1981).

Entrainment of the central pattern generating network for locomotion

Both the paralysed spinal dogfish (Grillner & Wallén, 1977, 1982) and the in vitro preparation of the lamprey spinal cord (Andersson et al., 1981; Grillner, McClellan & Perret, 1981) have been used to study the influence of movement-related proprioceptive input on the centrally generated rhythm. In both these cases the feedback loop is open, in that no movements are produced by the preparation. Artificial movements of chosen frequency and amplitude can then be imposed by the experimenter during fictive locomotor activity. Fig. 1 shows the experimental arrangement in the dogfish and Fig. 2 that in the lamprey. The efferent activity was recorded from the ventral roots in a region which was rigidly fixed by clamps (dogfish) or pinned through the notochord to the Sylgard lining of the bath (lamprey). Side-to-side sinusoidal

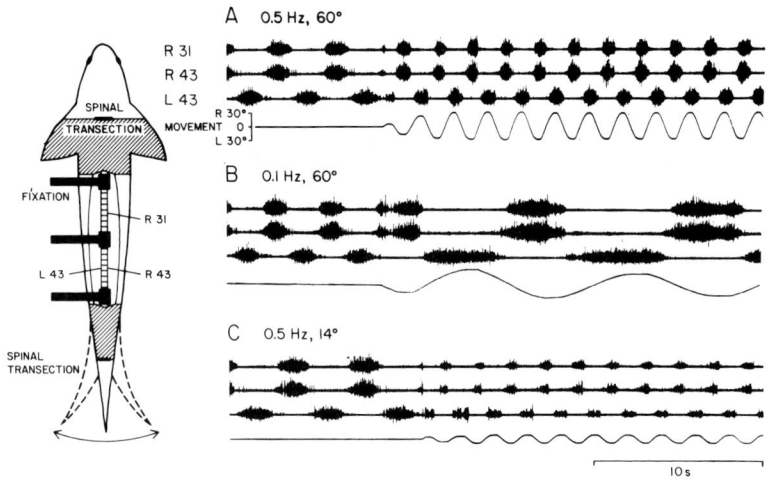

Fig. 1. Entrainment of spontaneous fictive locomotor rhythm in a paralysed spinal dogfish. A-C: ventral root recordings from segments indicated in A (L, left; R, right; numbering of segments starts with the first segment below the foramen magnum). Starting from the midposition, the tail fin was moved at different frequencies and amplitudes as indicated. In this example the intersegmental delay is caudal to rostral, as in backward swimming (see Grillner, 1974). Time and movement calibrations apply to all three sets of records. (From Grillner & Wallén, 1982).

movements of the caudal region, roughly mimicking swimming movements, were applied either by hand or with a motor.

Imposing such movements had an immediate and dramatic effect on the burst pattern, so that it became entrained in a 1:1 fashion by the movement, whether the frequency was slower or faster than that of the basic rhythm (Figs. 1 & 2). It can be seen that the effect was not restricted to the region nearest the tail, but that entrainment was apparent in all ventral root recordings. Furthermore, the activity became modified immediately at the onset of movement (Fig. 1B) and followed the movement quite precisely, cycle by cycle (Fig. 2, central panel).

Intrasegmental coordination was maintained, in that the burst durations became greater in the longer cycles and smaller in the shorter ones, with continued alternation between the left and right sides. Likewise the intersegmental coordination was maintained, the delay increasing or decreasing with the cycle duration. In addition, the burst amplitude was altered during entrainment, usually increasing, except at high frequencies or low movement amplitudes, when a decrease was seen (Grillner & Wallén, 1982).

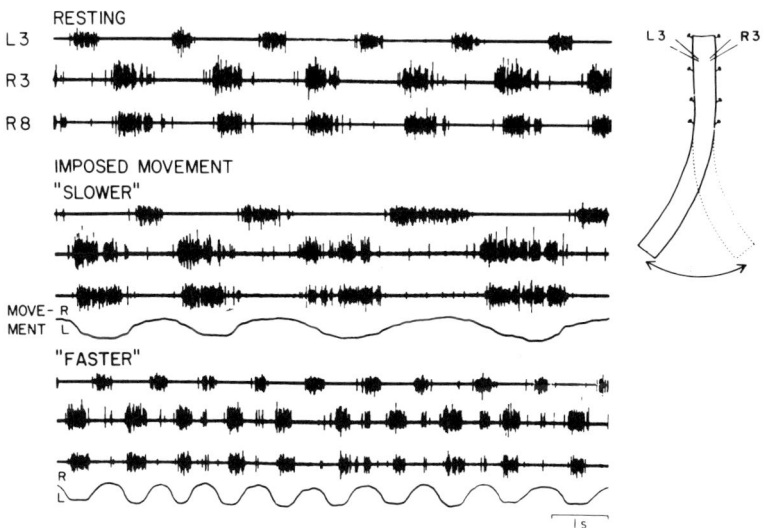

Fig. 2. Entrainment of fictive locomotor rhythm in lamprey spinal cord-notochord preparation, activated by D-glutamate. Ventral root recordings from segments indicated in upper panel (L, left; R, right; numbering starts with most rostral segment of preparation). Amplitude of movement approximately $40°$. Time calibration applies to all records. (From Andersson et al., 1981).

When movements were imposed in the absence of spontaneous rhythm in the paralysed spinal dogfish, activity was induced by the movement, and the rhythm persisted for some cycles after the movement had stopped (Fig. 3). Thus it is clear that the rhythm-generating network was activated rather than just the motorneurones, which all available evidence supports as being pure output elements (Wallén & Lansner, 1984).

In both preparations and at all frequencies, movement of the caudal portion towards one side was accompanied by a burst of activity on the opposite side. However, the precise phase relation between the bursts of activity and the movements depended on frequency, as generally occurs between coupled oscillators, whether biological or man-made (Pavlidis, 1973; Stein, 1976).

In both types of preparation, entrainment of the rhythm at integer ratios other than 1:1 could also occur, with e.g. a 2:1 or 1:3 relation between efferent bursts and movement cycles. Such ratios were seen when the movement frequency chosen was a great deal slower or faster than that of the basic rhythm, and especially when combined with a low movement

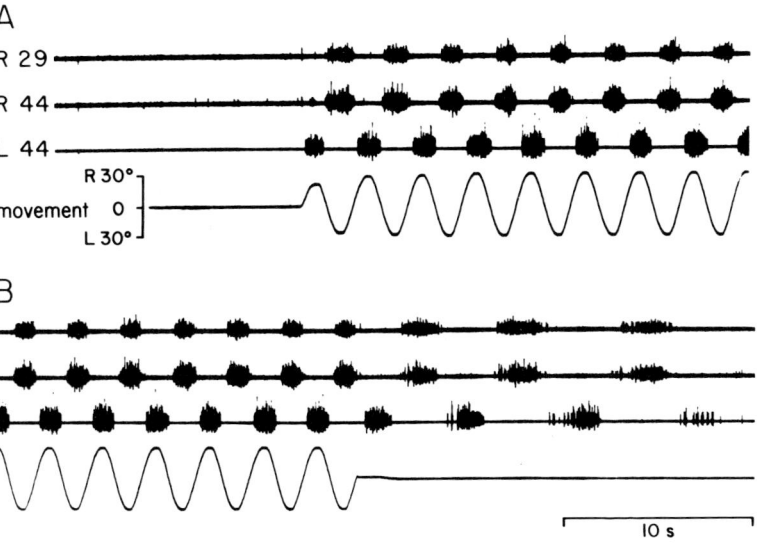

Fig. 3. Induction of fictive locomotion in a paralysed spinal dogfish with no initial spontaneous activity, by application of sinusoidal movements to the tail region (as in Figure 1). Segments indicated in A. The record in B is a direct continuation of that in A. (From Grillner & Wallén, 1982).

amplitude. These effects may reflect a partial entrainment of the central pattern generating network and a competing interaction between the movement-related input and the intrinsic rhythmicity of the pattern generator. Indeed, if the afferent input was enhanced by increasing the movement amplitude, complete 1:1 entrainment could occur. Also, if the same movement parameters that had given incomplete entrainment were employed in the absence of a spontaneous rhythm, activity could be induced and entrained in a 1:1 fashion.

To test whether the entrainment was mediated by purely position-sensitive elements, sinusoidal movements were superimposed on a sustained lateral curvature to one side or the other in the dogfish preparation (Grillner & Wallén, 1982). Entrainment by the sinusoidal movements still occurred, demonstrating a certain degree of dynamic sensitivity of the mechanosensitive elements. However, the burst durations were somewhat longer than normal on the side which was stretched by the sustained displacement, and somewhat shorter on the other. This indicates that the mechanosensitive elements also have static sensitivity.

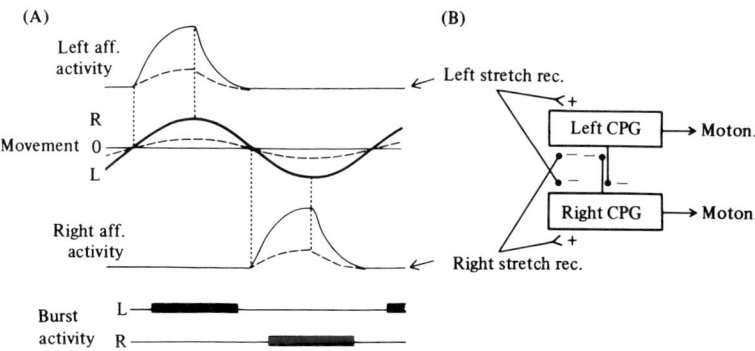

Fig. 4. Diagrammatic representation of proposed mechanism of entrainment of central pattern generators by stretch receptors. A: schematic form of afferent (aff.) activity caused by stretch of the left (L) and right (R) sides. Afferent discharge starts approximately when movement passes the midposition, and reaches its maximum at the extreme position, after which time activity decreases (dotted lines). Smaller movement amplitude causes a reduced afferent discharge (interrupted curves). Approximate timing of entrained bursts of activity is indicated below. B: Proposed central effects of afferent signals. Receptors (rec.) activated when left or right side is stretched reinforce and maintain generation of bursts in motor neurones (Moton.) by the central pattern generator (CPG) in the ipsilateral hemisegment, whereas they inhibit the generation of a burst by the contralateral CPG. (From Grillner & Wallén, 1982).

The various results from the two preparations are consistent with a mechanism for entrainment which is represented diagrammatically in Fig. 4 (Grillner & Wallén, 1982). In this scheme, stretch-sensitive elements with both static and dynamic sensitivity are able to reinforce the current state of the central pattern generating circuitry on the two sides of a segment. Ipsilaterally, burst activity is supported, whereas contralaterally silence is maintained, aided by mutual inhibition across the midline. Thus the rhythm

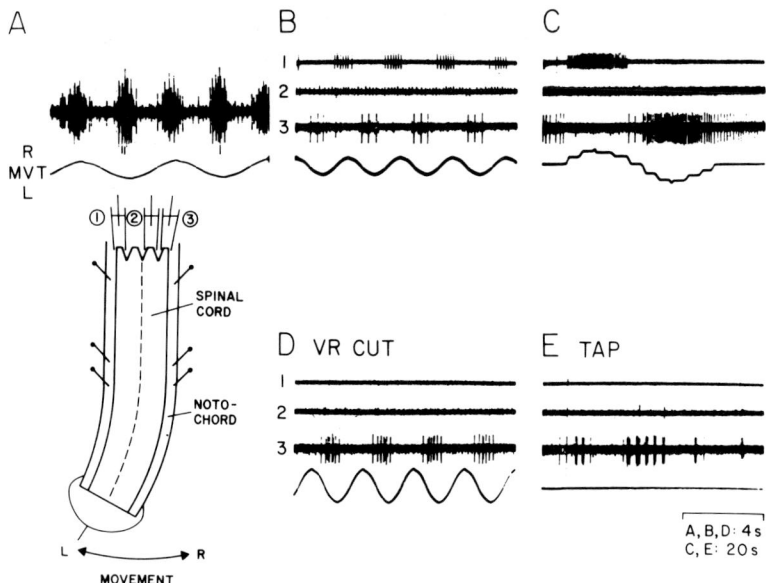

Fig. 5. Extracellular activity recorded from fascicles of the rostral fixed part of the lamprey spinal cord during imposed lateral movements (mvt) of the mobile caudal part of the preparation. Ca-free solutions used throughout. A, B, C: all dorsal roots cut but ventral roots intact. A: mass discharge in a left lateral fascicle of one preparation. B, C: single unit activity in lateral fascicles 1 and 3 (see drawing) of another preparation during sinusoidal (B) or stepwise (C) movements. D, E: same preparation as in B and C, except that the ventral roots were also cut. D: activation of the unit in fascicle 3 by sinusoidal bending. Loss of activation of unit in fascicle 1 may have resulted from disturbance during cutting of the ventral roots. E: response of unit in fascicle 3 to gentle tapping with a glass rod of the right lateral margin where bending occurs. (From Grillner et al., 1982).

can be entrained by movements at frequencies both above and below the spontaneous rate. There may also be a longer-lasting general facilitation of the pattern generator, since an initially silent preparation can be seen to remain rhythmically active for some cycles after the movement has ceased (Fig. 3).

Mechanosensitive elements within the lamprey spinal cord
In the lamprey preparation, section of all dorsal roots to remove any influence from mechanoreceptors which might exist in the notochord or meninges did not prevent entrainment, nor did section of all ventral roots except those from which recordings were made, thus ruling out any contribution from possible afferents in the ventral roots. Hence the conclusion was reached that there must exist within the spinal cord itself mechanosensitive elements which respond to lateral bending of the cord and which provide inputs to the locomotor pattern generating circuitry (Grillner, McClellan & Perret, 1981).

Experiments such as the one illustrated in Fig. 5 (Grillner et al., 1982) were designed to test whether mechanoreceptor activity could be detected in the ascending tracts of the isolated spinal cord. Suction electrodes were placed on dissected fascicles of the rostral end of the preparation. In the absence of excitatory amino acids (so that no fictive locomotion was occurring), the caudal end of the preparation was subjected to the same sort of lateral bending movements as had been used to entrain the fictive locomotor rhythm. During such movements, spike activity could be recorded from lateral fascicles of the spinal cord, as in Fig. 5A, but never in medial fascicles. Further dissection of the lateral fascicles allowed single unit activity to be recorded which responded preferentially to movement to one direction or the other, as in Fig. 5B.

During lateral bending, the regions of spinal cord maximally distorted are the very edges of the lateral margin. Fig. 5E illustrates that the unit recorded in fascicle 3 could also be activated by gentle tapping of the lateral edge of the spinal cord within a small area. The authors also found that the mechanoreceptor activity was not abolished in Ca^{++}-free or low Ca^{++}-high Mg^{++} Ringer solution, suggesting that no chemical synapse occurred between the mechanoreceptor and the ascending axon.

These results made one particular class of cells in the cyclostome spinal cord seem good candidates for mediating the observed mechanoreceptor effects. These cells were the Randezellen or edge cells, first noted by the early anatomists (e.g. Retzius, 1891; Kolmer, 1905) because of the position of their somata within the lateral tracts of the spinal cord. They have been further characterised by Rovainen (1974). They exhibit either an ipsilateral or contralateral axon extending rostrally for some segments in the lateral tracts. Their function has been unknown.

Recently, a study of the structure and the function of edge cells was undertaken by Grillner, Williams & Lagerbäck (1984), using both light and electron microscopy. Individual edge cells (visible under the dissecting microscope) were

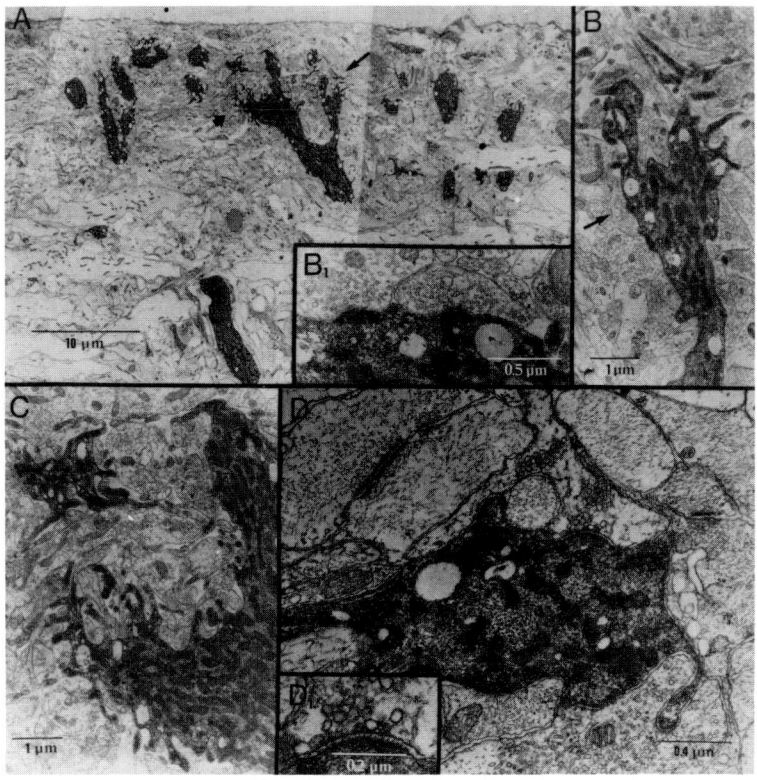

Fig. 6. Ultrastructure of lateral dendrites of edge cell injected with horseradish peroxidase (dark colour). A: Overview of lateral ramifications near lateral margin, which is upward. B & C: Higher magnification of two different lateral processes indicated by a short and a long arrow in A, showing abundance of fine fingerlike processes (dendritic tips). B_1: Two synapses in the left central part of B (arrow) shown under higher magnification. D: A bouton containing spherical synaptic vesicles in contact with the proximal part of an edge cell dendrite. D_1: The synaptic complex. (From Grillner et al., 1984).

Feedback in the control of locomotion in fish and lamprey

intracellularly injected with either Lucifer yellow, for light microscopy, or horseradish peroxidase (HRP), for electron microscopy. The shapes of the cell bodies were found to be quite variable, but each sent broad processes laterally which ended in very distinctive nestlike formations, aligned along the lateral margin of the cord. These can be seen in the reconstructions of Fig. 7. Lateral ramifications of an HRP-filled cell are shown at higher magnification in the electron micrograph of Fig. 6A, where the branches can be seen lined up close to the lateral margin. In B and C of this figure, one can see that in this region the cell membrane is thrown into many small fingerlike structures which are a few tenths of a μm in diameter. These resemble the 'dendritic tips' which occur in the stretch-sensitive region of the crayfish stretch-

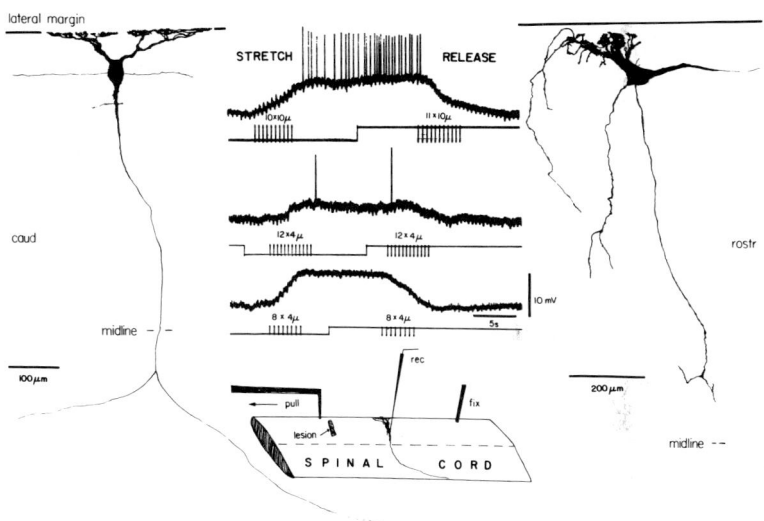

Fig. 7. Depolarising responses of lamprey edge cells to stretch of the spinal cord margin. General experimental arrangement shown at lower centre. Pull was applied rostro-caudally along the lateral margin with a stepping motor. The upper two sets of records in the centre panel are from the edge cell to the left (reconstruction from dorsal view of whole-mount preparation, Lucifer yellow); the lower set of records is from the cell to the right. Time and voltage calibrations apply to all three records. Upper trace in each case is intracellular potential; lower trace indicates stretch or release applied to the lateral margin, each vertical line signalling a 10μm or 4μm change in length. (From Grillner et al., 1984).

receptor neurone (Tao-Cheng et al., 1981).

Synapses from other cells onto the dendrites of injected edge cells are illustrated in Fig. $6B_1$ and 6D, and synapses have also been seen on edge cell somata. No synapses from the edge cells onto other neurones have been seen in the region of the lateral margin. The axon has not yet been traced in the electron microscope in search of outgoing synapses from the edge cells.

Calculations based on kinematic data of swimming lampreys (McClellan & Grillner, unpublished) have indicated that the distortion of the lateral margin of the spinal cord which occurs at the points of maximum curvature amounts to a change in length of up to around 5%. The extent of the margin occupied by the lateral ramifications of one edge cell (about 100-300 μm; see Fig. 7) is very much smaller than the minimum radius of curvature, and hence these processes are subjected to an almost purely rostrocaudal stretch, with negligible distortion in the mediolateral direction. Experiments were thus carried out to test the sensitivity of edge cells to stretch of up to 5% of a small region of the lateral margin of the spinal cord.

The lateral margin was stretched or released in small steps during intracellular recording from edge cells (Fig. 7). All edge cells tested depolarised when the margin was stretched and repolarised when the stretch was released. The depolarisation was not simply a response to injury of the edge cell soma by movement relative to the microelectrode, since hyperpolarisation was always the response to release, even when this was the first movement imposed after impalement of the cell.

The effects were graded with the degree of stretch, were reversible and reproducible, and they were not abolished in Ca^{++}-free Ringer. This evidence, together with the appropriateness of the edge cell structure to a stretch receptor function, speaks strongly in favour of a role for edge cells as mediators of movement-related feedback in the lamprey. A cell such as the one on the left in Fig. 7 may exert its influence primarily in other segments of the spinal cord, whereas a cell with dendritic branches in the cellular region, such as the one on the right, may play a segmental role as well.

Given the long, slender shape of the lamprey, any curvature of the body will be reflected in the spinal cord, and thus intraspinal mechanoreceptors seem quite appropriate. Neurones with somata near the lateral margin have also been described in reptiles and birds, but their function is unknown (Kappers et al., 1936). Whenever the body of a snake or the neck of a bird is arched, the spinal cord will also become curved. Thus it is tempting to speculate that intraspinal mechanoreceptors are not restricted to cyclostomes, but occur in other vertebrate groups as well.

What receptors are responsible for entrainment in the dogfish? Although muscle spindles have been observed in the jaw-closing muscles of a teleost fish (Maeda et al., 1983), they appear to be absent in the swimming musculature. In elasmobranchs, a stretch receptor in the fin muscles of rays has been described (Bone & Chubb, 1975) as well as a stretch-sensitive corpuscular ending in subcutaneous tissues of the dogfish (Roberts, 1969b; Bone & Chubb, 1976). Roberts (1969b) has shown that the subcutaneous receptors respond phasically to sinusoidal stretch of the body wall, so they may contribute to the entrainment. However, Grillner & Wallén (1982) found that entrainment still occurs after removal of both skin and muscle tissue. Thus there must be mechanoreceptors within the vertebral column and/or spinal cord with strong inputs to the central pattern generating circuitry.

Phase-dependent modulation of a tail fin reflex during locomotion

A brief pinch to the tail fin of a swimming fish is an example of an unexpected, short-lasting stimulus which might disrupt the locomotor rhythm. The reflex effects of such a stimulus have been studied in the spinal dogfish preparation (Grillner et al., 1977; Wallén, 1980).

If during swimming the tail fin is briefly touched or pinched, the ongoing movement is greatly augmented. This effect was studied systematically by applying brief (e.g. 50ms) electrical stimulus pulses to the tail fin during different phases of the movement cycle (Grillner et al., 1977). It was found that the reflex response varied with the phase of swimming, so that the response to a stimulus was always an increase in the electromyographic activity on whichever side was active, with little or no response on the opposite side. Thus the reflex response was reversed between the two sides of the segment according to the phase of the locomotor movement. The response was not restricted to the tail region, but rather involved the whole body, so that a well-organised reaction was produced. In functional terms the reflex response involves a quick withdrawal of the tail from the stimulus source and a momentary increase in swimming speed.

To investigate whether this modulation of the reflex in phase with the locomotor cycle depends on proprioceptive feedback, the corresponding experiment was also carried out on the paralysed spinal dogfish during fictive locomotion (Wallén, 1980). In the absence of movement, the reflex response, as recorded in the ventral roots, was still modulated in accordance with the locomotor cycle (Fig. 8). It is thus clear that phasic sensory feedback signals from the moving body are not necessary for this modulation. Therefore it was concluded that the phasic reflex modulation depends on central mechanisms. The varying excitability level of the motor neurones in phase with the fictive locomotor rhythm will clearly have a modulatory effect on their response, since a given reflex EPSP is more likely to generate an action potential during the depolarisation phase than during the phase of inhibition. The above findings could be explained by this mechanism alone, though the reflex modulation may also involve phasic gating of reflex arcs at a premotor level, exerted by the central pattern generating network for locomotion. Evidence for such a mechanism has been found in the cat (Edgerton, Grillner, Rossignol & Sjöström, 1976, cited in Forssberg et al., 1976; Andersson et al., 1978).

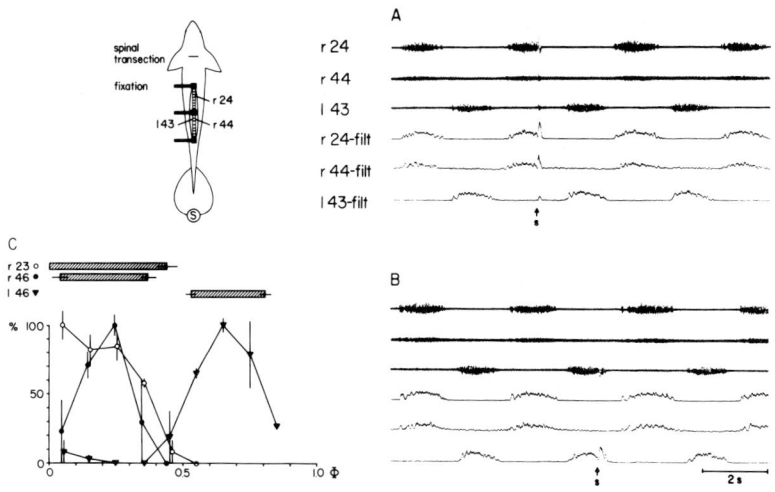

Fig. 8. Phase-dependent reflex modulation in the paralysed spinal dogfish. A, B: Ventral root recordings from segments indicated (r, right; l, left). Signals also shown after full-wave rectification followed by low-pass filtering. At arrow, a 50ms train of 1.0mA, 1ms pulses at 3.3ms intervals was delivered to the tail fin during activity on the right side (A), and a few seconds later during activity on the left (B). Time calibration applies to both. C: Phase dependence of response amplitude. Abscissa: fraction of the activity cycle at time of stimulation. Cycle duration normalised to 1.0 with the starting point defined as the onset of activity in the right rostral ventral root (r23). Ordinate: peak amplitude ($\pm$S.E.) from filtered record, from a series of 45 consecutive identical stimuli (as in A and B but only 0.5mA), expressed as percentage of the maximal response at that recording position. Symbols defined above, where the average timing ($\pm$S.D.) of the burst activity before stimulation is shown. (From Wallén, 1980).

Feedback in the control of locomotion in fish and lamprey

A marked difference was noted, however, in the reflex response pattern of the paralysed compared to the unparalysed preparation when the strength of the tail fin stimulation was increased (Wallén, 1980). The structure of the locomotor rhythm could become severely disturbed by the stimulation, including omission of bursts and rhythm resetting (Fig. 9, uppermost panel). This had not been seen in the actively swimming fish, suggesting that the proprioceptive feedback present under normal conditions might serve to stabilise the locomotor rhythm and make it less susceptible to disturbances. Indeed, when a side-to-side sinusoidal movement was artificially imposed on the tail region of the paralysed preparation, the timing of bursts was undisturbed by the application of a tail fin stimulus (Fig. 9). Thus it would appear that central mechanisms gate the reflex from the tail stimulus so that the response is appropriately integrated with the locomotor rhythm, and that in turn the movement-related feedback ensures that the central rhythm is not severely disrupted. This is a rather elegant interaction of 'peripheral' and 'central' mechanisms.

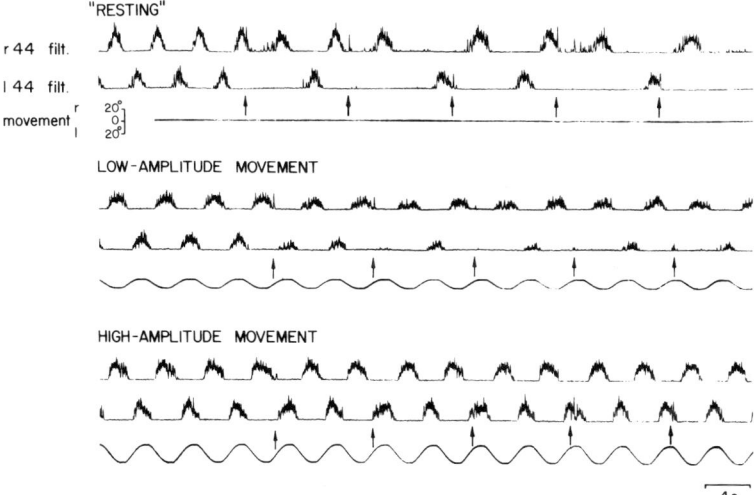

Fig. 9. Effect of imposed body movement on reflex response in paralysed spinal dogfish. Upper two traces in each example (filt.) are rectified and filtered ventral root recordings from the right (r) and left (l) sides of segment 44. At each arrow, a 50ms train (for timing see legend of Fig. 8) of 3mA pulses was delivered to the tail fin. The third trace in each case indicates imposed movement of the tail region. In middle panel, applied movement was about $14°$ peak-to-peak; in lower panel, about $29°$. Time and movement calibrations apply to all three panels. (From Wallén, 1980).

Functional considerations

As we have seen, movement-related feedback appears to have an important role in stabilising the locomotor rhythm during reflex responses to external disturbances. Feedback may also reduce the variability of the centrally generated pattern, as has been found by Droge & Leonard (1983a,b) in the stingray, in which both inter- and intrasegmental coupling become labile after paralysis of the decerebrate animal. As revealed in the entrainment experiments described above, however, movement-related feedback is potent enough to control the frequency of the centrally generated activity over a wide range. Can this phenomenon have any functional significance to the intact animal, where the swimming movements are caused not by the experimenter, but by the ventral root activity itself?

The actual movement of a particular portion of the body of a fish depends not only upon the muscle activation but also upon the passive inertial and viscoelastic properties of both the fish and the water. These are such that the mechanical wave travels more slowly down the body of the fish than the wave of muscle activation (Grillner & Kashin, 1976). Thus in the anterior portion of the body, the development of curvature is nearly synchronous with the muscle activation, whereas in the more posterior regions, the activation precedes the curvature formation by about one quarter cycle (as in Fig. 4, cf. also Grillner & Kashin, 1976). Such a phase advance is necessary if the muscle activation is to aid rather than oppose the thrust against the water by the leading edge of the travelling mechanical wave (see Blight, 1977). The entrainment experiments demonstrate that movement-related feedback is able to maintain the timing between activation and body curvature (Figs. 1 & 2). Thus it would be expected that in the intact animal, feedback can correct any discrepancy which might develop in this relationship: whenever the mechanical wave would tend to travel too slowly or too quickly, even if for only a fraction of a cycle, the movement-related feedback can be expected to slow down or speed up the wave of neural activation appropriately.

One would expect that during the course of evolution the central pattern-generating networks and the effects upon them of descending drive would have evolved so as to be well-matched to the mechanical properties of the body and the water, at least for uniform swimming in still water. Feedback may play a significant role during growth and development of the animal in this matching of neural activity to body form. Even during constant conditions in the adult it may provide the fine tuning required to optimise the system. It will become especially important, however, when conditions are changing, which is probably a more common situation than constancy for most species of fish. A swimming fish will turn, will accelerate and decelerate, and must cope with varying water currents. It may enter areas of seaweed or sandy water,

Feedback in the control of locomotion in fish and lamprey

or it may start to burrow into a bank of mud, in which case the hydromechanical conditions will become very different (cf. Grillner & Wallén, 1984). Any of these changes would tend to upset the relative timing between the neural and mechanical activity, and an optimally working feedback system, well integrated with descending modulatory signals, will thus be of critical importance for the precise control of the animal's movements.

Acknowledgement
We are grateful to Professor S. Grillner for very helpful comments on the manuscript.

REFERENCES

Andersson, O., Forssberg, H., Grillner, S. & Lindquist, M. (1978). Phasic gain control of the transmission in cutaneous reflex pathways to motorneurons during 'fictive' locomotion. Brain Res., 149, 503-507.

Andersson, O., Forssberg, H., Grillner, S. & Wallén, P. (1981). Peripheral feedback mechanisms acting on the central pattern generators for locomotion in fish and cat. Can. J. Physiol. & Pharmacol, 59, 713-726.

Bethe, A. (1899). Die Locomotion des Haifisches (Scyllium) and ihre Beziehungen zu den einzelnen Gehirntheilen und zum Labyrinth. Pflügers Arch. ges. Physiol., 76, 470-493.

Blight, A. R. (1977). The muscular control of vertebrate swimming movements. Biol. Rev., 52, 181-218.

Bone, Q. & Chubb, A. D. (1975). The structure of stretch receptor endings in the fin muscles of rays. J. mar. biol. Ass. U. K., 55, 939-943.

Bone, Q. & Chubb, A. D. (1976). On the structure of corpuscular proprioceptive endings in sharks. J. mar. biol. Ass. U. K., 56, 925-928.

Brodin, L. & Grillner, S. (1984). Excitatory amino acid receptors and the initiation of fictive swimming in the lamprey spinal cord in vitro. Neurosci. Lett. Suppl., 18, 386.

Brodin, L., Grillner, S. & Rovainen, C. M. (1985). NMDA, kainate and quisqualate receptors and the generation of fictive locomotion in the lamprey spinal cord. Brain Res. (In press).

Cohen, A. & Wallén, P. (1980). The neuronal correlate of locomotion in fish. 'Fictive swimming' induced in an in vitro preparation of the lamprey spinal cord. Expl. Brain Res., 41, 11-18.

Droge, M.H. & Leonard, R.B. (1983a). Swimming pattern in intact and decerebrated stingrays. J. Neurophysiol., 50, 162-177.

Droge, M. H. & Leonard, R. B. (1983b). Swimming rhythm in decerebrated, paralysed stingrays: normal and abnormal coupling. J. Neurophysiol., 50, 178-191.

Forssberg, H., Grillner, S., Rossignol, S. & Wallén, P. (1976). Phasic control of reflexes during locomotion in vertebrates. In Neural Control of Locomotion. Eds. Herman, R. M., Grillner, S., Stein, P. S. G. & Stuart, D. G. Plenum, New York, pp 647-674.

Gray, J. (1936). Studies in animal locomotion. IV. The neuromuscular mechanism of swimming in the eel. J. exp. Biol., 13, 170-180.

Gray, J. (1950). The role of peripheral sense organs during locomotion in the vertebrates. Symp. Soc. exp. Biol., 4, 112-126.
Gray, J. & Sand, A. (1936). The locomotory rhythm of the dogfish (Scyllium canicula). J. exp. Biol., 13, 200-209.
Grillner, S., (1974). On the generation of locomotion in the spinal dogfish. Expl. Brain Res., 20, 459-470.
Grillner, S. (1975). Locomotion in vertebrates: central mechanisms and reflex interaction. Physiol Rev., 55, 247-304.
Grillner, S. & Kashin, S. (1976). On the generation and performance of swimming in fish. In Neural Control of Locomotion. Eds. Herman, R. M., Grillner, S., Stein, P. S. G. & Stuart, D. G. Plenum, New York, pp 181-202.
Grillner, S., McClellan, A. & Perret, C. (1981). Entrainment of the spinal pattern generators for swimming by mechanosensitive elements in the lamprey spinal cord in vitro. Brain Res., 217, 380-386.
Grillner, S., McClellan, A. & Sigvardt, K. (1982). Mechanosensitive neurones in the spinal cord of the lamprey. Brain Res. 235, 169-173.
Grillner, S., McClellan, A., Sigvardt, K. Wallén, P. & Wilén, M. (1981). Activation of NMDA-receptors elicits 'fictive locomotion' in lamprey spinal cord in vitro. Acta physiol. Scand., 113, 549-551.
Grillner, S., Perret, C. & Zangger, P. (1976). Central generation of locomotion in the spinal dogfish. Brain Res., 109, 255-269.
Grillner, S., Rossignol, S., & Wallén, P. (1977). The adaptation of a reflex response to the ongoing phase of locomotion in fish. Expl. Brain Res., 30, 1-11.
Grillner, S. & Wallén, P. (1977). Is there a peripheral control of the central pattern generators for swimming in dogfish? Brain Res., 127, 291-295.
Grillner, S. & Wallén, P. (1982). On peripheral control mechanisms acting on the central pattern generators for swimming in the dogfish. J. exp. Biol., 98, 1-22.
Grillner, S. & Wallén, P. (1984). How does the lamprey CNS make the lamprey swim? J. exp. Biol., 112, 337-357.
Grillner, S., Wallén, P., McClellan, A., Sigvardt, K., Williams, T. & Feldman, J. (1983). The neural generation of locomotion in the lamprey: an incomplete account. In Neural Origin of Rhythmic Movements. Eds. Roberts, A. & Roberts, B. Symp. Soc. exp. Biol., 37, 285-303.
Grillner, S., Williams, T. & Lagerbäck, P.-A. (1984). The edge cell, a possible intraspinal mechanoreceptor. Science, N.Y., 223, 500-503.
Holst, E. von (1935a). Erregungsbildung und Erregungsleitung im Fischrückenmark. Pflügers Arch. ges. Physiol., 235, 345-359.
Holst, E. von (1935b). Uber den Process der zentralnervösen Koordination. Pflügers Arch. ges. Physiol., 236, 149-158.
Kappers, L. K. A., Huber, G. C. & Crosby, E. C. (1936). The Comparative Anatomy of the Nervous System of Vertebrates Including Man. Macmillan, London.
Kolmer, W. (1905). Zur Kenntnis des Rückenmarks von Ammocoetes. Arb. anat. Inst., Wiesbaden, 29, 165-214.
Lissmann, H. W. (1946a). The neurological basis of the locomotory rhythm in the spinal dogfish (Scyllium canicula, Acanthias vulgaris) I. Reflex behaviour. J. exp. Biol., 23, 143-161.

Lissmann, H. W. (1946b). The neurological basis of the locomotory rhythm in the spinal dogfish (Scyllium canicula, Acanthias vulgaris) II. The effect of deafferentation. J. exp. Biol., 23, 162-176.

Le Mare, D. W. (1936). Reflex and rhythmical movements in the dogfish. J. exp. Biol., 134, 429-442.

Maeda, N., Miyoshi, S. & Toh, H. (1983). First observation of a muscle spindle in fish. Nature, Lond., 302, 61-62.

Pavlidis, T. (1973). Biomechanical Oscillators: their Mathematical Analysis. Academic Press, New York.

Poon, M. L. T. (1980). Induction of swimming in lamprey by L-DOPA and amino acids. J. comp. Physiol., 136, 337-344.

Retzius, G. (1891). Zur Kenntnis des centralen Nervensystems von Myxine glutinosa. Biol. Unters., 11, 47-53.

Roberts, B. L. (1969a). Spontaneous rhythms in the motoneurons of spinal dogfish (Scyliorhinus canicula). J. mar. biol. Ass. U. K., 49, 33-49.

Roberts, B. L. (1969b). The response of a proprioceptor to the undulatory movements of dogfish. J. exp. Biol., 51, 775-785.

Rovainen, C. M. (1967). Physiological and anatomical studies on large neurons of central nervous system of the sea lamprey (Petromyzon marinus) I. Muller and Mauthner cells. J. Neurophysiol., 30, 1000-1023.

Rovainen, C. M. (1974). Synaptic interactions of identified nerve cells in the spinal cord of the sea lamprey. J. comp. Neurol., 154, 189-206.

Stein, P. S. G. (1976). Mechanisms of interlimb phase control. In Neural Control of Locomotion. Eds. Herman, R. M., Grillner, S., Stein, P. S. G. & Stuart, D. G. Plenum, New York, pp 465-487.

Steiner, I. (1886). Uber das Centralnervensystem der grünen Eidechse, nebst weiteren Untersuchungen über das des Haifisches. Sber. preuss. Akad. Wiss., 32, 539-543.

Tao-Cheng, J.-O., Hirosawa, K. & Nakajima, Y. (1981). Ultrastructure of the crayfish stretch receptor in relation to its function. J. comp. Neurol., 200, 1-21.

Wallén, P. (1980). On the mechanisms of a phase-dependent reflex occurring during locomotion in dogfish. Expl. Brain Res., 39, 193-202.

Wallén, P. & Lansner, A. (1984). Do the motoneurones constitute a part of the spinal network generating the swimming rhythm in the lamprey? J. exp. Biol., 113, 493-497.

Wallén, P. & Williams, T. L. (1984). Fictive locomotion in the lamprey spinal cord in vitro compared with swimming in the intact and spinal lamprey. J. Physiol., 347, 225-239.

Watkins, J. C. (1981). Pharmacology of excitatory amino acid transmitters. In Amino Acid Neurotransmitters. Eds. deFeudis, F. V. & Mandel, P. Raven Press, New York, pp 205-212.

Chapter Twenty-two

HOW LOCUSTS FLY STRAIGHT

C.H.F. Rowell, H. Reichert and J.P. Bacon

In this article we use the term locust to include both of the commonly studied genera Locusta and Schistocerca, and presume that our statements and generalisations cover acridid grasshoppers as a whole, at least those capable of flight. Insect flight in general and locust flight in particular has been the subject of a lot of work. At the risk of creating artificial divisions, we think it is useful to split up the neurophysiological aspects of locust flight into the following conceptual parts, most of which are represented in this symposium by one or more research workers:

1. The central rhythm generator (CRG) for flight, which is becoming known through the work of Robertson and Pearson (1982, 1983). It appears to be a distributed oscillator composed of interneurones spread over at least six thoracic and abdominal ganglia, not apparently containing intrinsic pacemaker cells but instead generating its cyclical output through circuit properties.

2. In close association with the CRG (and perhaps only mentally separable from it) is the sensory feedback occasioned by the wingbeat. This information arises from a multitude of sense organs, mostly proprioceptors, including the wing-hinge stretch receptor, the tegula, and the hairs and campaniform sensilla of the wings themselves (see e.g. Pabst & Schwarzkopf, 1962; Gettrup, 1962, 1963, 1965, 1966; Waldron, 1968; Wendler, 1974, 1983; Burrows, 1975; Altman & Tyrer, 1978; Möhl & Nachtigall, 1978; Möhl, 1979; Kien & Altman, 1979; Neumann et al., 1982; Pearson et al., 1983; Heukamp, 1983). It is typically phase-locked and modulated at wing-beat frequency. Sense organs that report windspeed and perhaps lift, such as the antennae (Gewecke, 1972a, 1975), may also fall into this category, though these differ from the rest in that phasic modulation at wing-beat frequency is less conspicuous.

3. The processes responsible for the initiation, maintenance and cessation of flight are very poorly understood. We know some of the sensory inputs which are

effective. For example, wind on the head, loss of tarsal contact, sudden sound and change in illumination all promote flight onset. We also have a few clues about the pharmacology of the units involved. For example, both poisoning with canavaline (Kammer et al., 1978) and the injection of octopamine into the neuropile (Sombati, 1983; Kinnamon et al., 1984) increase the probability of flight.

4. Course maintenance. In order to maintain a straight course in the face of deviations brought about either by faulty motor performance or by external influences (e.g. turbulence), the animal must be able to detect deviations and thereafter to carry out appropriate corrective steering. In short, it must have autopilot circuits. This is the subject of our chapter. Our knowledge has advanced considerably since Wilson's (1972) review of the subject.

5. Voluntary steering has obviously a close relationship with course maintenance. It is also a subject of our research, but for brevity we will ignore the topic in this review.

THE NEUROPHYSIOLOGY OF COURSE MAINTENANCE

The motor output
Once a deviation from course has been perceived as outlined below, the animal has a repertoire of at least two sorts of steering behaviour which it uses to correct its course. These are:

i. Alteration in wingbeat, brought about by asymmetric modulation of the flight motor neurone (FMN) discharge (Wilson & Weis-Fogh, 1962; Dugard, 1967; Koch, 1976; Zarnack & Möhl, 1977; Baker, 1979a,b; Taylor, 1981b; and others). Thüring in our laboratory has recently shown that these modulations of neural activity or wingbeat do indeed correlate with appropriate changes in torque, thus closing a logical loop which has remained open for more than two decades.

ii. Use of the hind legs and/or the abdomen as rudder surfaces (Gettrup & Wilson, 1964; Camhi, 1970a,b; Camhi & Hinkle, 1974; Taylor, 1981a). The aerodynamic effect of this has not been directly measured, but seems obvious enough.

Certainly the changes in wingbeat and probably also ruddering are first elicited by the direct action on the thoracic motor centres of the sensory signal generated by deviation from course (Taylor, 1981a,b; Rowell & Pearson, 1983). These behaviours can also be evoked more indirectly: the deviation signal elicits compensatory movements of the head to bring it back to the desired orientation. The positional mismatch between head and prothorax is detected by neck proprioceptors, and used secondarily to drive both sorts of steering behaviour (Haskell, 1959; Goodman, 1959, 1965; Taylor, 1981a).

The deviation signal

With what sense organs is the deviation from course perceived, and how is the signal encoded within the CNS?

i. The animal can detect <u>angular deviations from the direction of forward movement</u> by means of the directionally sensitive wind-hairs of the head (Weis-Fogh, 1949), body (Pflüger & Tautz, 1982) and cerci (Arbas, 1980; Altman, 1983). They respond to yaw, pitch, and even to roll if this is combined with one of the preceding angular movements. The antennae, which theoretically should be able to contribute information, do not seem to be used in this role (Gewecke, 1972b; Gewecke & Philippen, 1978; Heinzel & Gewecke, 1979; but see Arbas, 1980).

ii. Additionally, the flying locust can detect <u>angular deviation with respect to the surrounding visual world</u> by using the ocelli or the compound eyes. The ocelli are sensitive to nett luminosity over a wide field centred on the horizon, and thus to pitch and roll, but not to yaw (M. Wilson, 1978a,c; Taylor, 1981a; Thüring, unpublished). The compound eyes can report pitch, roll and – if a structured visual field is available – also yaw. Several different sorts of cues are available to the compound eyes; not only the luminosity difference between earth and sky and the position of landmarks (such as the horizon or vertical objects) on the retina can be used, but also the position of the poles of the flow fields on the retina, if the animal is flying past relatively near objects (Collett & King, 1975; Gibson, 1979; Wagner, 1982).

iii. A structured visual field also allows the compound eye to detect <u>translational movement with respect to the visual world</u>. The best studied aspect of this is the maintenance of a constant position by animals flying (or swimming) against a current – this is a situation which is rarely, if ever, encountered by a flying grasshopper. In a moth it has been shown convincingly (Preiss & Kramer, 1983) that the animals detect sinking or climbing through the expansion or contraction of the visual pattern on the ventral half of the eye, and alter their lift production accordingly (curiously, they do this only when seeking to remain in a pheromone plume). It seems very probable that something similar will be found in other insects, including locusts, but the necessary experiments have not yet been done.

Finally, there exist reports on tethered locusts (Wilson, 1968; Baker, 1975) which show that a structured visual field contributes to flight stability.

In summary, then, the animals use the wind-hairs, the ocelli and the compound eyes to detect their deviations from course; angular (as opposed to translational) deviations seem to be the most important, and are certainly the best understood.

All three sensory systems are represented in the ventral

How locusts fly straight

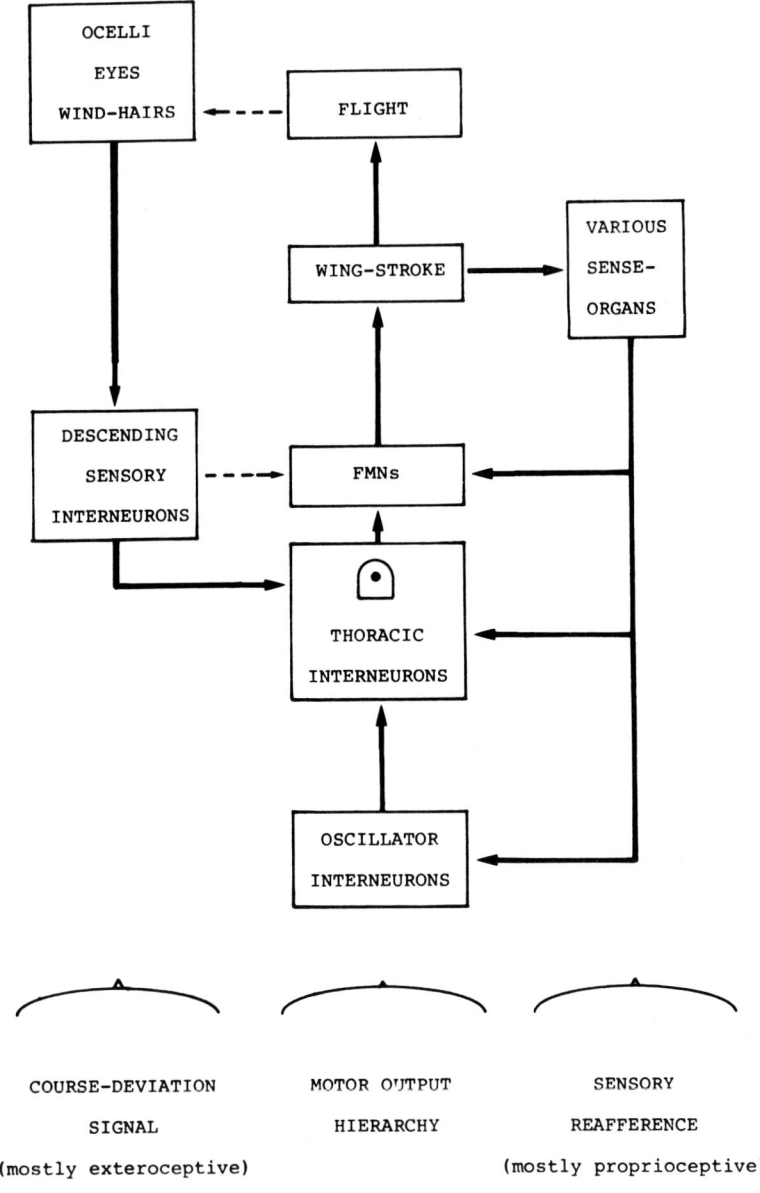

Fig. 1. Schematic representation of the organisation of the locust flight system. Dotted lines represent relatively weak interactions. The premotor interneurones of the thorax act as AND gates, mixing phasic

nerve cord by prominent descending interneurones which originate in the brain and carry the deviation signal to the thoracic motor centres responsible for steering behaviour. These descending interneurones and their thoracic connections will now be reviewed. First the three modalities will be considered in isolation; then convergence between them will be examined.

The ocellar interneurones
Whatever the other functions of the ocelli may be, their function in locust flight is unambiguous (M. Wilson, 1978a; Taylor, 1981a,b); they report angular deviations with respect to the horizon. The ocelli appear designed (M. Wilson, 1978a) to reject all information other than nett luminosity. This fact brings to their functioning within the locust flight system great sensitivity and speed of action. To the experimenter it also means that they can be stimulated in a biologically appropriate way with very simple apparatus (in contradistinction to the compound eye). We accordingly studied this system first, and know most about it. These results have been partially published (Rowell & Pearson, 1983; Reichert & Rowell, 1985) and recently summarised in a preceding SEB Symposium (Reichert, 1985); here we will give only a short resume.

The Hesse-Wilson theory of ocellar function in flight steering was confirmed experimentally by Taylor (1981a,b) and subsequently by Thüring (unpublished), who has shown that the ocelli alone (i.e. after removal of the compound eyes) can elicit not only modulation of the FMN firing pattern but also appropriate torque changes in flight. A number of workers (in locusts chiefly M. Wilson, 1978b,c; Simmons, 1981, 1982) have recently investigated the processing of visual information in the ocellar retina and in the first and second order interneurones driven by the retinula cells. Descending ocellar information, judging by the behaviour elicited, proceeds to the neck muscles via tonic second-order interneurones, which have not yet been characterised; it also goes by phasic second-order descending interneurones (described in detail by Simmons, 1980a; Rowell & Pearson, 1983; Reichert, Rowell & Griss, unpublished) to the pterothoracic ganglia. Here they make weak synaptic connections with flight motor neurones

---

input from the oscillator circuit with largely non-phasic input from the descending sensory interneurones which report course deviations. Only the most relevant and best-documented components are shown. For example, the following may be present as well; i) sensory interneurones in the reafference loop; ii) premotor interneurones postsynaptic to the oscillator which do not also receive exteroceptive information; iii) thoracic interneurones receiving exteroceptive information which do not also receive input from the oscillator.

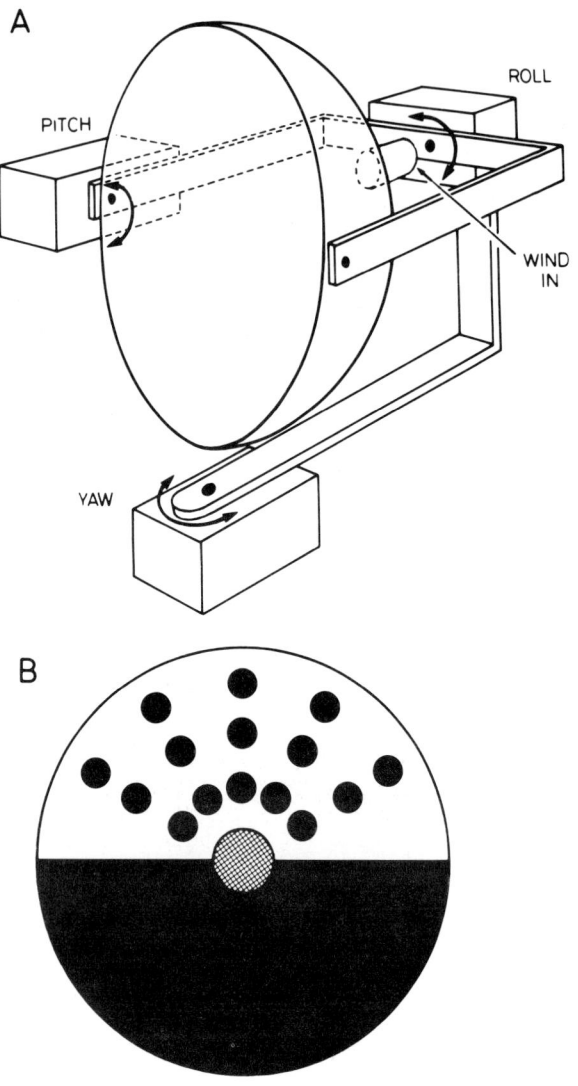

Fig. 2. A. Diagram of the artificial horizon used to simulate angular deviations from course during flight while recording electrophysiologically from the thoracic ganglia and connectives. Three motors move the hemisphere around the three axes of rotation. The locust's head is positioned in the centre of rotation. A current of air enters through a pipe set in the back of the hemisphere and simulates the wind caused by flight; it also serves to induce flight. The

(Simmons, 1980a; Rowell & Pearson, 1983) and strong ones with thoracic interneurones (Rowell & Pearson, 1983; Reichert, 1985; Reichert & Rowell, 1985). Typically, these thoracic interneurones are powerfully presynaptic to the FMNs, are partially depolarised during flight, and are strongly modulated at wing-beat frequency (apparently through being post-synaptic to the CRG for flight). This rhythmic modulation gates the ocellar signal, allowing it to generate spikes in the premotor interneurones only during discrete phases of the wing-beat cycle. The result is that the FMNs receive significant ocellar input only a) when the animal is flying and b) at an appropriate phase of the wing-beat cycle. Parallel subpopulations of thoracic interneurones (gated in either elevator or depressor phase and having either excitatory or inhibitory connections with either ipsi- or contralateral FMNs) ensure that all ocellar signals, regardless of their phase or time of arrival in the wing-beat cycle, are appropriately utilised for steering corrections. These relationships are summarised in Fig. 1.

This relatively well-understood system, whereby the ocellar information influences the flight motor output so as to produce corrective steering, can be taken provisionally as a model for the others. How closely do the information channels serving compound eye and wind-hairs conform to the ocellar plan?

Compound eye units
It has long been known (e.g. Parry, 1947) that descending units carrying compound eye information are to be found in the locust ventral nerve cord. Some attempts have been made to catalogue these (Catton & Chakraborty, 1969; Williamson & Burns, 1982), but these investigations have utilised "classical" optical stimuli (e.g. whole field or small field ON and OFF stimuli, contrasting edges, moving stripe patterns over a limited test area) and most have shown merely that these stimuli do not elicit much activity in the cord units. The problem with the compound eye units is clearly one of finding the appropriate stimulus. Perhaps the only descending compound eye unit in the locust which has been reasonably well characterised using biologically appropriate stimuli is the DCMD (descending contralateral movement detector) neurone, which responds selectively to the novel movement of small objects in the monocular visual field (for a review see O'Shea

---

hemisphere is made from translucent plastic and is illuminated diffusely from above and behind. A pattern (see B) is painted on the hemisphere with opaque black paint.
B. Formal representation of the pattern on the inside of the hemisphere. The edges of the circle represent 90° angular distance from the centre, where the pipe enters.

& Rowell, 1977). Although the DCMD makes synapses with FMNs (Pearson & C. Goodman, 1979; Simmons, 1980b), experiments have so far failed to disclose any effect on flight muscle activity during flight (Taylor & Rowell, unpublished); nor have we found any thoracic interneurones of the type associated with steering that are postsynaptic to it.

The task of simulating the visual stimuli received by a flying locust during deviations from course is complex, especially if it is assumed that the animal is flying close enough to the ground to be experiencing a "flow field" on the retina (Collett & King, 1975; Gibson, 1979; Wagner, 1982). If this requirement is relaxed and it be assumed that the animal is flying sufficiently high over a more or less featureless landscape to avoid significant flow field stimulation, then a simulation is easier. We have achieved this by mounting the animal for neurophysiological recording with its head at the centre of a vertically oriented translucent hemisphere (Fig. 2), into which the animal looks. (The back of the eye can be painted out, to ensure that the animal sees only the hemisphere, though this appears to be less necessary than we would have guessed). The inner surface of the lower half of the hemisphere is painted black, and the whole illuminated from behind and above. The clear "sky" half bears a few symmetrically placed black spots to provide some visual structure in the upper field. The hemisphere is mounted in gimbals permitting independent movement around all three rotational axes (i.e simulating roll, pitch and yaw, or combinations of these) and each axis is equipped with a remotely controlled servomotor. In the centre of the hemisphere air enters at a known velocity, providing a means of starting the animal "flying" (i.e., producing a rhythmic output in the nerves to the flight muscles reminiscent of that seen in flight, as described by Robertson & Pearson, 1982). Also, as described in the following Section, it provides a steerable flow of air over the head.

Using this artificial horizon, we were immediately able to find highly specific <u>descending visual interneurones</u> with remarkably complex properties, which provide detailed information about visually mediated deviations from the normal flight position and thus from course. They have the following general characteristics:

i. In comparison with the phasic ocellar responses, the latency of the compound eye signal is considerably greater (of the order of 80-100ms total delay to the thorax, compared with approximately 25ms for ocellar units at the same temperature) which is the expected consequence of the extra processing required in the optic lobe.

ii. The receptive field can be monocular or binocular.

iii. They are directionally sensitive to deviation - e.g., a unit may respond to a roll to the left, but a roll to the right would produce either no response or inhibition.

How locusts fly straight

iv. They are specific to complex combinations of movements, often combinations that are aerodynamically meaningful; e.g., we know a unit which responds to roll to the

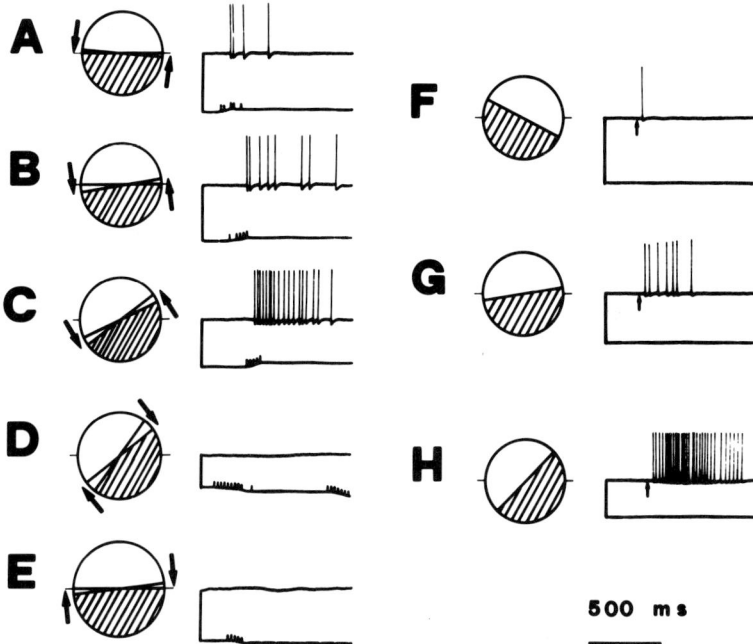

Fig. 3. Specimen response from the axon of an identified "absolute deviation unit", recorded intracellularly in the mesothoracic ganglion. The animal saw the artificial horizon shown in Fig. 2. In this Figure only the responses to roll are shown.

Anticlockwise movement of the horizon away from the horizontal, simulating a clockwise roll by the insect away from the normal flight position, evokes large responses which increase in frequency and become more tonic with increasing deviation (B,C). Movement in the same sector in the reverse direction produces no response (D,E). Movement in the preferred direction, but towards rather than away from the horizon, is ineffective; only when the horizon is very close to the horizontal (A) is there a small response. Calibration pulses on the stimulus monitor trace represent $9°$ of roll.

The position of the visual pattern also affects the response to wind. Close to the normal flying position there is a relatively weak response to frontal wind (G), the arrow marking the approximate onset of the wind stimulus. When however the pattern is stationary in a favoured position (i.e. after a simulated clockwise role), there is a vigorous response to frontal wind (H). When it is stationary in an unfavoured position, the same wind response is inhibited (F).

left, yaw to the left, and pitch up – that is, it responds optimally to a climbing banked turn to the left.

v. Most strikingly, they are sensitive to the absolute orientation of the visual pattern on the retina. We define as "normal" orientation the situation in which the horizon of the stimulator lies on the horizontal plane running through the centre of the two compound eyes. Movement of the hemisphere in the preferred direction <u>away</u> from the "normal" orientation results in <u>large and increasingly tonic</u> responses; movement in the same direction <u>towards</u> the normal orientation produces little or no response; movement in the opposite direction, as previously stated, gives no response or even inhibition (Fig. 3). These units are then <u>absolute deviation detectors</u>. They are not responsive to movement of the visual field as such, but only to movement which signals deviation of the animal from the normal flight position in a certain direction.

It is hard to imagine pure yaw units that would have this absolute deviation property, as they would have to employ some form of short-term memory of arbitrary visual landmarks to give a fixed reference point; the horizon, which fulfils this need for roll and pitch, is here of no help. Indeed, we have not as yet found any such units, but this may be simply because, for technical reasons, we have so far studied mainly the responses to pitch and to roll.

In the thoracic ganglia, we find a further population of <u>thoracic interneurones</u>. Their morphology is very similar to that of the corresponding ocellar thoracic neurones, and as we describe below, at least some individual neurones are common to both populations. They are postsynaptic to the descending compound-eye units just described. Like the ocellar units, these units are typically tonically depolarised during flight and rhythmically modulated at wing-beat frequency, and at least some are presynaptic to FMNs. In short, it appears so far that the ocellar model is good for the compound-eye units too.

<u>Wind-hair units</u>
The study of wind units lags behind that of the visual units. The wind-hairs themselves were early described and their significance in flight recognised (Weis-Fogh, 1949 and subsequent works; Guthrie, 1964; Camhi, 1969a,c; Smola, 1970; Tyrer et al., 1979), but the interneurones have been relatively neglected. Camhi (1969b) described a variety of descending wind-hair units, including a "recentre" unit, which had similar properties to the visual "absolute deviation units" we have discovered, but operating in the reverse direction. It (or they) provided information that the animal was now back on course, not (as in the visual units described above) that it had deviated from it. This finding is now even more interesting than it was originally, and must be confirmed. Camhi's pioneering work dates from before the era

of routine intracellular recording and dye-filling, so the identity of the units has not been established.

Since then, only one wind-hair unit has been characterised to modern standards: this is the tritocerebral commissure giant (TCG), first filled and described by Bacon & Tyrer (1978). The hairs driving this unit have been characterised, and their directionality and pattern of excitatory and inhibitory connections with the TCG shown (Mühl & Bacon, 1983). They explain the directionality of the interneurone, which prefers wind laterally displaced from centre (i.e., what the animal would experience during a yaw off course). Because of its unusually favourable anatomy, which makes it possible to stimulate it selectively without previous intracellular penetration, it has been possible to show (Bicker & Pearson, 1983) that under appropriate conditions onset of activity in the TCG alone can initiate flight; also that activity in one TCG during tethered flight will induce a modulation of the FMN firing pattern compatible with the theoretically expected yaw correction (Mühl & Bacon, 1983). Work with our new steerable horizon shows that the TCG is also responsive to downward pitch, via the wind-hairs, and not to roll stimuli. Some activity can be elicited by changes in whole-field illumination (Bacon & Tyrer, 1978) though this in itself appears not to impart any directional sensitivity to the neurone.

Although only the TCG has been intensively studied, it is known that other uni- or multimodal wind-hair neurones exist in the cord. Varanka & Sviderksy (1974a,b) estimated 10-15 wind-responsive units in the cord, and Williamson & Burns (1982) described 7 such descending interneurones, though without further characterisation other than their position in the cross-section of the connective. Several of the ocellar and compound eye units also carry wind information (see next Section). Pflüger (1984) describes a cord unit which receives most of its input from prothoracic hairs, but also is postsynaptic to a few cephalic hairs. Its role in flight is not yet known.

In common with ocellar and compound eye units, the TCG appears to synapse both with FMNs (Bacon & Tyrer, 1979; Tyrer, 1980) and with thoracic interneurones (Bicker, personal communication; Bacon, unpublished). Reichert & Rowell (unpublished) have filled thoracic interneurones with the usual physiological properties (i.e. depolarized during flight, modulated at wing-beat frequency, strongly affected by sensory input) and the morphology of steering interneurones which respond only to wind stimuli and not to visual inputs. The directionality of these units is, however, as yet unknown.

If the wind-activated thoracic units are indeed gated by the CRG for flight like the visual units (as appears to be the case), then the basic problem of how to feed randomly occurring sensory information into a rhythmically active system at the correct phase is basically solved. However, recent work (Bacon & Möhl, 1979, 1983; Horsmann et al., 1983) has made it clear that the wind-hair signal in real flight is far from random, and is actually rhythmically modulated at wing-beat frequency. This effect is due to modulation by the wing-stroke of the flow of air over the head. The wind-hair information is therefore modulated at source, and the phasic information that arrives at the thorax is then gated by the mechanism previously described.

In summary, study of the wind system is in its infancy. The advent of our new steerable stimulator should produce a great deal of much-needed data. The indications to date are that the same organisation applies as we have determined for the two visual systems. The plan derived originally from the ocellar units seems to be applicable here too.

Multimodality and convergence

So far, we have spoken of the sensory modalities as if they were three isolated entities. At least at the interneuronal level, this is emphatically not the case. There is extensive convergence, first obvious at the level of the descending interneurones in the connectives, and many of the descending and thoracic interneurones referred to above are actually multimodal. Most of the "ocellar" units (though not all) are responsive to the wind-hairs as well, and these two inputs interact in a manner first described by Simmons (1980a); wind plus light OFF produces a summation, and light ON inhibits an ongoing wind response. The complex units driven by the compound eye are also responsive to wind, and show a very strong interaction. First, the wind responses are behaviourally compatible with the visual responses; thus, units which respond best to a visually simulated yaw to the left prefer wind currents impinging on the right hand side of the head, and so on. In the "absolute deviation units", a negative visual input (e.g. a maintained roll in the antipreferred direction away from the normal orientation) will produce a maintained and virtually complete inhibition of the wind response. Thus, not only are the wind and visual responses spatially compatible, their interaction results in an editing process whereby the unit is silenced if it is not stimulated within a certain range of conditions, presumably corresponding to the range in which it can mediate appropriate corrective steering. Yet more strikingly, at least some of the descending units are trimodal, responding to wind, compound eyes and ocelli simultaneously. In these the ocellar input too is spatially compatible with the others; a visual roll-to-the-left unit, for example, responds to dimming of the left hand lateral ocellus, and the spikes so produced sum with the visual response. Illuminating this ocellus, however, powerfully inhibits the visual response, just as in the interaction with wind (Fig. 4). Exactly how general this trimodal convergence is remains to be seen, but it appears likely that many fewer descending units are involved in reporting course deviations than has previously been thought. Possibly the majority carry unique and very precise information derived from all three sensory modalities; the challenge will be to find the correct stimulus configurations. The specificity for the effective stimulus is remarkable; we have only recently determined, with the aid of the artificial horizon, that three particular descending "ocellus and wind" units, with which we have worked routinely for some years, are also powerfully driven by the compound eye – but only when confronted with the correct stimulus configuration.

At the level of the <u>thoracic interneurones</u>, we have already found all of the possible sorts of convergence; ocellar with wind, ocellar with compound eye, compound eye with wind, and units responding to all three modalities. This last group presumably will be found to include cells involved in correctional manoeuvres of which pitch is a component, because only pitch is signalled by all three modalities.

<u>Concluding remarks</u>
We appear to be currently in that gratifying situation which occasionally confronts experimenters; after a period of

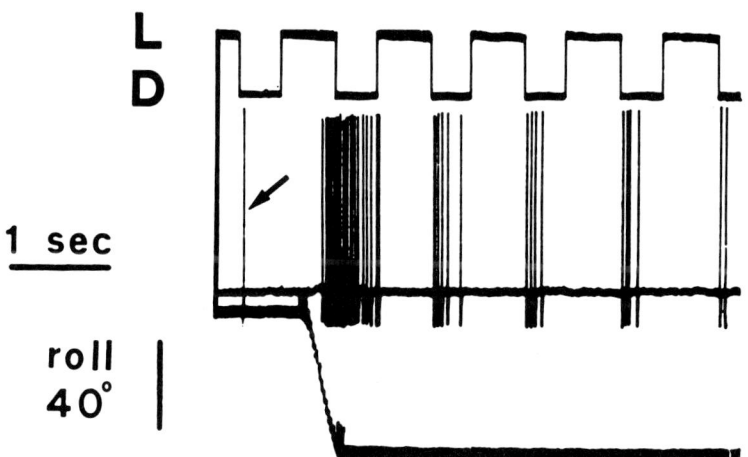

Fig. 4. Intracellular recording from the same unit as shown in Fig. 3 in a different individual. The upper trace (L, light; D, dark) shows step changes in illumination (1 log unit) delivered by the light pipe to that lateral ocellus which is connected to the neurone. Each OFF stimulus elicits a single spike (arrow). The lower trace shows a large and effective roll stimulus. The ensuing train of spikes (mediated by the compound eye) is modulated by the inhibition exerted by the ocellus during light ON.

demanding work it suddenly seems that the principles which have emerged (in this case from a study of the ocellar interneurones in the thoracic nervous system) are applicable to a much larger population of neurones, and that we are much closer to an understanding of how the locust autopilot functions than we would have guessed would be the case a year ago. This is largely due to the degree of high-order convergence found within the interneurones bringing information about deviations from course to the thorax. Much of the circuitry worked out for the ocellar neurones is therefore automatically valid for signals derived from the eye and wind-hairs. Once again, it appears that a sound working strategy for understanding the insect nervous system is to assume that little or no redundancy exists; the insect interneurones pack the maximum amount of information into the smallest number of units compatible with the basic evolutionary history of their nervous system. Although there are limitless opportunities for further work to confirm and refine the general scheme advanced here, we believe that the basic understanding of how grasshoppers fly straight has been achieved.

Acknowledgement
The work described was supported by the Schweizerische Nationalfonds.

REFERENCES

Altman, J.S. (1983). Sensory inputs and the generation of the locust flight motor pattern: from the past to the future. In Symposium on the Physiology and Biophysics of Insect Flight. Biona Report 2. Ed. Nachtigall, W.. Gustav Fischer, Stuttgart, pp. 127-136.

Altman, J.S. & Tyrer, N.M. (1978). The locust wing hinge stretch receptors. I. Primary sensory neurones with enormous central arborizations. J. comp. Neurol., 172, 409-430.

Arbas, E.A. (1980). A neuroethological study of flight loss and aerial manoeuvering in insects. Dissertation, University of Oregon, Eugene.

Bacon, J. & Möhl, B. (1979). Activity of an identified wind interneurone in a flying locust. Nature, Lond., 278, 638-640.

Bacon, J. & Möhl, B. (1983). The tritocerebral commissure giant (TCG) wind-sensitive interneurone in the locust. I. Its activity in straight flight. J. comp. Physiol., 150, 439-452.

Bacon, J. & Tyrer, N.M. (1978). The tritocerebral commisure giant (TCG): a bimodal interneurone in the locust, Schistocerca gregaria. J. comp. Physiol., 126, 317-325.

Bacon, J. & Tyrer, M. (1979). Wind interneurone input to flight motor neurones in the locust, Schistocerca gregaria. Naturwissenschaften, 66, 116-117.

Baker, P.S. (1975). Optomotor responses of flying locusts. Expl. Brain Res., 23, 13.

Baker, P.S. (1979a). The wing movements of flying locusts during steering behaviour. J. comp. Physiol., 131, 49-58.

Baker, P.S. (1979b). The role of forewing muscles in the control of direction in flying locusts. J. comp. Physiol., 131, 59-66.

Bicker, G. & Pearson, K.G. (1983). Initiation of flight by an identified wind sensitive neurone (TCG) in the locust. J. exp. Biol., 104, 289-294.

Burrows, M. (1975). Monosynaptic connexions between wing stretch receptors and flight motoneurones of the locust. J. exp. Biol., 62, 189-220.

Camhi, J.M. (1969a). Locust wind receptors. I. Transducer mechanisms and sensory response. J. exp. Biol., 50, 335-348.

Camhi, J.M. (1969b). Locust wind receptors. II. Interneurones in the cervical connective. J. exp. Biol., 50, 349-362.

Camhi, J.M. (1969c). Locust wind receptors. III. Contribution to flight initiation and lift control. J. exp. Biol., 50, 363-374.

Camhi, J.M. (1970a). Yaw-correcting postural changes in locusts. J. exp. Biol., 52, 519-532.

Camhi, J.M. (1970b). Sensory control of abdomen posture in flying locusts. J. exp. Biol., 52, 533-538.

Camhi, J.M. & Hinkle, M. (1974). Response modification by the central flight oscillator of locusts. J. exp. Biol., 60, 477-492.

Catton, W.T. & Chakraborty, A. (1969). Single neuron responses to visual and mechanical stimuli in the thoracic nerve cord of the locust. J. Insect Physiol., 15, 245-258.

Collett, T. & King, A.J. (1975). Vision during flight. In The Compound Eye and Vision of Insects. Ed. Horridge, G.A. Clarendon Press, Oxford, pp. 437-466.

Dugard, J.J. (1967). Directional change in flying locusts. J. Insect Physiol., 13, 1055-1063.

Gettrup, E. (1962). Thoracic proprioceptors in the flight system of locusts. Nature, Lond., 193, 498-499.

Gettrup, E. (1963). Phasic stimulation of a thoracic stretch receptor in locusts. J. exp. Biol., 40, 323-334.

Gettrup, E. (1965). Sensory mechanisms in locomotion: the campaniform sensilla of the insect wing and their function during flight. Cold Spring Harb. Symp. quant. Biol., 30, 615-621.

Gettrup, E. (1966). Sensory regulation of wing twisting in locusts. J. exp. Biol., 44, 1-17.

Gettrup, E. & Wilson, D.M. (1964). The lift-control reaction of flying locusts. J. exp. Biol., 41, 183-190.

Gewecke, M. (1972a). Antennen und Stirn-Scheitelhaare von Locusta migratoria L. als Luftströmungsorgan bei der Flugsteuerung. J. comp. Physiol., 80, 57-94.

Gewecke, M. (1972b). Bewegungsmechanismus und Gelenkrezeptoren der Antennen von Locusta migratoria L. (Insecta, Orthoptera). Z. Morph. Tiere, 71, 128-149.

Gewecke, M. (1975). The influence of the air-current organs on the flight behaviour of Locusta migratoria. J. comp. Physiol., 103, 79-96.

Gewecke, M. & Philippen, J. (1978). Control of the horizontal flight course by air current sense organs in Locusta migratoria. Physiol. Entomol., 3, 43-52.

Gibson, J.J. (1979). The Ecological Approach to Visual Perception. Houghton

Mifling, Boston.

Goodman, L.J. (1959). Hair receptors in locusts. Hair plates on the first cervical sclerites of the Orthoptera. Nature, Lond., 183, 1106-1107.

Goodman, L.J. (1965). The role of certain optomotor reactions in regulating stability in the rolling plane during flight in the desert locust, Schistocerca gregaria. J. exp. Biol., 42, 385-408.

Guthrie, D.M. (1964). Observations on the nervous system of the flight apparatus in the locust Schistocerca gregaria. Ql. J. microsc. Sci., 105, 183-201.

Haskell, P.T. (1959). Function of certain prothoracic hair receptors on the desert locust. Nature, Lond., 183, 1107.

Heinzel, H.-G. & Gewecke, M. (1979). Directional sensitivity of the antennal campaniform sensilla in locusts. Naturwissenschaften, 66, 212.

Heukamp, U. (1983). Die Rolle von Mechanorezeptoren im Flugsystem der Wanderheuschrecke (Locusta migratoria L.). Dissertation, Universität Köln.

Horsmann, U., Heinzel, H.-G. & Wendler, G. (1983). The phasic influence of self-generated air current modulations on the locust flight motor. J. comp. Physiol., 150, 427-438.

Kammer, A.E., Dahlman, D.L. & Rosenthal, G.A. (1978). Effects of the non-protein aminoacids L-canavanine and L-canaline on the nervous system of the moth Manduca sexta (L.). J. exp. Biol., 75, 123-132.

Kien, J. & Altman, J.S. (1979). Connections of the locust wing tegula with metathoracic flight motorneurons. J. comp. Physiol., 133, 299-310.

Kinnamon, S.C., Klaasen, L.W., Kammer, A.E. & Klaasen, D. (1984). Octopamine and chlordimeform enchance sensory responsiveness and production of the flight motor pattern in developing and adult moths. J. Neurobiol., 15, 283-294.

Koch, U.T. (1976). A miniature movement detector applied to recording of wingbeat in Locusta. Fortschr. Zool., 24, 327-332.

Möhl, B. (1979). High-frequency discharge of the locust wing hinge stretch receptor during flight. Naturwissenschaften, 66, 158-159.

Möhl, B. & Bacon, J. (1983). The tritocerebral commissure giant (TCG) wind-sensitive interneurone in the locust. II. Directional sensitivity and role in flight stabilisation. J. comp. Physiol., 150, 453-466.

Möhl, B. & Nachtigall, W. (1978). Proprioceptive input on the locust flight motor revealed by muscle stimulation. J. comp. Physiol., 128, 57-65.

Möhl, B. & Zarnack, W. (1977). Activity of the direct downstroke flight muscles of Locusta migratoria (L) during steering behaviour in flight. II. Dynamics of the time shift and changes in the burst length. J. comp. Physiol., 118, 235-247.

Neumann, L., Möhl, B. & Nachtigall, W. (1982). Quick phase-specific influence of the tegula on the locust flight motor. Naturwissenschaften, 69, 393-394.

O'Shea, M. & Rowell, C.H.F. (1977). Complex neural integration and identified interneurons in the locust brain. In Identified Neurons and Behavior of Arthropods. Ed. Hoyle, G. Plenum, New York, pp. 307-328.

Pabst, H. & Schwartzkopf, J. (1962). Zur Leistung der Flügel-gelenkrezeptoren von Locusta migratoria. Z. vergl. Physiol., 45, 396-404.

Parry, D.A. (1947). The function of the insect ocellus. J. exp. Biol., 24, 211-219.

Pearson, K.G. & Goodman, C.S. (1979). Correlation of variability of structure with the variability of synaptic connection of an identified interneuron in locust. J. comp. Neurol., 184, 141-166.

Pearson, K.G., Reye, D.N. & Robertson, R.M. (1983). Phase-dependent influence of wing stretch receptors on flight rhythm in the locust. J. Neurophysiol., 49, 1168-1181.

Pflüger, H.-J. (1984). The large fourth abdominal intersegmental interneuron: a new type of wind-sensitive ventral cord interneuron in locusts. J. comp. Neurol., 222, 343-357.

Pflüger, H.-J. & Tautz, J. (1982). Air movement sensitive hairs and interneurons in Locusta migratoria. J. comp. Physiol., 145, 369-380.

Preiss, R. & Kramer, E. (1983). Stabilization of altitude and speed in tethered flying gypsy moth males: influence of (+) and (-) disparlure. Physiol. Entomol., 8, 55-68.

Reichert, H. (1985). The cellular basis of sensorimotor coordination in the flight control system of the locust. In Coordination of Motor Behaviour. Eds. Bush, B.M.H. & Clarac, F. Cambridge University Press, Cambridge, pp. 121-140.

Reichert, H. & Rowell, C.H.F. (1985). Integration of non-phaselocked, exteroceptive information in the control of rhythmic flight in the locust. J. Neurophysiol. (In press).

Robertson, R.M. & Pearson, K.G. (1982). A preparation for the intracellular analysis of neuronal activity during flight in the locust. J. comp. Physiol., 146, 311-320.

Robertson, R.M. & Pearson, K.G. (1983). Interneurons in the flight system of the locust: distribution, connections and resetting properties. J. comp. Neurol., 215, 33-50.

Rowell, C.H.F. & Pearson, K.G. (1983). Ocellar input to the flight motor system of the locust: structure and function. J. exp. Biol., 103, 265-288.

Simmons, P. (1980a). A locust wind and ocellar brain neurone. J. exp. Biol., 85, 281-294.

Simmons, P. (1980b). Connexions between a movement-detecting visual interneurone and flight motoneurones of a locust. J. exp. Biol., 86, 87-97.

Simmons, P.J. (1981). Synaptic transmission between second- and third-order neurones of a locust ocellus. J. comp. Physiol., 145, 256-276.

Simmons, P.J. (1982). Transmission mediated with and without spikes at connexions between large second-order neurones of locust ocelli. J. comp. Physiol., 147, 401-414.

Smola, U. (1970). Rezeptor- und Aktionspotentiale der Sinneshaare auf dem Kopf der Wanderheuschrecke Locusta migratoria. Z. vergl. Physiol., 70, 335-348.

Sombati, S. (1983). Orchestrating behaviours: a central modulatory role of octopamine in the locust. Soc. Neurosci. Abstr., 9, 77.

Taylor, C.P. (1981a). Contribution of compound eyes and ocelli to steering of locusts in flight. I. Behavioural analysis. J. exp. Biol., 93, 1-18.

Taylor, C.P. (1981b). Contribution of compound eyes and ocelli to steering of locusts in flight. II. Timing changes in flight motor units. J. exp. Biol., 93, 19-31.

Tyrer, N.M. (1980). Transmission of wind information on the head of the locust to flight motor neurons. In Neurobiology of Invertebrates. Ed. Salánki, J. Adv. physiol. Sci., 23, 557-571.

Tyrer, N.M., Bacon, J.P. & Davies, C.A. (1979). Sensory projections from the wind-sensitive head hairs of the locust Schistocerca gregaria. Cell Tissue Res., 203, 79-92.

Varanka, I. & Svidersky, V.L. (1974a). Functional characteristics of the interneurons of wind-sensitive hair-receptors on the head in Locusta migratoria. L. I. Interneurons with excitatory responses. Comp. Biochem. Physiol., 48A, 411-426.

Varanka, I. & Svidersky, V.L. (1974b). Functional characteristics of the interneurons of wind-sensitive hair receptors on the head in Locusta migratoria. L. II. Interneurons with inhibitory responses. Comp. Biochem. Physiol., 48A, 427-438.

Wagner, H. (1982). Flow-field variables trigger landing in flies. Nature, Lond., 297, 147-148.

Waldron, I. (1968). The mechanism of coupling of the locust flight oscillator to oscillatory inputs. Z. vergl. Physiol., 57, 331-347.

Weis-Fogh, T. (1949). An aerodynamic sense organ stimulating and regulating flight in locusts. Nature, Lond., 163, 873-874.

Wendler, G. (1974). The influence of proprioceptive feedback on locust flight coordination. J. comp. Physiol., 88, 173-200.

Wendler, G. (1978). The possible role of fast wing reflexes in locust flight. Naturwissenschaften, 65, 65.

Wendler, G. (1983). The locust flight system: functional aspects of sensory input and methods of investigation. In Symposium on the Physiology and Biophysics of Insect Flight. Biona Report 2. Ed. Nachtigall, W. Gustav Fischer, Stuttgart, pp. 113-125.

Williamson, R. & Burns, M.D. (1982). Large neurones in the locust neck connectives. I. Responses to sensory inputs. J. comp. Physiol., 147, 379-388.

Wilson, D.M. (1961). The central nervous control of flight in a locust. J. exp. Biol., 38, 471-490.

Wilson, D.M. (1968). Inherent asymmetry and reflex modulation of the locust flight motor pattern. J. exp. Biol., 48, 631-641.

Wilson, D.M. (1972). Stabilizing mechanisms in insect flight. In Proceedings of an International Study Conference on Current and Future Problems in Acridology. Ed. Hemming, C.F. & Taylor, T.H.C. H. M. Stationery Office, London, pp. 47-52.

Wilson, D.M. & Weis-Fogh, T. (1962). Patterned activity of coordinated motor units, studied in flying locusts. J. exp. Biol., 39, 643-668.

Wilson, M. (1978a). The functional organisation of locust ocelli. J. comp. Physiol., 124, 297-316.

Wilson, M. (1978b). Generation of graded potential signals in the second order cells of the locust ocellus. J. comp. Physiol., 124, 317-331.

Wilson, M. (1978c). The origin and properties of discrete hyperpolarising potentials in the second order cells of locust ocellus. J. comp. Physiol., 128, 347-358.

Zarnack, W. & Möhl, B. (1977). Activity of the direct downstroke flight muscles of Locusta migratoria (L.) during steering behaviour in flight. I. Patterns of time shift. J. comp. Physiol., 118, 215-233.

Chapter Twenty-three

INTERACTIONS OF SEGMENTAL AND SUPRASEGMENTAL INPUTS WITH THE SPINAL PATTERN GENERATOR OF LOCOMOTION

S. Rossignol & T. Drew

An important question in the field of motor control is how the neural networks responsible for generating movements integrate the sensory inputs produced by these movements, so that the resulting feedback does not disrupt their progression but instead serves in their regulation. One way in which this question can be approached is to study the characteristics of the feedback signal by recording the discharge of specific types of peripheral afferents, or of central neurones, in relation to the various movement parameters. This approach is usually based on the assumption that the resulting effects of the recorded activity on the motor output can be derived from the effects that these particular neurones are known to have on motor neurones in motionless animal preparations. What if these effects are augmented, blocked or rerouted through pathways which are operative only in certain types of movements? Indeed, "what the cat's hindlimb tells the cat's spinal cord" and what the cat's spinal cord listens to may be different.

Another approach, which takes into account these changes in the effectiveness of transmission, consists in stimulating peripheral afferents or supraspinal centres during the movements themselves and studying the effects on motor neurones by recording electromyographic (EMG) responses. This approach has the advantage of revealing the underlying principles that may apply not only to "expected" inputs but also to "unexpected" inputs.

It should be realised that neural networks generating movements must be able to deal with unexpected inputs that may have a particular significance because they occur during the movement or because they arise in a particular phase of the movement. Indeed, cutaneous stimulation of the dorsum of the foot, which would be insignificant in an animal lying down, may, during the swing phase of locomotion, threaten the equilibrium of the animal. In this paper the principal findings on the responses of motor neurones to segmental and suprasegmental inputs during locomotion will be briefly reviewed and some new pertinent findings, obtained mainly in chronically implanted cats, will be used to discuss how those elements which generate locomotion and those elements which are involved in its control and regulation may usefully interact.

MODULATION OF REFLEX RESPONSES DURING LOCOMOTION

Ipsilateral hindlimb
The usual response to a stimulation of the skin of the hindlimb of a motionless cat is the so-called flexion reflex which activates muscles whose function is to withdraw the limb from the source of stimulation (Sherrington, 1910). However, at the time of the description of this reflex, it was also reported that reflex reactions to skin stimulation could exceptionally evoke excitatory responses in extensor muscles depending on the state of anaesthesia, the type of preparation, the position of the head or the position of the limbs (see Rossignol et al., 1981a,b for a review of these aspects). It is thus clear that cutaneous information can reach not only flexor motor neurones but also extensor motor neurones. What then determines the efficacy of transmission through one route or the other?

This question was initially studied in chronic spinal kittens walking on a treadmill with stimulation of cutaneous afferents produced by using either a light tap to the dorsum of the hindfoot or by a brief electrical shock through electrodes fixed to the dorsal surface of the foot (Forssberg et al., 1975, 1976, 1977). It was observed that the same stimulation given during mid-swing and in mid-stance evoked short latency responses in either flexor or extensor muscles respectively (latencies ranged from 10 to 25ms depending on muscles). This suggested the use of the expression "phase dependent reflex reversal" to indicate that the same stimulation gave a response in antagonistic muscles depending on the phase of the locomotor cycle. By delivering stimuli in all phases of the step cycle, it could be seen that this "reflex reversal" was in fact the result of a continuous modulation of the amplitude of the responses to flexor and extensor muscles as a function of the step cycle. This modulation was such that responses in flexors were maximal during their main period of locomotor activity (although the responses could still be large even after the locomotor burst) and extensor responses were maximal during the period of activation of extensor muscles in the step cycle. Thus, although there was not a strict relationship between the presence or absence of locomotor EMG activity and the time when the muscle was reflexly activated (period of reflex responsiveness), there was a general coincidence between the two.

These initial observations on the hindlimb were confirmed or extended in other preparations. For instance, in thalamic cats walking spontaneously on a treadmill (Duysens & Pearson, 1976; Duysens, 1977), stimulation of the plantar surface of the foot similarly altered the swing phase or the stance phase of the stimulated hindlimb. Although the stimulation of various cutaneous nerves could affect the step cycle differently it was always in a phase-dependent manner. Similar experiments in chronically implanted cats free to walk on a platform or on a treadmill (Duysens & Stein, 1978) revealed that several different pathways could be utilised and thus be controlled separately throughout the step cycle. For instance, Wand et al. (1980) showed that electrical stimulation of the skin of the hindfoot during swing evoked responses in knee and ankle flexors. During stance, these responses were absent and were replaced by a small and variable amount of inhibition and excitation in extensors. Furthermore, mechanical obstruction during swing revealed, besides the responses in the flexors, a large response in the ankle extensor gastrocnemius lateralis (GL) which was absent during the electrical stimulation at strengths up to three times threshold.

The various responses occurring at different latencies and in different muscle groups were particularly well studied by Forssberg (1979, 1981) and essentially confirmed by Duysens and Loeb (1980). In flexor muscles electrical stimulation of the foot may induce both a short latency (10ms) and a longer latency (25ms) response. These two responses appear to be transmitted through separate pathways since their amplitude was modulated differentially throughout the step cycle. In extensor muscles, there was generally a prominent inhibitory response followed by a late excitatory response. Moreover, in GL, there could be a short latency (10ms) excitatory response appearing only in swing. It thus seems that at least five different pathways are controlled: two to flexors, both excitatory, and three to extensors, two excitatory and one inhibitory. These pathways can all be activated in decerebrate and spinal cats so that they are essentially spinal although their relative importance depends on the type of preparation.

Other limbs
The amplitude of responses in the hindlimb contralateral to the cutaneous stimulation is also modulated in a phase dependent manner. This has been confirmed in chronic spinal cats (Julien et al., 1982), in decerebrate cats (Rossignol & Gauthier, 1978; Gauthier & Rossignol, 1981) and intact cats (Duysens et al., 1980).

Similar experiments have also been performed in the forelimbs with skin stimulation of the metacarpal region of decerebrate cats walking on a treadmill (Miller et al., 1977). Although there is clearly a phase dependent modulation of the amplitude of responses (their Fig. 6A), the relationship of the modulation and the locomotor activity of the tested muscles is not obvious. In intact cats, mechanical taps of the forefoot applied in different parts of the step cycle could evoke responses in both ipsilateral and contralateral flexors and extensors (Matsukawa et al., 1982).

All the above studies have looked at responses evoked ipsilaterally or contralaterally in the same girdle. Miller et al. (1977) also studied, in decerebrate walking cats, the effects of stimulation of cutaneous afferents in the limbs of one girdle on muscles of the limbs in the other girdle. With stimulation of the forelimb, a clear phase dependent organisation of responses was found in the hindlimbs. In the forelimbs, with stimulation of the other girdle, there was an alternation between the periods of reflex responsiveness of antagonistic muscles, although these periods did not strictly correspond to the respective periods of locomotor activity in the recorded muscles.

Fictive locomotion
A key question concerning the above findings is whether the actual movement of the limbs, with all the phasic afferent inputs, are essential for the modulation of the reflex responses. This problem was first studied in acute low spinal cats (Andersson et al., 1978) which, after paralysis and injection of Nialamide and DOPA (Jankowska et al., 1967) develop a rhythmical activity which can be recorded in the peripheral nerves and which has the typical reciprocal organisation and cycle structure of locomotion (fictive locomotion as documented by Grillner and Zangger, 1979). This preparation thus allows evaluation of central mechanisms in the absence of movement and cyclical afferent inputs. With skin stimuli similar to those which had been used before in the chronic spinal cat, it was found that, in flexor alpha motor neurones, the EPSPs were large early in the phase of flexor activity and were abolished during the stance phase (Andersson et al., 1978). Similar findings (i.e. largest EPSPs during activity) were made in extensor motor neurones although the modulation was not as deep as in flexors. Since no conductance changes were found in the motor neurones, these alterations in PSPs were attributed to changes in the transmission of reflex pathways occurring at a premotor neurone level. In one chronic spinal cat which was also acutely prepared, short latency (10ms) and long latency (25ms) PSPs were found in flexor and extensor motor neurones as in the chronically prepared but otherwise intact cat.

Again, in similar preparations, Schomburg and Behrends (1978) also found, at least in some cases, that cutaneous stimulation could induce EPSPs in flexor and extensor motor neurones and that these were maximal during their phase of activity. However, it appears that in extensors, IPSPs dominated and were greatest during their activity. This was later summarised in the following way: EPSPs dominate in flexors during the period of flexor activity and IPSPs dominate in extensors during the period of extensor activity (Schomburg et al., 1981). This was said to conform to the flexor reflex pattern. However it should be remembered that in these experiments there was a variety of EPSPs and IPSPs in different combinations, revealing once again that several pathways are involved.

Studies of cutaneous reflexes during fictive locomotion in decorticate cats (Perret & Cabelguen, 1980; Cabelguen et al., 1981) pointed to an even more complex picture. It was shown that some bifunctional muscles could be reflexly activated not only during the period of the step cycle in which they are normally active but also in periods where the muscles can be active only under certain conditions.

As far as the contralateral hindlimb is concerned, crossed flexion and crossed extension could also be found during locomotion in acute low spinal cats injected with Nialamide and DOPA. This central control, however, did not seem to be as complete as that found during actual locomotion, suggesting that afferent inputs during walking may participate in, or reinforce, the selection of some of these reflex responses (Rossignol et al., 1981a,b).

Stimulation of skin and muscle nerves in both forelimbs evoked EPSPs in hindlimb flexor and extensor motor neurones. The EPSPs were found to be maximal during the phase of activity of the respective motor neurones (Schomburg et al., 1977). Later, the same authors described that stimulation of group II and III fibres in forelimbs induced a mixture of EPSPs and IPSPs in the hindlimbs, much as did the cutaneous stimulation. There, the principle seemed to be that EPSPs dominate during motor neurone activity, while IPSPs dominate in the reciprocal phase (Schomburg et al., 1981).

The principles emerging from the preceding review can be summarised as follows:

1. Cutaneous inputs from different sources can give rise to responses in antagonistic muscle groups in different parts of the step cycle.

2. The magnitude of the responses in any particular group, flexor or extensor, is, in general, modulated throughout the step cycle.

3. Excitatory responses tend to be maximal during or close to the period of locomotor activity of a given muscle.

4. There may be several excitatory and inhibitory pathways that are differentially controlled during the step cycle.

5. This last conclusion together with the absence of conductance changes in motor neurones suggest that the control is exerted at a premotor neurone rather than a motor neurone level.

6. Since there is not generally a one to one relationship

between the phase of locomotor activity of the muscles and the period during which they are responsive, it would appear that at least some of the implicated interneurones are different from those interneurones of the central pattern generator which are responsible for driving the motor neurones in their characteristic locomotor pattern.

Comparison of responses in hindlimbs and forelimbs in intact cats

Some recent results obtained in chronically implanted cats trained to walk on a treadmill have suggested that the reflex control of at least some muscles may be still more complex.

Fig. 1 illustrates the overall movement of the hindlimb when it encounters an obstacle. Fig. 1A shows a reconstruction of the movement of the limb at all joints during the swing phase of the control step cycle which immediately precedes the swing phase in which the limb is perturbed (Fig. 1B). In this figure, after the moment of stimulation (see arrow aimed at the foot), there is a marked flexion of the limb which withdraws the foot upwards and backwards away from the obstacle. This withdrawal is seen more clearly when the trajectories of the tip of the foot in the control cycle (Fig. 1C) and in the perturbed cycle (Fig. 1D) are compared. Note that the trajectory of the knee in the perturbed cycle is much less altered than that of the foot and that the vertical excursion of the hip is unchanged, showing that most of the flexion occurs distally to the hip (i.e. at the knee and ankle).

This is confirmed by examination of Fig. 2A in which the angular movements at the hip, knee and ankle have been reconstructed for three step cycles with the perturbed step cycle falling in the middle of the display. Note that immediately after contact with the stimulus, represented by the thin vertical line, the first joint to flex is the knee, while the ankle and hip appear to be locked. Following this, the ankle flexes strongly and there is a small flexion at the hip. These actions cause the limb to be raised above and away from the obstacle. Following these flexions there is an extension at the knee and the ankle which restores the normal pattern of locomotion. It is noteworthy that this sequence of angular changes is exactly the same as that described earlier for the chronic spinal kitten (Forssberg et al., 1977).

Fig. 2B displays the electromyograms (EMGs) of the knee flexor, semitendinosus (St), on the perturbed side as well as the knee extensor, vastus lateralis (VL), on the perturbed (i) and contralateral side (co). The stimulation arrives just as the iSt burst would normally end (compare this burst with the preceding one). The stimulation evokes a large excitatory response (latency 11ms to onset, 40ms to peak) which has a peak amplitude more than twice that of the normal locomotor EMG. The overall response lasts for 150ms, that is slightly

Interactions of inputs with locomotor spinal pattern generator

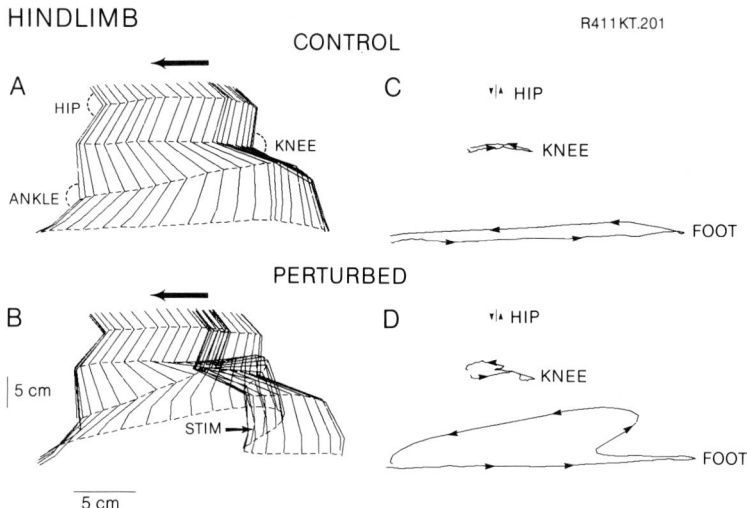

Fig. 1. Response of a hindlimb to a mechanical perturbation.
Chronically implanted cats were trained to walk on a treadmill in a plexiglass enclosure. Spots of light-reflective material were positioned over bony landmarks and the movements of the animal were recorded on video tape using a shutter video camera. The X-Y coordinates of the spots were measured in every field (16.7ms) from a television monitor using a manually controlled cursor and were transferred to a PDP 11/34 minicomputer in order to reconstruct the movement in the form of stick diagrams (A and B) or the trajectory of a point (C and D). Since, in the treadmill situation, the animal is practically stationary relative to the camera, the movement (reconstructed as stick diagrams) may be difficult to visualise because of superpositions of points unless each frame is displaced relative to the preceding one by a certain amount. This, in the cases of A and B, is the distance travelled by the foot between each frame. Consequently, the horizontal calibration is twice that of the vertical one. Although the problem of superposition does not arise when displaying the trajectory of a point (as in C and D), the calibrations were kept the same to allow a direct correspondence between the trajectories and stick diagrams. Note that for these reconstructions the values were filtered with a moving window of five consecutive values.
In A and B, the arrows pointing to the left indicate the direction of the movement of the limb. In B, STIM indicates the exact video field on which contact of the foot with a metal bar equipped with a microswitch was observed.
In C and D, the small arrows on the trajectories indicate the direction of the movement. Note that the hip is fixed to zero in the X axis so that it only moves along the Y axis.

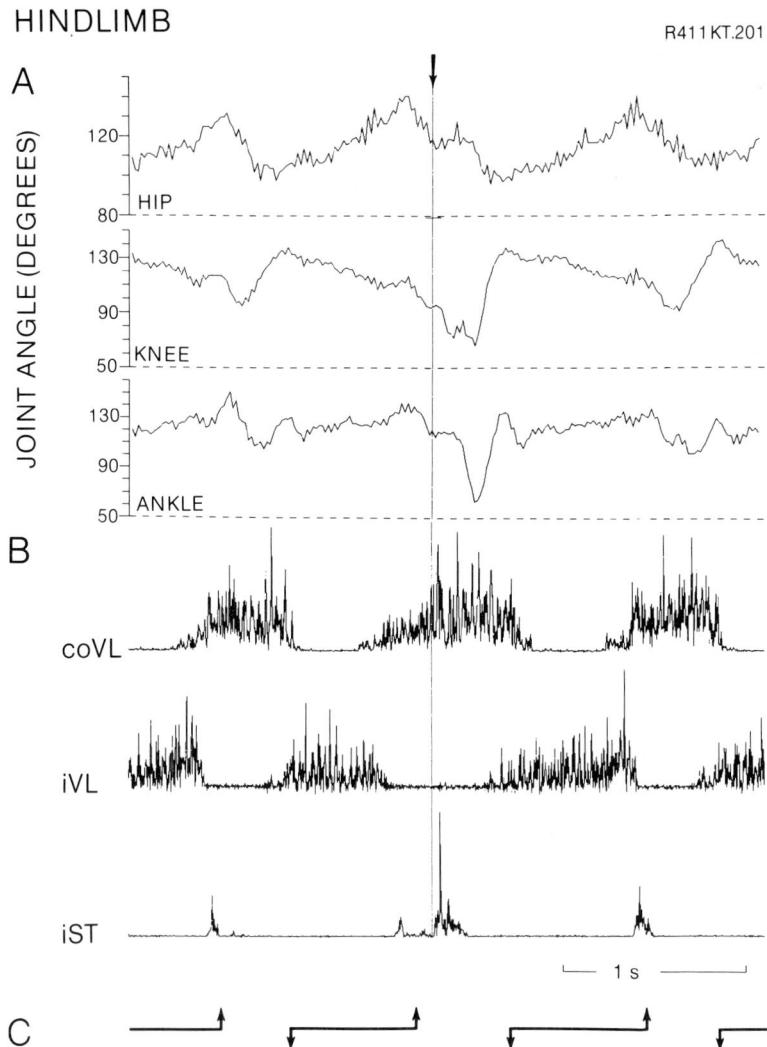

Fig. 2. Angular movements and EMG traces corresponding to Fig. 1.
In A, the X-Y coordinates of the points used in Fig. 1 were used to reconstruct the angular movements of the hip, knee and ankle (see Fig. 1A) using trigonometry. Three continuous cycles are displayed. The thin vertical line indicates the moment of contact of the metal rod. Flexion is indicated by a downward deflection of the traces. Note that the peak knee flexion is earlier than both the peak ankle and hip flexion. The values were filtered with a moving window of three consecutive values.
In B are shown EMGs, chronically recorded by teflon coated stainless

longer than the time of a normal St locomotor burst. Note that the following iVL burst is retarded by an equivalent time so that, effectively, the step cycle is being reset by the stimulation. On the contralateral side, there is also an initial excitation of VL (26ms latency) and a prolongation of the burst to match the overall prolongation of the cycle on the other side.

Thus, when the hindlimb hits the obstacle during swing, the cat initially flexes the knee to withdraw the foot and then flexes the ankle and hip to bring the foot above and in front of the obstacle. The knee flexion is accomplished by a strong activation of, among other muscles, semitendinosus at a time when it is either active or immediately after its period of normal locomotor activation. During this hyperflexion of the perturbed limb, the extensor of the contralateral limb augments its activity to sustain the added weight. In this particular case, it then appears that the classical flexion reflex and crossed extension reflex (Sherrington, 1910) are perfectly well integrated into the locomotor pattern.

Figs. 3 and 4 display in a similar fashion the response of a forelimb when the foot hits an obstacle just as it leaves the treadmill belt. Fig. 3A illustrates the swing of the forelimb in the control cycle. In Fig. 3B, at the arrow, the foot hits the metal rod and is withdrawn backwards and upwards. Again, this withdrawal is more clearly seen in Fig. 3C and 3D by comparing the trajectories of the foot during the perturbed cycle and during the control cycle. The initial action (see Fig. 4A) was a strong flexion at the shoulder, while the wrist, and to a lesser extent the elbow, were locked. Thus, the foot is initially withdrawn from the obstacle mainly through a hyperflexion of the shoulder. Following this, the elbow and wrist are both hyperflexed to bring the foot over the obstacle. Note, in Fig. 3D, that the trajectory of the elbow is changed as much as is that of the foot and that the vertical excursion of the shoulder increased in the perturbed cycle as compared to the control cycle, indicating that much of the flexion occurred proximally to the elbow joint.

---

steel wires sewn into the bellies of the muscles and led subcutaneously to multipin connectors cemented to the skull of the animal. The EMGs have been digitised at 1KHz and displayed after rectification and filtration using a PDP 11/34 computer. The synchronisation of A and B was achieved by simultaneously recording an SMPTE time code on the video image and EMG recordings.

In C, the duty graph indicates the periods of contact of the ispilateral foot with the belt (stance, solid line) and the periods of lack of contact with the belt (swing, space). Arrows indicate the moment of contact and lift of the foot in relation to the treadmill belt.

A comparison between the responses evoked in the fore- and hindlimb reveals both similarities and differences between the adopted strategies. The initial responses to the perturbation of the hindlimb is to flex the limb at the knee to raise the foot above the obstacle, while, when the forelimb is obstructed, the whole limb appears to be withdrawn from the obstacle before the elbow is flexed. This apparent difference in strategy between homologous joints (knee and elbow) must result from the difference in their direction of movement. However, it is of interest to note that the first reaction of the cat, for both limbs, is to lock the joints distal to the portion which is actively moved. Thus, in the hindlimb, the ankle is locked while the knee flexes (see also Wand et al., 1980) and in the forelimb the wrist, and to a lesser extent the elbow, are locked while the shoulder flexes. Both movements raise the distal limb away from the obstacle, while the subsequent flexion of the more distal joints then allows the limb to clear the obstacle. The similarities between shoulder and knee action, as well as between elbow and ankle, have been pointed out previously (Cabelguen et al., 1981).

The most striking observation in the EMG recording of Fig. 4B is the large, short latency excitation (12-15ms) of the ipsilateral long head of triceps (Tri) in a period when this muscle is inactive. As can be seen from its pattern of firing and the associated duty graph in Fig. 4C, the long head

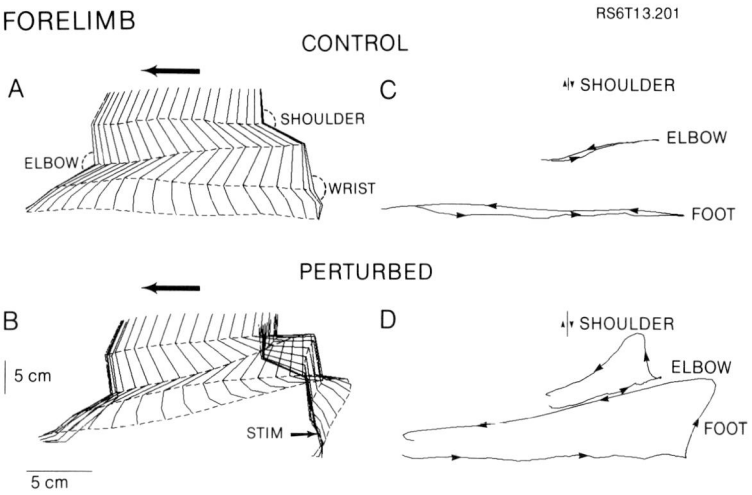

Fig. 3. Response of a forelimb to mechanical perturbation.
Details as in Fig. 1. Note that a closure of the shoulder angle is described in the text as a shoulder flexion.

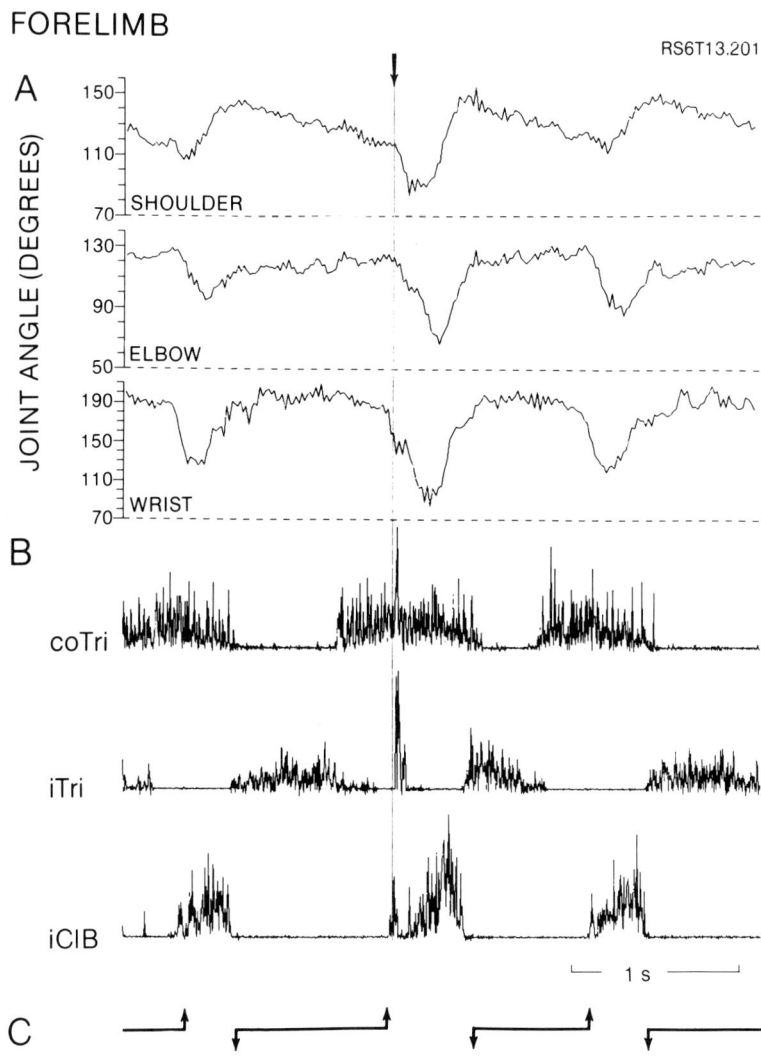

Fig. 4. Angular movements and EMG traces corresponding to Fig. 3. Details as in Fig. 2. Refer to Fig. 3A for orientation of the angles. Note that the shoulder angle closes (flexes) during the stance and flexion phases of the locomotion cycle whereas it opens (extends) during the extension phase of swing.

of triceps normally discharges during stance. Because of the anatomical attachments of this muscle, it could serve both as an elbow extensor and a shoulder flexor during stance. It is presumed here that the large EMG response occurring out of phase would actually participate in the flexion of the shoulder. In fact, in a different cat, muscles acting only at the shoulder and active during the swing phase (such as teres major) were recruited together with triceps during the responses. Furthermore, other experiments showed that the pure elbow flexor brachialis (Br, see also burst in cleidobrachialis ClB) and the pure elbow extensor lateral head of triceps (Tril) were also excited approximately at the same time during these responses. The resulting co-contraction serves to lock, or at least partially retard the flexion of the elbow (see small plateau in Fig. 4A). A strong contraction of the long head of triceps in these conditions would almost necessarily lead to a shoulder flexion. There is also presumably a co-contraction of the wrist muscles as movement of this joint appears to be completely blocked.

The presence of responses in the long head of triceps only in the period of the step cycle in which the muscle is inactive (see Fig. 5 and Fig. 7) is in contrast with almost all previous situations where muscles tended to have their period of maximal reflex responsiveness at around the time of their normal locomotor activation (see however Wand et al., 1980). Thus the pathway leading to the short latency activation of triceps during swing must be controlled by interneurones which are different from those responsible for driving the triceps motor neurones in the locomotion cycle. This situation bears a striking resemblance to the modulation of the jaw opening reflex during mastication in the rabbit. In this preparation, when strong stimuli are applied to the lower lip or the gum during the chewing cycle the activation of the digastric muscle which opens the mouth is maximal during the period of jaw closure, that is during the period of inactivity of the muscle in the chewing cycle (Lund & Rossignol, 1981; Lund et al., 1981; see also Lund & Olsson, 1983).

It should also be noted, in Fig. 4B, that there is a longer latency (21ms) response in the contralateral triceps. This response is akin to crossed extension responses described for the hindlimb and suggests that it represents the activation of a pathway different from that involved in evoking out of phase responses in the ipsilateral limb.

The previous sections have considered the reaction of the limbs to a mechanical stimulation. In order to quantify the responses, experiments were also performed using weak electrical stimulation of a cutaneous nerve innervating the dorso-lateral aspect of the forearm and the dorsum of the foot. In Fig. 5, the amplitude of different types of responses recorded in the long head and lateral head of triceps has been plotted for electrical stimuli given to the superficial radial

Interactions of inputs with locomotor spinal pattern generator

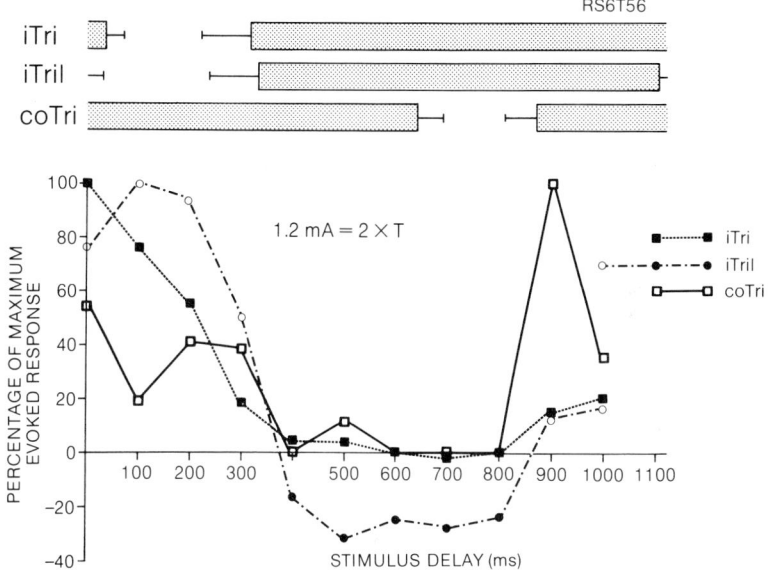

Fig. 5. Modulation of different types of responses in triceps grouped in latency ranges.

Cats were also implanted with electrodes placed around the superficial radial nerve, held in place by a polymer cuff surrounding the nerve (see Julien & Rossignol, 1982). Single pulses of 0.2ms were delivered every three step cycles and in different parts of the step cycle using a delay circuit triggered from the onset of iClB. The intensity of the current was adjusted to be a multiple of the threshold for evoking the smallest flexion response in the forelimb of the resting animal. Here the stimulation was twice threshold. The digitised EMG activity before and after the stimuli was averaged, and the background EMG was subtracted from the response. The true responses obtained (averages of between 7 to 20 responses per point; total number = 164) are displayed as a percentage of the maximal response recorded in any muscle (see Drew & Rossignol, 1984, for further methodological details). The shaded rectangles above the graph indicate the times (±1SD) when the various triceps muscles were active during the step cycle. Note that although the curve for the lateral head of triceps (Tril) is continuous, two different symbols are used above and below the zero line to indicate that these excitatory and inhibitory responses are mediated by different pathways.

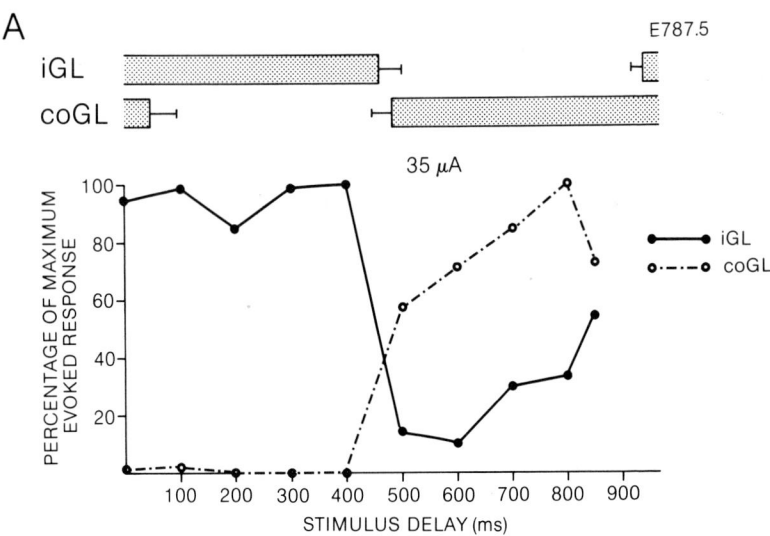

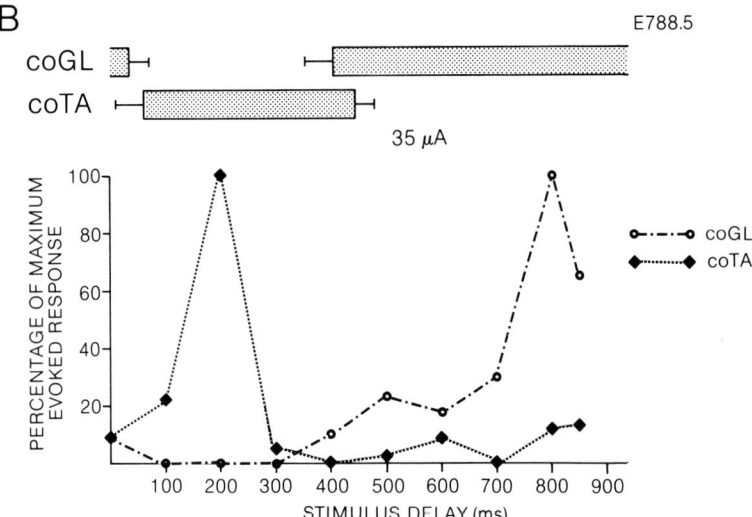

Fig. 6. Responses of hindlimb muscles to MRF stimulation during locomotion. Cats were acutely decerebrated, and muscles were implanted percutaneously with copper wires. The MRF stimulation (11 pulses, 0.2ms duration, 300Hz, at 35µA) was delivered by a glass coated tungsten microelectrode lowered through the cerebellum. The stimulus train was triggered by the activity of the ipsilateral vastus lateralis and was

nerve at different delays in the step cycle relative to the onset of the iClB burst.

For the lateral head of triceps there are, in the phase of inactivity, short latency (10-12ms) responses which largely resemble the short latency responses in the long head of triceps. Whereas in the long head there is no response to the stimulus during the main period of activity of the muscle, in the lateral head during that period there is a primary inhibitory response at latency 17-19ms. Moreover, as also shown in Fig. 4B, there is also a longer latency response (21-26ms) in the long head of triceps when it is recruited as a crossed extensor (CoTri, Fig. 5). At different strengths of stimulation, other excitatory (but less prominent) responses could also be seen in the ipsilateral triceps during the period of activity, but these will not be further detailed at the present.

It should also be noted that the out of phase response is seen in both the long and lateral heads of triceps with either mechanical or cutaneous stimulation, whereas in the hindlimbs the out of phase response in GL was seen only to a mechanical perturbation (Wand et al., 1980).

In summary, it appears that, at least for some muscles of the forelimb, some excitatory reflex pathways are only activated out of phase with the period of locomotor activity of the muscles, while other excitatory or inhibitory pathways are open only during the period of activity. All these pathways appear to play distinct roles and to be controlled differentially.

MODULATION DURING LOCOMOTION OF LIMB RESPONSES TO INPUTS OF SUPRASEGMENTAL ORIGIN

Much less is known about the modulation of responses to stimulation of suprasegmental structures during walking. Pioneering work in this direction was done by Orlovsky (1972) who stimulated the vestibular nucleus, the medullary reticular formation (MRF), the red nucleus and the pyramidal tract in high decerebrate cats walking on a treadmill. The main conclusion, in the words of the author, was "that transient increase in descending pathway activity does not affect the

---

given at every third step cycle. The axes are as for Fig. 5.

In A, the curves show that the MRF can induce excitatory responses in the GL of either side depending on the phase of the step cycle.

In B, the responses obtained in TA and GL are plotted as a function of the stimulus delay within the step cycle. The stimulation site was different in A and B. (Further methodological details and complete results to be found in Drew and Rossignol, 1984).

timing of onset and cessation of muscular activity in a locomotor cycle, but does determine the degree of motor unit activity". For instance, it was seen that stimulation of the MRF augmented the amplitude of the ankle flexor tibialis anterior (TA) during the time it was active in the step cycle (swing) and inhibited the ankle extensor (GL), at the time when it was active (stance). These changes in the output of motor neurones did not, however, affect the timing of the discharges of the respective muscles. Similar principles were seen with the other descending systems (except for the pyramidal tract), although the changes in EMG amplitude corresponded to the suggested specific connectivity of each system.

Other findings suggested that different principles may be involved. Russell and Zajac (1979) stimulated Deiters' nucleus and the medial longitudinal fasciculus (MLF) and found that electrical stimulation of these descending systems could alter not only the amplitude but also the timing of the discharges of motor neurones recorded in peripheral nerves of paralysed decerebrated cats injected with Nialamide and DOPA (fictive locomotion).

In a more recent study in decerebrated cats (Drew & Rossignol, 1984), it was found that stimulation of the MRF could indeed reset the locomotor rhythm in the decerebrate cat walking on a treadmill. It was also found that stimulation of the MRF could activate flexors or extensors of both the forelimbs and the hindlimbs depending on the phase of stimulation. Fig. 6A displays the amplitude of the responses recorded in the ipsilateral GL (latency 9-14ms) and the contralateral GL (latency 12-17ms) as a function of the phase of the step cycle. Similarly, Fig. 6B (same animal but a different location than for Fig. 6A) shows in the same hindlimb (contralateral to the stimulation) that there can be responses in TA as well as GL, depending on the phase of the step cycle in which the stimulation is delivered. These results are less surprising when one considers that indeed the MRF has known connections not only with flexor motor neurones but also extensor motor neurones (see Drew and Rossignol, 1984, for a discussion of this point).

Thus again there seems to be a variety of detailed descending pathways converging on spinal elements and each of these pathways appears to be selectively controlled.

A further development of this concept was obtained recently in chronic cats, implanted as for the experiments described in Figs. 1 to 5 with the addition of a chamber positioned over the cerebellum which allowed stimulation of the MRF with a conventional glass coated tungsten microelectrode. Fig. 7 compares the result of stimulating the superficial radial nerve and the MRF in the same animal. In Fig. 7A, the cutaneous nerve stimulation given on the right (R) side during swing evoked an out of phase excitatory

Interactions of inputs with locomotor spinal pattern generator

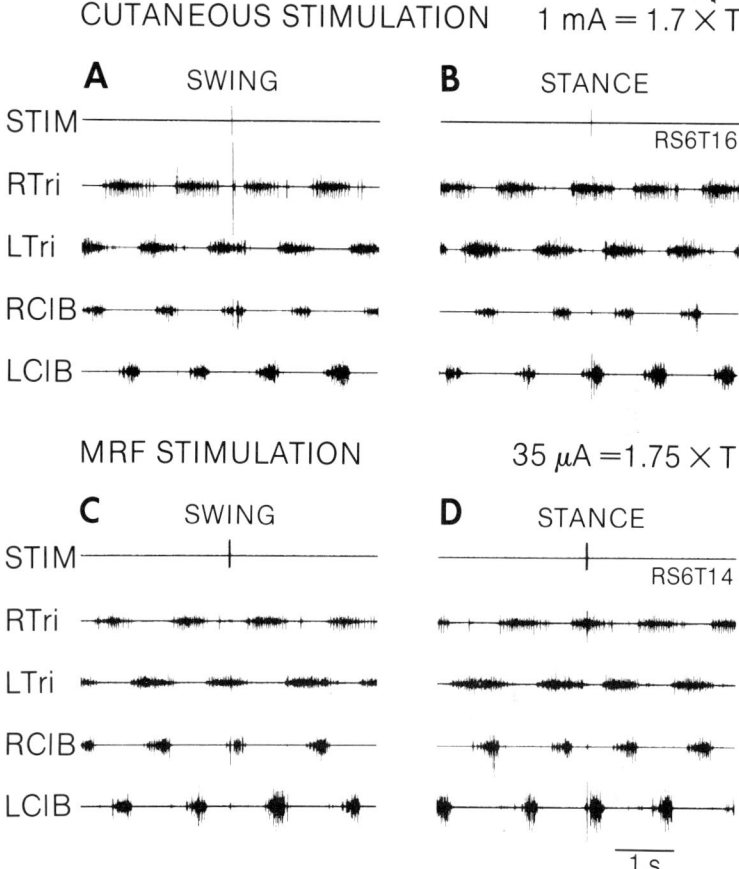

Fig. 7. Comparison of forelimb responses to cutaneous stimulation and MRF stimulation.
Cutaneous stimulation was delivered to the right superficial radial nerve during either swing (A) or stance (B). Note that a short latency excitatory response occurs in RTri during swing (when the muscle is normally inactive) but not during stance (when the muscle is active). During a different session in the same cat, the MRF was stimulated through a glass-coated tungsten microelectrode lowered through the cerebellum with a small microdrive fixed to the skull of the animal. At a stimulation strength of 35µA, there was no appreciable response in RTri during swing (C), though there was a short latency response in RTri during stance (D). Note that the stimulation in A and B was a single pulse, while in C and D it was a train of stimuli (11 pulses, 0.2ms at 300Hz).

response in the ipsilateral RTri (latency 10ms). Note in LTri the smaller crossed extension response at a longer latency. The same stimulation in stance (Fig. 7B) was ineffective in relation to RTri, although it did evoke longer latency responses in both RClB and LClB. Stimulation of the MRF on the left side (Figs. 7C and 7D) caused short latency (10ms) responses primarily when the muscles were already active. Thus in Fig. 7C, there was a reciprocal activation of LTri and RClB when they were active and, in Fig. 7D, reciprocal activation of the RTri and LClB.

This indicates that in those selected instances, two different pathways leading to the short latency excitation of RTri have been activated in two different parts of the cycle; out of phase by cutaneous stimulation and in phase by MRF stimulation. Whether or not these two pathways share common elements in this situation remains to be studied.

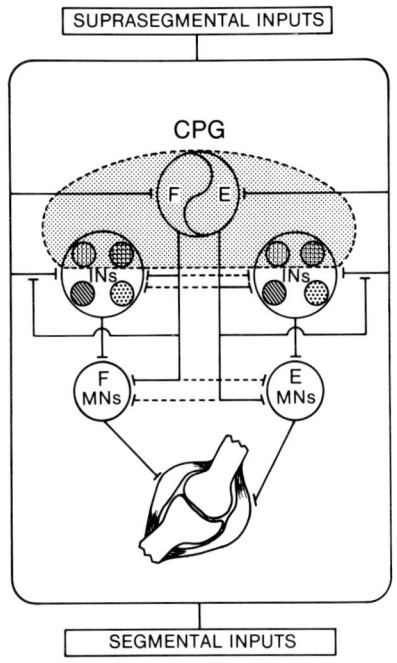

Fig. 8. General scheme summarising the modes of interaction between the central pattern generator for locomotion and the inputs from segmental or suprasegmental origins. For details see text.

Interactions of inputs with locomotor spinal pattern generator

GENERAL MODEL

Fig. 8 represents the main mechanisms that appear to regulate the interactions between segmental and suprasegmental inputs with the spinal central pattern generator (CPG).

In this diagram both types of inputs are shown to project directly to the CPG because in both cases they have been demonstrated to reset the step cycle. The present review was also restricted to inputs reaching motor neurones through interneuronal pathways and does not treat the known monosynaptic pathways.

The flexor and extensor portions of the CPG (F and E) are shown to send a command signal (locomotor drive) to the motor neurones (MNs). It is assumed here that there are interneurones within the CPG whose specific task is to provide this locomotor drive and that these interneurones, although preferentially connected to certain functional groups of motor neurones may also send signals to motor neurone pools of the antagonists as suggested by Perret and Cabelguen (1980) and Cabelguen et al. (1981). Since a similar dual command may also be sent to groups of interneurones which are part of specific segmental or suprasegmental pathways, there is also the possibility that a motor neurone group may be driven in the locomotor cycle in one phase of the step cycle and be mainly responsive to a segmental or to a suprasegmental input in the reciprocal phase of the step cycle.

The interneurones (INs) on each side are grouped together because they are mainly connected to synergistic groups of motor neurones. They would represent the premotor neurone elements. All the observations presented in this paper suggest that there are many different interneuronal pathways involved (represented by several interneurones on each side) and that each of these pathways may be selectively controlled at different times of the step cycle. Some interneurones may be shared by segmental and suprasegmental pathways as shown by Lundberg (1969).

An unanswered question is whether or not to include these interneurones within the CPG or outside the CPG. This depends ultimately on the definition of the CPG. In the present context the CPG has been defined as those elements which provide a drive to the motor neurones in the characteristic pattern of locomotion. All other controlling elements have been placed outside the CPG, although under its influence. However, one might well include these INs within an extended or expanded definition of the CPG, which would thus include not only the rhythmical driving elements but also all elements which are under the control of the CPG except the motor neurones. This, then, is a functional concept of the CPG whereby some elements may participate in, or may be controlled by, different CPGs at different times. Thus, the dashed line includes the rhythmical interneurones as well as some control interneurones. It is, however, important to separate the rhythmical units and the control units, since it has been shown that the period of reflex responsiveness of motor neurones does not necessarily match their period of locomotor activity.

Finally, the CPG or INs may control presynaptic elements. This mechanism, which has received some attention (Bayev & Kostyuk, 1982), may be of importance in the control of segmental and also suprasegmental inputs. There are probably, however, a variety of mechanisms used simultaneously to control selectively the interactions between the CPG and elements carrying information about the moving limbs or other more complex descending influences.

SUMMARY

The group of neurones responsible for generating the characteristic pattern of locomotion (central pattern generator or CPG) has not only to interact with neurones carrying peripheral and supra-spinal information concerning normal locomotor movements but should also be prepared to compensate rapidly for unexpected inputs that may perturb any of the limbs in any phase of the step cycle.

Towards this end, it appears that the CPG independently controls several reflex pathways, which may be excitatory or inhibitory, to both flexor and extensor motor neurones of each limb. The transmission in certain reflex pathways may thus be facilitated during one phase of the step cycle and reduced in the reciprocal phase. For example, in some animal preparations, cutaneous stimulation of the foot may evoke a response in flexor muscles during the swing phase but not in the stance phase. However, during the stance phase, the same stimulus may evoke an excitation of extensor muscles. This modulation of reflex transmission during walking appears to be similar in ipsilateral, contralateral and propriospinal cutaneous pathways and seems to be achieved largely by the CPG itself since it can also be found during fictive locomotion. As a general rule, the maximal excitatory responses evoked in a motor neurone pool tend to occur during or close to its period of locomotor activity. However, as the periods of locomotor activity and reflex responsiveness may be dissociated, the principal control of reflex modulation would appear to be not at the motor neurone but rather at the premotor neurone level. Indeed, as an example, electrical stimulation of the superficial radial nerve causes a large excitatory response in the long head of triceps when given during swing i.e. when the muscle is inactive, and evokes only a small response during the stance phase when the muscle is normally active.

Specific control of transmission also seems to be exerted

on descending pathways. Stimulation of the medullary reticular formation, for instance, may evoke, on the ipsilateral or contralateral side, flexion or extension responses in decerebrate cats walking on a treadmill depending on the phase of stimulation within the step cycle. Taken together, these facts suggest that the same muscle may be activated in one phase by a suprasegmental pathway and in another phase by a segmental pathway.

Thus, during locomotion, the CPG controls selectively the transmission of interneurones of several distinct excitatory and inhibitory pathways of segmental and suprasegmental origin so that motor neurone responses are best suited to compensate for an unexpected perturbation at any time during the step cycle.

Acknowledgements

This work was supported by a group grant from the Canadian MRC. T. Drew was awarded a post-doctoral fellowship also from the MRC. The assistance of J. Provencher, G. Blanchette, S. Bergeron, R. Bouchoux, E. Rupnik and S. Philippe is most gratefully acknowledged. Dr J.M. Cabelguen is gratefully thanked for his comments on this manuscript.

REFERENCES

Andersson, O., Forssberg, H., Grillner, S. & Lindquist, M. (1978). Phasic gain control of the transmission in cutaneous reflex pathways to motoneurones during "fictive" locomotion. Brain Res., 149, 503-507.

Bayev, K.V. & Kostyuk, P.G. (1982). Polarization of primary afferent terminals of lumbosacral cord elicited by the activity of spinal locomotor generator. Neuroscience, 7, 1401-1409.

Cabelguen, J.M., Orsal, D., Perret, C. & Zattara, M. (1981). Central pattern generation of forelimb and hindlimb locomotor activities in the cat. In Regulatory Functions of the CNS. Motion and Organisation Principles. Eds. Szentagothai, J., Palkovits, M. & Hamori, J.. Adv. Physiol. Sci., 1, 199-211.

Drew, T. & Rossignol, S. (1984). Phase dependent responses evoked in limb muscles by stimulation of the medullary reticular formation during locomotion in thalamic cats. J. Neurophysiol., 52, 653-675.

Duysens, J. (1977). Reflex control of locomotion as revealed by stimulation of cutaneous afferents in spontaneously walking premammillary cats. J. Neurophysiol., 40, 737-751.

Duysens, J. & Loeb, G.E. (1980). Modulation of ipsi- and contralateral reflex responses in unrestrained walking cats. J. Neurophysiol., 44, 1024-1037.

Duysens, J., Loeb, G.E. & Weston, B.J. (1980). Crossed flexor reflex responses and their reversal in freely walking cats. Brain Res., 197, 538-542.

Duysens, J. & Pearson, K.G. (1976). The role of cutaneous afferents from the distal hind-limb in the regulation of the step cycle of thalamic cats. Expl. Brain Res., 24., 245-255.

Duysens, J. & Stein, R.B. (1978). Reflexes induced by nerve stimulation in walking cats with implanted cuff electrodes. Expl. Brain Res., 32, 213-224.

Forssberg, H. (1979). Stumbling corrective reaction: a phase-dependent compensatory reaction during locomotion. J. Neurophysiol., 42, 936-953.

Forssberg, H. (1981). Phasic gating of cutaneous reflexes during locomotion. In Muscle Receptors and Movement. Eds. Taylor, A. & Prochazka, A.. MacMillan, London, pp. 403-412.

Forssberg, H., Grillner, S. & Rossignol, S. (1975). Phase dependent reflex reversal during walking in chronic spinal cats. Brain Res., 85, 103-107.

Forssberg, H., Grillner, S. & Rossignol, S. (1977). Phasic gain control of reflexes from the dorsum of the paw during spinal locomotion. Brain Res., 132, 121-139.

Forssberg, H., Grillner, S., Rossignol, S. & Wallén, P. (1976). Phasic control of reflexes during locomotion in vertebrates. In Neural Control of Locomotion. Eds. Herman, R.M., Grillner, S., Stein, P.S.G. & Stuart, D.G.. Plenum Press, New York, pp. 647-674.

Gauthier, L. & Rossignol, S. (1981). Contralateral hindlimb responses to cutaneous stimulation during locomotion in high decerebrate cats. Brain Res., 207, 303-320.

Grillner, S. & Zangger, P. (1979). On the central generation of locomotion in the low spinal cat. Expl. Brain Res., 34, 241-261.

Jankowska, E., Jukes, M.G.M., Lund, S. & Lundberg, A. (1967). The effect of DOPA on the spinal cord. 5. Reciprocal organization of pathways transmitting excitatory action to alpha motoneurones of flexors and extensors. Acta physiol. scand., 70, 369-388.

Julien, C., Barbeau, H. & Rossignol, S. (1982). Gain changes in cutaneous reflexes during locomotion in the adult chronic spinal cat. Soc. Neurosci. Abstr., 8, 168.

Julien, C. & Rossignol, S. (1982). Electroneurographic recordings with polymer cuff electrodes in paralyzed cats. J. Neurosci. Methods, 5, 267-272.

Lund, J.P. & Olsson, K.A. (1983). The importance of reflexes and their control during jaw movement. Trends Neurosci., 6, 458-463.

Lund, J.P., Rossignol, S. & Murakami, T. (1981). Interactions between the jaw opening reflex and mastication. Can. J. Physiol. & Pharmacol., 59, 683-690.

Lund, J.P. & Rossignol, S. (1981). Modulation of the amplitude of the digastric jaw opening reflex during the masticatory cycle. Neuroscience, 6, 95-98.

Lundberg, A. (1969). Convergence of excitatory and inhibitory action of interneurones in the spinal cord. In The Interneurone. Ed. Brazier, M.A.B.. University of California Press, Los Angeles, pp. 231-265.

Matsukawa, K., Kamei, H., Minoda, K. & Udo, M. (1982).Interlimb coordination in cat locomotion investigated with perturbation. I. Behavioral and electromyographic study on symmetric limbs of decerebrate and awake walking cats. Expl. Brain Res., 46, 425-437.

Miller, S., Ruit, J.B. & van der Meche, F.G.A. (1977). Reversal of sign of long spinal reflexes dependent on the phase of the step cycle in the high decerebrate cat. Brain Res., 128, 447-459.

Orlovsky, G.N. (1972). The effect of different descending systems on flexor and extensor activity during locomotion. Brain Res., 40, 359-371.

Perret, C. & Cabelguen, J.-M. (1980). Main characteristics of the hindlimb locomotor cycle in the decorticate cat with special reference to bifunctional muscles. Brain Res., 187, 333-352.

Rossignol, S. & Gauthier, L. (1978). Patterns of contralateral limb responses to nociceptive stimuli during locomotion. Soc. Neurosci. Abstr., 4, 304.

Rossignol, S., Julien, C. & Gauthier, L. (1981a). Stimulus-response relationships during locomotion. Can. J. Physiol. & Pharmacol., 59, 667-674.

Rossignol, S., Julien, C., Gauthier, L. & Lund, J.P. (1981b). State dependent responses during locomotion. In Muscle Receptors and Movement. Eds. Taylor, A. & Prochazka, A.. MacMillan, London, pp. 389-402.

Russell, D.F. & Zajac, F.E. (1979). Effects of stimulating Deiters' nucleus and medial longitudinal fasciculus on the timing of the fictive locomotor rhythm induced in cats by DOPA. Brain Res., 177, 588-592.

Schomburg, E.D. & Behrends, H.B. (1978). Phasic control of the transmission in the excitatory and inhibitory reflex pathways from cutaneous afferents to alpha motoneurones during fictive locomotion in cats. Neurosci. Lett., 8, 277-282.

Schomburg, E.D., Behrends, H.B. & Steffens, H. (1981). Changes in segmental and propriospinal reflex pathways during spinal locomotion. In Muscle Receptors and Movements. Eds. Taylor, A. & Prochazka, A.. MacMillan, London, pp. 413-425.

Schomburg, E.D., Roesler, J. & Meinck, H.M. (1977). Phase-dependent transmission in the excitatory propriospinal reflex pathway from forelimb afferents to lumbar motoneurones during fictive locomotion. Neurosci. Lett., 4, 249-252.

Sherrington, C.S. (1910). Flexion-reflex of the limb, crossed extension reflex, and reflex stepping and standing. J. Physiol., 40, 28-121.

Wand, P., Prochazka, A. & Sontag, K.H. (1980). Neuromuscular responses to gait perturbations in freely moving cats. Expl. Brain Res., 38, 109-114.

Chapter Twenty-four

STEPPING REFLEXES AND THE SENSORY CONTROL OF WALKING IN CRUSTACEA

F. Clarac

In recent years, studies of locomotion in decapod Crustacea have been concerned mainly with the organisation of the motor output, the coordination of the legs, and the role played by the sensory input in regulating the stepping pattern (Clarac, 1982; Evoy & Ayers, 1982).
   Studies of the different proprioceptors operating in walking permit their classification into four different types (Clarac, 1977): <u>movement</u> and <u>position</u> <u>receptors</u>, the chordotonal organs spanning each joint; <u>neuromuscular structures</u>, the thoraco-coxal muscle receptor organ at the base of the leg (see Chapter 9 by Drs Bush and Cannone) and the myochordotonal organ sensitive not only to movements of the mero-carpopodite joint but also to the contraction of the accessory flexor muscle; <u>tension receptors</u>, the apodeme tension receptors located within muscles in the region of the apodemes; and <u>cuticular receptors</u> such as the cuticular stress detectors in the basipodite/ischiopodite region.
   The purpose of this paper will be to review how peripheral feedback can act on the motor output, and will particularly consider the effect of disturbances upon the walking pattern. Most of the studies discussed here have been performed on the rock lobster, <u>Jasus lalandii</u>, walking on a treadmill, because this permits manipulation of the walking pattern and hence evaluation of the role of proprioceptive feedback (Barnes et al., 1972; Ayers & Davis, 1977a; Clarac & Chasserat, 1983).
   This review will consider the following topics: firstly, the control of the stepping cycle in a single leg; secondly, the effects of feedback on the overall pattern of walking; and finally, a computer model that has been constructed to simulate stepping sequences and is used to test hypotheses on the possible uses of feedback during walking.

CONTROL OF THE STEPPING CYCLE IN A SINGLE LEG

<u>Joint function</u>
Most decapod crustaceans are able to walk in all directions by

Sensory control of walking in Crustacea

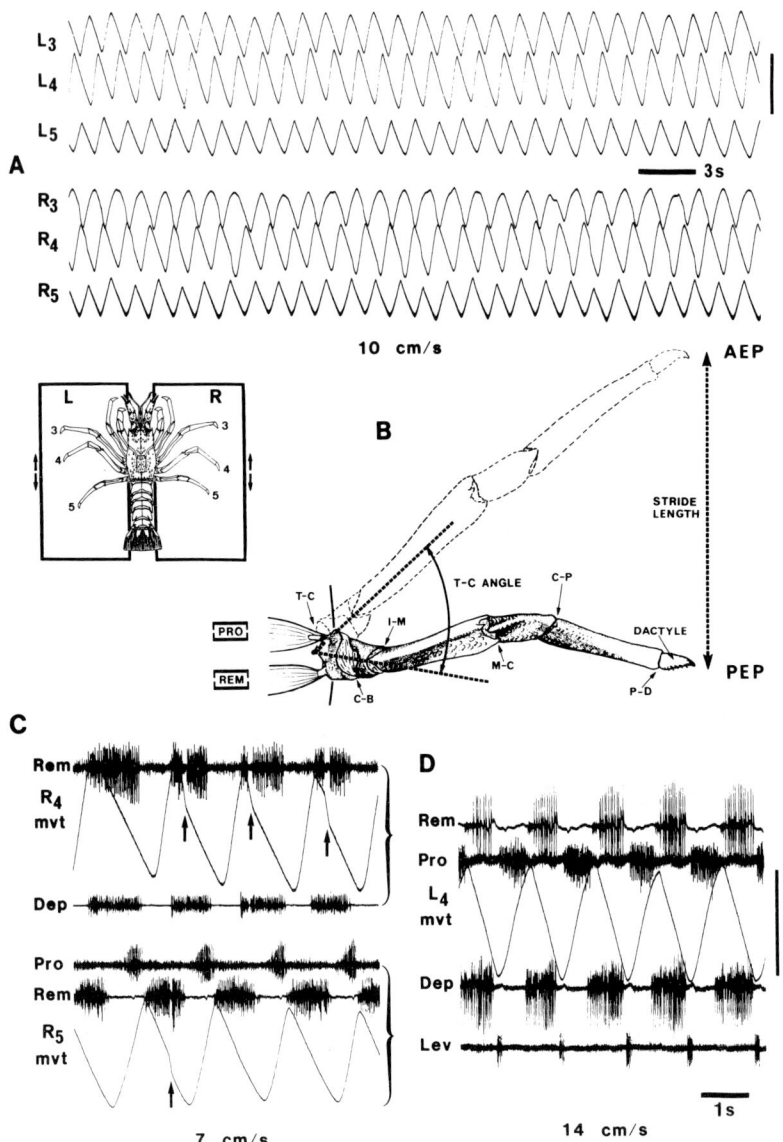

Fig. 1. Movements and EMGs in the rock lobster Jasus lalandii walking forwards on a treadmill.

A. Continuous recording of movements of the 6 rear legs, labelled as in B, during walking. The vertical bar represents 10 cm of stride amplitude

changing the relative timing of contractions in muscles operating at different leg joints.

Even though there are slight anatomical differences between the different groups of decapods, the functions of the different joints (see Fig. 1) during a step cycle can be summarised as follows (Clarac & Coulmance, 1971; Ayers & Davis, 1977a; Evoy & Ayers, 1982):

The coxo-basipodite joint (C-B) is moved by levator and depressor muscles and is mainly responsible for moving the limb in the vertical plane. The levator is active during the return stroke and the depressor during the power stroke, whatever the direction of walking. The discharges of these muscles thus define the two parts of the step cycle. Since the levator appears to be the most centrally determined of all leg muscles (Ayers & Davis, 1977a), its discharges are an accurate measure of step frequency.

The thoraco-coxal joint (T-C) and the mero-carpopodite joint (M-C) are "directional joints". During forwards and backwards walking, the promotor and remotor muscles of the T-C joint are largely responsible for the forward and backward movement of the legs, while the M-C joint is mainly used as a strut. In lateral walking, however, the extensor and flexor muscles of the M-C joint are used to extend and flex the leg sideways while T-C joint angle hardly changes.

The other joints, the ischio-meropodite (I-M), carpo-propodite (C-P) and propo-dactylopodite (P-D) are of secondary

---

in the anterior to posterior plane, forward movements being indicated by upward movement of the traces.
B. On the left is the experimental arrangement of the animal walking on the split belt of the treadmill with the 3 back legs numbered (L, left; R, right). On the right are shown details of a.4th right leg viewed from above with the different joints labelled: T-C, thoraco-coxal; C-B, coxo-basipodite; I-M, ischio-meropodite; M-C, mero-carpopodite; C-P, carpo-propodite; P-D, propo-dactylopodite. The promotor muscle (PRO) and the remotor muscle (REM) govern the T-C angle and are of primary importance in determining the stride length, which is limited by the anterior extreme position (AEP) and the posterior extreme position (PEP).
C. Simultaneous EMGs and traces of the leg movement of the two right back legs R4 and R5 during forwards walking. Two power stroke muscles, remotor (Rem) and depressor (Dep), are recorded for leg R4; the two T-C muscles, promotor (Pro) and remotor, are recorded for leg R5. The arrow indicates slippage of the leg on the treadmill corresponding to gaps in the remotor discharge.
D. Simultaneous EMGs of the four main muscles of the two proximal joints, promotor and remotor of the T-C joint, levator (Lev) and depressor of the C-B joint. The vertical bar represents 10 cm of stride amplitude, and refers to C as well as D.
In A, C and D the speed of the belt is indicated below each trace.

Sensory control of walking in Crustacea

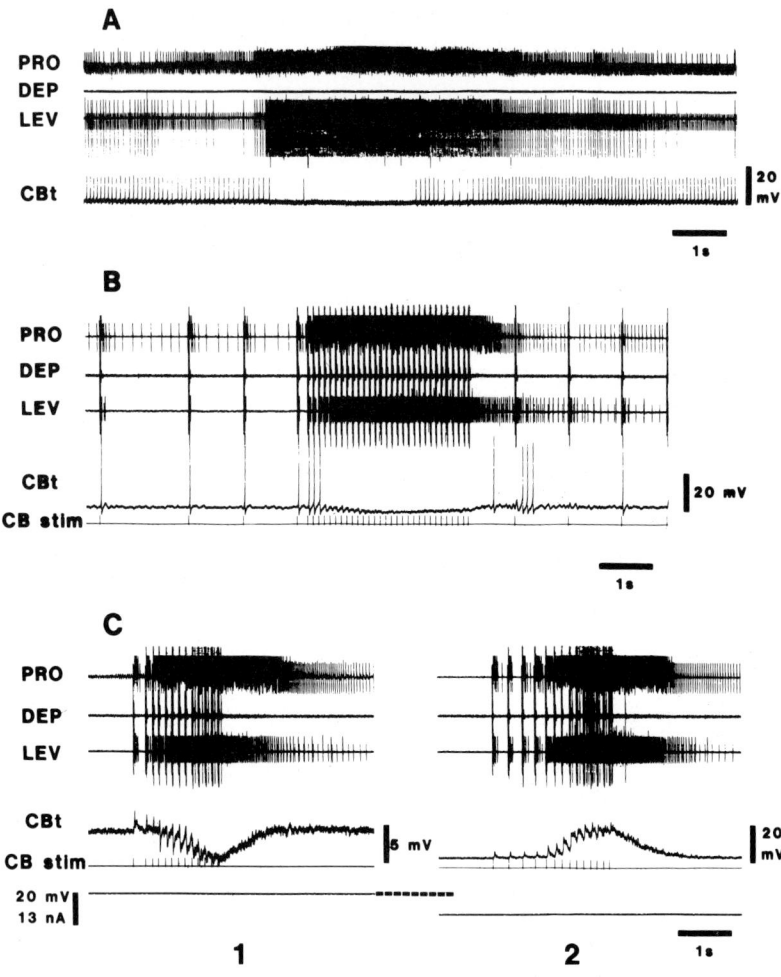

Fig. 2. Preliminary evidence for presynaptic inhibition of primary chordotonal organ afferents in the isolated thoracic ganglion of Carcinus maenas (Simmers & Clarac, unpublished). A single afferent terminal from the chordotonal organ spanning the coxo-basipodite joint (CBt) is recorded in the central nervous system together with the motor nerves of the promotor (PRO), levator (LEV) and depressor (DEP) muscles.
A. During spontaneous bursting of the promotor and levator axons, action potentials recorded in the chordotonal organ terminals are inhibited.
B. During stimulation of the chordotonal organ nerve, action potentials are initially recorded in the chordotonal organ terminal but during a series of stimulations, which elicit reflex bursts in the motor nerves, activity in the terminal is inhibited by IPSPs.

importance in walking of the rock lobster. Most of the time they act as struts, thus giving the leg a degree of stability. However, this is not true of all decapods during free walking. For example, in the crab Carcinus maenas the P-D joint operates in lateral walking, while the C-P joint plays an important role in adjusting the leg when the animal walks over an uneven surface.

In Figs. 1C and D, sequences of forward walking are presented where the levator discharges with the promotor during the return stroke, and the depressor with the remotor during the power stroke. In backward walking these relationships are altered, the levator being active at the same time as the remotor during the return stroke. The strengths of these muscle discharges are important in determining the total stride length of the leg (see Fig. 1B), defined as the difference between the anterior extreme position (AEP) and the posterior extreme position (PEP) of the leg tip on the treadmill. Systematic studies at different imposed speeds of walking on a power driven treadmill have demonstrated the stability of the stride during normal walking. This is probably a general feature of multilegged animals as previously suggested by Grillner (1981). Different stride lengths are, however, associated with different legs, as shown in Fig. 1A.

Intra-leg reflexes
In the last twenty years, a variety of reflexes have been described in the decapod Crustacea in several different types of preparation. Bush (1962) first defined the resistance reflex, analogous to the myotatic reflex of vertebrates, that depends upon sensory activation of chordotonal organs and produces facilitation of motor neurone discharge of passively stretched muscles. We now have some evidence that this type of reflex is monosynaptic (Wiens, 1976; Blight & Llinás, 1980). It has been demonstrated that the resistance reflex can be switched into an assistance reflex in some preparations at a certain degree of high nervous excitability (Vedel, 1980, 1982; DiCaprio & Clarac, 1981). However, as the preparation was immobilised in these experiments, the relevance of these results to walking is unclear.

In a series of experiments on crabs and crayfish, Barnes et al. (1972) and Barnes (1977) demonstrated that the resistance reflex does not operate during walking for normal unimpeded movement produced by the animal. Barnes thus proposed that resistance reflexes were inhibited centrally by

---

C. The chemical nature of the inhibition is demonstrated when the polarisation of the IPSPs is reversed by hyperpolarisation of the afferent terminal (in 2).

an efference copy signal generated by the central command that precisely matched the sensory input "expected" from any central command. If the leg slipped on the treadmill, the sensory input would be increased and would overcome the inhibition, producing a resistance reflex as in fact occurred.

Cruse et al. (1983), in a study of driven walking in the rock lobster, concluded that the resistance reflex was part of a servo-mechanism continuously relating the central command to the actual movement. Studying the force exerted above the treadmill during the power stroke, it was found that at slow speeds the forces were greater than at faster speeds. This paradoxical result may be explained by the modulation of motor output by resistance reflexes. When the reference input representing the "desired" leg position takes on a value posterior to the actual leg position, the resulting error signal increases the force directed posteriorly. On the contrary, if the treadmill belt is going too fast, the resulting error signal decreases the same force. Such an effect has been demonstrated by Barnes (1977) in crayfish walking over a slippery patch on the surface of the treadmill, and is illustrated here in Fig. 1C. As soon as the leg goes faster and until it recovers its "normal" position, remotor activity is completely abolished. This sort of position controlling servo-mechanism operates during the power stroke but not the return stroke. This type of regulation has often been presented in a slightly different way as the concept of "load compensation"; an animal exerts more propulsive force and has a stronger motor output when it carries load.

Very little information concerning the central thoracic connections of sensory elements has yet been published in Crustacea. I would like to present here preliminary data from experiments in which motor and sensory activity was recorded in an isolated thoracic ganglion of Carcinus maenas (Simmers & Clarac, unpublished). During promotor and levator bursts, spiking activity of chordotonal afferents recorded from their terminals in the ganglion was inhibited (Fig. 2A). In addition, during repetitive stimulation of the whole chordotonal nerve, discrete IPSPs are recorded in the afferent terminal. This inhibition appears to be of a chemical nature due to its reversal upon hyperpolarisation. This result, which must be confirmed, can be interpreted as a presynaptic inhibition of some chordotonal organ afferent terminals by "motor neuronal" or "interneuronal" bursting activity. This could be a mechanism of filtering sensory input during motor commands. Such action could explain the inhibition of resistance reflexes during normal unimpeded walking (see also Chapter 9 by Drs Bush and Cannone).

In addition to resistance reflex studies, we have also demonstrated other effects of proprioceptive feedback that depend upon the phase of walking as Rossignol and Gauthier (1980) showed in the cat. The step cycle is organised into two sub-systems, corresponding to power and return strokes. The character of these two states is different and elicits activity in different types of proprioceptor. Switching between the two states occurs at the AEP and PEP, where the leg may pause for a short time. Since chordotonal afferents are maximally activated at extreme joint positions (Clarac, 1977), they may play a particularly important role in determining when the leg switches from one state to the other, as was first suggested by Davis (1969) for lobster swimmerets.

An example of such a phase dependant effect is illustrated in Fig. 3 where the movement of a leg during the step cycle is blocked by a rod during either power or return stroke (Fig. 3). If the block is inserted during a power stroke, stepping of the leg is inhibited, and the leg exerts force against the rod until it is removed (Fig. 3A). Thus a leg never starts a new return stroke before reaching its PEP. If we make the same interruption during a return stroke, the

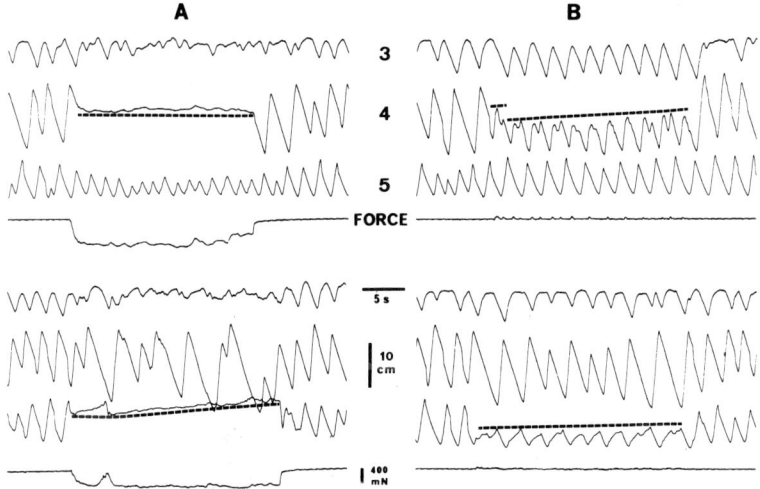

Fig. 3. Differential effect of blocking the movement of a leg during forward walking of the rock lobster. The movement of the three ipsilateral legs (3, 4 and 5) are recorded simultaneously with the force exerted against the rod. In A, a leg is blocked during the power stroke (above, leg 4 blocked; below, leg 5 blocked). In B, a leg is blocked during the return stroke (above, leg 4 blocked; below, leg 5 blocked). During the power stroke, a large force is exerted against the rod and cycling of the leg is interrupted; movements of the other ipsilateral legs are also disturbed. During the return stroke, no significant force is recorded and cycling of the leg continues at a reduced amplitude, the leg in front compensating by making a longer stride.

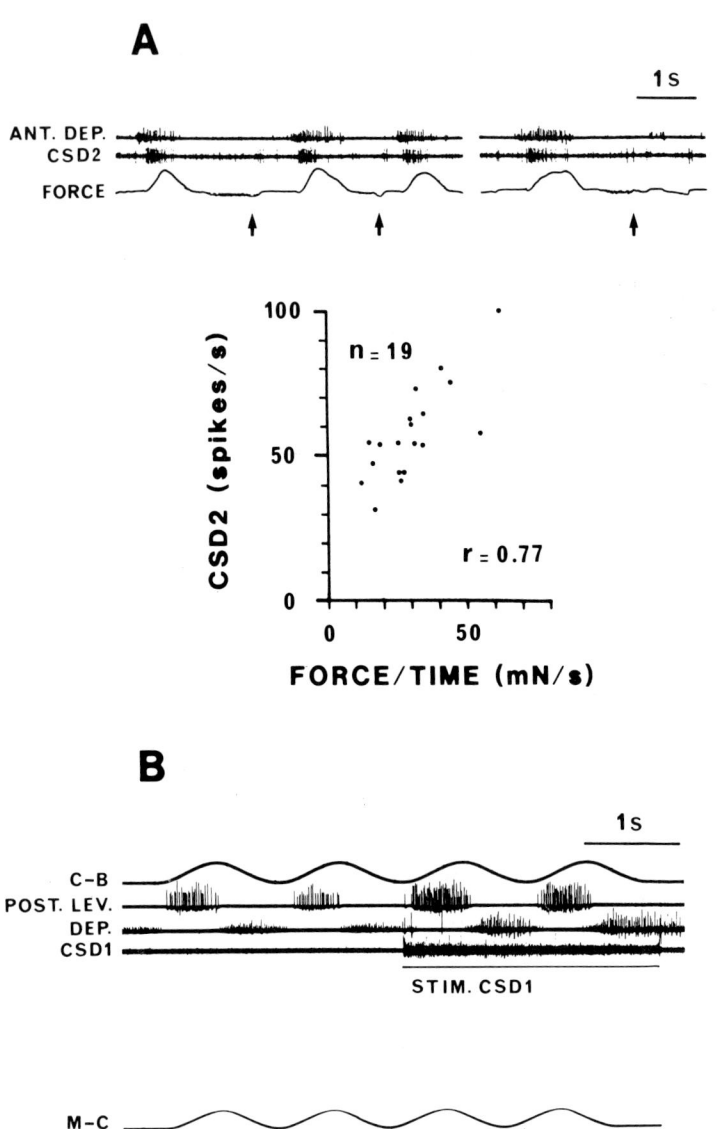

effect is different; sometimes the leg stays against the rod without exerting force, but often the response is as in Fig. 3B, the leg continuing to step but with a reduced stride. This confirms that sensory information controls the two parts of the step cycle in different ways and emphasises the importance of the force component in the regulation of stepping.

The effects of disruption of sensory input have also been shown to depend upon the direction of walking. In the crab, Carcinus, there is close coupling between the M-C and C-B joints during lateral walking (Clarac & Coulmance, 1971), that is modulated as the crab changes its direction of walking. When, during lateral walking, the M-C joint of a leg on the trailing side of the body is immobilised in an extended position, the cyclic levation of the leg at the C-B joint is exaggerated, while M-C immobilisation on the leading side inhibits C-B levation completely. These effects are absent during forward walking and present to a small degree during oblique walks. It has been hypothesised that this reflex is under the control of interneurones coordinating the direction of locomotion such as those found in the circumoesophageal connectives of crayfish by Bowerman (1977). Tight coupling between the different joints of a single leg is presumably due to complex interjoint reflexes such as those described by Ayers and Davis (1977b) and Clarac et al. (1978).

One of the most significant of recent results concerns the role of the cuticular stress detectors. These two receptors (CSD1 and CSD2) are situated in the basipodite/ischiopodite region of the leg, one on each side of the autotomy plane. They consist of a connective tissue strand innervated by several dozen bipolar sensory neurones, which is attached to a zone of soft cuticle. Despite their location

Fig. 4. A. Responses of a cuticular stress detector (CSD2) during walking. Simultaneous recording of activity in CSD2, the anterior depressor muscle (ANT. DEP) and the force exerted by the leg during forward walking in the crayfish. CSD2 discharges correspond to power strokes. The arrows show slight firing of CSD2 during dragging of the leg. The graph below demonstrates the close relation of the force exerted by the leg and CSD2 activity during walking. The positive slope of the force curve being positively correlated with firing frequency in CSD2 during this time. (Klärner & Barnes, 1985a,b).

B. Modulation of reflex effect of a chordotonal organ by CSD1 in the rock lobster. CSD1 stimulation increases the resistance reflex discharge elicited in the posterior levator (POST. LEV) and depressor (DEP) muscles by sinusoidal movement of the coxo-basipodite (C-B) joint. A similar effect occurs in the levator (LEV) muscle during sinusoidal movement of the mero-carpopodite (M-C) joint, while a continuous tonic discharge is produced in the depressor muscle. (From Vedel & Clarac, 1979).

distant from the tip of the leg, Klärner and Barnes (1985a,b) have shown them to be able to record leg contact with the ground at the AEP and monitor forces during the power stroke. Experiments in which chronic recordings were obtained from CSD2 in the legs of walking crayfish showed that the highest frequencies of CSD2 firing occurred when the leg was placed on the ground at the onset of the power stroke, though smaller bursts of sensory activity occurred if the leg dragged along the ground during the return stroke (Fig. 4A). As the graph of Fig. 4A shows, the overall discharge of CSD2 was positively correlated with the force exerted by the leg on the ground during the power stroke. In a previous experiment on a restrained non-walking rock lobster, Vedel and Clarac (1979) demonstrated that CSD1 is able to modulate reflex muscle discharge within the proximal region of the leg. It can not only increase the reflex activation of the C-B and M-C joints but also, more specifically, modify the discharge of one or two levator or depressor units (Fig. 4B). These results emphasise the importance of the simultaneous activation of different sensory receptors that occurs naturally during walking and provide insight into how power stroke and return stroke control mechanisms could be different (see Zill et al., 1985).

PERIPHERAL CONTROL OF INTERLEG COORDINATION

Most decapod Crustacea walk by adopting a pattern in which a leg alternates both with the adjacent ipsilateral legs and the contralateral leg of the same segment. Also, a leg does not normally start its return stroke before the leg behind has touched the ground (see Clarac & Barnes, 1985, for more details). Usually the overall pattern of interleg coordination corresponds to a metachronal wave. This may be anteriorly or posteriorly directed, depending on whether the mean phase relationship of adjacent ipsilateral legs is greater or smaller than 0.5.

Several theories have been presented to explain how these patterns of interleg coupling are produced. For instance, the strong preference for a particular phase relationship between legs was hypothesised by von Holst (1973) to be due to a "magnet effect" between coupled leg oscillators with slightly different inherent frequencies.

"In phase" and "alternating" patterns
Although an alternating pattern of leg coordination during walking is usually dominant, an altogether different pattern of coordination, in which the movements of the legs quickly follow each other in descending sequence, can occur when legs are not generating propulsive forces (Fig. 7A). Such "in phase" coordination has been recorded in two different

Sensory control of walking in Crustacea

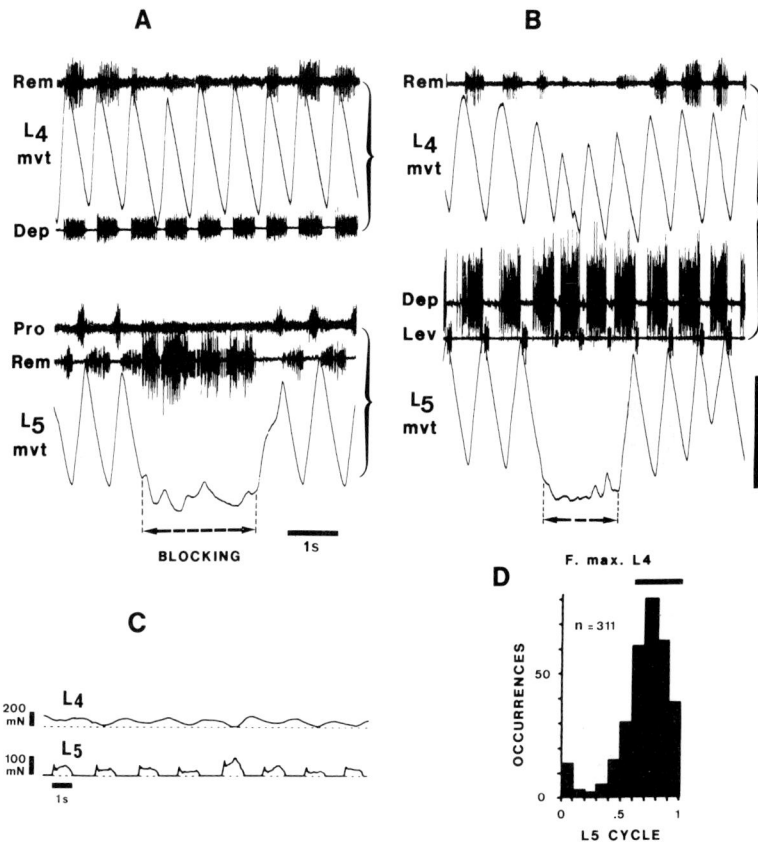

Fig. 5. Ipsilateral ascending interleg effects during forwards walking.
A & B. Simultaneous recording of EMG activity and leg movements in legs 4 and 5. Blocking leg 5 during power stroke produces a sustained discharge in the leg 5 remotor muscle (Rem) and inhibits activity in the leg 4 remotor. The effect is specific to the remotor as the two C-B muscles (depressor and levator) continue to discharge in the step cycle.
C. If a leg is held immobile on a platform and leg 5 continues to walk, measurements of the forces produced by the legs indicate that the minimum force exerted by leg 4 (L4) occurs during the power stroke of leg 5 (L5).
D. A plot of the force exerted by L4 during L5 step cycles indicates that maximum force occurs during the return stroke of L5 (indicated by the bar).
The vertical bar represents 10 cm of stride amplitude in the anterior to posterior plane, forward movement being indicated by upward movement of the traces.
C & D from Cruse et al. (1983).

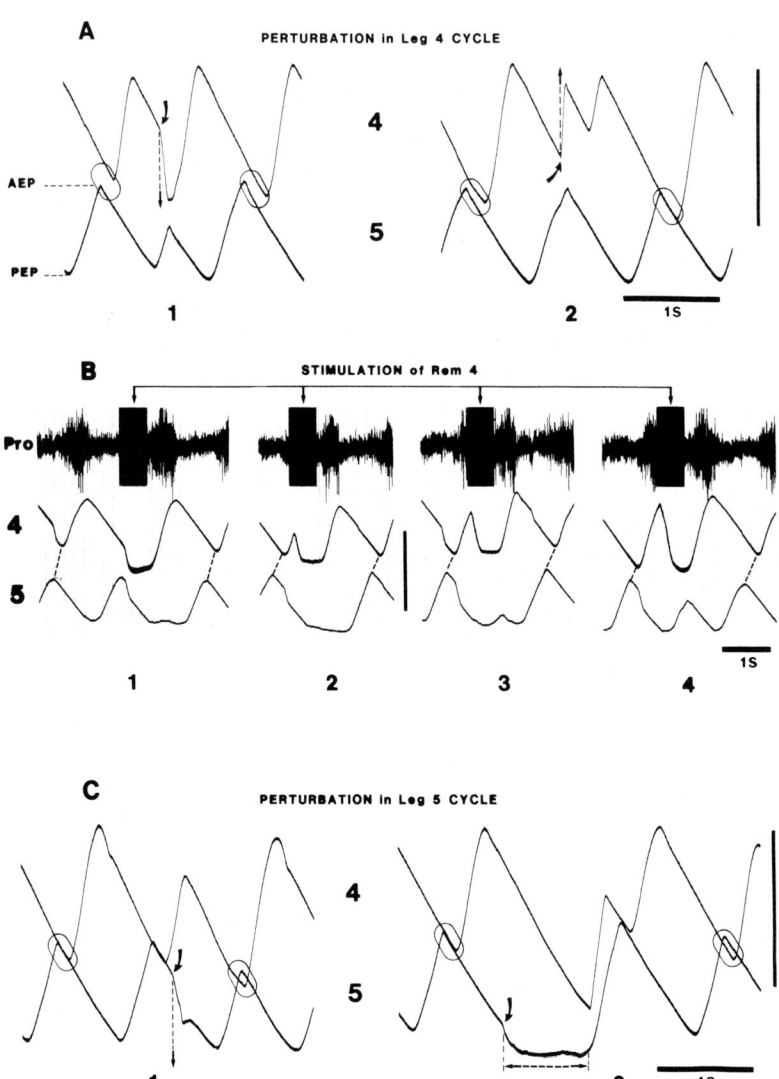

Fig. 6. Spatio-temporal coordination between ipsilateral legs 4 and 5 during forwards walking. Systematic disturbances of step cycling have been produced by passively pushing a leg (A and C) or by stimulating the remotor muscle (B).

A. Pushing leg 4 posteriorly with a rod elicits an early power stroke in leg 5, while pushing anteriorly evokes an additional power stroke in leg 4. AEP, anterior extreme position; PEP, posterior extreme position.

experimental situations.

The rock lobster, like most other decapod crustaceans, is able to autotomise its legs at the level of the ischiopodite. If a leg is autotomised, the stump continues to oscillate when the animal walks, but its movements are now in phase with the leg in front. If legs are progressively autotomised until just one leg is intact, all the stumps behind it move in phase with the walking cycle of the intact leg, while stumps in front stay immobile. Thus an "in-phase" pattern appears after autotomy.

Such synchronous motor activity has also been recorded in intact animals in a pattern of behaviour recently described as waving (Pasztor & Clarac, 1983). In this behaviour legs 1 to 4 are held up off the ground and, together with the third maxillipeds, oscillate in phase over wide arcs. This waving can sometimes extend over the mid-plane of the animal. In such cases the two sides usually wave in antiphase. No significant forces are exerted by limbs during these movements.

These two observations suggest that substrate contact and/or power stroke forces are necessary for production of the normal "alternating" pattern of leg coordination. In their absence, legs exhibit an "in phase" pattern of coordination produced by a descending excitatory influence of presumably central origin.

Ipsilateral ascending interleg effects

When a rock lobster is walking on the treadmill, placing a rod in the path of either leg 4 or leg 5 during the power stroke substantially inhibits the walking pattern of the other legs. Myographic recordings during the blocking of leg 5 indicate that the most important effect is that a sustained discharge in the leg 5 remotor muscle is accompanied by inhibition in the leg 4 remotor (Fig. 5A,B). Leg 4 movement does not stop as the leg is passively moved by the belt. In contrast, there is no effect on other power stroke muscles such as the depressor and only a slight effect on return stroke muscles such as the levator.

A direct dependence of leg 4 force upon the leg 5 power stroke has been demonstrated by placing leg 4 on a force

---

B. Stimulation of the remotor muscles of leg 4 produces a sustained power stroke and delays the return stroke of leg 4. Leg 5 is not promoted until after the onset of the next leg 4 power stroke. Pro, promotor muscle EMG.
C. Pushing leg 5 posteriorly elicits an early power stroke in leg 4 while a maintained leg 5 remotion increases the leg 4 power stroke. The alternation of the two legs recovers only when a leg 5 power stroke is initiated.
In A, B and C, the vertical bar represents 10 cm of stride amplitude.

platform while leg 5 continues to step. In this situation, leg 4 does not step but exerts cyclical forces, which are maximal during leg 5 return strokes and minimal during leg 5 power strokes (Cruse et al., 1983).

These two observations suggest that the normal "alternating" mode of coordination is brought about, at least in part, by an anteriorly directed inhibitory influence in which leg remotion is inhibited by the power stroke of the leg behind.

Cruse (1983) has constructed a simulation model of leg coordination using these data in which the force component is the major element of regulation. This model reproduces perfectly the alternating gait of the intact animal and the "in phase" coordination of animals with autotomised legs. According to this model, "in phase" coordination occurs between a stump and an immediately anterior intact leg because the normally dominant antiphase influence of the rearward leg is reduced when its distal segments are removed by autotomy.

Spatio-temporal coordination
Spatio-temporal adjustments are involved in the generation of the alternating gait as well as force. For example, the PEP of a leg is always controlled by the leg to the rear, so that if a leg cannot reach its AEP, the leg in front makes a longer stride (Fig. 3B).

The spatial component of the coordination of legs 4 and 5 have been particularly well studied. Systematic modification of the step cycle can be produced by passively pushing a leg or stimulating a leg muscle and observing the effects on the other leg. As shown in Fig. 6, these experiments have demonstrated that the coordinating influence can work both ways. Leg 4 modification affects leg 5 as readily as leg 5 action affects leg 4. All disturbances are corrected within one step and subsequent steps are normal.

During normal walking, there is some evidence that the leg behind dominates the leg in front (Chasserat & Clarac, 1983). In particular, the onset of leg 5 contact with the ground at the start of the power stroke seems to elicit the initiation of the return strokes of leg 4 after a variable delay. This has also been found with crayfish (Klärner & Barnes, 1985a). Because of the absence of a leg posterior to it, the PEP of leg 5 is a very stable point, defined by the postural geometrical arrangement of all the legs with the body.

Contralateral coordination
Descriptive studies of the phase relationships of legs during walking have demonstrated that in all decapod Crustacea the coordination of contralateral legs is much more labile than that of ipsilateral legs (Clarac, 1982). For instance, it is not unusual to see a decapod use only the legs of one side of

the body during turning. It is thus appropriate to ask whether the rules governing the coordination of ipsilateral legs can also be applied to contralateral leg coordination. To answer this question we have split the treadmill into two belts, the speeds of which can be varied independently. In theory, when the belts are moved at different speeds, the legs on each side of the body might either move at frequencies appropriate to the belt speeds or both sides might maintain the same frequency of stepping. Since the lobster is on a power driven treadmill, the velocity of the power strokes are controlled by the belt speeds. Thus absolute coordination across the body could only be maintained but adjusting stride length and/or leg velocity during the return stroke.

In the particular sequence shown in Fig. 8, the two belts were initially moving at 6cm s^{-1} and the 4th pair of legs operated more or less in opposition (mean phase, 0.36) with nearly identical strides (7 and 6.5cm s^{-1} for the two sides). The belt on the right side was then slowed down to 4.5cm s^{-1}. This resulted in a reduction in the frequency of stepping, not only on the right side but also on the left, even though the belt speed on that side was unchanged. Absolute coordination was maintained due to a longer stride on the unchanged belt and a shorter stride on the slower belt. When the right hand belt speed was increased again to its initial rate, stride lengths and mean phase returned to their original values.

These results demonstrate that contralateral legs may be coordinated as accurately as ipsilateral legs and that their preferred mode of coordination is also an "alternating" pattern. I conclude that peripheral information plays a predominant role in maintaining the alternation of legs during walking and that similar signals may be used in both ipsilateral and contralateral coordination.

Summary of interleg effects

Fig. 7 summarises the major features of interleg coordination. The "alternating" pattern of coordination seen in normal walking animals and the "in phase" pattern seen both during waving behaviour and when all legs are autotomised (stump preparation) are illustrated in Fig. 7A. Primary roles in the production of the "alternating" pattern are played by the power stroke force component and by leg contact at the AEP (Fig. 7B). Klärner and Barnes (1985a,b) have demonstrated that CSD2 records leg contact with the ground and influences the coordination of neighbouring ipsilateral legs. In crayfish (as in the rock lobster), the PEP of a leg closely follows the AEP of the leg behind. If a wire is bound round CSD2 in the latter leg so that it can no longer signal the AEP, the coordination of the two legs is significantly more variable. The force produced by a leg during the power stroke is often at a maximum early in the power stroke, decreasing more slowly to zero at the PEP (Cruse et al., 1983; Klärner & Barnes, 1985b).

Sensory control of walking in Crustacea

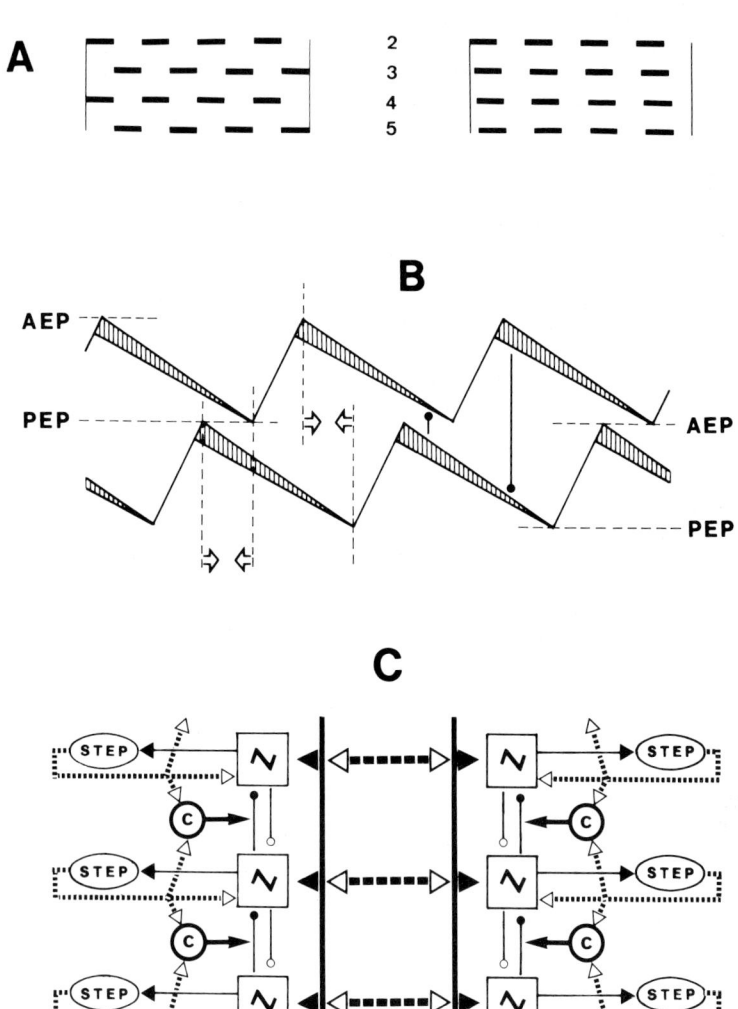

Fig. 7. Schematic representation of ipsilateral and contralateral coordination.
A. In walking ipsilateral legs are used in an alternating pattern (left), while in waving or when all legs are autotomised (stump preparation) all legs move in phase (right). Black bars represent the power strokes.

It exerts an inhibitory influence on neighbouring legs (Fig. 5). This effect, together with the increased input from chordotonal organs at extreme joint positions (spatial component of interleg coordination), appears to be sufficient to switch a leg from power stroke to return stroke, with fine control of the timing of the PEP being provided by neighbouring leg CSD input. Similar mechanisms appear to control interleg coordination in both cockroaches and cats (Pearson & Duysens, 1976).

Fig. 7C is a speculative wiring diagram of the mechanism controlling interleg coordination which is derived from experiments on both leg autotomy and manipulation of sensory feedback during walking. According to this model, sensory inputs from each leg feed into a comparator which activates connections between the central rhythm generators of neighbouring legs. This model considers only ipsilateral coordination since contralateral coordination is more labile and therefore difficult to define.

In the rock lobster, back legs impose their rhythm on the legs in front, presumably because anteriorly directed interleg effects are more powerful than posteriorly directed ones. No such dominance can be seen between contralateral pairs of legs during normal walking. However, use of a split treadmill has shown that legs on a more slowly moving belt seem to impose their rhythm on the legs on the faster side. During normal walking, coordination can be maintained by adjusting both power and return stroke components of the step (though on the power drawn treadmill power stroke velocity is determined by the belt speed). If such changes are insufficient to maintain coordination, a leg can make an extra step. Such extra steps are commonly seen during normal walking sequences as well as following experimental manipulations (Fig. 6).

B. Diagram of the movement of two ipsilateral legs in walking indicating the critical factors of leg coordination; the white arrows represent the coordination of movement based upon the onset of the power stroke of one leg inducing the onset of the return stroke of the other. The black dots represent the inhibition of forces exerted at the end of the power stroke of one leg by the initiation of remotion of the other. AEP, anterior extreme position; PEP, posterior extreme position.

C. Wiring diagram of the connections producing coordination of the three posterior pairs of legs by interaction of sensory feedback with a central oscillating element which commands the step activity. Coexistence of the "in phase" and the "alternating" patterns is explained by a descending "in phase" connection. The inhibitory connections between each segmental rhythmic element are under the control of a sensory comparator (C). Black ascending dots emphasise the greater influence of posterior legs on anterior legs in comparison with the opposite direction (white dots).

Sensory control of walking in Crustacea

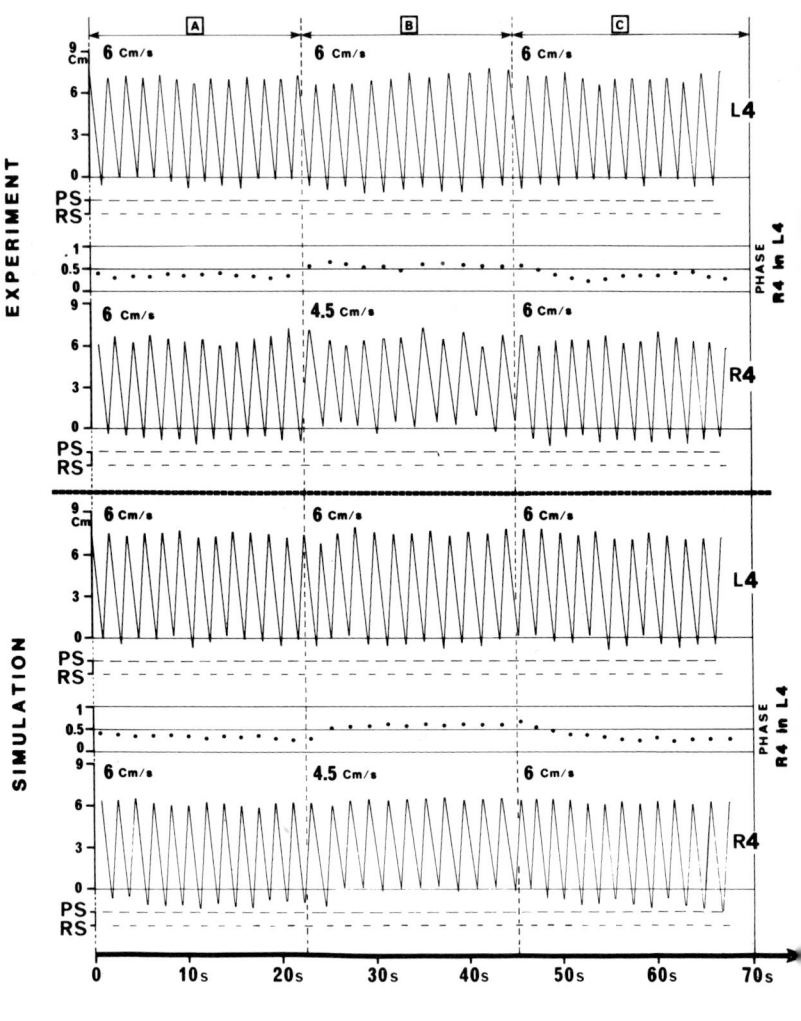

Fig. 8. Comparison of an experimental and a simulated sequence of walking on a split belt treadmill. Leg movements of a bilateral pair of legs (R4 and L4) have been recorded with durations of power (PS) and return strokes (RS) and the instantaneous phase relationship of the two legs (R4 in L4). The two belts are driven at 6 cm s^{-1} (A and C), while in B the right belt is slowed down to 4.5 cm s^{-1}. To obtain the simulated sequence, mean stride (7 and 6.5 cm) and mean return stroke duration (0.65s) have been given with specific coefficients for coupling, hierarchy and strategy.

Sensory control of walking in Crustacea

COMPUTER SIMULATION OF LOBSTER WALKING

Our knowledge of step control for driven walking in the rock lobster has led us to develop a computer model able to simulate stepping sequences (Chasserat & Clarac, unpublished). Consecutive steps are drawn on a plotter as "leg" position versus elapsed time. For a single leg, the calculated data include both invariant parameters (mean stride, return stroke duration and pause time) and imposed parameters (belt speed and number of steps). The variability observed in real walking is reproduced by using a random function to compute the successive PEP values. Any given step is partly dependent upon the parameters of the preceding one, and partly influenced by random factors. To simulate the interactions between adjacent legs, it was necessary to add three more control parameters, <u>coupling</u> (the strength of coupling), <u>hierarchy</u> (the dominating action of one leg upon another) and <u>strategy</u> (whether power or

TABLE 1

Detailed comparison of experimental and simulated values from Fig. 8.

		Experiment				Simulation		
L4	Total 35 steps	A	B	C	Total	A	B	C
Stride in cm	7 ±0.43	6.8 ±0.3	7.4 ±0.3	6.8 ±0.3	7.1 ±0.4	7 ±0.3	7.2 ±0.4	7.1 ±0.3
RS duration in s	0.67 ±0.07	0.64 ±0.05	0.75 ±0.05	0.64 ±0.05	0.66 ±0.02	0.64 ±0.01	0.68 ±0.02	0.65
Phase L4 in R4	0.42 ±0.12	0.36 ±0.07	0.56 ±0.06	0.36 ±0.08	0.43 ±0.11	0.34 ±0.03	0.57 ±0.05	0.38 ±0.05
R4	Total	A	B	C	Total	A	B	C
Stride in cm	6.3 ±0.5	6.5 ±0.4	5.8 ±0.4	6.6 ±0.3	6.4 ±0.4	6.6 ±0.3	5.9 ±0.3	6.7 ±0.2
RS duration in s	0.66 ±0.09	0.69 ±0.08	0.66 ±0.11	0.61 ±0.04	0.65 ±0.01	0.65	0.63	0.65

return stroke parameters are adjusted to produce coordination). Boundary conditions prevent stepping parameters assuming impossible values.

The mechanism selected here to control interleg coupling is the "magnet effect" concept (von Holst, 1973). The model continuously creates a balance between full adjustment (complete absolute coordination) and random occurrences (i.e. no coordination). It works as if reciprocal attractive forces were applied between a given extreme position of one leg (AEP) and the opposite extreme position of the other leg (PEP). Contralateral phase relationships are usually simulated with a weak coupling coefficient and no leg dominance. Ipsilateral phase simulation needs a stronger coupling coefficient and a dominating influence of a posterior leg upon its next anterior neighbour.

Fig. 8 presents real and simulated sequences of the movements of a pair of contralateral legs of a rock lobster walking on a split belt treadmill, while Table 1 gives means and standard deviations of stepping parameters of both these sequences of walking. In order to reproduce the behaviour of the real animal, we have used the following coefficients in our simulation. During sequences A and C we have assumed that the left side is dominant, but that during the B sequence, when the two belts are moving at different speeds, the hierarchy is reversed so that legs on the slower belt dominate. A low coefficient of coupling is sufficient when the two belts are moving at the same speed, while during the B sequence stronger coupling is necessary. The adopted strategy for the left side of the body is adjustment of return stroke parameters, while for the right side power stroke values are regulated. The simulation of patterns of leg coordination that replicate real walking rather accurately does not imply that our hypotheses are totally valid but does confirm that a locomotor system can work with this minimum of regulatory information.

As in insects (Bässler, 1983; see also Chapter 12 by Dr Zill) sensory information is always interacting with the motor commands to adjust the stepping cycle. The controls are complex, but even though we are still very far from having a complete understanding of the effects of proprioceptive feedback, I am optimistic that this experimental approach to behaviour will continue to be very productive in elucidating the neuronal processes generating locomotion.

REFERENCES

Ayers, J.L. & Davis, W.J. (1977a). Neuronal control of locomotion in the lobster Homarus americanus. L Motor programs for forward and backward walking. J. comp. Physiol., 115, 1-27.
Ayers, J.L. & Davis, W.J. (1977b). Neuronal control of locomotion in the

lobster Homarus americanus. II. Types of walking leg reflexes. J. comp. Physiol., 115, 29-46.
Barnes, W.J.P. (1977). Proprioceptive influences on motor output during walking in the crayfish. J. Physiol., Paris, 73, 543-564.
Barnes, W.J.P., Spirito, C.P. & Evoy, W.H. (1972). Nervous control of walking in the crab Cardisoma guanhumi. II. Role of resistance reflexes in walking. Z. vergl. Physiol., 76, 16-31.
Bässler, U. (1983). Neural Basis of Elementary Behavior in Stick Insects. Springer Verlag, Berlin.
Blight, A.R. & Llinas, R. (1980). The non-impulsive stretch-receptor complex of the crab: a study of depolarization-release coupling at a tonic sensorimotor synapse. Phil. Trans. R. Soc. B, 290, 219-276.
Bowerman, R.F. (1977). The control of arthropod walking. Comp. Biochem. Physiol., 56A, 231-247.
Bush, B.M.H. (1962). Proprioceptive reflexes in the legs of Carcinus maenas L.. J. exp. Biol., 39, 89-105.
Chasserat, C. & Clarac, F. (1983). Quantitative analysis of walking in a decapod crustacean, the rock-lobster Jasus lalandii. II. Spatial and temporal regulation of stepping in driven walking. J. exp. Biol., 107, 219-243.
Clarac, F. (1977). Motor coordination in crustacean limbs. In Identified Neurons and Behavior of Arthropods. Ed. Hoyle, G. Plenum, New York, pp. 167-187.
Clarac, F. (1982). Decapod crustacean leg coordination during walking. In Locomotion and Energetics in Arthropods. Eds. Herreid, C.F. & Fourtner, C.. Plenum, New York, pp. 31-71.
Clarac, F. & Barnes, W.J.P. (1985). Peripheral influences on the coordination of the legs during walking in decapod crustaceans. In Coordination of Motor Behaviour. Eds. Bush, B.M.H. & Clarac, F.. Society for Experimental Biology Seminar Series 24. Cambridge University Press, Cambridge, pp. 249-269.
Clarac, F. & Chasserat, C. (1983). Quantitative analysis of walking in a decapod crustacean, the rock-lobster Jasus lalandii. I. Comparative study of free and driven walking. J. exp. Biol., 107, 189-217.
Clarac, F. & Coulmance, M. (1971). La marche laterale du crabe (Carcinus). Z. vergl. Physiol., 73, 408-438.
Clarac, F., Vedel, J.P. & Bush, B.M.H. (1978). Intersegmental reflex coordination by a single joint receptor organ (CB) in rock lobster walking legs. J. exp. Biol., 73, 29-46.
Cruse, H. (1983). The influence of load and leg amputation upon coordination in walking crustaceans: a model calculation. Biol. Cybern., 49, 119-125.
Cruse, H., Clarac, F. & Chasserat, C. (1983). The control of walking movements in the leg of the rock lobster. Biol. Cybern., 47, 87-94.
Davis, W.J. (1969). Reflex organization in the swimmeret system of the lobster. I. Intrasegmental reflexes. J. exp. Biol., 51, 547-563.
DiCaprio, R.A. & Clarac, F. (1981). Reversal of a walking leg reflex elicited by a muscle receptor. J. exp. Biol., 90, 197-203.
Evoy, W.H. & Ayers, J.L. (1982). Locomotion and control of limb movements. In The Biology of Crustacea, Vol. 4, Neural Integration and Behavior. Eds. Sanderman, D.C. & Atwood, H.L.. Academic Press, New York, pp. 61-105.
Grillner, S. (1981). Control of locomotion in bipeds, tetrapods and fish. In

Handbook of Physiology, Section 1, The Nervous System, Vol. 2, Motor Control. Ed. Brooks, V.B. American Physiological Society, Bethesda, pp. 1179-1236.

Holst, E. von (1973). The Behavioural Physiology of Animals and Man, Vol. 1. (Translated by Robert Martin). University of Miami Press, Miami.

Klärner, D. & Barnes, W.J.P. (1985a). Physiology and function of a cuticular stress detector (CSD2) in the crayfish Astacus leptodactylus. I. CSD2 activity during walking and its influence on leg coordination. (In preparation).

Klärner, D. & Barnes, W.J.P. (1985b). Physiology and function of a cuticular stress detector (CSD2) in the crayfish Astacus leptodactylus. II. Forces produced by crayfish legs during walking. (In preparation).

Pasztor, V.M. & Clarac, F. (1983). An analysis of waving behaviour: an alternative motor programme for the thoracic appendages of decapod Crustacea. J. exp. Biol., 102, 59-77.

Pearson, K.G. & Duysens, J. (1976). Function of segmental reflexes in the control of stepping in cockroaches and cats. In Neural Control of Locomotion. Eds. Herman, R.M., Grillner, S., Stein, P.S.G. & Stuart, D.G.. Plenum, New York, pp. 519-537.

Rossignol, S. & Gauthier, L. (1980). An analysis of mechanisms controlling the reversal of crossed spinal reflexes. Brain Res., 182, 31-45.

Vedel, J.P. (1980). The antennal motor system of the rock lobster: competitive occurrence of resistance and assistance reflex patterns originating from the same proprioceptor. J. exp. Biol., 87, 1-22.

Vedel, J.P. (1982). Reflex reversal resulting from active movements in the antenna of the Rock lobster. J. exp. Biol., 101, 121-133.

Vedel, J.P. & Clarac, F. (1979). Combined reflex actions by several proprioceptive inputs in the rock lobster walking legs. J. comp. Physiol., 130, 251-258.

Wiens, T.J. (1976). Electrical and structural properties of crayfish claw motoneurons in an isolated claw-ganglion preparation. J. comp. Physiol., 112, 213-233.

Zill, S.N., Libersat, F. & Clarac, F. (1985). Single unit sensory activity in free walking crabs: force sensitive mechanoreceptors of the dactyl. Brain Res. (In press).

Section 7
Feedback and Motor Control in Man

Chapter Twenty-five

INTRODUCTION

J.A. Stephens

Each of the following chapters is concerned with the study of afferent input and its importance for motor control in man. In the first, Prof Vallbo describes how afferent input can be recorded directly in human subjects at the level of activity of single identified afferent units. Dr Matthews describes experiments designed to compare the reflex effects of Group I and Group II muscle spindle endings. Dr Forssberg considers the importance of afferent input in modifying the step cycle in man and finally Prof Marsden and Dr Rothwell provide evidence to suggest that afferent input plays an important functional role at the level of the cerebral cortex as well as in the spinal cord in man.

Proprioceptive activity from human finger muscles
 In Chapter 26 Vallbo discusses the role of proprioceptors in motor control and the nature of the information provided by spindles and tendon organs, particularly to what extent motor parameters such as joint position, velocity of movement, and force output of the active muscles are described in the afferent discharge. Conclusions are drawn from visual inspection of microneurographic recordings made in man of the behaviour of afferent units with endings in the finger muscles, the long flexors or extensors in the forearm, when subjects performed voluntary contractions on instruction.
 It is concluded that tendon organ afferent fibres provide two types of information. One is said to describe accurately the total force output of the muscle, the other to monitor the exact moment of recruitment of individual motor units at low levels of contraction strength. This conclusion is based on the observation that during smoothly increasing ramp contractions some afferent units show step like increments in their firing (Fig. 1A of Chapter 26), while others increase their firing smoothly (Fig. 1B of Chapter 26). To my mind the information carried by the two types of unit is identical, namely that they each convey a sample of the total force being generated by the muscle. The fact that one shows step like

changes in firing while the output of the other is smooth merely reflects the different sample of muscle fibres that contribute force in series with the different receptors. One can suppose that the 'smooth' type has relatively few fibres in series, the 'step' type many more; the more fibres in series the more steps in firing will be recorded as the parent motor units are recruited. To conclude that there is physiological significance in what is simply the consequence of an expected and predictable difference in anatomical arrangement between muscle fibres and sensory capsule seems unlikely. For without other evidence to the contrary one would suppose this arrangement to vary in some continuous manner around some mean value of muscle fibres per capsule. A systematic study in animal experiments is required before any conclusion can be reached that there exist a real subpopulation of receptors that by their anatomy provide a special signal to the CNS. If on the other hand the population of muscle fibres connected in series with the population of receptor capsules is indeed random then the ensemble afferent input from tendon organs would contain an unbiased representation of the total force output generated by the receptor bearing muscle and extremes of individual unit behaviour would be adventitious.

The behaviour of muscle spindle afferent units during voluntary movements is described. It is concluded that within the same muscle some spindle endings increase their firing during lengthening but not during active shortening, while others increase their firing during active shortening but do not fire during lengthening. If this is an adequate summary description of spindle afferent behaviour, then the information contained in the muscle afferent signal can aptly be described as ambiguous. Faced with such a 'no sense' signal one is led to wonder whether the true information content of the firing pattern is being properly recognised.

One possibility is that the analysis of records by visual inspection may be inadequate and thus conclusions drawn from simple visual inspection of afferent spike trains as in Figs. 5-8 of Chapter 26 may be misleading. Some processing of the signal may be necessary before its information content can be recognised. For example the overwhelming impression gained from Fig. 6 is that of an increase in firing of the afferent ending associated with and therefore signalling shortening of the receptor bearing muscle. This may not be the case. Suppose for example the central nervous decoder of this signal acts like a differentiator. In this case slow steady changes in spindle firing rate would be filtered out and ignored. The information contained in the signal would then be revealed as being contained in the rate of change of impulses sec^{-1} of afferent input rather than the absolute number of impulses sec^{-1} per se. If this were the case then the changes in steady firing produced by fluctuation in fusimotor activity during slow contractions as in Fig. 6 would be disregarded by the CNS; the information of significance would be the moment to moment fluctuations in afferent firing rate as illustrated for the same afferent in Fig. 8. Applying this type of decoding to the afferent discharge in Fig. 8 it is apparent that short term fluctuations in firing are in the direction of muscle lengthening being associated with an increase and muscle shortening a decrease in firing frequency. This being the case, the apparent ambiguity of information contained in the two examples of afferent firing shown in Figs. 5-8 would disappear. In summary some processing of the afferent signals may be necessary before either the experimenter or the CNS can understand what information is being conveyed.

In his discussion Vallbo suggests that muscle afferent input may have a special importance for the control of weak contractions. To support this view he points out that spindle afferents exert particularly powerful effects on small motor neurones which dominate weak contractions. This is the case but a careful analysis of this problem by Harrison, Taylor and Chandler (1980) indicated that the relationship between Group Ia EPSP amplitude and motor unit force output was such that the increment of total force output of a muscle per unit increment in Group Ia firing is proportional to the initial force. In other words contrary to the intuitive view expressed by Vallbo the reciprocal relationship between Group Ia EPSP amplitude and motor unit force output is such that the gain of the monosynaptic Group Ia reflex is not in fact more powerful at low rather than high initial force levels.

Human long-latency stretch reflexes - a new role for the secondary ending of the muscle spindle?
In Chapter 27 Matthews describes experiments whose results lead him to the view that Group II spindle afferent endings are responsible for the long latency components of the reflex response to muscle stretch in human flexor pollicis longus (FPL) muscle. The experimental observations are straightforward. Muscle stretch produces short and long latency components. Muscle vibration produces only a short latency response. On the assumption that vibration excites predominantly muscle spindle primary endings he concludes that activity in these afferent fibres does not produce a long latency response. Muscle stretch on the other hand can be expected to excite both muscle spindle primary and secondary endings, and elicits a long latency response. He therefore concludes that the long latency response is mediated by Group II spindle afferents. On the cessation of vibration, emg activity is reduced at short latency but with no obvious second long latency component of action. On release of muscle stretch there is a powerful long latency reduction of activity but often little apparent short latency effect. Muscle release can be expected to reduce the firing of both muscle spindle

primary and secondary endings. He therefore concludes that the long latency response to muscle release is due to the cessation of the firing of the slower conducting Group II muscle spindle afferent fibres.

Central to these conclusions are two suppositions. The first rests on the argument that because stimulation of Group Ia afferents alone does not produce a long-latency response then this same input also does not produce a long-latency response to stretch when many other different types of afferent are active simultaneously. I find this hard to accept in view of the extensive convergence that must exist in the central nervous pathways involved. The second supposition is that the 'intensity' of the Group Ia input generated by the vibration is comparable to that produced by muscle stretch. It is argued that this is the case in the situation where the short latency response to stretch and vibration is the same. This is not a safe assumption because under the conditions of stretch a very different and much more complex muscle afferent input is generated, whose reflex effects are summated at the motor neurone pool to generate a certain time course of net excitation and hence change in motor unit firing probability and change in recorded averaged rectified emg. The recorded change in muscle electrical activity can say nothing of the intensity of the Group Ia afferent component alone. For this reason it is my view that the finding of the same sized short latency response to stretch and vibration is an inadequate guide to the relative intensity of the Group Ia input produced by the two types of stimulation. The likelihood is, in fact, that following muscle stretch the short latency effects of Group Ia input are much reduced by the concomitant reflex effects of Group Ib activity accompanying rapid muscle stretch. This being the case, in the situation where vibration evokes the same sized short latency response as muscle stretch, a much less intense Group Ia input can be expected to be involved. This could account for the lack of long latency response to vibration under these conditions. Any long latency reflex effects of more powerful vibration are unavoidably hidden by the reflex discharge of motor units produced by the short latency component.

The role of Group II muscle spindle afferents in the response of FPL to muscle stretch and release depends on the conclusion that Group Ia input generated by muscle stretch does not produce a long latency response. This in turn depends on a comparison between the reflex effects of muscle vibration and stretch. This comparison would appear to be insecurely based.

During discussion of this paper at the meeting, Prof Marsden reminded participants that long latency responses to muscle stretch were reduced or absent in patients with signs of dorsal column disease or damage to motor cortex. There followed some lively discussion between Marsden and Matthews about the interpretation and significance of these clinical data. Could they be reconciled with the hypothesis favoured by Matthews that Group II muscle afferents were responsible for long latency responses or did they remain telling evidence that conduction along long ascending and descending central nervous pathways were involved? Participants listened with rapt attention as Matthews defended his case bravely against eloquent scientific riposte from Marsden. Parsimony prevailed.

Phase dependent step adaptations during human locomotion

Forssberg describes experiments in which subjects were asked to push or pull on a fixed support while walking on a treadmill. Activity in the arm muscles was found to be preceded by changes in the activity of muscles in both legs. The spatial and temporal pattern of this anticipatory response varied according to the timing of the arm muscle activity within the step cycle. When the instruction to push/pull was given on heel strike, the pattern in the stance leg turned out to be similar to the anticipatory responses that had been described earlier by Cordo and Nashner (1982) when they performed the same experiment in standing subjects. The same spatio-temporal pattern of leg muscle activity also follows externally applied disturbances such as can be produced if the surface under the supporting leg is withdrawn backwards during locomotion (Nashner, 1980) or if a rocking platform is used to rotate the ankle during fixed stance (Nashner, 1977). The similarity between the 'postural synergies' (proprioceptive or anticipatory) during fixed stance and locomotion leads Forssberg to conclude that the same central programme is used. He postulates that 'two separate systems exist, one for locomotion, and another for postural reactions, where activity from the two systems is summated at a late pre-motor or motor neurone level'.

To my mind the fact that the pattern of response produced by a number of different procedures is similar does not mean that the same mechanisms are involved but merely indicates that the same set of mechanical corrections are required to meet the different circumstances. To use the word 'system' in place of 'mechanism' is misleading for without evidence it conjures up a vision of a set of neurones whose function is 'postural' rather than 'locomotor'. The computer analogy is likewise misleading. The fact that the final motor command is similar under different circumstances is no evidence that the same sequence of central nervous events has been responsible. Study of the pattern of emg bursts is traditional but fundamentally phenomenological. What would be more illuminating would be to use these data to identify the physiological controlled variables in these experiments. Once this had been done studies of emg bursts would become studies of mechanism rather than as at present descriptions of pattern and timing, rhymes without reason.

Abnormal feedback and movement disorders in man, with particular reference to cortical myoclonus

In Chapter 29 Marsden and Rothwell begin by reviewing the general categories of movement disorders in man that can be attributed to abnormal feedback. The categorisation of movement disorder according to underlying physiological mechanism is characteristic of Marsden's group and is valuable because it establishes a framework of hypothesis for logical treatment that is open to experimental test. The main topic of their review is cortical reflex myoclonus.

In their introduction Marsden and Rothwell discuss the evidence that long latency muscle reflex responses which follow muscle stretch or cutaneous stimuli in man employ transcortical pathways and a number of criticisms of the evidence are presented and discussed. The authors anticipate the significance of their clinical observations by concluding that the existence of transcortical reflex pathways in normal subjects is dramatically revealed in patients with cortical reflex myoclonus. An example is shown in Figs. 1 and 2 of Chapter 29.

Fig. 1 shows for patient J.B. that a large positive-negative wave over the contralateral scalp precedes each myoclonic jerk by about 22ms. This is taken to indicate that the myoclonic jerks can be attributed to electrical discharges arising from the sensori-motor cortex and then travelling by the direct corticospinal pathway to the spinal cord and lower motor neurones. Fig. 2 shows for the same patient scalp-recorded somatosensory evoked potentials and reflex muscle jerks in the hypothenar muscle following electrical stimulation of the cutaneous nerves of the affected side. The latency to onset of the first negative component of the cortical evoked response on the contralateral side is about 20ms, followed by a positive peak at about 34ms. The interval between the latter positive peak and the onset of the peak in the average rectified reflex electromyogram is about 20ms.

Taken together Figs. 1 and 2 are used to argue that the reflex myoclonus emg peak in Fig 2 evoked by electrical stimulation of the digital nerves is produced as the result of a transcortical pathway. We are invited to assume that the cutaneous input has activated or triggered the same cortical mechanism that was responsible for the spontaneous myoclonic jerk shown in Fig. 1. We are led to conclude by inference that cutaneous input has directly or indirectly activated the sensori-motor cortex to produce a reflex myoclonic jerk by way of the direct cortico-spinal pathway. Whatever the explanation we should consider whether the result can safely be used as evidence to support the view that the same pathways are active, albeit much less powerfully, in normal subjects. The evidence is circumstantial and some will criticise on that basis alone. If accepted we are left with further evidence to support the view that transcortical reflexes do exist, and that their disruption in disease may underly recognisable indicators of dysfunction.

REFERENCES

Cordo, P.J. & Nashner, L.M. (1982). Properties of postural adjustments associated with rapid arm movements. J. Neurophysiol., 47, 287-302.
Harrison, P.J., Taylor, A. & Chandler, R.B. (1980). The significance of motor unit size relative to Ia excitation for the gain of the stretch reflex. Adv. Physiol. Sci., 1, 109-112.
Nashner, L.M. (1977). Fixed patterns of rapid postural responses among leg muscles during stance. Expl. Brain Res., 30, 13-24.
Nashner, L.M. (1980). Balance adjustments of humans perturbed while walking. J. Neurophysiol., 44, 650-664.

Chapter Twenty-six

PROPRIOCEPTIVE ACTIVITY FROM HUMAN FINGER MUSCLES

A.B. Vallbo

Theories concerning the role of intramuscular proprioceptors in motor control and perception were for a long time based on recordings of afferent activity from reduced preparations, e.g. anaesthetised, spinal, or decerebrate animals. Such preparations have provided a wealth of data on the physiological properties of sense organs and their capacity to measure movements and forces, as well as information on their central connections and synaptic actions on segmental and suprasegmental neurones. However, reduced preparations are not very suitable for direct analysis of sensory impulses in natural movements, except in simple reflexes. Admittedly, some studies of muscle spindle discharge have been done during natural and more complex movements, i.e. treadmill walking, respiratory movements, and jaw movements in the decerebrate or anaesthetised cat (e.g. Critchlow & von Euler, 1963; Davey & Taylor, 1967; Shik et al., 1968).

Recordings of afferent activity in truly intact and behaving organisms were first reported in 1966 when a microneurographic method was introduced which allows the study of unitary nerve impulses in intact human subjects (Vallbo & Hagbarth, 1967, 1968). Some years later a method of recording afferent activity from behaving mammals was presented (Matsunami & Kubota, 1972; Cody & Taylor, 1973) and several other techniques have followed (Prochazka et al., 1975; Loeb, 1976; Schieber & Thach, 1980).

Unique advantages are offered by the approach of recording afferent activity in intact and behaving animals or human subjects. It is possible to assess the properties of proprioceptive signals during complex motor activities which are truly physiological and entail important features which cannot be guaranteed in reduced preparations, e.g. an undistorted activation of the muscle with regard to pattern of motor unit recruitment, adequately working feedback loops, and an alpha-gamma balance which is proper for the particular movement studied. In humans, voluntary contractions have been explored when movements are performed on instruction, implying that the subject's consciousness is engaged in planning the motor activity. In behaving animals, on the other hand, the emphasis has been on locomotor movements. Thus, more automatic contractions have probably been studied in behaving animals than in man (Jackson, 1932), although in both preparations they are referred to as voluntary.

A crucial test of a theory concerning the control function exerted by a group of sense organs in a particular motor task may be obtained by recording from non-reduced preparations because it can be directly assessed whether the properties of the afferent signal during the task is consistent with the theory. An example illustrating this approach bears on the follow-up length servo hypothesis (Merton, 1951) which predicts that the afferent discharge from muscle spindles should precede the electromyographic activity by some 20ms in the finger muscles at the onset of a voluntary contraction. This hypothesis can be refuted for this type of contraction, because actual recordings from human afferents clearly demonstrated that increased spindle firing consistently lagged the onset of the electromyographic activity by at least 10ms (Vallbo, 1971).

A good reason for studying afferent discharge from muscle spindles is that it may be used to deduce the nature and properties of the fusimotor activity during natural movements in non-reduced preparations. Admittedly, it has lately been demonstrated that such inferences must be done with considerable caution, particularly during active movements when the effect of changing muscle lengths greatly complicates the interpretation (Hulliger & Prochazka, 1983).

The present report will be concerned mainly with certain aspects of the role of proprioceptors in motor control and the nature of the information provided by spindles and tendon organs; in particular, to what extent motor parameters such as joint position, velocity of movement, and force output of the active muscles are described in the afferent discharge. Other aspects of proprioceptive functions studied with the microneurographic technique have been summarised in recent surveys (Vallbo et al., 1979; Hagbarth, 1981; Vallbo, 1981a,b).

The data were all extracted from units with endings in the extensor digitorum and flexor digitorum muscles on the forearm, when the subjects performed voluntary contractions on instruction. The method of recording afferent activity from single nerve fibres in human subjects has been described in several reports and will not be reiterated here (see Vallbo et al., 1979).

GOLGI TENDON ORGANS

For a long time, the Golgi tendon organs were thought to have mainly a protective role. By their reflex effects they were assumed to account for the well-known clasp knife phenomenon (Ballif et al., 1925; Fulton & Pi-Suner, 1928) and prevent the muscle from developing excessive tensions which might damage the muscle itself. This hypothesis was based on the finding that the receptors do not respond to passive stretch unless the tension reaches very high levels (Matthews, 1933). However, this view came into doubt when Jansen and Rudjord (1964) demonstrated high sensitivity to active forces. When the muscle contracted, the discharge rate was much higher than with passive stretch at corresponding tensions recorded at the main tendon. The function of the Golgi tendon organs in mammals has since been further explored by a number of groups using electrical stimulation of single motor nerve fibres. Houk and Henneman (1967) demonstrated that a strong discharge may be produced by the contraction of a single motor unit in spite of its force output being quite small. They also showed that a limited number of motor units (4-15) insert upon the individual sense organ and that distinct levels of afferent impulse rates may result when one, two, or three of these motor units are artificially activated by tetanic stimulation of their motor nerve fibres. Moreover, it has lately been demonstrated that the clasp knife phenomenon is probably not dependent on tendon organs but on other types of sense organs (Rymer et al., 1979).

In microneurographic studies on human subjects only a few recordings of afferent discharge from Golgi tendon organs have been published (Vallbo, 1970, 1974a; Burg et al., 1973; Burke, Hagbarth & Löfstedt, 1978). The reason is no doubt that interest has been focused on muscle spindle afferents. However, some interesting features have been reported and they are supported by a fair amount of unpublished observations.

One question is whether Golgi tendon organs produce a smoothly changing impulse rate during natural contractions when the force is modulated or if the rate is altered in steps as successive motor units are recruited. It seems that both types of responses do occur. Fig. 1A shows the discharge of a unit in the finger flexor muscles during an isometric contraction when the subject was pressing his finger against a strain gauge. The contraction intensity was low; it reached only about 5 per cent of the maximal voluntary force of the receptor-bearing muscle and the afferent response started at less than 0.5 per cent. It may be seen that the rate changed in steps to attain two separate levels, although the rate declined slightly with time after the initial peak. A similar time course was reported by Houk and Henneman (1967) when they electrically stimulated single motor units. A third and higher level was also seen with this unit when the contraction force was further increased (not illustrated).

The separate levels of discharge are most reasonably ascribed to the activation of different numbers of motor units inserting upon the Golgi tendon organ, possibly one, two, and three motor units. In a recent study similar responses have been demonstrated with reflex contractions in the decerebrate cat when the contractions were weak (Crago et al., 1982). With stronger contractions it was a regular finding in this preparation that the discharge was smoothly graded with muscular force.

Whether the step response of Golgi tendon organs has a functional role in motor control remains an open question. If the interpretation given above is accepted, it follows that a step in impulse rate specifies the exact time when an additional motor unit is recruited. This might be of significance in the design of the motor output. If the autogenetic inhibition is working effectively, the step increase of impulse rate might, by suppressing motor units

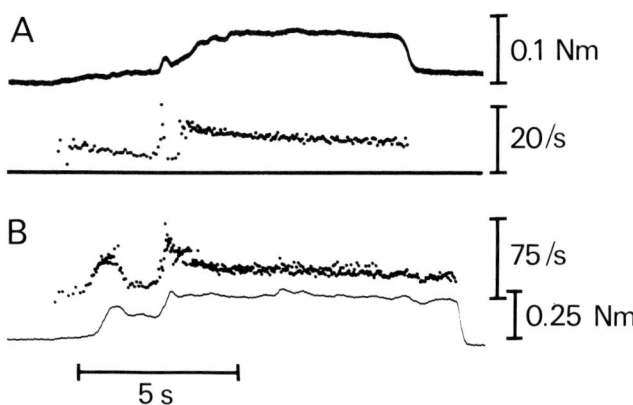

Fig. 1. Contrasting patterns of impulse discharge during isometric voluntary contractions in two different units classified as Golgi tendon organs. Both endings were located on the forearm and in the finger flexor muscles. The subject pressed his finger against a strain gauge and was instructed how to vary the force. The amount of muscular contraction is indicated by the analog signals which give the torque around the meta-carpo-phalangeal joint of the appropriate finger. The contractions in A and B did not exceed 5 and 15 per cent, respectively, of the maximal voluntary force of the receptor-bearing muscle. Nm, Newton-metre.
The unit in A exhibited a step response when the contraction force was smoothly rising, suggesting that individual motor units were recruited when the impulse rate changed. In B the rate changed smoothly with the force signal, providing a fairly accurate description of the total contractile force output of the whole muscle. (From Vallbo, 1970, 1974a).

which are close to firing level, retard their activation so that synchronisation of motor unit recruitment is minimised. This might decrease physiological tremor at low contraction levels (Hagbarth & Young, 1979).

A different pattern is seen in Fig. 1B which illustrates the discharge from another Golgi tendon organ during a weak isometric contraction. The discharge started at a torque below 0.5 per cent of the maximum voluntary force. It may be seen that the rate was then modulated in very close parallelism with the details of force changes. These records demonstrate that there are sense organs which may provide an accurate description of the total force output of the muscle, at least in the low range.

In contrast to their high sensitivity to active force, the responses of these two units to passive stretch were small; only a few impulses were elicited even with very fast stretches (not illustrated). It should be noted that recordings of Group IB afferents in man are limited to isometric and weak contractions, i.e. the range below about 20 per cent of maximal voluntary force. In behaving cats recordings during movements have been published but the forces were not monitored (Prochazka & Wand, 1980).

Thus available recordings from human subjects during voluntary contractions suggest that Golgi tendon organs may provide two kinds of information. One describes accurately the total force output of the muscle. The other type of response monitors the exact moment of recruitment of the individual motor unit at low levels of contraction. Similarly two kinds of responses have recently been reported in reflex contractions of the decerebrate cat (Crago et al., 1982).

MUSCLE SPINDLES

Afferent activity in relaxed muscles
It has been repeatedly emphasised that there are no signs of fusimotor activity to spindles in relaxed muscles of human subjects. This statement is based on the finding that responses grossly like those of de-efferented mammalian spindles are seen when the EMG of the receptor-bearing muscle is quiet and no indication of active force around the joint is seen. As illustrated by the example of Fig. 2 the discharge rate is low (e.g. Vallbo, 1974a) and broadly time-invariant as long as the joint position is kept constant. When ramp stretches are applied the rate increases in the same manner as that of de-efferented spindles. When the test is repeated, almost identical responses appear (not illustrated). However, if the subject does not manage to remain relaxed, the response to ramp stretch is usually markedly altered from one test to another.

The interpretation that fusimotor activity is lacking in

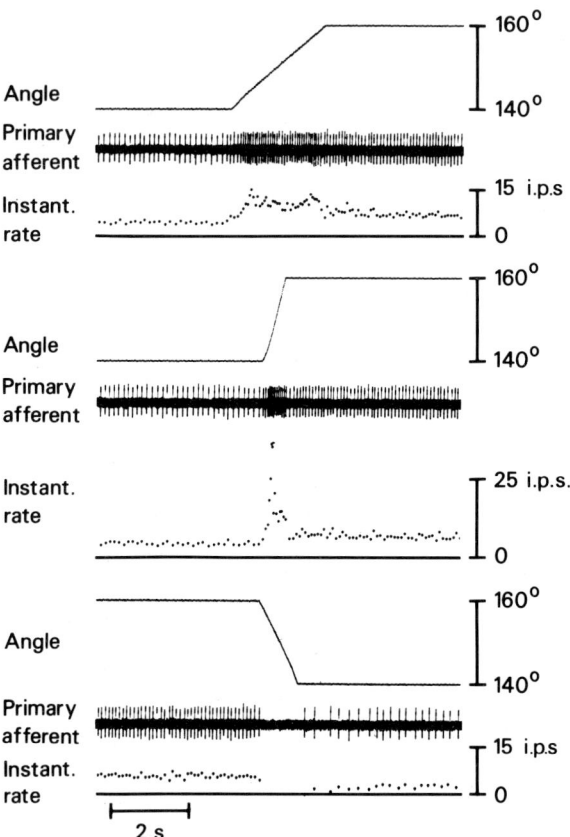

Fig. 2. Muscle spindle response to ramp stretch and release while the receptor-bearing muscle is relaxed. A manipulator moved the four fingers in concert while the subject was instructed to remain passive. EMG and torque recordings (not shown) revealed no indications of muscle activation.
The unit was classified as a muscle spindle primary afferent and originated from the finger flexor muscles on the forearm. The afferent exhibited several characteristics of a muscle spindle primary ending, i.e. initial burst, high dynamic sensitivity, deceleration response at the end of the stretch phase (middle run) and silence during the release of the stretch (lower run). Muscle lengthening is indicated upwards in this and all other records. (Vallbo, unpublished).

relaxed muscles is corroborated by the finding that spindle afferent discharge remains largely unchanged when the nerve is blocked with local anaesthetics proximal to the recording site (Hagbarth et al., 1975; Burke et al., 1976b). Had the spindles been subjected to a substantial fusimotor drive, one would expect a decrease of impulse rate because the block would stop the fusimotor impulses from reaching the spindles.

Whether there is absolutely no fusimotor activity to relaxed muscles can of course not be answered on the basis of these tests. The possibility remains that a minimal gamma drive is present which might contribute to the small and time-dependent variations in interspike intervals that can be demonstrated in statistical tests of spike trains (Nordh et al., 1983).

The lack of a substantial fusimotor drive is interesting in the light of earlier findings with reduced preparations which indicated that the gamma system is readily activated by a number of different stimuli, often prior to the skeletomotor system and at a lower threshold (see Granit, 1970). Such findings suggest that the fusimotor system would be continuously active in the intact preparation and modulated by the numerous reflex mechanisms known to influence the excitability of the fusimotor neurones (e.g. Appelberg et al., 1983a,b,c), even when the muscles are non-contracting. There has been no convincing evidence of such effects in humans, suggesting that reflex effects onto the fusimotor system do not come through until the fusimotor neurones are recruited by synaptic actions from other sources.

Isometric contractions and position holding

When the subject contracts the receptor-bearing muscle there are clear indications of an increased fusimotor drive. Fig. 3 illustrates the response from a unit classified as a muscle spindle primary ending during two isometric contractions. It may be seen that the discharge was elevated during the main part of the contractions and that the stronger contraction to the right was associated with a higher impulse rate. Thus, there was a positive relation between contraction force and impulse rate although it was not very close. Marked deviations are seen particularly at the beginning and end of contraction in the left-hand record. In this respect the contrast to the record from the Golgi tendon organ of Fig. 1B is striking. A crude relation between force and spindle discharge rate was found not only with isometric contractions (Vallbo, 1974b), but also during position holding against a constant load as illustrated in the plot of Fig. 4 (Vallbo & Hulliger, 1982). It may be seen that there is a positive relation between the two variables although the scatter is considerable. It should be noted that this force-rate relation has been explored within the low range of contraction intensities only, up to about 30 per cent of maximal voluntary force.

These findings seem to justify the conclusion that there is a dependence between fusimotor drive and skeletomotor activity. Both are activated simultaneously and an increase in the latter is often associated with an increase in the former. Since this conclusion is based, not on a direct analysis of efferent activity in fusimotor fibres, but on spindle afferent recordings while the muscle length was held constant, it is not possible to differentiate between the separate types of fusimotor neurones (gamma, beta, static, dynamic). However, some inferences in this respect are probably justified. The high irregularity seen in many spindle afferents (Fig. 3) lends some support for the involvement of static fusimotor neurones (Emonet-Dénand et al., 1977). With regard to gamma versus beta activity, there is some evidence from blocking experiments (Burke et al., 1979). When the large fibres in the nerve were blocked with a pressure block so that the muscle was totally paralysed, the subject was still able to produce spindle contraction indicating that voluntary contractions engage gamma motor neurones and not only alpha and beta motor neurones.

Whether the conclusion that a dependence exists between alpha and gamma activity, particularly the static gamma, should be extrapolated to other kinds of motor activities which have not yet been studied is a matter of opinion. The conclusion is based on observations with voluntary isometric contractions without a very clearly defined goal as well as

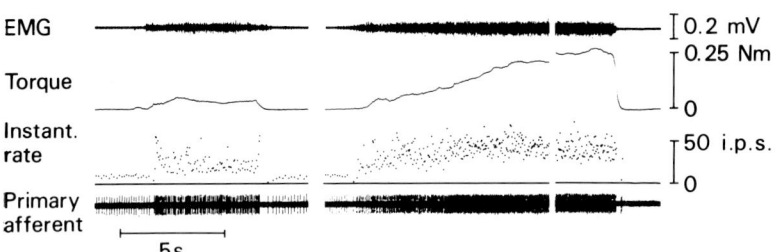

Fig. 3. Muscle spindle responses during isometric contractions. Recording from a finger flexor unit while the subject pressed his finger against a strain gauge and varied the strength of the contraction according to verbal instructions.
A positive relation between impulse rate and torque is suggested by the rate being higher with the stronger contraction to the right, although in contrast to the Golgi tendon organ discharge of Fig. 1B the relation is not very intimate. Same unit as in Fig. 2. The analog signals give the torque at the meta-carpo-phalangeal joint of the appropriate finger. EMG recorded with surface electrodes in this and all other figures. Six seconds are cut out at the gap in the right hand records. Nm, Newton-metre. (Modified from Vallbo, 1981a,b).

position holding with visual feedback. The simplistic view which has been presented here and in several previous articles is a reasonable interpretation of the findings available so far from human studies. Whether it is the final view or not seems immaterial as long as it includes clear and testable statements. On the basis of later studies on lightly anaesthetised or behaving animals, different views have been advanced. In some preparations, the relation between the amount of static gamma activity and muscle contraction seems to be relatively close. This has led to the suggestion that the static gamma motor neurones "provide a template of the intended movement" (Gottlieb & Taylor, 1983). Such an interpretation has been rejected in other investigations (e.g.

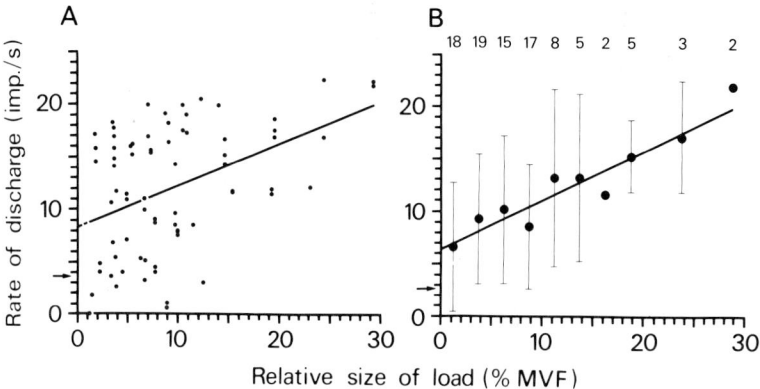

Fig. 4. Relation between spindle discharge rate and size of load during position holding against a constant torque in a visual tracking task. A shows data from individual tasks collected from 7 finger extensor units classified as muscle spindle primary afferents. B gives the impulse rate averaged for loads of comparable size [class width 2.5 per cent of maximal voluntary contraction force (MVF); the bars indicate ±1 S.D. from the mean]. Data from 11 finger extensor units including the 7 units of A. Figures at the top of the diagram give number of tasks behind each point. The horizontal arrows at the ordinates indicate level of discharge in relaxed muscles, averaged for the sample of units represented in the diagram. Straight lines are linear regression lines fitted to the pooled data of each diagram. Their slopes were 0.40 and 0.47 i.p.s./%MVF in A and B, respectively. The impulse rate was measured over 5s, while the load is expressed as a percentage of the MVF.
The diagrams illustrate the positive relation between impulse rate and force which is often found in the low range of contraction intensities. (From Vallbo & Hulliger, 1982).

Prochazka & Wand, 1981). With regard to dynamic fusimotor activity, on the other hand, there seems to be some agreement that it is not well related to muscular contraction (e.g. Gottlieb & Taylor, 1983; Hulliger & Prochazka, 1983).

Active movements

About 15 years ago, examples of proprioceptive activity in human nerves during voluntary movements and position holding as opposed to isometric contractions were published (Hagbarth & Vallbo, 1968), and further material has been presented in

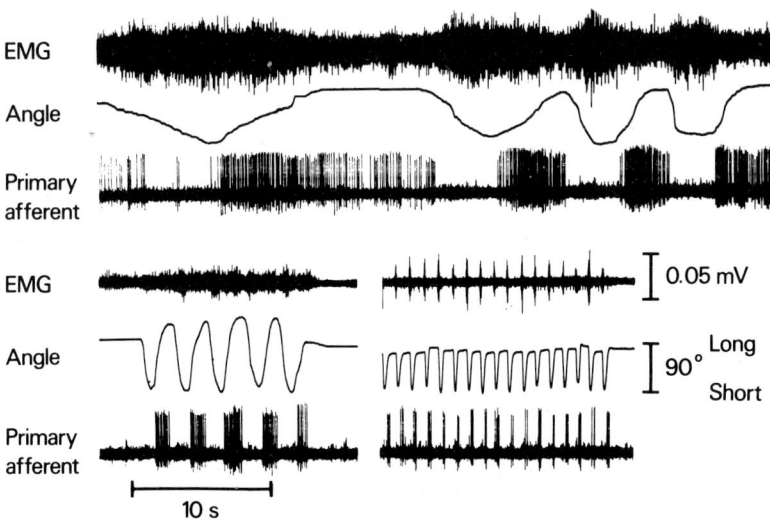

Fig. 5. Discharge from a muscle spindle responding mainly during the lengthening phases of voluntary movements. The unit was classified as a primary afferent and originated from the flexor digitorum muscle. Lengthening movements (extensions) are indicated upwards and shortening movements downwards in these and all other records. The subjects performed, on instruction, finger movements engaging all the three joints of the appropriate finger. Amplitude of movement is, for the sake of simplicity, given in degrees of metacarpophalangeal joint excursion corresponding to the change in muscle length when all joints were moved. Amplitudes were approximated on the basis of estimates of joint radii. With the contractions at the top the muscles were working against a small weight, corresponding to less than 10 per cent of the maximal voluntary force, whereas the faster movements in the lower records were unloaded. Minimal and maximal speeds during shortening were about 0.05 and 4.5 muscle belly resting lengths per second. (Vallbo, unpublished).

later publications (Vallbo, 1973, 1981a,b; Burke, Hagbarth & Löfstedt, 1978; Hulliger & Vallbo, 1979; Roll & Vedel, 1982). However, more studies of a systematic nature are definitely needed, because a clear picture of spindle afferent discharge during voluntary movements in man has not yet emerged.

There seems to be a considerable variation between individual spindle afferents with regard to their response patterns during active movements (Vallbo, 1973, 1981b; Burke, Hagbarth & Löfstedt, 1978). Two contrasting examples are shown in Figs. 5 and 6. The endings of the two units were located in the finger flexor muscles. They were classified as muscle spindle primary, rather than secondary, afferents on the basis of their high dynamic sensitivity and high variability of discharge rate which were qualitatively estimated. For the unit of Fig. 6, these two features are illustrated in Figs. 2 and 3.

The unit of Fig. 5 responded almost exclusively during lengthening movements when the antagonists were stretching the receptor-bearing muscle whereas there was almost no discharge while the muscle was actively shortening. This pattern was seen with very slow movements as well as with somewhat faster movements. Also when the finger was passively moved by the experimenter while the subject was instructed to remain relaxed, the unit responded vigorously during phases of muscle stretch (not illustrated). This suggested that the spindle was subjected to only minimal fusimotor drive or none at all during the active movements. However, there was no doubt that

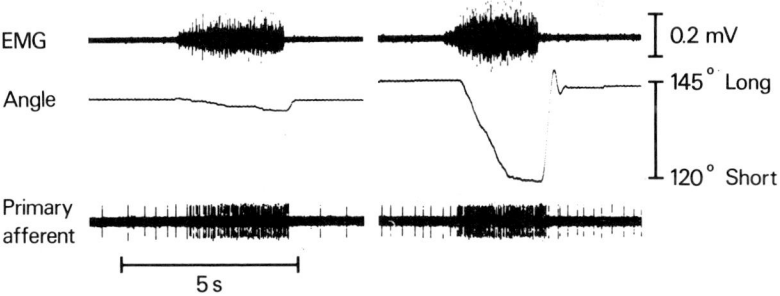

Fig. 6. Discharge from a muscle spindle exhibiting responses mainly during shortening phases of voluntary movements. Same unit as in Figs. 2 and 3. The movements were performed with the metacarpophalangeal joint exclusively, and against a small load, corresponding to less than 3 per cent of the maximal voluntary force. Minimal and maximal speeds of shortening were about 0.05 and 1.0 muscle belly resting lengths per second. (Modified from Vallbo, 1981a).

Proprioceptive activity from human finger muscles

this unit was within reach of fusimotor activity because it responded readily during isometric contractions (not illustrated).

The unit of Fig. 6, on the other hand, exhibited an opposite response pattern in that its impulse rate was fairly high when the receptor-bearing muscle was shortening and moving the joint. However, it was not particularly active during phases of active lengthening movements, regardless of whether these were very slow or of moderate speed. This was not due to an inherent lack of length sensitivity, because it responded readily to passive movements produced by a manipulator if the muscle was given a short resting period beforehand, as illustrated in Fig. 2. Thus, there seemed to

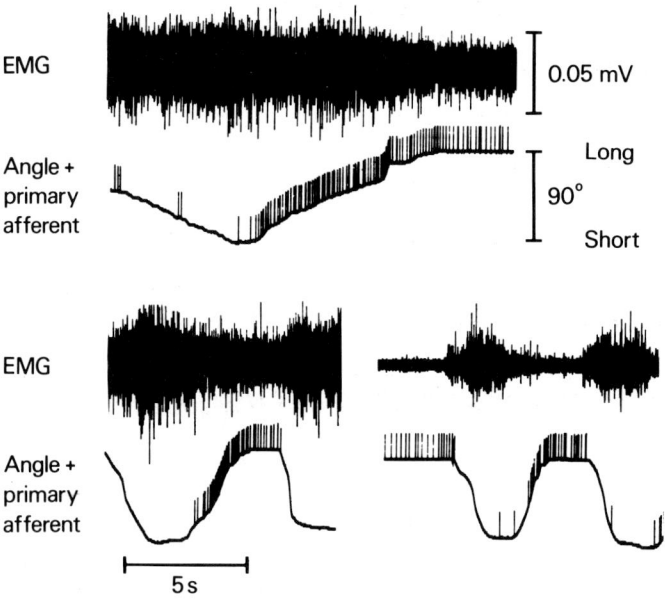

Fig. 7. Responses of a muscle spindle afferent to variations in speed of lengthening movements. The lower records show time course of voluntary finger movements with superimposed standard pulses triggered by nerve impulses to highlight the relation between impulse discharge and details of movements. A small load, less than 10 per cent of maximal voluntary force, was opposing finger movements in the top records and at lower left, whereas the contraction at lower right was unloaded. Same unit as in Fig. 5 and partly same tests. Movement recordings and calibration as in Fig. 5. (Vallbo, unpublished).

Proprioceptive activity from human finger muscles

be, in the same muscle, some spindle endings which respond during lengthening and others which respond during shortening when the subject performed slow voluntary movements. Hence, all phases of movement would be covered by the population of spindles in the muscle.

Although the two units of Figs. 5 and 6 exhibited different discharge patterns whose significance in motor control is not immediately obvious, there is one common feature in their responses. Figs. 7 and 8 were constructed to illustrate that both units were highly sensitive to small irregularities of movement. It may be seen that the unit of Fig. 7 which was active during lengthening movements fired at a higher rate the faster the movements, whereas it often paused when the movement was slower. With the unit which was active during shortening movements, the rate was higher the slower the shortening movements and vice versa (Fig. 8).

A lengthening of the muscle may be regarded as a movement in the positive direction and a shortening as a movement in the negative direction. If the velocities are defined accordingly, it follows that a change of muscular velocity in the positive direction resulted in an increase of the impulse rate with both units, whereas a change in the negative direction resulted in a fall with both, although one of them covered the negative range of velocities and the other covered the positive range.

Figs. 7 and 8 illustrate that the modulations in impulse rate which occurred during slow movements may be quite pronounced. It is interesting to compare them to responses elicited by external disturbances which gave rise to substantial reflex contractions of the receptor-bearing muscle. An example is illustrated in Fig. 9, which shows the

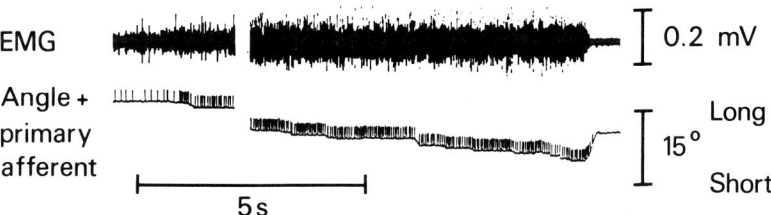

Fig. 8. Responses of a muscle spindle afferent to variations in speed of shortening movements. The muscle was working against a small load corresponding to less than 10 per cent of maximal voluntary force. Same unit as in Figs. 2, 3 and 6. Display of movement and nerve impulses as in Fig. 7. At gap in records, 6.5s is cut out. (Modified from Vallbo, 1981a).

discharge from the particular afferent of Fig. 7 as well as the EMG activity of the parent muscle. In these tests, externally applied force pulses were delivered which tended to stretch the contracting muscle. It may be seen that small stretching of about 15 degrees, which elicited only 4 afferent impulses, were associated with large reflexes as seen in the EMG records. In these tests, the subject was asked to resist the movement while the experimenter applied a force on the finger tip.

DISCUSSION

Recordings from human spindle afferents in non-isometric contractions have largely been concerned with position holding and slow movements of small amplitudes. Further, the opposing forces have been fairly low. Hence, with regard to dynamic and mechanical parameters, only a limited range of the full motor repertoire has been tested. However, the data collected seem to be of particular interest in relation to precision movements which are often performed within this range. The proprioceptors may have a prominent role in the fine control of such movements and in exact position holding, whereas learned movements of large amplitudes seem to be less dependent on proprioceptive input (Bizzi et al., 1978; Rothwell et al., 1982; Sanes & Evarts, 1984). Moreover, the fact that spindle afferents exert particularly powerful effects on the small motor neurones (Eccles et al., 1957; Burke, Rymer & Walsh, 1976) which dominate in weak

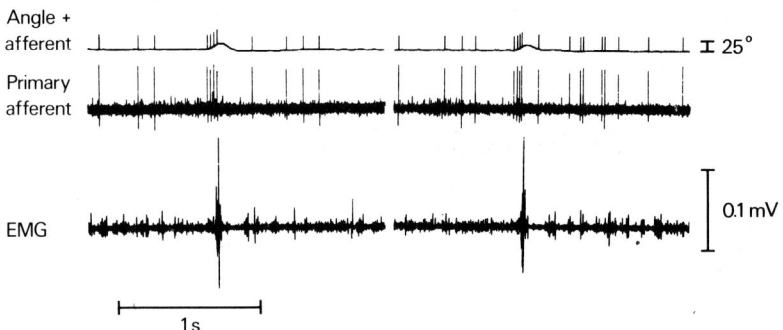

Fig. 9. Responses of a spindle afferent from the finger flexor muscles to imposed movements while the subject was instructed to resist. Force pulses in the direction of stretching the receptor-bearing muscle were applied on the finger tip. The second trace shows movement and time of occurrence of nerve pulses as in Figs. 7 and 8. Same unit as in Figs. 5 and 7. (Vallbo, unpublished).

contractions, suggests that muscle spindles have an important role in the control of such contractions.

Examples have been presented of two kinds of spindle responses, occurring during slow voluntary movements performed as a result of a balanced co-contraction of agonists and antagonists. Some afferents were active mainly while the muscle was being stretched by the antagonists whereas others were discharging mainly while the muscle was shortening. The basis of these two kinds of responses has not been clarified, but the findings may be interpreted in terms of different amounts of fusimotor drive to the individual spindle. The unit responding during lengthening but not when the muscle was actively shortening (Fig. 5 and 7) would be subjected to insufficient fusimotor drive when the muscle was contracting to compensate for the unloading effect of the shortening movement. The response of the other unit (Figs. 6 and 8) would be accounted for by a substantial increase of fusimotor drive while the receptor-bearing muscle was actively shortening. The intrafusal contraction would more than compensate for the mechanical unloading and keep the impulse rate above resting level. At the same time the receptor was retaining a fairly high dynamic sensitivity. This interpretation fits well with the demonstration by Burke, Hagbarth & Skuse (1978) that separate spindle receptors have different threshold sensitivities during active contraction, i.e. some spindle afferents increase their firing rate at minimal levels of muscle contraction, whereas others exhibit an accelerated firing rate only when the contraction reaches higher levels under isometric conditions.

In the execution and control of voluntary movements it seems that units with the two response patterns may cooperate to achieve the desired performance. It is reasonable to assume that spindles with lengthening responses and spindles with shortening responses are present in flexors as well as extensors. The afferents from mutually antagonistic muscles with opposite responses in this respect would constitute a pair which, through the excitatory effects of the afferents on the homonymous motor neurones, would exert a similar effect in the control of speed of joint movement. Consider a slow flexion of a finger. Some variations in speed of movement usually occur even if the subject is asked to perform a smooth movement. Two sets of spindles would respond to these variations, i.e. the spindles in the flexors which are discharging during shortening and the spindles in the extensors which are discharging during lengthening of the parent muscle. When the speed of flexion increases above the desired speed the flexor spindles would fire at a lower rate and withdraw some excitation from the "accelerating" motor neurones, whereas simultaneously, the extensor spindles would fire at a higher rate and increase the excitation of the "braking" motor neurones. Hence, these two sets of spindles would cooperate because they would simultaneously produce opposite effects on the accelerator and the brake of the movement controlling system and both effects would tend to reduce variations in speed of movement. During extensions of the finger two other sets of spindles would exert corresponding effects. It is possible that reflex effects via the IA inhibitory interneurones (Eccles & Lundberg, 1958; Hultborn et al., 1976) may be added to the direct effects on the motor neurones and therefore contribute to the elimination of tremor and irregularities of movements.

These two kinds of responses have been seen during slow movements with small loads. It seems that the control effects discussed above would be particularly beneficial under these conditions when the delay in the reflex loop would be acceptable in relation to the speed of movements (Rack, 1981).

Spindle and muscular responses to externally applied stretch as illustrated in Fig. 9 suggest that modulations of the impulse rate during voluntary movements were large enough to exert a powerful control effect of the muscular activity. However, several points may complicate an interpretation in this direction. It is not known to what extent the particular afferent shown in Fig. 9 is representative of all the units accounting for the reflex contraction. It may be argued that other spindles in the muscle may respond much more strongly to the stretches which elicited reflex contractions, but we have no observation to support such a presupposition. Moreover, spindles in synergists which were not explored may be more responsive, particularly those in small hand muscles which have strong heteronymous effects in the primate (Clough et al., 1968). Second, it cannot be taken for granted that muscle spindles are the only sense organs of significance for the reflex effects. For instance, skin afferents from the finger may be involved as well (Marsden et al., 1977; Torebjörk et al., 1978; Garnett & Stephens, 1980; Westling & Johansson, 1984). Moreover, the reflex gain in the centre may be set to different values in the two test situations. Still, the findings seem to indicate that the impulse modulations during slow precision movements in afferents from the dynamically sensitive spindle endings are large in relation to the modulations elicited by small external disturbances which give rise to substantial reflex effects. Thus it seems that a basic requisite for a powerful control of precision movements from spindle afferents has been demonstrated, in that it has been shown that there exist spindles which are deeply modulated by small variations in velocity during lengthening as well as during shortening movements.

Acknowledgement
This work was supported by the Swedish Medical Research Council (projects 2075 and 3548) and by Gunvor and Josef Anérs Stiftelse.

REFERENCES

Appelberg, B., Hulliger, M., Johansson, H. & Sojka, P. (1983a). Actions on γ-motoneurones elicited by electrical stimulation of group I muscle afferent fibres in the hind limb of the cat. J. Physiol., 335, 237-253.

Appelberg, B., Hulliger, M., Johansson, H. & Sojka, P. (1983b). Actions on γ-motoneurones elicited by electrical stimulation of group II muscle afferent fibres in the hind limb of the cat. J. Physiol., 335, 255-273.

Appelberg, B., Hulliger, M., Johansson, H. & Sojka, P. (1983c). Actions on γ-motoneurones elicited by electrical stimulation of group III muscle afferent fibres in the hind limb of the cat. J. Physiol., 335, 275-292.

Baliff, L., Fulton, J.F. & Liddell, E.G.T. (1925). Observations on spinal and decerebrate knee-jerks with special reference to their inhibition by single break-shocks. Proc. R. Soc. B, 98, 589-607.

Bizzi, E., Dev, P., Morasso, P. & Polit, A. (1978). Effect of load disturbances during centrally initiated movement. J. Neurophysiol., 41, 542-556.

Burg, D., Szumski, A.J., Struppler, A. & Velho, F. (1973). Afferent and efferent activation of human muscle receptors involved in reflex and voluntary contraction. Expl Neurol., 41, 754-768.

Burke, D., Hagbarth, K.-E. & Löfstedt, L. (1978). Muscle spindle activity in man during shortening and lengthening contractions. J. Physiol., 277, 131-142.

Burke, D., Hagbarth, K.-E., Löfstedt, L. & Wallin, B.G. (1976a). The response of human muscle spindle endings to vibration of non-contracting muscles. J. Physiol., 261, 673-693.

Burke, D., Hagbarth, K.-E., Löfstedt, L. & Wallin, B.G. (1976b). The responses of human muscle spindle endings to vibration during isometric contraction. J. Physiol., 261, 695-711.

Burke, D., Hagbarth, K.-E. & Skuse, N.F. (1978). Recruitment order of human spindle endings in isometric voluntary contractions. J. Physiol., 285, 101-112.

Burke, D., Hagbarth, K.-E. & Skuse, N.F. (1979). Voluntary activation of spindle endings in human muscles temporarily paralysed by nerve pressure. J. Physiol., 287, 329-336.

Burke, R.E., Rymer, W.Z. & Walsh, J.V. Jr. (1976). Relative strength of synaptic input from short latency pathways to motor units of defined type in cat medial gastrocnemius. J. Neurophysiol., 39, 447-458.

Clough, J.F.M., Kernell, D. & Phillips, C.G. (1968). The distribution of monosynaptic excitation from the pyramidal tract and from primary spindle afferents to motoneurones of the baboon's hand and forearm. J. Physiol., 198, 145-166.

Cody, F.W.J. & Taylor, A. (1973). The behaviour of spindles in the jaw-closing muscles during eating and drinking in the cat. J. Physiol., 231, 49-50P.

Crago, P.E., Houk, J.C. & Rymer, W.Z. (1982). Sampling of total muscle force by tendon organs. J. Neurophysiol., 47, 1069-1083.

Critchlow, V. & von Euler, C. (1963). Intercostal muscle spindle activity and its motor control. J. Physiol., 168, 820-847.

Davey, M.R. & Taylor, A. (1967). The activity of jaw muscle spindles recorded together with active jaw movements in the cat recovering from anaesthesia. J. Physiol., 190, 8-9P.

Eccles, J.C., Eccles, R.M. & Lundberg, A. (1957). The convergence of monosynaptic excitatory afferents on to many different species of alpha motoneurones. J. Physiol., 137, 22-50.

Eccles, R.M. & Lundberg, A. (1958). Integrative pattern of Ia synaptic actions on motoneurones of hip and knee muscles. J. Physiol., 144, 271-298.

Emonet-Dénand, F., Laporte, Y., Matthews, P.B.C. & Petit, J. (1977). On the subdivision of static and dynamic fusimotor actions on the primary ending of the cat muscle spindle. J. Physiol., 268, 827-861.

Fulton, J.F. & Pi-Suner, J. (1928). A note concerning the probable function of various afferent end-organs in skeletal muscle. Am. J. Physiol., 83, 554-562.

Garnett, R. & Stephens, J.A. (1980). The reflex responses of single motor units in human first dorsal interosseus muscle following cutaneous afferent stimulation. J. Physiol., 303, 351-364.

Gottlieb, S. & Taylor, A. (1983). Interpretation of fusimotor activity in cat masseter nerve during reflex jaw movements. J. Physiol., 345, 423-438.

Granit, R. (1970). The Basis of Motor Control. Academic Press, London.

Hagbarth, K.-E. (1981). Fusimotor and stretch reflex functions studied in recordings from muscle spindle afferents in man. In Muscle Receptors and Movement. Eds. Taylor, A. & Prochazka, A. Macmillan, London. pp. 277-285.

Hagbarth, K.-E. & Vallbo, A.B. (1968). Discharge characteristics of human muscle afferents during muscle stretch and contraction. Expl Neurol., 22, 674-694.

Hagbarth, K.-E., Wallin, G. & Löfstedt, L. (1975). Muscle spindle activity in man during voluntary fast alternating movements. J. Neurol. Neurosurg. Psychiat., 38, 625-635.

Hagbarth, K.-E. & Young, R.R. (1979). Participation of the stretch reflex in human physiological tremor. Brain, 102, 509-526.

Houk, J.C. & Henneman, E. (1967). Responses of Golgi tendon organs to active contractions of the soleus muscle of the cat. J. Neurophysiol., 30, 466-489.

Hulliger, M. & Prochazka, A. (1983). A new simulation method to deduce fusimotor activity from afferent discharge recorded in freely moving cats. J. Neurosci. Methods, 8, 197-204.

Hulliger, M. & Vallbo, A.B. (1979). The responses of muscle spindle afferents during voluntary tracking movements in man. Load-dependent servo assistance? Brain Res., 166, 401-444.

Hultborn, H., Illert, M. & Santini, M. (1976). Convergence on interneurones mediating the reciprocal Ia inhibition of motoneurones. Disynaptic Ia inhibition of Ia inhibitory interneurones. Acta physiol. scand., 96, 193-201.

Jackson, J.H. (1932). Selected writings of John Hughlings Jackson. Vol. II. Ed. Taylor, J.. Hodder & Stoughton, London.

Jansen, J.K.S. & Rudjord, T. (1964). On the silent period and Golgi tendon organs of the soleus muscle of the cat. Acta physiol. scand., 62, 364-379.

Loeb, G.E. (1976). Somaesthetic unit activity during normal walking in the cat. Soc. Neurosci. Abstr., 6, 545.

Marsden, C.D., Merton, P.A. & Morton, H.B. (1977). The sensory mechanism of servo action in human muscle. J. Physiol., 265, 521-535.
Matsunami, K. & Kubota, K. (1972). Muscle afferents of trigeminal mesencephalic tract nucleus and mastication in chronic monkeys. Jap. J. Physiol., 22, 545-555.
Matthews, B.H.C. (1933). Nerve endings in mammalian muscle. J. Physiol., 78, 1-53.
Merton, P.A. (1951). The silent period in a muscle of the human hand. J. Physiol., 114, 183-198.
Nordh, E., Hulliger, M. & Vallbo, A.B. (1983). The variability of inter-spike interval of human spindle afferents from relaxed muscles. Brain Res., 271, 89-99.
Prochazka, A. & Wand, P. (1980). Tendon organ discharge during voluntary movements in cats. J. Physiol., 303, 385-390.
Prochazka, A. & Wand, P. (1981). Independence of fusimotor and skeletomotor systems during voluntary movements. In Muscle Receptors and Movement. Eds. Taylor, A. & Prochazka, A.. Macmillan, London. pp. 229-243.
Prochazka, A., Westerman, R.A. & Ziccone, S.P. (1975). Spindle units recorded during voluntary hindlimb movements in the cat. Proc. Aust. Physiol. Pharmacol. Soc., 6, 101-102.
Rack, P.M.H. (1981). Limitations of somatosensory feedback in control of posture and movement. In Handbook of Physiology, Section 1, The Nervous System, Vol. 2, Motor Control. Ed. Brooks, V.B. American Physiological Society, New York, pp. 229-256.
Roll, J.P. & Vedel, J.P. (1982). Kinaesthetic role of muscle afferents in man, studied by tendon vibration and microneurography. Expl. Brain Res., 47, 117-190.
Rothwell, J.C., Traub, M.M., Day, B.L., Obeso, J.A., Thomas, P.K. & Marsden, C.D. (1982). Manual motor performance in a deafferented man. Brain, 105, 515-542.
Rymer, W.Z., Houk, J.C. & Crago, P.E. (1979). Mechanisms of the clasp-knife reflex studied in an animal model. Expl. Brain Res., 37, 93-113.
Sanes, J.N. & Evarts, E.V. (1984). Motor psychophysics. Hum. Neurobiol., 2, 217-225.
Schieber, M.H. & Thach, W.T. (1980). Alpha-gamma dissociation during slow tracking movements of the monkey's wrist: preliminary evidence from spinal ganglion recording. Brain Res., 202, 213-216.
Shik, M.L., Orlovskii, G.N. & Severin, F.V. (1968). Locomotion of the mesencephalic cat elicited by stimulation of the pyramids. Biophysics, 13, 143-152.
Torebjörk, H.E., Hagbarth, K.-E. & Eklund, G. (1978). Tonic finger flexion reflex induced by vibratory activation of digital mechanoreceptors. In Active Touch. Ed. Gordon, G.. Pergamon, Oxford. pp. 197-203.
Vallbo, A.B. (1970). Slowly adapting muscle receptors in man. Acta physiol. scand., 78, 315-333.
Vallbo, A.B. (1971). Muscle spindle response at the onset of isometric voluntary contractions in man. Time difference between fusimotor and skeletomotor effects. J. Physiol., 318, 405-431.
Vallbo, A.B. (1973). Muscle spindle afferent discharge from resting and contracting muscles in normal human subjects. In New Developments in Electromyography and Clinical Neurophysiology, Vol.3. Ed. Desmedt, J.E.

Karger, Basel. pp. 251-262.

Vallbo, A.B. (1974a). Afferent discharge from human muscle spindles in non-contracting muscles. Steady state impulse frequency as a function of joint angle. Acta physiol. scand., 90, 303-318.

Vallbo, A.B. (1974b). Human muscle spindle discharge during isometric voluntary contractions. Amplitude relations between spindle frequency and torque. Acta physiol. scand., 90, 319-336.

Vallbo, A.B. (1981a). Basic patterns of muscle spindle discharge in man. In Muscle Receptors and Movement. Eds. Taylor, A. & Prochazka, A.. Macmillan, London. pp. 263-275.

Vallbo, A.B. (1981b). A critique of the papers by Loeb and Hoffer; and Prochazka and Wand. Differences in muscle spindle discharge during natural movements in cat and man. In Muscle Receptors and Movement. Eds. Taylor, A. & Prochazka, A. Macmillan, London. pp. 249-255.

Vallbo, A.B. & Hagbarth, K.-E. (1967). Impulses recorded with microelectrodes in human muscle nerves during stimulation of mechanoreceptors and voluntary contractions. Electroenceph. clin. Neurophysiol., 23, 392.

Vallbo, A.B. & Hagbarth, K.-E. (1968). Activity from skin mechanoreceptors recorded percutaneously in awake human subjects. Expl Neurol., 21, 270-289.

Vallbo, A.B., Hagbarth, K.-E., Torebjörk, H.E. & Wallin, B.G. (1979). Somatosensory, proprioceptive and sympathetic activity in human peripheral nerves. Physiol. Rev., 59, 919-957.

Vallbo, A.B. & Hulliger, M. (1982). The dependence of discharge rate of spindle afferent units on the size of the load during isotonic position holding in man. Brain Res., 237, 297-307.

Westling, G. & Johansson, R.S. (1984). Factors influencing the force control during precision grip. Expl. Brain Res., 53, 277-284.

Chapter Twenty-seven

HUMAN LONG-LATENCY STRETCH REFLEXES - A NEW ROLE FOR THE SECONDARY ENDING OF THE MUSCLE SPINDLE?

P.B.C. Matthews

The stretch reflex in which the stretched muscle pulls back so as to resist a disturbing force is such a widespread response in the animal kingdom that it might be thought to be worked out as an object for study. The neural circuitry responsible for such an apparently simple example of negative feedback might well have been expected to have achieved an optimum configuration early in evolution and remained relatively constant thereafter. At least in higher mammals it might be expected to be invariant. Indeed it is still sometimes taught that this is so, and that all there is to the reflex is excitation of the group Ia afferents from the primary ending of the muscle spindle leading to monosynaptic activation of the motor neurones of the stretched muscle and its close synergists. Unfortunately, things have proved to be a great deal more complex, and this circuit is now widely recognized as providing only a part of the mechanism of the stretch reflex. Debate continues, with no obvious end in sight, as to how much additional neural circuitry is involved and what functional advantages accrue to the body by its introduction (cf. Houk & Rymer, 1981).

The particular matter that I shall be concerned with is the later components of response to a continuing stretch, following on from the initial monosynaptically evoked response, the so-called long-latency stretch reflex. It can, of course, be suggested that there is nothing special here to get excited about, since as long as the stretch continues so also will the group Ia excitation and with it an enhanced monosynaptically-evoked motor discharge. Nobody now doubts that this contributes at least part of the response. The question is whether the additional time available from the beginning of the stretch has allowed the initial afferent volleys to reach the motor neurones by more circuitous routes as well and thus excite them yet further. The most radical hypothesis, which has been favoured and expounded by many in the last decade, is that in man the motor cortex itself has become involved to provide a major transcortical contribution to the stretch reflex, as illustrated in Fig.1. On the basis of my own recent work on the problem, which has already been published in detail (Matthews, 1984a,b,c), I hope to persuade you of two main things. First, that in spite of some recent doubts there is indeed a special long-latency stretch reflex for at least some human muscles, well in advance of any 'voluntary' response. Second, that such a reflex depends, at least in part, on the group II afferents from the secondary endings of the muscle spindles in the stretched muscle producing a spinal reflex excitation of its motor neurones, with the delay occurring because of the relatively slow conduction velocity of these afferents rather than because of involvement of the cortex. In many situations these afferents discharge as vigorously as the group Ia afferents. Whether this leaves any room for a transcortical stretch reflex as well must be a matter for debate, but I do not want to take issue on that now since there seems to be insufficient evidence to decide the matter either way.

Historical survey
Before defending my propositions it seems appropriate to survey the growth and recent modifications of the transcortical hypothesis. Among other things this may help expose the frailty of much of the evidence which has been adduced in favour of the cortex being uniquely responsible for the later components of the stretch reflex and so help

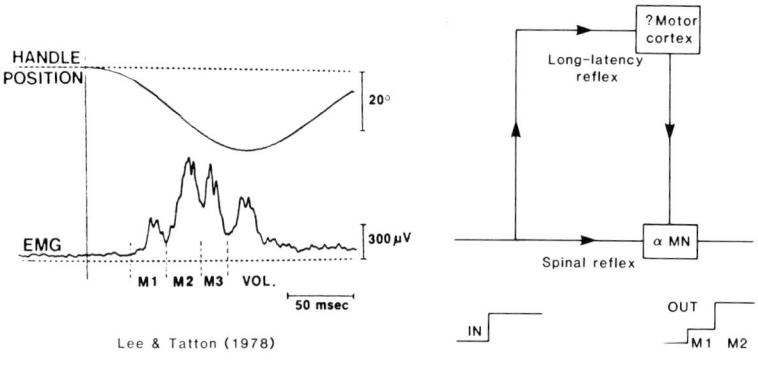

Fig. 1. Left, an example of the complex electromyographic response evoked from a muscle on stretching it by displacing the joint at which it acts. This example is for the wrist extensors. VOL., voluntary. Right, schema illustrating how the later M2/M3 wave might arise from the activation of a 'long-loop' reflex involving the cortex.

open the way for the present alternative. Marsden, Rothwell and Day (1983) have recently reviewed the matter from a diametrically opposed point of view and as they provided a rather full bibliography a minimum of references will be provided now.

The story starts in 1960 with Hammond who performed an experiment which has been reproduced almost <u>ad</u> <u>infinitum</u> by his successors, but who have been far less cautious in their interpretation of very much the same findings. What Hammond did was to ask a human subject to flex his elbow against a rigid stop to achieve a preset target force and then at an unexpected moment the arm was rapidly forced into extension, thereby stretching the already contracting flexor muscles. What Hammond found was that the initial electromyographic response from biceps, at the time expected for group Ia monosynaptic action, was surprisingly modest and without appreciable mechanical action. The main EMG response, followed by a rapid build up of contractile force, occurred with about twice the latency of the initial response. What Hammond concluded was simply that 'The long delay of this activity as compared with monosynaptic transmission time suggests either that it is carried by slower afferents, or that it takes a longer route in the central nervous system'. The former idea, which I am now resurrecting, has since been almost entirely neglected in favour of the latter in spite of a very considerable dearth of evidence bearing on the problem. This happened, it would appear, largely because of the successful development of recording from the cortex, especially in conscious animals performing a preset 'voluntary' motor task. These showed that muscle afferent activity does reach the cortex, and with a surprisingly short latency. Once there, it influences the firing of pyramidal tract neurones with great rapidity. Thus there is little doubt of the existence of the neural components of the circuitry required for the mediation of a 'transcortical stretch reflex'. But whether the circuits are actually organised to achieve this particular action is quite another matter. Among other things, the gain of the system might be low or activity might be distributed inappropriately to flexors and extensors, so animal work cannot be taken to resolve the human situation.

A major contribution to the development of thought about a possible transcortical routing of the stretch reflex was provided by Phillips in 1969 in an influential review. In relation to work on monkeys he specifically raised the question, in far clearer terms than had Hammond, and asked 'Is the CM (corticomotoneuronal) projection itself the efferent limb of a transcortical spindle circuit?' In favour of the existence of such 'a transcortical servo-loop which could dominate the segmental servo loop' he made two cogent points. First, in experiments on the baboon he had found that the overall effect produced monosynaptically on muscles of the hand was greater on activation of pyramidal tract fibres than it was for the appropriate group Ia afferents, especially when repetitive activity was considered. Second, he extended to the primate the finding already known for the cat that excitation of group Ia afferents evoked a short-latency activation of neurones in sensory area 3a adjacent to the motor cortex. However, when this latter sensory work was written up in full only very shortly afterwards (Phillips et al., 1971) the transcortical servo hypothesis was explicitly denied, since by then they had failed to find any sign of the expected functional connection from neurones in area 3a to those in area 4, the primary motor cortex. Nonetheless most subsequent authors writing in favour of the transcortical hypothesis have continued to quote the earlier paper, often as supporting their own position, while ignoring the more formed opinion of the second paper and its apparently definitive statement that 'The possibility raised in the introduction of a dense three neurone pathway leading from spindle primaries to the baboon's hand area and forming the afferent limb of a transcortical feedback loop (Phillips, 1969), can now be ruled out'. But, of course, this retraction was valid only for a route involving area 3a and was premature in relation to the basic concept of a transcortical reflex, since other pathways could be involved in the transmission of the activity to the pyramidal tract neurones. This indeed appears to be the case, explaining perhaps why so many have drawn a veil over the conclusions of the second paper.

Evarts (1981) and most recently Cheney and Fetz (1984) have now amply documented an early and powerful excitation of pyramidal tract neurones by joint movement and the accompanying muscle stretch. In the monkey this is appropriately timed to contribute to a 'long-latency' response of the muscle to the disturbance. In the cat, however, things have not worked out this way (Tatton et al., 1983). Although the cortical pyramidal tract cells are again excited with a short latency, their discharge is now too early for the transcortical hypothesis since it would influence the motor neurones when they are temporarily quiescent following their initial short-latency discharge, and appreciably in advance of their giving what might have been supposed to be an analogous 'long-latency' response. Thus species differences are now coming to the fore. Even for the monkey the problem remains whether the observed muscular responses really represent a 'stretch reflex' as classically understood with some sort of continuous relation between stimulus and response, or whether they sometimes represent some more complex type of 'triggered' pre-programmed activity as suggested for certain human responses (Houk & Rymer, 1981). For example, in the recent exemplary study by Cheney & Fetz (1984) a given muscle regularly showed long-latency excitation irrespective of whether it was being stretched or released, so that flexors and extensors of the wrist normally co-contracted in response to a perturbation in either direction. On the evidence presented these responses would indeed seem to have been mediated via the cortex, but it seems premature to equate them with the classical stretch reflex in which reciprocal activation of antagonistic muscles occurs. Thus, in spite of certain confident pronouncements to be found in the literature, animal work has yet to show a proven cortical stretch reflex in operation.

Attacks on the hypothesis

The transcortical hypothesis has recently come under determined attack from a variety of sources, coupled in most cases with the suggestion that there is no special long-latency reflex requiring an explanation. Perhaps the foremost of these has been that mounted by Hagbarth and his colleagues (Eklund et al., 1982). They began by inserting tungsten electrodes into various nerves in human subjects and recording the pattern of afferent discharge set up by the standard types of stretch. The implicit assumption of many previous workers seems to have been that the group Ia afferents would fire at a more or less constant rate from the beginning of stretch so that any late increase in the reflex response, or the appearance of partially separated bursts of motor activity, could be attributed to the action of the reflex centres. In fact, however, the afferent discharges were almost invariably found to be 'segmented' into separable bursts, each of which was normally followed by a short-latency motor response. The afferent segmentation seemed to be due to the mechanical oscillations which were seen to occur in the muscle even when the mechanical device applying the stretch operated completely smoothly. These oscillations were attributed to muscular resonances arising from its non-linear mechanical properties and considered to be inescapable in all experiments of this type. On this 'resonance' hypothesis there is no such thing as a special long-latency stretch reflex, simply a succession of separate short-latency responses leading to segmented EMG activity similar to that to be expected from the activation by the initial volley of a series of parallel reflex pathways with different reflex delays. But, of course, acceptance of the validity of such a peripheral origin for some of the observed EMG segmentation does not directly attack the transcortical hypothesis per se. All that has been demonstrated is that short-latency reflexes continue in operation at the time of any postulated long-latency response and so superimpose a finely segmented pattern upon it, as can also be demonstrated by vibrating the muscle while it is being stretched (Matthews et al., 1982). In other words, if the long-latency reflex exists it operates in parallel with continuing short-latency responses, and the overall response at the time of its supposed occurrence depends upon both pathways. Lee and Tatton's (1982) observations on the effect of shortening the duration of the stretch would appear to support the view that the long-latency response of the wrist flexors is so dependent on continuing short-latency action that it entirely fails to manifest itself as a separate entity when this is withdrawn, though this is not quite how these authors themselves put the matter. At any rate, probably nobody would now feel that on its own the mere observation of segmentation in a reflex response provides any sort of argument for the occurrence of a long-latency reflex. This is especially so, because various well known spinal mechanisms can be expected to transform even a continuous afferent input into a segmented motor output; notable among these are the rhythmic behaviour of motor neurones acting individually (cf. Matthews, 1984c) and their collective inhibition by Renshaw feedback the moment any appreciable number discharge.

A further objection to the transcortical hypothesis has been provided by the observation that in both monkeys and cats supposed long-latency responses (as shown by segmentation of the reflex output) of muscles acting at the elbow may persist after lesions of the cortex. They could even be seen in spinal animals thus also excluding all long-loop pathways involving higher centres from being responsible for the responses then studied. This conflicted with the pre-existing observations on man that lesions along the supposed transcortical route normally grossly interfered with the late responses of the long flexor of the thumb. It was recognised, of course, that any such abolition did not prove the operation of the transcortical route. The lesion could act equally by altering the activity of some descending pathway which controlled the excitability of any interneurones in the spinal cord involved in mediating the spinal stretch reflex. A subsequent study on muscles acting at the wrist and on the fingers of the monkey agreed with the human finding, and contrasted with the effects in the same animals on biceps, which acts at the elbow, for which somewhat similar late responses persisted after making the cortical lesion (Lenz et. al., 1983). Thus superficially similar responses may behave quite differently and may therefore be presumed to have different origins. Differences may occur not only for different species but also for different muscles of the same species, presumably depending upon their functional role. The late responses of various muscles controlling the primate hand seem to be particularly well developed; of these the human flexor pollicis longus provides a notable example as shown by the work of Marsden, Merton and Morton (1976).

The last stand?
Thus the proponents of the transcortical hypothesis have achieved an effective fighting retreat to the defensible

position that, in spite of a variety of difficulties confusing the issue and invalidating certain types of evidence, genuine transcortical stretch reflexes are still suggested to exist in the situation for which Phillips (1969) had originally proposed them, namely the muscles of the primate hand. This view was succintly expressed by Marsden, Rothwell and Day (1983) as follows 'It is our belief that the transcortical stretch reflex mechanism is a system which is evolved to its greatest extent for the control of the human hand, but which is much less evident, if not rudimentary, in most other parts of the human body.' Thus, the recent attacks on the transcortical hypothesis have indeed had a major impact on the early somewhat overenthusiastic presumption that almost any late component of the response evoked by forcible movement of a joint was likely to be attributable to the cortex. This shift of opinion can be seen, for example, by the gradual replacement in the literature of the term 'long-loop' by the more neutral one 'long-latency' when discussing the experimental observations; even this latter term still

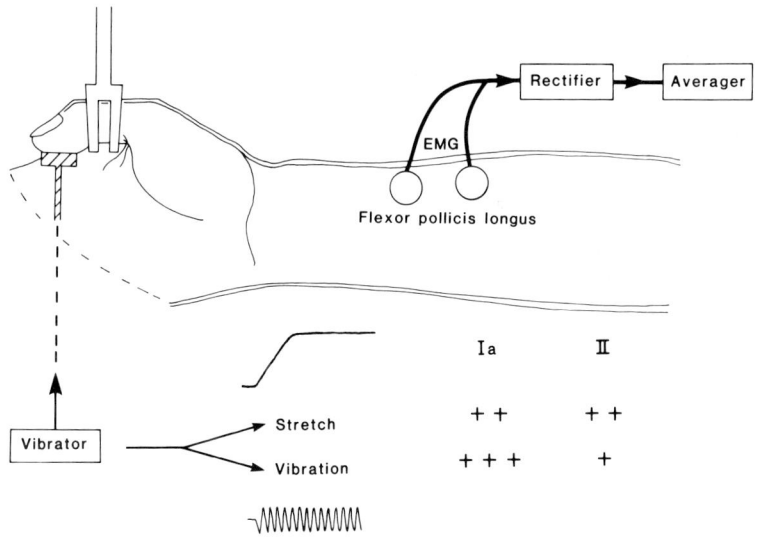

Fig. 2. The experimental arrangements used to study the long flexor of the thumb; this is the sole muscle responsible for flexion of the terminal phalanx. A large vibrator was used both to apply ramp and hold displacements to the thumb of about $10°$ ('stretch') and to move it sinusoidally by a smaller amount at 143 Hz (vibration). The subject maintained a steady force of 6N between stimuli and avoided reacting to them voluntarily. The schema at the bottom shows the relative excitatory effects of these two modes of stimulation upon the two types of spindle afferent.

somewhat begs the issue. But the transcortical hypothesis itself is by no means dead.

Present experiments
These were initiated in the hope of clarifying the situation by using short periods of vibration as the stimulus to reflex action as well as the conventional displacement of a joint with its accompanying gross muscle stretch. Fig. 2 illustrates the experimental arrangement. The subject exerted a force of about 20% maximal between stimuli and avoided making a voluntary response of any kind to them; this was achieved by exerting a constant effort throughout, which proved to be a surprisingly natural thing for even a naive subject to do. The point about using vibration as the stimulus is that it has a powerful excitatory action upon the group Ia afferents from the primary endings of the muscle spindles so that, irrespective of their latency, any reflex actions elicited by these afferents should be evoked by vibration as much as by stretch. In contrast, the secondary endings of the muscle spindles are relatively much less influenced by vibration while still responding strongly to stretch. Thus with vibration their reflex actions will be largely in abeyance, whereas with stretch they will be present in conjunction with

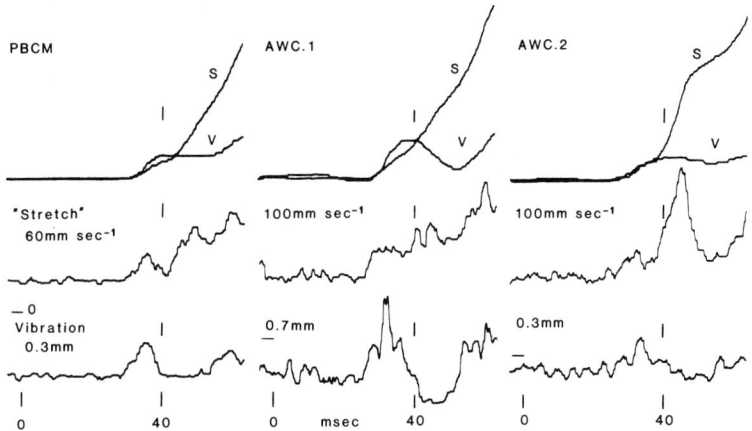

Fig. 3. Responses of the long flexor of the thumb to stretch (middle) and to vibration (bottom) from three separate experiments. The traces show the average value of the rectified EMG, based on 128 or 256 separate applications of each stimulus. Top, the 'cusums' for each pair of records superimposed upon each other to emphasize the differences in the long latency responses evoked by the two types of stimuli; the cusum is the integral of the rectified EMG after subtracting the initial level throughout. S, stretch; V, vibration. (From Matthews, 1984a.)

those from the group Ia afferents; the conventional view would seem to have been that the reflex action of the spindle group II afferents was too weak to affect the issue. In essence, if the long-latency response is due to group Ia action it should be seen equally for vibration as for stretch, but if it is due to group II action it should be very much smaller for vibration.

As illustrated in Fig. 3 the experimental findings have regularly supported the group II hypothesis. The middle and bottom sets of traces show the averaged electromyographic activity (rectified) for stretch and for vibration from three separate experiments. Both types of stimuli began at 0ms on the time scale and both elicit a short latency response at about 30ms. The responses illustrated have been selected from among others for different amplitudes of vibration and velocities of stretching, obtained on the same occasions, so as to match very approximately the sizes of the short-latency responses evoked by the two modes of stimulation; in the right hand records these initial responses were by good fortune virtually identical. In contrast, the pronounced activity evoked over the period from 40 to 50ms from the beginning of stretch simply does not occur for vibration. The top records confirm this difference in the time course of the responses to the two types of stimuli by superimposing their 'cusums', in which any deviation of the rectified EMG from its initial level is summed arithmetically from the beginning of the trace onwards. The yet later response seen at about 60ms with vibration seems attributable to a continued short-latency excitation, since in those cases in which the matter was tested it was abolished (after an appropriately short interval) when the period of vibration was suitably shortened.

Taken at face value such findings seem incompatible with the transcortical hypothesis as an explanation for the stretch response at 40 to 50ms. On the evidence of the similarity of the short-latency responses the two types of stimuli evoked similar initial bursts of group Ia firing. Since these self-same initial discharges are also those responsible for any transcortical reflex, on this hypothesis the later discharges should also have been similar. Since they were not, either some other mechanism must have been involved or some crucial factor has been neglected in reducing the argument to its logical essentials. If, however, the spindle group II afferents were to produce a spinal autogenetic excitation the findings would at once fall into place without any further saving hypothesis. They are undoubtedly excited well by stretch but poorly by vibration, and the delay is that to be expected if their conduction velocity is about half of that of the group Ia afferents as in cat limb muscles.

Possible complications
Vibration inevitably also excites numerous cutaneous afferents

in the pad of the thumb as well as Pacinian corpuscles over a wide area. It might be suggested that one or other of these produced an inhibition which was powerful enough to obscure the excitatory effects of a group Ia transcortical reflex, or even on occasion to over-ride it totally and produce a net inhibition as in the middle experiment of Fig. 3. This possibility was excluded for the cutaneous afferents by two separate controls. First, the vibration response persisted virtually unchanged when the thumb was anaesthetised by a ring block at its base. Second, the usual phasic responses were obtained when the vibration was applied directly to the tendon in the forearm close to its junction with the muscle by a separate small vibrator with a fine tip. Pacinian corpuscles seem an improbable source of such a powerful inhibition since the stretch response was not reduced when the two types of stimuli were combined. It might be suggested none-the-less that the rather phasic response to vibration is still somehow abnormal and not that to be expected from an afferent input which suddenly rises to a new steady level. Theoretical analysis shows, however, that this is just what is to be expected (Matthews, 1984c) with a step increase in afferent activity impinging upon a pool of discharging motor neurones, first because of the rhythmic firing properties of individual motor neurones, and second because of Renshaw feedback. Rather, it is the stretch response with its continued activity after the initial discharge which demands that the pool receive a continually increasing excitation, and one which is probably more than can be reasonably attributed to any progressive increase in group Ia activity acting via the short-latency pathway. Such continued excitation could in principle have been produced by a group Ia transcortical reflex; but it should then also have been seen for vibration, and is thus excluded as producing an appreciable effect under the present conditions.

'Off' responses
However, once the transcortical hypothesis is abandoned and a spinal mechanism sought for the various findings, then the group II hypothesis can no longer be defended as providing a uniquely suitable explanation solely on the basis of the responses at the onset of stimulation. This is because the precise pattern of afferent firing evoked by the stimuli is unknown. Hence it cannot be excluded that the absence of a reflex response at 40-50ms with vibration is simply because the vibration fails to provide an effective continued stimulus to the group Ia afferents, the reflex deficit thus being simply due to a deficit of group Ia short-latency action. The first cycle of vibration might excite the group Ia afferents far more powerfully than any subsequent cycle (i.e. failure of 1:1 driving) so that the overall afferent discharge would then fall, whereas with stretch the afferent discharge might be

suggested to build up progressively. If this were the case, the argument based on matching their initial short-latency responses would be simply irrelevant. However, the group II hypothesis also predicts that equivalent differences between the effects of stretch and vibration should be seen on the termination of a period of stimulation. Again, because of the peripheral conduction delay, any excitatory effect of group II activity should be withdrawn from the motor neurone later than that of group Ia excitation. The requisite late effects are indeed observed for release of stretch but not for termination of vibration (Fig. 4). But now the different effects of the two types of stimuli can no longer be reasonably attributed to aberrations in group Ia afferent firing. Even if vibration were to fail to maintain the afferent discharge at its initial high level throughout, the delayed decline in the response at the cessation of stretch would not be explained.

However, there has been the minor unexpected complication that for flexor pollicis longus the effects of withdrawing the maintained group Ia excitation have been surprisingly slight, and on release of a stretch they have usually been lost in the noise. This is not so for all muscles and may perhaps be due to a particularly high level of autogenetically-elicited presynaptic inhibition of the group Ia afferents. Fig. 4 shows the typical findings. On cessation of vibration there is an obvious short-latency withdrawal of excitation and no obvious second long-latency component of action. On release of stretch there is a powerful long-latency reduction of activity and no apparent short-latency effect, though it has been indubitably observed for flexor pollicis longus on certain other occasions. The latency of this delayed off response was not appreciably altered by changing the velocity of release which, along with other observations, argues that it is not simply due to mechanical lags.

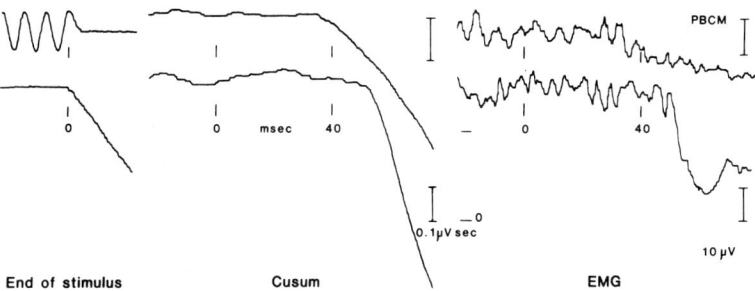

Fig. 4. Effects of terminating stimulation seen in the averaged rectified EMG of the long flexor of the thumb, on release of stretch (lower traces) and on cessation of vibration (upper traces). Details as in Fig. 3.

Neither is the delayed reduction of activity seen on release from stretch attributable to an inhibition, as could be evoked by the concomitant stretch of the antagonist, rather than to a withdrawal of excitation. This was shown by comparing the response to short stretches lasting 15-30ms with that to continued stretch. The reduction in activity on release again occurred with a long latency. During the period between the expected short-latency off effect (as given by stopping vibration) and the observed long-latency off effect, motor activity was enhanced. Such continuing excitation did not occur with short periods of vibration, and is presumed to represent the response to activation of group II afferents, seen on its own by virtue of continuing for a brief period after short-latency excitation had ceased.

Behaviour of a wrist flexor

The application of vibration to the tendon of flexor carpi radialis elicits a similar phasic reflex response to that seen for flexor pollicis longus. Once again this contrasts with the response elicited at the onset of stretch when the evoked electromyographic activity continues to increase well after it has come to an end when elicited by vibration. The difference in the action of the two modes of stimulation is particularly apparent in patients with Parkinson's disease, many of whom show enhanced long-latency (M2) responses to stretch (Lee & Tatton, 1978). As illustrated in Fig. 5 such enhancement is unaccompanied by any sign of a similar long-latency response to vibration (Cody et al., 1984). Two conclusions follow. First, it seems likely that for this wrist flexor, as for the long flexor of the thumb, the long-latency stretch response depends upon the spindle group II input, both in the normal

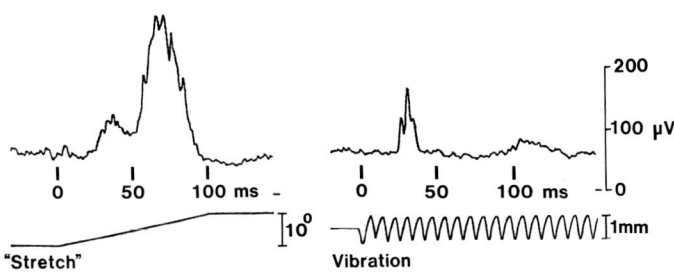

Fig. 5. Responses of flexor carpi radialis to 'stretch' (produced by wrist displacement) and to vibration (applied to its tendon proximal to the wrist) in a patient with Parkinson's disease. (From Cody et al., 1984).

and in Parkinson's disease. Second, the reflex pathways mediating their action would appear to be facilitated in Parkinson's disease, presumably by tonic activity descending from unknown higher centres to the spinal cord. How far this is responsible for the characteristic rigidity of this condition remains to be decided, but on the limited current evidence it cannot be held uniquely responsible.

Behaviour of the short flexor of the thumb

On the transcortical hypothesis the transmutation of the stretch reflex from a simple spinal response to the predominantly long-latency, presumed transcortical, response shown by the long flexor of the thumb might be expected to have occurred also for the other thumb muscles. These must be equally under voluntary control and would thus likewise appear to benefit by the anatomical shifting of the centres mediating their automatic stretch responses from the spinal cord to the motor cortex. Any such encephalisation can be reasonably suggested to help unify the reflex and voluntary controlling mechanisms by locating them at a single site, and so simplify the task of the 'will' in producing and regulating fine voluntary movements. In fact, for the moderate rates of

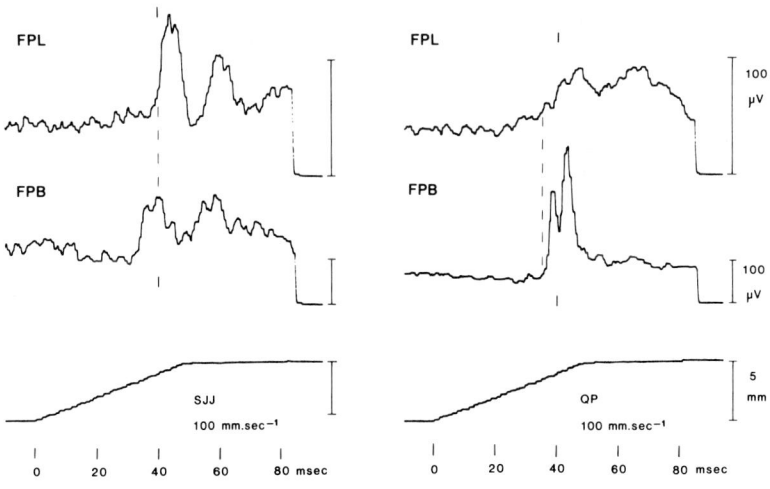

Fig. 6. The contrasting electromyographic responses of the long (FPL) and short (FPB) flexors of the thumb when it is moved in its entirety (i.e. not clamped as in Fig. 2) thereby producing comparable stretching of both muscles. Results from two separate subjects. The short flexor lies about 15cm further away from the spinal cord. (From Matthews, 1984b).

stretch at which the long flexor gives a nearly pure long-latency-response the short flexor commonly still shows a very pronounced short-latency response, as illustrated in Fig. 6 (Matthews, 1984b). In these two examples, from two different subjects, the two muscles respond to one and the same disturbance in near synchrony, in spite of their different separation from the spinal cord. The initial response was a long-latency response for the more proximal long flexor and a short-latency response for the more distal short flexor. The motor discharges responsible for this latter response must have left the spinal cord before the arrival of any putative descending signal from the cortex responsible for the former response, thus rendering impossible any cortical integration of the initial responses of the two muscles, although in mechanical terms they must be co-operating closely. Thus the rationale for the encephalisation of the stretch reflex would appear to be seriously undermined, and at the very least there seems no obvious reason why it should not have occurred similarly for all the muscles of the thumb.

It should be noted, however, that the relative strengths of the long-latency mechanisms of the two thumb flexors cannot as yet be determined. On the basis of findings like those in Fig. 6 it is tempting to believe that it is the stronger for the long flexor, but this is not necessarily so since there is a difference in the potency of their short-latency mechanisms. The absolute amount of motor discharge elicited by a given amount of activity in the long-latency pathway can be expected to vary inversely with the size of any initial short-latency response, since because of refractoriness etc. the motor neurone pool will be less ready to be excited by any subsequent input.

DISCUSSION

The present findings are all explicable on the basis of the single assumption that under the present conditions the spindle group II afferents act within the spinal cord to produce autogenetic excitation. The additional delay, above that for the short-latency pathway, is readily attributable to the slower peripheral conduction of these afferents; in the cat their conduction velocity is about half that of the group Ia afferents. The neural circuitry responsible for their postulated action remains obscure, but as in animal work the autogenetic monosynaptic action of group II afferents is weak so polysynaptic pathways seem likely to be involved. Such pathways have yet to be clearly delimited in the cat, the species for which we have the most detailed knowledge of spinal cord function. But given our present limited understanding of the complexities of interneuronal connections this is hardly surprising, and does not militate seriously against the present hypothesis. It may be that the correct preparation with the correct interneuronal set has yet to be investigated. It should be emphasised that the earlier suggestion (Matthews, 1969) for a similar group II action in the decerebrate cat is an essentially different case, and the disproof of that evidence, as several have claimed, would not disprove the present case, and vice versa.

It might be suggested that since the muscles presently studied in man are all labelled anatomically as 'flexors', the present findings are a simple corollary of the work on spinal and anaesthetised cats in which spindle group II afferents have been suggested to belong to a widespread class of 'flexor reflex afferents' that produce excitation of flexors and inhibition of extensors. It seems unlikely, however, that muscles controlling the human hand would be organised in such a simple manner and it is believed that a more specific reaction is being studied. This view is supported by the finding, in the single subject studied, that when the activity of the short flexor was recorded with the proximal phalanx clamped so that it was not stretched then, although it was tonically active, it was no longer excited on moving the thumb and eliciting the usual long-latency response from the long flexor, which was of course then stretched. In addition, as illustrated in Fig. 1, Lee and Tatton (1978) have recorded long-latency responses from the wrist extensors which they considered to be essentially the same as those of the wrist flexors.

At the present stage of investigation there is no claim that the group II hypothesis provides a unique explanation; the wide variety of human observations can doubtless also be dealt with by a series of somewhat ad hoc suggestions tailored to cover each particular instance. But it provides a sufficient, single, unifying explanation as discussed in greater detail elsewhere (Matthews, 1984a,b,c). It offers also a more rational view of the functional operation of the stretch reflex than the classical view that it receives its entire excitatory input from the group Ia afferents. If instead it depends upon both types of spindle afferent, then their co-operative action would ensure the initial stretch response was of high speed and high sensitivity, as provided by the group Ia contribution, but this would only be maintained to produce an appreciable overall reflex effect when the more sluggish spindle secondary endings signalled that muscle length had actually deviated significantly from its controlled value. In addition, a ready opportunity is provided for the higher centres to control the early and the later parts of the response somewhat independently by adjusting the excitability of the relevant interneurones.

Credibility of Long loop action?
The final question is 'Where does the transcortical hypothesis

now stand?; does it remain credible?' This must remain a matter of opinion, but various points merit discussion.

First, the present experiments support earlier work arguing that there is indeed a genuine, distinct, long-latency component of the stretch reflex which thus offers itself as a candidate for transcortical mediation. The results with release, in particular, contradict the view that all the effects can be attributed to segmentation of the group Ia discharge coupled with inhibitory interactions within the spinal cord.

But, second, the results with vibration make it unreasonable to attribute the excitation seen at 40-50ms from the onset of stretch to group Ia action, thus excluding a group Ia-initiated transcortical reflex as uniquely responsible for the marked excitation that then normally occurs.

Third, has a transcortical reflex been finally excluded as making any contribution to the later responses? Of course not. To begin with, the present analysis has paid rather little attention to effects occurring beyond 50ms from the beginning of the stimulus, when among other things a group II transcortical response begins to become a possibility. Also, even within the period 40-50ms a weak group Ia initiated transcortical response might have been present and served merely to counteract an inhibition which would otherwise have occurred instead of producing an excitation, since no such excitation was seen with vibration.

Fourth, the present work has been primarily concerned with establishing a significant role for the spindle group II afferents. Acceptance of this view carries no implications as to what other neural routes might be also concerned with mediating a given response; on the present hypothesis the response at 40-50ms indubitably depends upon group Ia as well as upon group II activity. Only for the earliest times can the contributions of different pathways that operate in parallel be disentangled by virtue of any differences in their latency.

Finally, to prove one thing is not necessarily to disprove another. And at some stage, especially when a subject is told to respond actively to a stimulus, the cortex and other higher structures can be expected to intervene in the control of a muscle which is being voluntarily contracted. The question then becomes yet more complex, and partly entangled by semantics, when one asks whether the first effects of higher activity should be seen as a servo-type proportional response, a pre-programmed triggered response, or a 'voluntary' response, and how far these can each be influenced by the subject's prior intent.

But if the transcortical hypothesis had not been introduced over a decade ago, on the tacit supposition that the group II afferents could be ignored, then on the basis of the current evidence there would now seem little need to introduce it de novo to explain the human findings. Indeed, it now falls to its proponents to rescue the transcortical hypothesis, whether by demolishing the present conclusion that much that has hitherto been attributed to group Ia transcortical action is better attributed to spinal group II action, or by adducing new evidence that the transcortical loop operates in parallel with at least two separate spinal loops.

SUMMARY

When the length of a muscle which a subject is already contracting voluntarily is changed by forcibly moving the relevant joint it shows a 'stretch reflex' type of response, with activity increasing on stretch and decreasing on release. EMG recordings suggest that the reflex may consist of separate short and long latency components. Both of these are normally attributed to the central action of the group Ia afferents from the spindle primary endings of the muscle studied, but there is continuing disagreement as to the central pathways involved. The favoured hypothesis of the last decade has been that the late response is a 'long loop' reflex, with the group Ia excitatory action routed via the motor cortex. Recently, however, influential voices have claimed that the late response is not a separate reflex entity, but is simply due to the continued short-latency action of a continued group Ia input; any 'segmentation' of the response is suggested to arise adventitiously, perhaps from a similar patterning of the group Ia discharge resulting from intramuscular resonances or from the complex interplay of excitatory and inhibitory mechanisms within the spinal cord.

The experiments described favour a third possibility. They support the view that at least for the human flexor pollicis longus there is indeed a genuine, distinct, long-latency stretch reflex. But the findings seem incompatible with the idea that it is due to the group Ia input acting transcortically. This is because activation of group Ia afferents by short periods of vibration, rather than by stretch, fails to elicit a late response, including when the short-latency response is similar to that elicited by stretch. This new finding suggests that under the present conditions the group II afferents from the spindle secondary endings are acting spinally to exert a powerful autogenetic excitatory action (presumably via interneurones in addition to any monosynaptic effect). The spindle secondary endings are excited much more powerfully by stretch than they are by vibration. On this view the delay of the long-latency stretch response is simply attributable to the slower peripheral conduction of the group II afferents, as compared with the group Ia afferents. In propounding this hypothesis particular attention has been paid to the reflex behaviour of the muscle on its release, since the pattern of afferent discharge then seems more reliably predictable than that occurring on stretch. It should be emphasised that this new suggestion is also entirely compatible with any segmentation of the reflex response arising from a variety of other peripheral and spinal mechanisms. Nor has the possibility of a transcortical contribution been excluded, but the evidence of others for the occurrence of any appreciable action being produced by this route now seems far from compelling.

REFERENCES

Cheney, P.D. & Fetz, E.B. (1984). Corticomotoneuronal cells contribute to long-latency stretch reflexes in the rhesus monkey. J. Physiol., 349, 249-272.
Cody, F.W.J., MacDermott, N., Matthews, P.B.C. & Richardson, H.C. (1984). Electromyographic responses of Parkinsonian patients to wrist extension and to flexor tendon vibration. J. Physiol., 351, 19P.
Eklund, G., Hagbarth, K.E., Hagglund, J.V. & Wallin, E.U. (1982). The 'late' reflex responses to muscle stretch: the 'resonance hypothesis' versus the 'long-loop hypothesis'. J. Physiol., 326, 79-90.
Evarts, E.V. (1981). Role of motor cortex in voluntary movements in primates. In Handbook of Physiology, Section 1, The Nervous System, Vol. 2, Motor Control. Ed. Brooks, V.B. American Physiological Society, New York, pp. 1083-1119.
Hammond, P.H. (1960). An experimental study of servo action in human muscular control. In Proc. III International Conference in Medical Electronics. Butterworths, London, pp. 190-199.
Houk, J.C. & Rymer, W.Z. (1981). Neural control of muscle length and tension. In Handbook of Physiology, Section 1, The Nervous System, Vol. 2, Motor Control. Ed. Brooks, V.B. American Physiological Society, New York, pp. 257-323.
Lee, R.G. & Tatton, W.G. (1978). Long loop reflexes in man: clinical applications. In Cerebral Motor Control in Man: Long Loop Mechanisms. Ed. Desmedt, J.E. Karger, Basel, pp. 167-177.
Lee, R.G. & Tatton, W.G. (1982). Long latency reflexes to imposed displacements of the human wrist: dependence on duration of movement. Expl. Brain Res., 45, 207-216.
Lenz, F.A., Tatton, W.G. & Tasker, R.R. (1983). The effect of cortical lesions on the electromyographic response to joint displacement in the squirrel monkey forelimb. J. Neurosci., 3, 795-805.
Marsden, C.D., Merton, P.A. & Morton, H.B. (1976). Servo action in the human thumb. J. Physiol., 257, 1-44.
Marsden, C.D., Rothwell, J.C. & Day, B.L. (1983). Long latency automatic responses to muscle stretch in man: their origins and their function. In Motor Control Mechanisms in Man: Electrophysiological Methods and Clinical Applications. Ed. Desmedt, J.E. Raven, New York, pp. 509-539.
Matthews, P.B.C. (1969). Evidence that the secondary as well as the primary

endings of the muscle spindles may be responsible for the tonic stretch reflex of the decerebrate cat. J. Physiol., 204, 365-393.

Matthews, P.B.C. (1984a). Evidence from the use of vibration that the human long-latency stretch reflex depends upon spindle secondary afferents. J. Physiol., 348, 383-415.

Matthews, P.B.C. (1984b). The contrasting stretch reflex responses of the long and short flexor muscles of the human thumb. J. Physiol., 348, 545-558.

Matthews, P.B.C. (1984c). Observations on the time course of the electromyographic response reflexly elicited by muscle vibration in man. J. Physiol., 353, 447-461.

Matthews, P.B.C., Bawa, P. & Matthews, H.R. (1982). Synchronisation of motor firing by vibration during stretch evoked responses of human wrist flexors. Expl. Brain Res., 45, 313-316.

Phillips, C.G. (1969). Motor apparatus of the baboon's hand. Proc. R. Soc. B, 173, 141-174.

Phillips, C.G., Powell, T.P.S. & Wiesendanger, M. (1971). Projection from low-threshold muscle afferents of hand and forearm to area 3a of baboon's cortex. J. Physiol., 217, 419-446.

Tatton, W.G., North, A.G.E., Bruce, I.C. & Bedingham, W. (1983). Electromyographic and motor cortical responses to imposed displacements of the cat elbow: disparities and homologies with those of the primate wrist. J. Neurosci., 3, 1807-1817.

Chapter Twenty-eight

PHASE DEPENDENT STEP ADAPTATIONS DURING HUMAN LOCOMOTION

H. Forssberg

The muscular contractions of a walking man are much weaker than those produced during voluntary leg movements of comparable amplitude (Basmajian, 1967). They are particularly weak if the subject is allowed to choose a comfortable speed, which coincides with the speed with the lowest energy expenditure (Ralston, 1976). The weakness of the muscle contractions has been suggested to be due to a specific oscillatory behaviour of the skeleto-muscular system (Winter & Robertson, 1978). The position of the body in front of the foot support also allows the force of gravity to accelerate the body forwards (Pedotti, 1977). Although the EMG-amplitudes are small, they vary considerably during consecutive step cycles even during straight locomotion on a flat surface. EMGs and joint angle accelerations vary much more than the leg movements and the reaction forces (Fomin et al., 1976; Herman et al., 1976; Pedotti, 1977). Hence, the basic human locomotor rhythm is very efficient with a low intensity motor pattern, while adaptive mechanisms with signals of relatively higher amplitude continuously modify the stereotyped pattern to produce regular and smooth locomotor movements. This paper describes some adaptive mechanisms during human locomotion focusing on two studies done in collaboration with Dr Lew Nashner.

Proprioceptive postural reflexes
Nashner (1980) reported a brisk increase in the activity of the extensor muscles of the ankle occurring about 95-100ms after the surface under the supporting leg was withdrawn backwards during human locomotion. A 75% increase in ankle dorsiflexion during the stance phase, produced either by a backward movement or a toe up rotation of the supporting surface, elicited 2-3 times more activity in these muscles than occurs during a normal step. A forward movement or toe down rotation of the supporting surface, on the other hand, which reduced the dorsiflexion, evoked instead a burst of activity in the dorsiflexing muscles.

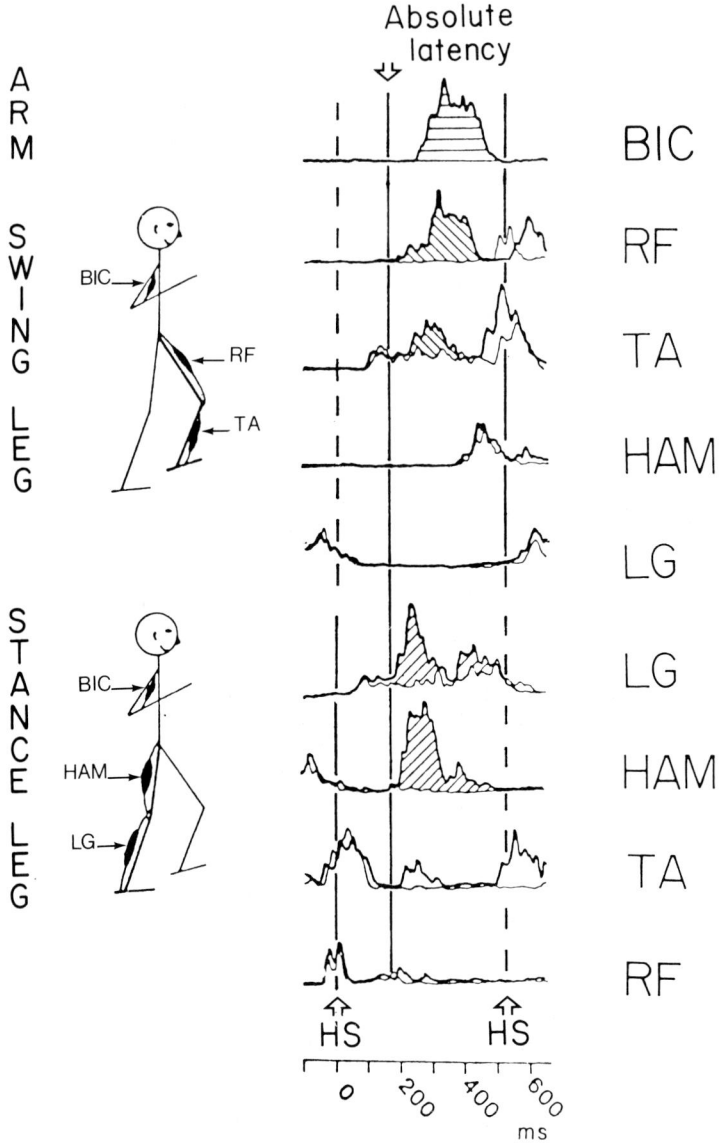

Fig. 1. Ensemble average of EMG-recordings from the arm, the swing leg and the stance leg during 8 normal steps (fine lines) of one subject compared to 8 steps in which an audio signal to pull a handle was delivered at heel strike (heavy lines). Subjects walked on a treadmill holding a handle and were instructed to pull as soon as they heard the signal. The signal was triggered by foot switches. EMGs were recorded by

These ankle responses during walking had the same latencies and directional specificity as the postural sway programmes that are elicited during a forward or backward sway around the ankle during fixed stance (Nashner, 1977). An imposed forward sway elicits a counteracting backward sway synergy, which consists of a wave of muscular contractions radiating proximally in the calf muscles, the hamstring muscles and the back muscles. A backward sway, on the other hand, elicits a forward sway synergy with contractions in the muscles of the ventral aspect of the leg and body. Due to their fixed temporal and spatial pattern, Nashner suggested that a postural synergy was generated by a central programme set in advance, rather than by each muscle being individually controlled. It appears that these same synergies are initiated and added to the locomotor activity when forward or backward sways are imposed during locomotion, although only the distal leg muscles were monitored during walking (Nashner, 1980).

The amplitude of the EMG-adjustments in both extensor and flexor muscles varied during the stance phase. Both the extensor response during increased dorsiflexion and the flexor response during decreased dorsiflexion were largest at the beginning of the stance phase. They were smaller when the perturbation occurred later and almost absent toward the end of stance. Both the position of the foot (support on heel, whole foot or toe) and the intensity of muscular activity were eliminated as causes of the variation in the responses by testing the same perturbation in different fixed postural positions (Nashner, 1980). These phase dependent modifications were therefore probably due to the ongoing locomotor rhythm.

Anticipatory postural programmes

If the body is moved forward or backward by voluntary arm contractions against an external object, the activity in the arm muscles is preceded by postural contractions in leg and

---

surface electrodes. Vertical lines indicate from left to right: 1) heel strike (HS) of stance leg triggering the audio signal 2) the absolute latency, i.e. the first muscular activity in arm pull step cycles that deviated from the normal locomotor bursts 3) heel strike of the swing leg. The EMGs of the swing leg and of the stance leg were recorded from the same leg during different trials, but are combined in this figure to show the integrated stance-swing pattern. EMG-activity associated with arm pull and referred to in the text is highlighted by diagonal lines. Stick figures demonstrate the anatomical configuration of contracting muscles. The muscles recorded were the biceps muscle (BIC), rectus femoris of the quadriceps muscle (RF), the anterior tibialis muscle (TA), the hamstring muscles (HAM), and the lateral gastrocnemius muscle (LG). Zero in the time scale below the traces indicates the trigger signal. (From Nashner & Forssberg, 1985).

Phase dependent step adaptations during human locomotion

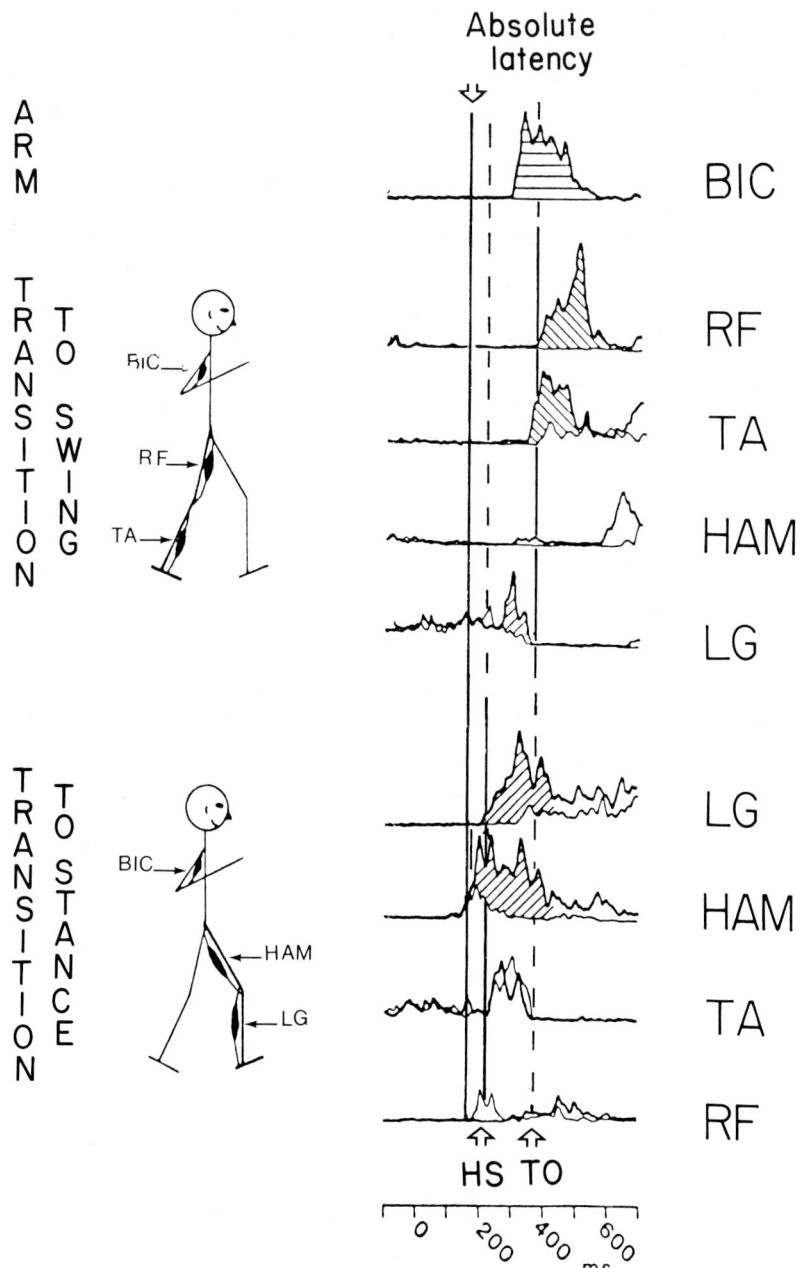

trunk (Marsden et al., 1977; Cordo & Nashner, 1982). The temporal and spatial pattern of the postural contractions is the same as in the backward and forward sway synergies induced by rotation around the ankle. Cordo and Nashner (1982) proposed that the same postural programmes are used, whether induced by proprioceptive signals or triggered prior to a voluntary movement.

When subjects walking on a treadmill hold a handle and are instructed to pull, they anticipate the destabilising force by first inducing postural contractions in the legs (Nashner & Forssberg, 1985). The first muscular activity that deviates from the normal (unperturbed) locomotor activity occurs in the lateral gastrocnemius muscle (LG) of the stance leg about 203±32ms (n=5) after an audio trigger signal. The postural contractions then radiate proximally to the hamstring muscles (HAM) after an additional delay of 36±30ms (see Fig. 1). At about the same time a postural contraction is evoked in rectus femoris (RF) of the swing leg followed by activity in the anterior tibialis muscle (TA) 12±8ms later (Fig. 1). The temporal pattern of the postural contractions in the stance leg is thus similar to a backward sway synergy with a distal to proximal sequence. It is followed by a forward swing synergy in the swing leg with a proximal to distal sequence, which will increase the forward swing. Contraction of the biceps muscle (BIC) in the arm is delayed until the postural contractions are initiated and follows about 39±27ms after the first LG-contraction.

A push against the handle initiates a forward sway synergy in the stance leg with activity in TA 186±32ms after the trigger signal, followed by activity in the knee extensor vastus lateralis (VL) and RF. The sway synergy is followed by a backward swing synergy in the swing leg with activity in HAM 45±30ms after the initial stance leg TA-activity. In addition, activity in the swing leg TA is initiated 63±22ms after stance leg TA. Again the contraction of the muscle in the arm, in this case the triceps, is delayed until the postural synergies are initiated.

Fig. 2. Ensemble averaged EMG-traces of 8 normal steps (thin lines) and 8 steps in which the signal to pull was delivered about 200ms before heel strike (heavy line). This timing of the trigger signal evoked postural contractions during the transition phases between swing and stance. The EMGs during transition to swing and during transition to stance were recorded from the same leg in different trials. The timing is therefore not comparable between the two legs. Vertical lines indicate from left to right: 1) absolute latency 2) heel strike (HS, the transition from swing to stance) 3) toe off (TO, the transition from stance to swing). At the bottom; time scale counted from the trigger signal. Abbreviations as in Fig. 1.

This distinct pattern was disrupted when the <u>pull</u> occurred during the transition phases between the swing and the stance phases. The proximal part of the forward swing synergy (i.e. RF), activated during the end of swing, changed to the backward sway synergy at heel strike. HAM was activated about 50ms before heel strike and LG just before heel strike (10±10ms). If the signal was delivered somewhat later, the RF contraction disappeared and only the backward sway muscles were activated with the same proximal to distal sequence and with the same temporal coupling to heel strike (Fig. 2). Instead of a fixed temporal pattern (distal to proximal) synchronised to the trigger signal, the contractions of the backward swing synergy were phase locked to heel strike. When the trigger signal was delivered closer to heel strike than the absolute latency, the backward sway synergy was induced after heel strike, with a distal to proximal sequence, synchronised to the signal.

During transition from stance to swing only the distal part of the backward sway synergy was activated during the end of stance (Fig. 2). The postural activity was then switched to a forward swing synergy at toe off. Also during this period of the step cycle the otherwise fixed proximal to distal sequence of the swing synergies was broken and the contractions instead locked to toe off.

Subjects who were allowed to time their handle pulls without any external trigger signal preferred to initiate the pull at heel strike without any ipsi- or contralateral preference. They used the same strategy with a sway and swing synergy initiated prior to the arm contraction, although they could plan the movement in advance. We wondered whether the latency to the arm contraction when triggered by a signal would be shorter during these preferred periods of the step cycle. However, the latency to the arm contraction as well as the absolute latency (i.e. the latency to the first detectable postural response) was constant during the whole step cycle.

The postural contractions during the stance phase are identical to the sway synergies activated in the same subject during fixed stance as judged by the temporal pattern and the relation of contractile strength between the involved muscles (see Nashner & Forssberg, 1985). The swing synergies have not been found during fixed stance. The concept of central programmes might seem contradicted by the disruption of the synergies during the transition phases. The dropping out of muscle contractions occurred, however, in an all-or-none mode suggesting that a fraction of the programme was suddenly blocked. The observed fractionation of the pattern is thus consistent with a central programme organised in subunits which can be separately activated. Such an organisation of central pattern generators has earlier been suggested for locomotion (Edgerton et al., 1976) and scratch reflexes (Mortin et al., 1982).

Visual corrections

Normally man is heavily dependent on vision to orient the body in space and to plan and coordinate movements in a precise way. The perception of space and of movement in space has been studied and discussed by several investigators (see Johansson et al., 1980). Lee and Lishman (1977) suggested a theoretical model in which the flow of the image of the peripheral environment over the retina during locomotion could be used to control stepping, balance and foot placement by programming several step cycles ahead. The study reported here examines the dependence of motor patterns of visual corrections upon how long in advance a change of step length was signalled. A final manuscript is in preparation.

Adult volunteers (n=8) walked on a walkway constructed of transparent plates under which lamps could be lit to mark a path which the subject should walk on. Microswitches under each plate monitored foot contact and correct foot placement. The light pattern on the walkway could suddenly be changed to a new path. The switching of the light was triggered from the microswitches and the duration of the warning time before the correction had to be made was controlled by a delay unit. A normal stride length was 66cm while a lengthened one was 99cm. The subjects performed 3-5 normal steps before the long step. By varying normal steps, long steps and short steps in the right or left leg with different warning times, the subject could never predict the change in advance.

No systematic analysis was made to test how late before touch down the light pattern could be switched and the subject still manage to correct the step properly. If the switch occurred 200ms after the heel strike of the contralateral leg, i.e. approximately 100ms after ipsilateral lift off and about 350-400ms before touch down of a normal step, all subjects succeeded in stepping on the correct plate. However, all subjects missed occasionally, which indicates that this is close to the time when the information comes too late.

During such a late warning all subjects had a fixed pattern of forceful contractions in the proximal leg muscles. The EMG-bursts of the corrective contractions were several times larger than the locomotor bursts (Fig. 3). The hip extensors of the stance leg (GM, HAM) were activated simultaneously (±10ms) with the hip flexor in the swing leg (RF; Fig. 3A). In most subjects (7/8) these contractions occurred 30-50ms after increased contractions in the stance ankle extensors (LG; Fig. 3A). The EMG-increase in LG, which occurred 150-200ms after the change of the light pattern, was much smaller in relation to the locomotor activity in LG than the relative responses in the proximal muscles. In several subjects (5/8) the dorsiflexor of the ankle (TA) was co-contracted, although it was initiated 10-50ms after the activity in LG. In one subject there was activity only in TA. In the swing leg the burst in the bi-articular HAM (hip

Phase dependent step adaptations during human locomotion

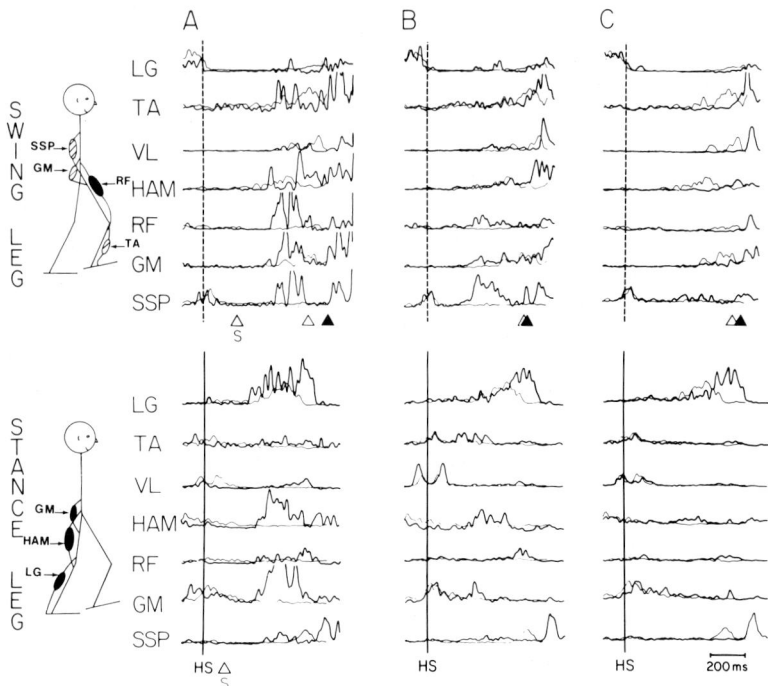

Fig. 3. Ensemble average of EMG-recordings from 8 steps with normal length (66 cm; thin line) and from 4-6 steps in which the subject was instructed to lengthen the step (step length 99 cm; heavy line). Subjects walked on a walkway with lighted plates on which the feet were placed. The signal to lengthen the step was a switching of the light pattern. A: Switching of pattern occurred 100ms (stance leg) and 200ms (swing leg) after heel strike of the stance leg. B: Switching of pattern occurred at heel strike. C: Switching of pattern occurred several steps in advance. EMGs from the swing phase and the stance phase were recorded in the same leg during different trials and are combined in this figure. Note the time difference of the stimulus in A. The vertical lines indicate heel strike (HS) in the stance leg. Arrows with an S beneath in A indicate the switch of the light pattern. Open arrows mark heel strike in the swing leg during a normal step and filled arrows mark heel strike in a lengthened step. Stick figures demonstrate the postural contractions after a late warning (A). Time scale at lower right. LG - lateral gastrocnemius; TA - anterior tibialis; VL - vastus lateralis; HAM - hamstring; RF - rectus femoris; GM - gluteus medius; SSP - supraspinatus.

extensor-knee flexor) was followed by activity in the pure knee extensor VL after a 5-15ms delay (in 4/7 recorded subjects; not seen in Fig. 3). In one subject (seen in Fig. 3A) there was instead a significant increase in TA, 20-30ms after the activity in RF. The hip extensor (GM) and the back muscle (SSP) were also activated on the swing side 30-40ms and 15-25ms respectively after the RF-activity in the two subjects in which these muscles were recorded. No recordings were made from muscles on the ventral aspect of the trunk.

The hip muscles were usually activated in a triphasic pattern at a late warning. During a lengthened step the hip extensors in the stance leg (HAM and GM) and the hip flexor in the swing leg (RF) and, with a small delay, the knee flexor in the swing leg (VL) acted as agonists and were activated together. They were followed by activity in the antagonists (RF in the stance leg and HAM in the swing leg) about 50-150ms later. This was followed by a second burst of activity in the agonists about 200-300ms after the first agonist bursts. The triphasic pattern was not obligatory. All subjects could also activate the agonists in one longer burst, which was the pattern used with a slightly longer warning time.

When the change of the light pattern occurred 200ms earlier, i.e. at contralateral heel strike about 100ms before toe off, there were also stimulus synchronised EMG-bursts added to the normal locomotor activity. The proximal muscles of both legs were activated with the same temporal pattern as at a later warning but with much smaller intensities and without a triphasic pattern (Fig. 3B). In the distal stance leg muscles, however, the pattern was reversed. In 5 of 8 subjects TA was activated first and with a larger relative amplitude than LG. This is functionally important since a dorsiflexing torque at the beginning of stance will rotate the body forwards around the ankle, while a plantarflexion at the end of stance, with only the toe contacting ground and the centre of body in front of the foot, will thrust the body forward. In the swing leg VL was still activated, but now with a longer delay, phase locked to the end of swing. A knee extension at the end of swing is meaningful, prolonging the step, while a too early knee extension might have caused a stumble.

When the switching of the light pattern was delivered several steps in advance, no extra EMG-bursts could be detected. The only significant change of the EMG-pattern, except a prolongation of the duration of the locomotor bursts, was an increase of the total LG locomotor activity in 4 of 8 subjects (comparison of integrated EMG-activity of the locomotor bursts; $p<0.05$). In a few subjects (3/8) a non-significant increase of the small RF-burst in the beginning of swing could be seen. No detailed analysis was made of the motor activity in the preceding cycles.

The CNS has thus at least two different mechanisms to

correct a step. With a long prediction time no extra contractions are needed. Very slight, hardly detectable, increases in ordinary locomotor activity are sufficient. This can probably be achieved by just increasing the drive on the locomotor circuits, with a proportionately larger potentiation on some parts of the network (e.g. plantar flexors). This mechanism, which probably is much more energy saving, can be used up to 500ms before contralateral heel strike. Corrections planned after this time use additional contractions with delays synchronised from the light switch. In the proximal muscles these contractions have a fixed time pattern, which is triphasic at a very late warning, when the intensity of the contractions is greatest. In the distal stance muscles too the corrective contractions are coordinated to a fixed temporal pattern, but here they are locked to certain phases of the step cycle and do not occur at other times. The fixed temporal pattern of the visual corrections supports the existence of central programmes initiating sets of muscular contractions. The increase of the intensity at later warnings suggests an interaction between the two different mechanisms at longer warning times, with a possible summation of a late increased locomotor drive and corrective synergies with weaker intensities. It might also be that a corrective synergy is more effective early in the swing phase. The phase locking of some contractions, with a reversal of the pattern in the ankle of the stance leg, reflects an interaction between the corrective programmes and the locomotor activity (see last section).

Cutaneous reflexes

A tactile stimulus to the dorsum of the cat hind paw during the swing phase of locomotion elicits a corrective reaction. The leg will be lifted over a perturbing obstacle by short latency reflexes in the flexor muscles. Simultaneous reflexes to the contralateral extensor muscles induce an increased extension (Forssberg et al., 1977; Forssberg, 1979; Duysens & Loeb, 1980). These reflex contractions are superimposed on the ordinary locomotor activity in a similar way as are the corrective motor patterns during human locomotion discussed above. The phase dependence of the stumbling corrective reaction in the cat might thus help us to understand the central mechanisms of phase controlled adaptations in humans.

The same tactile stimulus that gives a flexor response during the swing phase elicits a late extensor response during the stance phase of locomotion in spinal cats. There is thus a reversal from a flexor response to an extensor response depending on the phase of the step cycle. By intracellular recordings from motor neurones in locomoting chronic spinal cats in which the movements had been blocked by curare (fictive locomotion), it was found that the transmission in the reflex pathways from the dorsum of the paw to extensor and flexor motor neurones was phasically controlled by the central pattern generator for locomotion (Andersson et al., 1978; Forssberg, 1981). Early and late reflex pathways to the flexors and early reflex pathways to the extensors were open during flexor activity and late reflex pathways to the extensor motor neurones were open during extensor activity. Central gating has also been found in several other motor systems (e.g. mastication; Lund et al., 1981).

Several investigators have tried to evoke similar corrective reactions during human locomotion by applying non-noxious stimuli to the foot and leg (Kanda & Sato, 1983; Berger & Dietz, 1983; Forssberg, unpublished). In no case has it been successful. It seems that man does not have tactile corrective synergies that compensate for a stumble. However, there is evidence that cutaneous reflexes in the legs are phase dependent in man. Lisin, Frankstein and Rechtmann (1973) applied noxious stimuli to a hemiplegic man and noticed a reversal from a flexor reflex during rest to an extensor reflex during locomotion, while Kanda and Sato (1983) elicited EMG-responses by non-noxious stimuli. Although no movement reaction was noted, the size of these EMG-responses varied in phase with the locomotor activity.

Mechanisms of phase control in man

The appearance of stepping movements in human infants at birth indicates that locomotor-like activity i) is innate, ii) can be induced by peripheral influence (support on the feet, extension of hips), and iii) does not involve neural mechanisms above the brain stem (walking anencephalic infants). Studies on the ontogeny of human locomotion have shown that children still use an infantile, digitigrade pattern when they start to walk without support, and that there is a gradual transformation of the locomotor pattern to a plantigrade gait during the following years (Forssberg, 1982, 1985). The slow and gradual progress to a mature, plantigrade pattern suggests that adult man uses the same circuits as the infants, but that these are influenced by some other mechanism to produce a more efficient pattern for bipedal gait. It is reasonable to assume that the original locomotor activity during infancy is generated by a central network in the spinal cord in a similar way as experimentally demonstrated in animals (c.f. Grillner, 1981). This implies that locomotor rhythm generation in adult man is also produced by spinal central networks.

The similarity between the postural synergies (proprioceptive or anticipatory) during fixed stance and locomotion suggests that the same central programmes are used. Thus it is postulated that there are two separate systems, one for locomotion and another for postural reactions, where activity from the two systems is summated at a late premotor or motor neurone level. By analogy with this, visual corrective programmes are transmitted in systems separate to locomotion and summated to the locomotor activity. It is possible that the postural synergies and the visual corrections actually are controlled by the same programmes or parts of the same programmes, since the temporal and spatial patterns are similar, although not identical.

Although the postural synergies and the visual corrective programmes are separate from the locomotor generating circuits, they are obviously controlled by locomotor activity. There is a switch to a completely different pattern between the swing and the stance phase. The postural synergies are fractionated during the transition phases so that only a part of the synergy is activated. In both the postural synergies and the visual corrections, some contractions are locked to certain phases of the step cycle. For example, the contraction of LG in the backward sway synergy is locked to heel strike (Fig. 2). During a visual correction a TA-response is locked to early stance and a LG-response to late stance and a VL-response to late swing (Fig. 3B). The locomotor activity thus selects between different programmes (swing or stance) but can also control the timing of individual contractions within a programme. This could be done by blocking a subunit of a programme during a certain period of the phase.

Certain parts of the CNS (e.g. the cerebellum) receive continual information about the movements of the body from peripheral receptors as well as efference copy signals from the movement generators, informing them of which movements are coming next (Arshavsky et al., 1972a,b). A central motor coordinating system would therefore have the information needed to control the postural and visual programmes on a feed forward basis and could directly activate the proper programme. Another possible, and more efficient mechanism, however, would be a gating system similar to that controlling cutaneous reflexes in cats. The innate spinal locomotor circuits could continuously control the excitability level of the corrective programmes. A central motor coordinating system could then select pairs of swing or sway programmes. Due to the excitability level of the programmes, controlled by the locomotor generator, the proper swing or sway synergy would be initiated. The locomotor circuits could also by a similar mechanism block a subunit of a selected programme during a certain phase when the rest of the programme is activated.

Acknowledgements
This work was supported by the Swedish Medical Research Council (4X-5925, 4P-6745), Norrbacka-Eugenia Stiftelsen and the Swedish Sport Research Council. The valuable comments of Dr T. Williams are gratefully acknowledged.

Phase dependent step adaptations during human locomotion

REFERENCES

Andersson, O., Forssberg, H., Grillner, S. & Lindquist, M. (1978). Phasic gain control of the transmission in cutaneous reflex pathways to motoneurons during "fictive" locomotion. Brain Res., 149, 503-507.

Arshavsky, Y.I., Berkinblit, M.B., Gelfand, I.M., Orlovsky, G.N. & Fukson, O.I. (1972). Activity of the neurons of the dorsal spinocerebellar tract during locomotion. Biophysics, 17, 506-514.

Arshavsky, Y.I., Berkinblit, M.B., Gelfand, I.M., Orlovsky, G.N. & Fukson, O.I. (1972). Activity of the neurons of the ventral spinocerebellar tract during locomotion. Biophysics, 17, 926-941.

Basmajian, J.V. (1967). Muscles Alive: their Function revealed by Electromyography. Williams & Wilkins, Baltimore.

Berger, W. & Dietz, V. (1983). Corrective reactions to disturbance of stance and gait studied on man. J. Physiol., 341, 33P.

Cordo, P.J. & Nashner, L.M. (1982). Properties of postural adjustments associated with rapid arm movements. J. Neurophysiol., 47, 287-302.

Duysens, J. & Loeb, G.E. (1980). Modulation of ipsi- and contralateral reflex responses in unrestrained walking cats. J. Neurophysiol., 44, 1024-1037.

Edgerton, V.R., Grillner, S., Sjöström, A. & Zangger, P. (1976). Central generation of locomotion in vertebrates. In Neural Control of Locomotion. Eds. Herman, R.M., Grillner, S., Stein, P.S.G. & Stuart, D.G.. Plenum Press, New York, pp. 439-464.

Fomin, S.V., Gurfinkel, V.S., Feldman, A.G. & Shtilkind, T.I. (1976). Movements of the joints of human legs during walking. Biophysics, 21, 572-577.

Forssberg, H. (1979). Stumbling corrective reaction: a phase-dependent compensatory reaction during locomotion. J. Neurophysiol., 42, 936-953.

Forssberg, H. (1981). Phasic gating of cutaneous reflexes during locomotion. In Muscle Receptors and Movement. Eds. Taylor, A. & Prochazka, A., Macmillan, London, pp. 403-412.

Forssberg, H. (1982). Spinal locomotor functions and descending control. In Brain Stem Control of Spinal Mechanisms. Eds. Sjölund, B. & Björklund, A.. Elsevier Biomedical Press, Amsterdam, pp. 253-271.

Forssberg, H. (1985). Ontogeny of human locomotor control. I. Infant stepping, supported locomotion and transition to independent locomotion. Expl. Brain Res., 57, 480-493.

Forssberg, H., Grillner, S. & Rossignol, S. (1977). Phasic gain control of reflexes from the dorsum of the paw during spinal locomotion. Brain Res., 132, 121-139.

Grillner, S. (1981). Control of locomotion in bipeds, tetrapods, and fish. In Handbook of Physiology, Section I, The Nervous System, Vol. II, Motor Control. Ed. Brooks, V.B. American Physiological Society, Bethesda, pp. 1179-1236.

Herman, R., Wirta, R., Bampton, S. & Finley, F.R. (1976). Human solutions for locomotion. I. Single limb analysis. In Neural Control of Locomotion. Eds. Herman, R.M., Grillner, S., Stein, P.S.G. & Stuart, D.G.. Plenum, New York, pp. 13-49.

Johansson, G., von Hofsten, C. & Jansson, G. (1980). Event perception. A. Rev. Psychol., 31, 27-63.

Kanda, K. & Sato, H. (1983). Reflex responses of human thigh muscles to non-noxious sural stimulation during stepping. Brain Res., 288, 378-380.

Lee, D.N. & Lishman, R. (1977). Visual control of locomotion. Scand. J. Psychol., 18, 224-230.

Lisin, V.V., Frankstein, S.I. & Rechtmann, M.B. (1973). The influence of locomotion on flexor reflex of the hind limb in cat and man. Expl Neurol., 38, 180-183.

Lund, J.P., Rossignol, S. & Murakami, T. (1981). Interactions between the jaw-opening reflex and mastication. Can. J. Physiol. & Pharmacol., 59, 683-690.

Marsden, C.D., Merton, P.A. & Morton, H.B. (1977). Anticipatory postural responses in the human subject. J. Physiol., 275, 47-48P.

Mortin, L.I., Keifer, J. & Stein, P.S.G. (1984). Three forms of the turtle scratch reflex. Soc. Neurosci. Abstr., 8, 159.

Nashner, L.M. (1977). Fixed patterns of rapid postural responses among leg muscles during stance. Expl. Brain Res., 30, 13-24.

Nashner, L.M. (1980). Balance adjustments of humans perturbed while walking. J. Neurophysiol., 44, 650-664.

Nashner, L.M. & Forssberg, H. (1985). Anticipatory postural reactions induced by arm movements while walking. J. Neurophysiol. (In press).

Pedotti, A. (1977). A study of motor coordination and neuromuscular activities in human locomotion. Biol. Cybern., 26, 53-62.

Ralston, H.J. (1976). Energetics of human walking. In Neural Control of Locomotion. Eds. Herman, R.M., Grillner, S., Stein, P.S.G. & Stuart, D.G.. Plenum, New York, pp. 77-98.

Winter, D.A. & Robertson, D.G.E. (1978). Joint torque and energy patterns in normal gait. Biol. Cybern., 29, 137-142.

Chapter Twenty-nine

ABNORMAL FEEDBACK AND MOVEMENT DISORDERS IN MAN, WITH PARTICULAR REFERENCE TO CORTICAL MYOCLONUS

C.D. Marsden and J.C. Rothwell

The question raised by this title is, what types of human movement disorders may be attributed to abnormal feedback? Disorders of movement in man comprise three major categories:
1. Those in which a central lesion causes loss of normal movement, as with damage to the cortico-motor neurone pathways or to the cerebellum. Such lesions cause distortion of the effects of peripheral feedback to the spinal cord and brain, which may contribute to some of the features of pyramidal and cerebellar deficit, for example producing changes in muscle tone. These matters will not be considered here.
2. Those in which there is loss of peripheral feedback by partial or complete deafferentation. The effects of deafferentation in adult man are complex. On the one hand there is a reasonable preservation of a range of complex motor patterns, indicating a rich store of learned motor programmes in the adult central nervous system which can be executed in the absence of information from the periphery. On the other hand, a number of specific deficits can be identified in deafferented man, including difficulties in maintaining a constant position or force of contraction, and in executing a sequence of motor actions in the absence of visual feedback. There is also a question as to what extent peripheral feedback is required to learn motor programmes. The effect of deafferentation in man has been described by Rothwell, Traub, Day, Obeso, Thomas and Marsden (1982) and the subject has been reviewed by Marsden, Rothwell and Day (1984) and so will not be discussed here.
3. Those in which there is an exaggerated response to feedback, which is the topic of this review. The most obvious example of this category of abnormality is myoclonus, and in particular that category of myoclonus known as "cortical reflex myoclonus" (Hallett et al., 1979). In this group of conditions there is excessive cerebral cortical response to peripheral input, such that sensory stimuli or motor action cause muscle jerking. The system involved appears to be that which contributes to the long-latency stretch reflexes

described in Chapter 27 by Dr Matthews, and earlier by Marsden, Merton and Morton (1972, 1973) and Lee and Tatton (1975), after Phillips' (1969) original hypothesis of the existence of a transcortical stretch reflex pathway.

LONG-LATENCY REFLEXES AND TRANSCORTICAL REFLEX PATHWAYS

Late responses to cutaneous stimuli (Jenner & Stephens, 1982) and muscle stretch (Marsden et al., 1976a,b) of the human arm are now well-established. These responses occur much later than would be expected for simple fast spinal reflexes. Both the long-latency component of the cutaneous finger reflex (Jenner & Stephens, 1982) and the long-latency stretch reflex from the long flexor of the thumb or wrist (Marsden et al., 1977a,b; Lee & Tatton, 1975) are abolished by lesions of the posterior column pathways, the sensori-motor cortex and the cortico-motor neurone pathway, suggesting (although not proving) that they employ transcortical pathways. Short latency reflexes are unaffected, or even increased after such lesions.

The transcortical stretch reflex hypothesis has come under strong attack in recent years. First, long-latency stretch reflexes have been recorded in spinal animals (cat - Ghez & Shinoda, 1978; primate - Tracey et al., 1980). However, it is doubtful whether these residual long-latency events in the spinal animal are similar to the long-latency stretch reflexes that can be obtained from the human arm. Second, human muscle stretch can evoke repetitive primary spindle discharge (Hagbarth et al., 1981), so late responses may represent only a spinal response to a second or later spindle burst. However, it is difficult to see how lesions of the ascending sensory pathways to the brain would selectively influence such events whilst leaving the first spinal response intact. Third, Matthews (1984, and Chapter 27) puts forward the hypothesis that long-latency stretch reflexes are due to a spinal response to secondary spindle input. This notion is based primarily on the finding that vibration, a powerful primary spindle stimulus, does not produce long-latency responses of the type evoked by muscle stretch, which engages both primary and secondary spindle endings. However, there is no proof that vibration and stretch evoke similar afferent volleys into the proposed transcortical system. Some of these criticisms of the long-latency stretch reflex concept have been reviewed at length by Marsden, Rothwell and Day (1983).

Whatever the criticism raised, there is little doubt that a transcortical stretch (and cutaneous) pathway exists. Anatomy and electrophysiology indicate an ascending pathway for low-threshold muscle afferents to the sensori-motor cortex in animals (see Weisendanger & Miles, 1982 for review). Studies in man indicate that muscle afferents from both the leg (Starr et al., 1981) and hand (Gandevia et al., 1984) gain access to the sensori-motor cortex, and we (Abbruzzese et al., 1985) also have been able to demonstrate a projection of muscle afferents from the wrist flexors in man to the brain. Thus a transcortical stretch (and cutaneous) pathway is likely to exist in man. How it is employed in the intact subject, in conjunction with spinal mechanisms, is open to debate. However, the existence of this pathway is dramatically revealed in patients with cortical reflex myoclonus.

CORTICAL REFLEX MYOCLONUS

Patients with myoclonus have involuntary muscle jerks. Myoclonus may arise in spinal cord, brainstem or cerebral cortex. We are concerned here with those with cortical myoclonus. Such patients may have jerks triggered by peripheral stimuli such as touch or muscle stretch (or both) (reflex myoclonus), by movement itself (action myoclonus), or spontaneously (spontaneous myoclonus), or any combination of these phenomena.

That myoclonus is arising in the cerebral cortex can be established by demonstration of greatly enhanced cerebral potentials evoked by peripheral stimuli which also provoke reflex myoclonic jerks, and/or back-averaging of a cortical potential prior to and time-locked to spontaneous myoclonic jerks.

Evidence for the participation of the cortex in the generation of some types of myoclonus was provided first by Dawson (1947), who studied a patient with generalised jerks provoked either by muscle stretch, or by electrical stimulation of a peripheral nerve. The electrical stimuli not only caused muscle jerking, but also produced giant somatosensory evoked potentials (SEPs). Further evidence for the existence of myoclonus arising in the motor cortex was provided by Kugelberg and Widen (1954) who described a patient with spontaneous, repetitive jerks of the right foot, preceded by focal EEG spikes recorded by electrocorticography from the medial surface of the opposite motor cortex. Subsequent surgical excision of this small zone abolished the myoclonus of the foot. Pagni, Marossero, Cabrini, Ettore and Infuso (1971) briefly reported a patient with stimulus-sensitive myoclonus who was studied by electrodes implanted into the thalamus, sensory cortex, motor cortex, and internal capsule. They concluded that the jerks were caused by afferent impulses in the spinal dorsal columns, a subsequent sensori-motor cortex discharge, and then an efferent volley passing down the cortico-motor neurone pathway.

Halliday (1967), in an influential review on myoclonus, distinguished "pyramidal" myoclonus with origin in motor cortex, from myoclonus arising in other sites. He drew attention to the frequent description of patients with spontaneous myoclonus associated with focal spike discharges arising from the opposite sensori-motor cortical area. In addition, he mentioned the patient described by Dawson as an example of the uncommon situation in which peripheral stimuli could evoke myoclonic jerking. Subsequently, Sutton and Mayer (1974), and Rosen, Fehling, Sedgwick and Elmquist (1977) each described individual patients with similar stimulus-sensitive myoclonus of cortical origin. Hallett, Chadwick and Marsden (1979) added a detailed electro-physiological description of such stimulus-sensitive myoclonus in three further cases and introduced the term cortical reflex myoclonus to describe the syndrome. Meanwhile, Shibasaki and Kuroiwa (1975) established that spontaneous myoclonus might be due to cortical discharge even though the routine surface EEG showed no abnormality. They employed the technique of back-averaging the EEG prior to the myoclonus by triggering from the EMG burst responsible for the jerk. By this means, they were able to extract from the averaged EEG a potential arising over the sensori-motor cortex preceding the muscle jerk by a short interval appropriate to conduction in cortico-motor neurone pathways. Chadwick, Hallett, Harris, Jenner, Reynolds and Marsden (1977) independently developed the same technique to show not only the existence of a motor potential in averaged records that was not apparent in simple surface EEG recordings in some cases of cortical myoclonus, but also the lack of correspondence of surface EEG spikes to myoclonic jerks in other patients. Thus, in a case of what they described as reticular reflex myoclonus, a large EEG spike, which superficially appeared linked to the myoclonic twitches, disappeared on back-averaging, indicating that it was not time-locked to the muscle discharges (Hallett et al., 1977). The use of this technique of back-averaging went a long way to clarify the mystery of the relationship of the EEG to myoclonic jerks. Many previous observers had found it difficult to interpret the significance of the presence or absence of spike discharges in the EEG, and their time relation to myoclonic twitches. It was not until it was possible to establish whether such EEG spikes were time-locked to the muscle jerks, or whether any other form of EEG abnormality could be averaged in relation to muscle twitches, that the true significance of cortical discharge in relation to myoclonus could be appreciated.

Recently, we have described a series of patients with different types of cortical myoclonus studied electro-physiologically (Obeso et al., 1985). Their myoclonus was attributed to electrical discharges arising from the sensori-motor cortex because either they had an abnormal cortical potential preceding spontaneous or action-induced jerks (Fig. 1), or because sensory stimuli evoked giant cortical potentials preceding their reflexly-evoked myoclonic jerks

Abnormal feedback and movement disorders in man

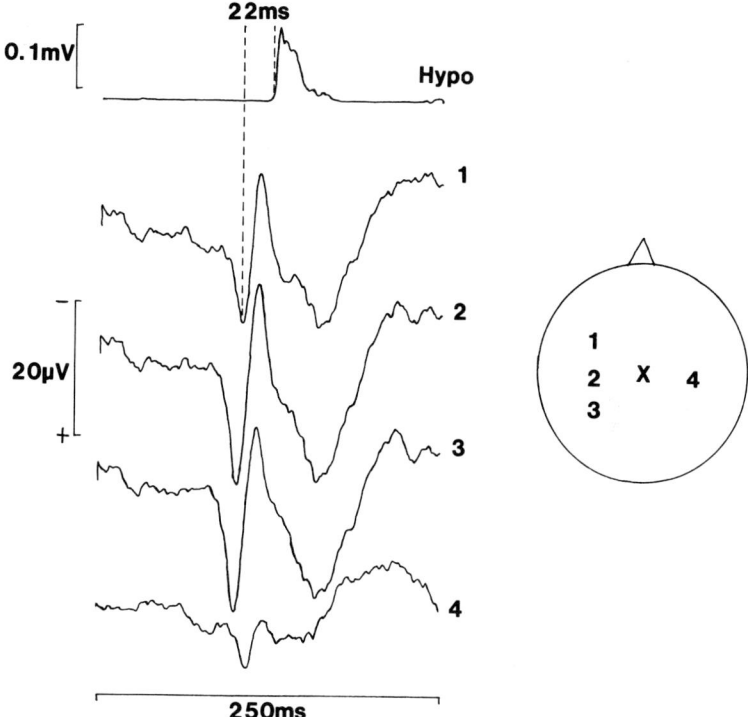

Fig. 1. Back-averaged (of 64) EEG potentials preceding spontaneous myoclonic jerks of the hypothenar muscles (Hypo) in a patient who had both spontaneous and reflex myoclonus together with occasional grand mal epileptic seizures. The averaging programme was triggered by the onset of muscle jerks in the rectified EEG record. A large positive-negative wave over the contralateral scalp precedes the myoclonic jerk by some 22 ms. EEG records are monopolar to a linked mastoid reference with high and low pass filters at 0.16 and 2.5kHz respectively. Electrode 2 is over the somatosensory hand area (a point 7cm lateral on a line joining the external auditory meatus to a point 2cm posterior to the vertex).

JB, a 34-year-old woman, had an undiagnosed left hemisphere lesion. For some six years she had experienced rare grand mal seizures, but continuous flexor jerking of the right hand and forearm while awake. These jerks occurred spontaneously on action, or in response to touch of the fingers or a tap with a tendon hammer. On examination, apart from the focal myoclonus of the right hand, there were no other neurological signs. Computer tomography scans repeatedly were normal, as were a left arteriogram and cerebrospinal fluid examination. Routine EEG showed a focal abnormality over the region of the left sensorimotor area, with spike/sharp wave discharges.

(Fig. 2). In either case, the cortical potentials were localised to, or were maximal over, the contralateral sensorimotor cortical region. The shape of the cortical potential preceding spontaneous or action-induced jerks was similar to that of the P1-N2 component of the SEP, as has been noted previously by Shibasaki, Yamashita and Kuroiwa (1978), and

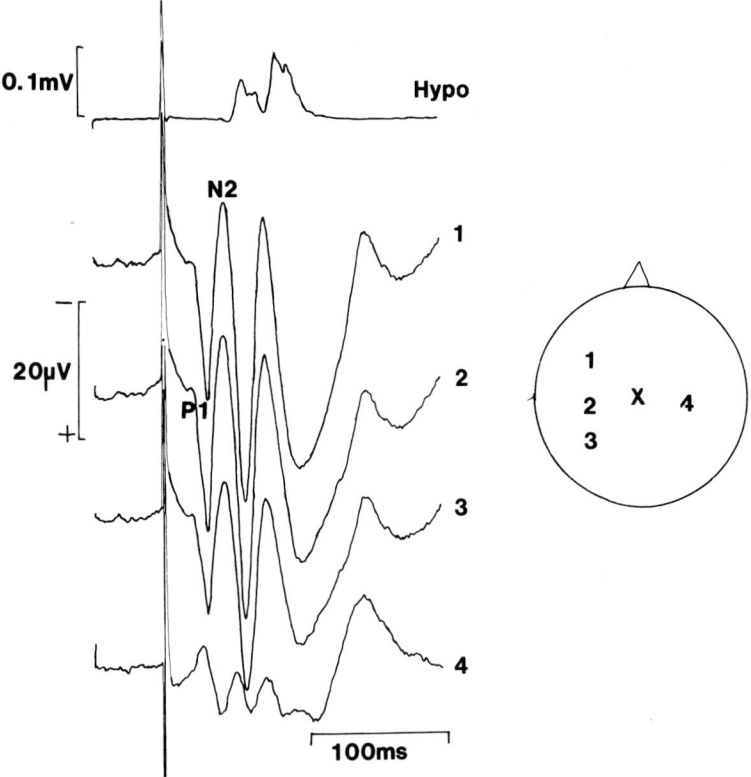

Fig. 2. Average (of 128) scalp-recorded somatosensory evoked potentials and reflex muscle jerks in the hypothenar muscles (Hypo) following electrical stimulation of the cutaneous nerves of the right index finger in patient JB. EEG montage as Fig. 1. The artefact at 50ms indicates the stimulus onset. A large P1-N2 potential precedes the first reflex muscle jerk by 22ms. In this patient a second positive-negative wave also precedes a second muscle jerk. The apparent response in the ipsilateral cortex is due to activity in the contralateral hemisphere being picked up by the mastoid reference.

Rothwell, Obeso and Marsden (1984). The interval between the positive peak of the cortical potential and the onset of spontaneous or reflex myoclonus was of the order of 20ms in the arm. This is consistent with rapid conduction down a direct corticospinal pathway (Pagni et al., 1964; Merton et al., 1982), and is compatible with involvement of motor cortex and pyramidal tract mechanisms in the myoclonus. At first sight, the cortical potentials preceding spontaneous or action-induced jerks, and the giant SEPs, might be considered to represent an abnormal synchronous discharge of motor cortical output neurones. However, elsewhere (Rothwell et al., 1984) we have drawn attention to evidence against this hypothesis. The source of the abnormal SEPs may not correspond accurately to the motor cortex, but often appears to lie posterior to the central sulcus. In such patients it may be that the motor cortex itself is normal, but that it is driven by an abnormal input from the sensory cortex. Indeed, a range

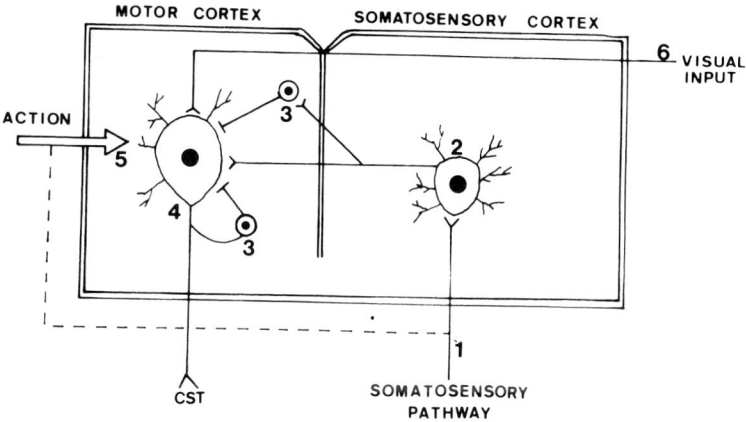

Fig. 3. Schematic diagram indicating the variety of different possible mechanisms which could produce cortical myoclonus in the patients studied. Reflex myoclonus could be due to abnormalities in a) the somatosensory (1) or visual input (6) to the cortex; b) transmission of the input from sensory receiving areas to motor cortex (2) (although not necessarily as a monosynaptic connection between somatosensory and motor cortex as shown here; more likely as a complex, cortico-cortical loop); c) the intrinsic interneurones of the motor cortex (3); or d) the final motor cortical output cells (4). Action myoclonus might arise from abnormal input to motor cortex from "voluntary motor centres" (5), or from mechanisms intrinsic to motor cortex (3,4). Spontaneous myoclonus might arise from motor cortical mechanisms (3,4) or from abnormal spontaneous input to the motor area from other regions of the brain (e.g. 2). CST, corticospinal tract.

of abnormalities of sensori-motor cortical function can be conceived as causing different types of cortical myoclonus (Fig. 3).

Some of these patients exhibited pure reflex myoclonus, with muscle jerks only in response to peripheral stimuli, either cutaneous and/or stretch. Others had myoclonus only on action. Others had spontaneous myoclonus, and some had epilepsia partialis continua or the frank Jacksonian motor epilepsy. All combinations of the varieties of cortical reflex myoclonus, action or spontaneous cortical myoclonus, and motor epilepsy were seen in individual patients. Cortical reflex myoclonus is of particular interest, for it shows that afferent input, from a variety of sources, may trigger myoclonic jerks, epilepsia partialis continua, and even Jacksonian seizures. Afferent information from skin joints and muscles (see e.g. Friedman & Jones, 1981) projects to the sensori-motor cortex, so it is not perhaps surprising that there is a wide variety of clinical types of cortical reflex myoclonus, differing in their necessary provocative stimuli, and in their association with other epileptic phenomena.

The spectrum of cortical myoclonus may be conceived as follows (Fig. 4). Some patients may have unstable sensori-motor cortical responses to a variety of somatic afferent inputs, from skin, joints or muscle, and so exhibit cortical reflex myoclonus. Many of these patients also exhibit unstable responses to those signals responsible for driving the sensori-motor cortex to produce willed movement, so causing action myoclonus. Such signals, associated with voluntary

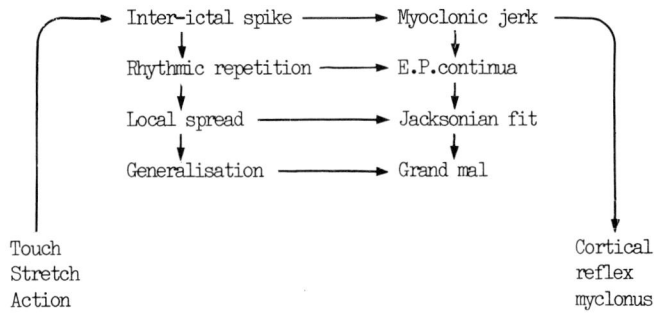

Fig. 4. Summary of relationships between cortical myoclonus and epilepsy. These events may occur in the motor cortex, as labelled, but as emphasised in the text, there may be patients in whom motor cortical mechanisms are normal. In these cases, the motor strip may be driven to discharge and produce myoclonic jerks by abnormal input from other areas of the brain. E.P. continua, epilepsia partialis continua. (From Marsden, 1980).

movement, may be those arriving at the sensori-motor cortex from other cortical areas, the basal ganglia, or the cerebellum. In addition, the spindle discharge that accompanies willed human muscle contraction (Vallbo, 1974) could trigger such sensori-motor cortical discharge to cause myoclonic jerks. Other patients may have spontaneous discharges of sensori-motor cortical neurones, causing spontaneous myoclonic jerks, with or without afferent triggers. In other patients the sensori-motor cortical neurones may discharge spontaneously and repetitively, with or without afferent triggers, to produce epilepsia partialis continua. If such discharges spread within the sensori-motor cortex, Jacksonian motor seizures then occur, with or without secondary generalisation. Viewed in this light, cortical myoclonus, whether reflex or spontaneous, is a part of the spectrum of epilepsy, as illustrated by some of the patients described here who all have manifestations of this spectrum.

CONCLUSIONS

The clinical phenomenon of cortical reflex myoclonus dramatically illustrates the effect of disordered cortical processing of afferent input from the periphery. Both cutaneous and stretch reflex-induced cortical myoclonus can occur, highlighting the existence of transcortical cutaneous (Jenner & Stephens, 1982) and stretch (Marsden et al., 1973) reflex pathways in man. A major question now is how these transcortical mechanisms interact with the spinal reflex machinery in the normal control of movement.

REFERENCES

Abbruzzesse, G., Berardelli, A., Rothwell, J.C. Day, B.L. & Marsden, C.D. (1985). Cerebral potentials and electromyographic responses evoked by stretch of wrist muscles in man. Expl. Brain Res., 58, 544-551.
Chadwick, D., Hallett, M., Harris, R., Jenner, P., Reynolds, E.H. & Marsden, C.D. (1977). Clinical, biochemical and physiological features distinguishing myoclonus responsive to 5-hydroxy-tryptophan, tryptophan with a monoamine oxidase inhibitor, and clonazepam. Brain, 100, 455-487.
Dawson, G.D. (1947). Investigations on a patient subject to myoclonic seizures after sensory stimulation. J. Neurol. Neurosurg. Psychiat., 10, 141-162.
Friedman, D.P. & Jones, E.G. (1981). Thalamic input to areas 3a and 2 in monkeys. J. Neurophysiol., 45, 58-85.
Gandevia, S.C., Burke, D. & McKeon, B. (1984). The projection of muscle afferents from the hand to cerebral cortex in man. Brain, 107, 1-13.
Ghez, C. & Shinoda, Y. (1978). Spinal mechanisms of the functional stretch reflex. Expl. Brain Res., 32, 55-68.
Hagbarth, K.-E., Hagglund, J.V., Wallin, E.U. & Young, R.R. (1981). Grouped

spindle and electromyographic responses to abrupt wrist extension movements in man. J. Physiol., 312, 81-96.

Hallett, M., Chadwick, D., Adam, J. & Marsden, C.D. (1977). Reticular reflex myoclonus: a physiological type of human post hypoxic myoclonus. J. Neurol. Neurosurg. Psychiat., 40, 253-264.

Hallett, M., Chadwick, D. & Marsden, C.D. (1979). Cortical reflex myoclonus Neurology, 29, 1107-1125.

Halliday, A.M. (1967). The electrophysiological study of myoclonus in man. Brain, 90, 241-284.

Jenner, J.R. & Stephens, J.A. (1982). Cutaneous reflex responses and their central nervous pathways studied in man. J. Physiol., 333, 405-419.

Kugelberg, E. & Widen, L. (1954). Epilepsia partialis continua. Electroenceph. clin. Neurophysiol., 7, 341-356.

Lee, R.G. & Tatton, W.G. (1975). Motor responses to sudden limb displacements in primates with specific C.N.S. lesions and in human patients with motor system disorders. Can. J. Neurol. Sci., 2, 285-293.

Marsden, C.D. (1980). The pathophysiology of myoclonus and its relation to epilepsy. Res. & Clin. Forums, 2, 31-46.

Marsden, C.D., Merton, P.A. & Morton, H.B. (1972). Servo action in human voluntary movement. Nature, Lond., 238, 140-143.

Marsden, C.D., Merton, P.A. & Morton, H.B. (1973). Is the human stretch reflex cortical rather than spinal? Lancet, i, 759-761.

Marsden, C.D, Merton, P.A. & Morton, H.B. (1976a). Servo action in the human thumb. J. Physiol., 257, 1-44.

Marsden, C.D., Merton, P.A. & Morton, H.B. (1976b). Stretch reflex and servo action in a variety of human muscles. J. Physiol., 259, 531-560.

Marsden, C.D., Merton, P.A., Morton, H.B. & Adam, J. (1977a). The effect of posterior column lesions on servo responses from the human long thumb flexor. Brain, 100, 185-200.

Marsden, C.D., Merton, P.A., Morton, H.B. & Adam, J. (1977b). The effect of lesions of the sensorimotor cortex and the capsular pathways on servo responses from the human long thumb flexor. Brain, 100, 503-526.

Marsden, C.D., Rothwell, J.C. & Day, B.L. (1983). Long-latency automatic responses to muscle stretch in man: origin and function. In Motor Control Mechanisms in Health and Disease. Ed. Desmedt, J.E.. Raven Press, New York, pp. 509-539.

Marsden, C.D., Rothwell, J.C. & Day, B.L. (1984). The use of peripheral feedback in the control of movement. Trends Neurosci., 7, 253-257.

Matthews, P.B.C. (1984). Evidence from the use of vibration that the human long-latency stretch reflex depends upon spindle secondary afferents. J. Physiol., 348, 383-415.

Merton, P.A., Morton, H.B., Hill, D.K. & Marsden, C.D. (1982). Scope of a technique for electrical stimulation of human brain, spinal cord, and muscle. Lancet, ii, 597-600.

Obeso, J.A., Rothwell, J.C. & Marsden, C.D. (1985). The spectrum of cortical myoclonus: from focal reflex jerks to spontaneous motor epilepsy. Brain, 108, 193-224.

Pagni, C.A., Ettore, G., Infuso, L. & Marossero, F. (1964). EMG responses to capsular stimulation in the human. Experientia, 20, 691-692.

Pagni, C.A., Marossero, F., Cabrini, G., Ettore, G. & Infuso, L. (1971). Pathophysiology of stimulus sensitive myoclonus: a stereo-EEG study.

Electroenceph. clin. Neurophysiol., 31, 176.
Phillips, C.G. (1969). Motor apparatus of the baboon's hand. Proc. R. Soc. B, 173, 141-174.
Rosen, T., Fehling, C., Sedgwick, M. & Elmquist, D. (1977). Focal reflex epilepsy with myoclonus; electrophysiological investigation and therapeutic implications. Electroenceph. clin. Neurophysiol., 42, 95-106.
Rothwell, J.C., Traub, M.M., Day, B.L., Obeso, J.A., Thomas, P.K. & Marsden, C.D. (1982). Manual motor performance in a deafferented man. Brain, 105, 515-542.
Rothwell, J.C., Obeso, J.A. & Marsden, C.D. (1984). On the significance of giant somatosensory evoked potentials in cortical myoclonus. J. Neurol. Neurosurg. Psychiat., 47, 33-42.
Shibasaki, H. & Kuroiwa, Y. (1975). Encephalographic correlates of myoclonus. Electroenceph. clin. Neurophysiol., 39, 455-463.
Shibasaki, H., Yamashita, Y. & Kuroiwa, Y. (1978). Electroencephalographic studies of myoclonus. Myoclonus-related cortical spikes and high amplitude somatosensory evoked potentials. Brain, 101, 447-460.
Starr, A., McKeon, B., Skuse, N. & Burke, D. (1981). Cerebral potentials evoked by muscle stretch in man. Brain, 104, 149-166.
Sutton, G.G. & Mayer, R.F. (1974). Focal reflex myoclonus. J. Neurol. Neurosurg. Psychiat., 37, 207-217.
Tracey, D.J., Walmsley, B. & Brinkman, J. (1980). Long-loop reflexes can be obtained in spinal monkeys. Neurosci. Lett., 18, 59-65.
Vallbo, A.B. (1974). Human muscle spindle discharge during isometric voluntary contractions: amplitude relations between spindle frequency and torque. Acta. physiol. scand., 90, 319-336.
Wiesendanger, M. & Miles, T.S. (1982). Ascending pathway of low threshold muscle afferents to the cerebral cortex and its possible role in motor control. Physiol. Rev., 62, 1234-1270.

AUTHOR INDEX

Figures in bold type indicate pages on which references are listed

Abbruzzesse, G. 467, **473**
Abbs, J.H. 313, **313**
Abraham, L.D. 175, **184**
Adam, J. 466, 468, **474**
Akeson, W.H. 221, **229**
Alexander, R.McN. 219, 221, **228**
Alexandrowicz, J.S. 145, **165**
Alstermark, B. 303, **305**
Altman, J.S. 52, **55**, 237, **247**, 308, 310, **313**, **314**, 337, 339, **350**, **352**
Altner, H. 237, **248**
Alverdes, F. 290, **294**
Amiel, D. **221**, 229
Anderson, D.J. 87, **90**
Anderson, J.H. 269, **274**
Anderson, M.E. 271, **273**
Andersson, O. 9, **11**, 42, 44, 45, 46, 47, 50, **53**, 301, **305**, 312, 313, **314**, 320, 330, **333**, 358, **375**, 461, **463**
Angaut-Petit, D. 5, **11**
Appelberg, B. 417, **427**
Appenteng, K. 78, 83, 85, 87, **90**, **91**, 117, 118, **119**, **120**, 171, **171**, 180, **184**
Arbas, E.A. 339, **350**
Arbuthnott, E.R. 126, 128, **142**
Arikawa, K. 62, **74**
Arridge, R.G.C. 221, **228**
Arshavsky, Y.I. 37, **53**, 462, **463**
Atwood, H.L. 239, **247**
Ayers, J.L. 277, **295**, 379, 381, 387, **398**, **399**

Bacon, J.P. 302, 337-54, 346, 347, 348, **350**, **352**, **354**

Baer, E. 221, **228**
Bailey, C.H. 239, **247**
Bak, M.J. 175, 176, **186**
Baker, P.S. 338, 339, **350**, **351**
Baker, R. 78, **90**, 265, 266, 269, 271, **273**, **274**, **275**
Bakker, D.A. 177, **186**
Baldissera, F. 302, 303, **305**
Baliff, L. 413, **427**
Ballard, K.J. 126, 128, **142**
Bampton, S. 451, **463**
Banks, R.W. 128, 137, **142**
Bannatyne, B.A. 239, **248**
Barbeau, H. 40, **55**, 357, **376**
Bard, P. 35, **53**
Barker, D. 123, 128, 137, 139, **142**
Barlow, D. 254, **258**
Barnes, W.J.P. 9, **11**, 50, **53**, 188, 199, **206**, 211-16, 253-8, 254, 256, **258**, 289, 290, 291, **294**, **296**, 379, 383, 384, 387, 388, 392, 393, **399**, **400**
Basmajian, J.V. 451, **463**
Bässler, U. 188, 194, 199, **206**, **207**, 311, 312, **314**, 398, **399**
Bawa, P. 435, **449**
Bayev, K.V. 374, **375**
Becht, G. 199, **206**
Bedingham, W. 434, **449**
Behrends, H.B. 358, 359, **377**
Bennet-Clark, H.C. 221, **228**
Bennett, M.V.L. 245, **250**
Bentley, D.R. 58, **74**
Berardelli, A. 467, **473**
Berger, C.S. 147, **165**
Berger, W. 461, **463**

Author index

Bergman, F. 105, **111**
Berkinblit, M.B. 462, **463**
Berthoz, A. 259-75, 260, 263, 264, 265, 266, 267, 269, 270, 271, 272, **273, 274, 275**
Bessou, P. 128, **142**
Bethe, A. 318, **333**
Bicker, G. 347, **351**
Binder, M.D. 177, **185, 186**
Bishop, G.A. 262, **274**
Bizzi, E. 270, **274**, 424, **427**
Bjursten, L.-M. 35, **53**
Blanchette, G. 39, **56**
Blight, A.R. 156, **165**, 332, **333, 373, 399**
Bone, Q. 329, **333**
Botterman, B.R. 177, **185**
Bowerman, R.F. 286, **294**, 387, **399**
Boyd, I.A. 115, 123-44, 123, 124, 126, 127, 128, 129, 130, 131, 132, 138, 140, 141, **142, 143**, 176, 179, 180, **184**
Brink, E. 177, **185**
Brinkman, J. 466, **475**
Brodfuehrer, P. 199, 206, **206**
Brodin, L. 319, **333**
Brown, A.G. 215, **216**, 239, **248**
Brown, T.I.H. 218, **228**
Bruce, I.C. 434, **449**
Brunelli, M. 239, **248**
Bryan, J.S. 239, **248**
Buchthal, F. 4, **11**
Budakova, N. 40, **53**
Buddenbrock, W. von 57, **74**
Bullock, T.H. 280, **295**
Burg, D. 413, **427**
Burgess, P.R. 176, **185**
Burke, D. 118, **119**, 181, **185**, 413, 417, 418, 421, 425, **427**, 467, **473, 475**
Burke, R.E. 37, **55**, 78, 90, 239, **249**, 424, **427**
Burns, M.D. 189, **206**, 214, **216**, 289, 291, **296**, 343, 347, **354**
Burrows, M. 6, **12**, 63, 70, **75**, 109, 110, 215, **215, 216**, 231-50, 233, 234, 235, 237, 239, 241, 242, 243, 246, **247, 248, 249, 250**, 291, **294**, 303, **305**, 337, **351**
Buser, P. 118, **120**

Bush, B.M.H. 115, 117, 118, **119**, 145-66, 145, 146, 147, 148, 149, 150, 151, 153, 156, 157, 158, 159, 162, **165**, 285, **295**, 379, 383, 384, 387, **399**,
Byrne, J. 239, **248**

Cabelguen, J.-M. 37, **56**, 118, **119**, 359, 364, 373, **375, 377**
Cabrini, G. 467, **474**
Calabrese, R.L. 93, **110**, 239, **248**
Cameron, W.E. 177, **185**
Camhi, J.M. 9, 93-111, 93, 95, 97, 99, 101, 102, 103, 105, 106, 107, 109, **110, 111**, 194, 195, **208**, 302, 338, 343, 346, **351**
Cannone, A.J. 118, **119**, 145-66, 146, 149, 151, 153, 156, 157, 159, 162, **165**, 379, 384
Carlstead, M.K. 280, 284, **297**
Carmel, P.W. 116, **119**
Castellucci, V. 239, **248**
Cattaert, D. 286, **295**
Catton, W.T. 343, **351**
Cavagna, G.A. 227, **228**
Chadwick, D. 465, 468, **473, 474**
Chakraborty, A. 343, **351**
Chandler, R.B. 405, **409**
Chasserat, C. 379, 384, 389, 392, **393, 399**
Chen, M.C. 239, **247**
Cheney, P.D. 434, **448**
Chubb, A.D. 329, **333**
Clarac, F. 47, 50, 164, **165**, 169, **171**, 188, **206**, 289, 290, **295, 297**, 302, 379-400, 379, 381, 382, 383, 384, 385, 387, 388, 389, 391, 392, **393, 399, 400**
Clark, F.J. 176, **185**
Clough, J.F.M. 426, **427**
Cody, F.W.J. 411, **427**, 442, **448**
Cohan, C.S. 27, **31**
Cohen, A. 319, **333**
Cohen, M.J. 215, **216**, 280, **295**
Collett, T. 339, 344, **351**
Collewijn, H. 255, **258**
Collins, W.F. 245, **247**
Comer, C. 106, **110**
Cope, T.C. 177, **186**
Cordo, P.J. 407, **409**, 455, **463**

Author index

Corvisier, J. 260, 269, 270, 271, 272, **274**, **275**
Coulmance, M. 381, 387, **399**
Crago, P.E. 89, **90**, 413, 414, 415, **427**, **429**
Critchlow, V. 411, 414, **427**
Croll, R.P. 15, 18, 19, 20, 21, 26, 27, 28, 30, **31**, **32**, **33**
Crommelinck, M. 270, **274**
Crosby, E.C. 328, **334**
Crowe, A. 128, 138, **143**
Cruse, H. 384, 389, 392, 393, **399**
Cullheim, S. 245, **248**
Czarkowska, J. 81, **90**, 177, **185**

Dagan, D. 65, **75**
Dahlman, D.L. 338, **352**
Daley, D.L. 93, 105, 107, 108, **110**
Dando, M.R. 22, **31**
Darlot, C. 265, 270, **274**
Davenport, A. 72, **74**
Davey, M.R. 169, **171**, 411, **427**
Davies, C.A. 346, **354**
Davis, K.B. 20, **32**
Davis, W.J. 10, **11**, 13-33, 13, 15, 17, 18, 19, 20, 21, 22, 26, 27, 28, 30, **31**, **32**, **33**, 50, 51, 52, **53**, **54**, 107, **110**, 284, 286, **295**, 379, 381, 385, 387, **398**, **399**
Dawson, G.D. 467, **473**
Day, B.L. 424, **429**, 437, **448**, 465, 466, 467, **473**, **474**, **475**
Delcomyn, F. 52, **54**, 93, 99, 107, 108, **110**, 187, 188, 189, 191, 193, 204, **206**, 307, **314**
Dellow, P.G. 83, **90**
Denise, P. 270, **274**
Descartes, R. 211, 212, **215**
Dev, P. 424, **427**
Diamant, J. 221, **228**
DiCaprio, R.A. 164, **165**, 188, **206**, 383, **399**
Dietz, V. 461, **463**
Dorsett, D.A. 3-12, 7, 8, **11**, 72, **74**
Downer, R.G.H. 72, **74**
Dreissman, G. 279, 280, 281, 282, **297**
Dresden, D. 197, **207**
Drew, T. 302, 355-77, 369, 370, **375**
Drewes, C. 93, **110**
Droge, M.H. 332, **333**

Droulez, J. 270, **274**
Dugard, J.J. 338, **351**
Dum, R.P. 37, **55**
Dutia, M.B. 135, 137, 138, **143**
Duysens, J. 47, **54**, **55**, 175, 176, **185**, **186**, 357, **375**, **376**, 395, **400**, 460, **463**

Eaton, R.C. 93, **110**
Eccles, J.C. 233, 234, 239, **247**, 424, 425, **428**
Eccles, R.M. 302, **305**, 424, 425, 426, **428**
Edgerton, V.R. 456, **463**
Egger, M.D. 239, **249**
Eggleston, A.C. 286, **295**
Eisen, J.S. 22, **32**, 245, **247**
Eklund, G. 118, **120**, 426, **429**, 435, **448**
Ellaway, P.H. 118, **120**
Elmquist, D. 468, **475**
Elsner, N. 59, **74**
Emonet-Dénand, F. 129, 139, **142**, **144**, 418, **428**
Enomoto, S. 83, **91**
Ettore, G. 467, 471, **474**
Euler, C. von 118, **120**, 411, **427**
Evans, P.D. 72, **74**, 243, **249**
Evarts, E.V. 424, **429**, 434, **448**
Evinger, C. 266, 269, **275**
Evoy, W.H. 6, **12**, 188, 199, **206**, 277, **295**, 379, 383, **399**
Eyring, H. 221, **228**

Fatt, P. 233, **247**
Fay, R.R. 280, 283, **295**
Fehling, C. 468, **475**
Feinstein, B. 242, **247**
Feldman, A.G. 451, **463**
Feldman, J. 9, **12**, 319, **334**
Fetz, E.B. 434, **448**
Field, L.H. 60, 66, 72, **75**
Fields, H.L. 6, **12**, 145, **166**
Finley, F.R. 451, **463**
Fomin, S.V. 451, **463**
Forman, R.R. 188, **207**, **208**
Forssberg, H. 35, 39, 40, 41, 49, 50, **54**, 175, **185**, 187, **207**, 301, **305**, 312, 313, **314**, 320, 330, **333**, **356**, 357, 358, 360, **375**, **376**, 451-64, 453, 455, 456, 460, 461, **463**, **464**

479

Author index

Fourtner, C.R. 191, 194, 199, 206, 206, 207, 233, 249
Fraenkel, G. 283, 295
Frank, E. 245, 250
Frankstein, S.I. 461, 464
Fraser, P.J. 284, 295
Freedman, W. 254, 258
Friedman, D.P. 472, 473
Fuchs, A.F. 260, 274
Fukami, Y. 125, 143
Fukson, O.I. 462, 463
Fukushima, K. 271, 275
Fullerton, B.C. 242, 247
Fulton, J.F. 413, 427, 428
Fyffe, R.E.W. 239, 247, 248,

Gandevia, S.C. 467, 473
Garcia-Rill, E. 36, 54
Garfin, S.S. 221, 229
Garnett, R. 426, 428
Gauthier, L. 47, 56, 356, 357, 359, 376, 377, 384, 400
Geisert, B. 237, 248
Gelfand, I.M. 37, 53, 462, 463
Getting, P.A. 8, 11, 22, 32, 33
Gettrup, E. 309, 315, 337, 338, 351
Gewecke, M. 337, 339, 351, 352
Ghez, C. 466, 473
Gibson, J.J. 339, 344, 351
Gillette, R. 15, 17, 18, 19, 20, 21, 28, 30, 32, 93, 107, 110
Gladden, M.H. 123, 124, 126, 128, 129, 130, 133, 135, 142, 143
Glantzman, D.L. 60, 74
Godden, D.H. 150, 151, 165
Gogan, P. 243, 248
Goldberg, J. 269, 275
Gomez, M.A. 221, 229
Goodman, C.S. 60, 74, 344, 353
Goodman, L.J. 338, 352
Goodwin, G.M. 118, 120
Goslow, G.E. 83, 91
Gottlieb, S. 9, 77-91, 117, 118, 120, 121, 419, 420, 428
Gracco, V.L. 313, 313
Graham, D. 188, 207
Graham Brown, T. 37, 53, 188, 207
Granit, R. 118, 120, 181, 185, 417, 428
Grantyn, A. 270, 271, 274
Grantyn, R. 271, 274

Gray, J. 318, 333, 334
Greene, P.R. 214, 216
Gregory, G.E. 233, 235, 250
Grigg, P. 176, 185
Grillner, S. 9, 11, 12, 35-56, 35, 36, 37, 38, 39, 40, 41, 42, 43, 44, 45, 46, 47, 49, 50, 51, 53, 53, 54, 55, 83, 90, 118, 120, 175, 185, 187, 207, 301, 304, 305, 312, 313, 314, 318, 319, 320, 321, 322, 323, 324, 325, 326, 327, 329, 330, 332, 333, 333, 334, 356, 358, 360, 375, 376, 383, 399, 456, 460, 461, 463
Gueritaud, J.P. 237, 243, 248
Guinan, J.J. 242, 247
Gunn, D.L. 283, 295
Gurfinkel, V.S. 451, 463
Guthrie, D.M. 199, 207, 346, 352

Hackett, J.T. 93, 110
Hagbarth, K.-E. 118, 120, 169, 172, 181, 186, 411, 412, 413, 415, 417, 418, 420, 421, 425, 426, 427, 428, 429, 430, 435, 448, 460, 466, 473
Hagglund, J.V. 435, 448, 466, 473
Halbertsma, J.M. 39, 40, 41, 47, 49, 50, 54, 55
Hallett, M. 465, 468, 473, 474
Halliday, A.M. 467, 474
Hammond, P.H. 79, 90, 225, 228, 433, 448
Harris, R. 468, 473
Harris, S. 239, 249
Harris-Warrick, R.M. 60, 75
Harrison, P.J. 78, 79, 90, 177, 185, 405, 409
Haskell, P.T. 338, 352
Hassan, Z. 179, 185
Hawkins, R.D. 239, 247
Head, S.I. 162
Hedwig, B. 309, 314
Heinzel, H.-G. 109, 110, 309, 314, 339, 348, 352
Heitler, W.J. 243, 248, 303, 305, 313, 314
Henneman, E. 4, 12, 413, 428
Herman, R. 451, 463
Hermkind, W.F. 280, 295
Heukamp, U. 337, 352
Heyer, C. 7, 12
Higuchi, T. 279, 280, 295

480

Author index

Hikosaka, O. 264, **274**
Hildebrand, C. 38, **56**
Hill, D.K. 471, **474**
Hinkle, M. 338, **351**
Hirai, N. 221, **228**
Hirosawa, K. 328, **335**
Hirth, C. 59, **74**
Hisada, M. 279, 280, 284, 285, 286, 288, 290, **295, 297**, **298**
Hoffer, J.A. 118, **120**, 182, **185**
Hofsten, C. von 457, **463**
Holst, E. von 38, **55**, 318, **334**, 388, 398, **400**
Homma, S. 78, **90**
Hongo, T. 118, **120**
Horcholle-Bossavit, G. 243, **248**
Horridge, G.A. 235, **247**, 254, **258**, 291, **294**
Horsmann, U. 109, **110**, 309, **314**, 348, **352**
Houk, J.C. 89, **90**, 177, 179, 181, **185**, 413, 414, 415, **427, 428, 429**, 431, 434, **448**
Hoy, M. 39, **56**
Hoyle, G. 6, 8, **11, 12**, 57-75, 58, 59, 60, 61, 62, 63, 64, 65, 66, 67, 68, 69, 70, 71, 72, **74, 75**, 188, 202, **207**, 242, 243, **248**
Huber, G.C. 328, **334**
Hughes, G.M. 191, **207**
Hulliger, M. 116, 117, 118, 119, **120**, 179, 180, **185**, 412, 417, 419, 420, 421, **427, 428, 429, 430**
Hultborn, H. 233, **248**, 302, 303, **305**, 426, **428**
Hume, R.I. 22, **32, 33**
Hunt, C.C. 129, **144**, 152, **166**

Iles, J.F. 9, **12**, 193, **207**, 310, 311, **314**
Illert, M. 302, 303, **305**, 426, **428**
Infuso, L. 467, 471, **474**

Jackson, J.H. 412, **428**
Jackson, M.B. 36, **54**
Jacquin, M. 239, **249**
Jami, L. 132, 142, **143**
Jankowska, E. 81, **90**, 177, 179, **185**, **186**, 233, 234, 242, **248, 249**, 301-5, 302, 303, **305**, 358, **376**
Jansen, J.K.S. 413, **428**

Jansson, G. 457, **463**
Jazak, M.M. 257, **258**, 283, 284, **296**
Jenner, J.R. 89, **91**, 466, **474**
Jenner, P. 468, **473**
Johannisson, T. 81, **90**, 177, **185**
Johansson, G. 457, **463**
Johansson, H. 417, **427**
Johansson, R.S. 426, **430**
Jones, E.G. 472, **473**
Jones, K.A. 293, **296**
Jordan, L. 36, **55**
Joseph, M.P. 242, **247**
Jukes, M.G.M. 358, **376**
Julien, C. 40, **55**, 356, 357, 359, 367, **376, 377**

Kalil, R.E. 270, **274**
Kamei, H. 37, **56**, 358, **376**
Kammer, A.E. 338, **352**
Kanda, K. 461, **463**
Kandel, E.R. 239, **247, 248**
Kappers, L.K.A. 328, **334**
Kashin, S. 318, 332, **334**
Kater, S.B. 7, **12**
Katoh, M. 83, **91**
Katsuki, Y. 280, **295**
Kawakami, T. 264, **274**, 275
Keifer, J. 456, **464**
Keller, A. 221, **228**
Kellerth, J-O. 245, **248**
Kennedy, D. 6, **12**, 93, **110**, 239, **248**, 284, **295**
Ker, R.F. 221, **228**
Kernell, D. 426, **427**
Kidokoro, Y. 83, **91**
Kien, J. 22, **32**, 52, **55**, 310, **314**, 337, **352**
Kinnamon, S.C. 338, **352**
King, A.J. 339, 341, 344, **351**
King, D.G. 246, **248**
Kitai, S.T. 262, **274**
Klärner, D. 387, 388, 392, 393, **400**
Klassen, D. 338, **352**
Klassen, L.W. 338, **352**
Koch, U.T. 338, **352**
Kolmer, W. 325, **334**
Kostyuk, P.G. 374, **375**
Kovac, M.P. 15, 17, 18, 19, 20, 21, 22, 26, 27, 28, 30, **31, 32, 33**, **107, 110**
Kramer, E. 339, **353**

481

Author index

Krasne, F.B. 60, **74**, 93, 107, **111**, 239, **248**, 286, **298**
Krauthamer, V. 191, **207**
Kravitz, E.A. 60, **75**, 215, **216**
Kristan, W.B. 9, **12**, 22, **23**, **33**
Kubo, Y. 83, **91**
Kubota, K. 83, **91**, 411, **429**
Kucera, J. 123, 124, **144**
Kuei, S.C. 221, **229**
Kugelberg, E. 467, **474**
Kulagin, A.S. 47, **55**
Kupfermann, I. 26, **33**, 53, **55**
Kuroiwa, Y. 468, 470, **475**
Kutsch, W. 59, **75**, 308, **314**

Lagerbäck, P.-A. 38, **55**, 242, 245, **248**, 326, 327, **334**
Lan-Couton, D. 132, **143**
Landgren, S. 233, **247**
Langberg, J.J. 105, 106, 107, 109, **111**
Lansner, A. 322, **335**
Laporte, Y. 129, 139, **142**, **144**, 418, **428**
Larimer, J.L. 286, **294**, **295**
Laverack, M.S. 145, **165**
Le Mare, D.W. 318, **335**
Lee, D.N. 457, **464**
Lee, R.G. 225, **228**, 436, 442, 445, **448**, 466, **474**
Lennard, P.R. 22, **32**, **33**
Lenz, F.A. 436, **448**
Leonard, R.B. 332, **333**
Lestienne, F. 270, **274**
Libersat, F. 169, **171**, 388, **400**
Liddell, E.G.T. 413, **427**
Light, A.R. 239, **247**, 249
Lindegard, B. 242, **247**
Linden, R.W.A. 83, **91**
Lindquist, M. 330, **333**, 358, **375**, 461, **463**
Lindstrom, S. 233, 234, **248**
Lipski, J. 81, **90**
Lishman, R. 457, **464**
Lisin, V.V. 461, **464**
Liske, E. 199, **207**
Lissmann, H.W. 187, **207**, 318, **334**, **335**
Litt, M. 221, **228**
Livingstone, M.S. 60, **75**
Llinás, R. 156, **165**, 383, **399**

Loeb, G.E. 117, 118, **120**, 171, **172**, 173-86, 175, 176, 177, 179, 182, 183, **184**, **185**, 186, 357, **375**, 411, **428**, 460, **463**
Löfstedt, L. 413, 417, 421, **427**, **428**
Lopez-Barneo, J. 265, **274**
Lorenz, K.Z. 57, **75**
Lorton, E.D. 280, 283, **295**
Lucas, S.M. 177, **186**
Lund, J.P. 83, 85, 87, **90**, 356, 359, 366, **376**, **377**, 461, **464**
Lund, S. 118, **120**, 358, **376**
Lundberg, A. 37, **55**, 81, 82, **91**, 169, **172**, 177, **186**, 302, 303, **305**, 358, 373, **376**, 424, 425, 426, **428**
Luschei, E.S. 118, **120**

McClellan, A. 9, **11**, 319, 320, 324, 325, **334**
McCrea, D.A. 177, 179, **185**, **186**
McCrea, R.A. 260, 262, 263, 264, 266, 269, **274**, **275**
MacDermot, N. 442, **448**
Macht, M. 35, **53**
Mackel, R. 179, **185**, **186**
McKeon, B. 467, **473**, **475**
McMahan, T.A. 214, **216**
McMahan, U.J. 215, **216**, 237, **249**
MacMillan, D.L. 52, **55**
McWilliam, P.N. 124, **143**
Madrid, J. 239, **249**
Maeda, M. 266, **275**
Maeda, N. 329, **335**
Malamed, S. 239, **249**
Malmgren, K. 81, 82, **91**, 177, **186**
Mann, C. 131, 132, **143**
Marder, E. 22, **32**, 245, **247**
Marks, W.B. 117, **120**
Marossero, F. 467, 471, **474**
Marsden, C.D. 79, 87, **91**, 213, 215, 225, **228**, 424, 426, **429**, 433, 436, 437, **448**, 455, **464**, 465-75, 465, 466, 467, 468, 471, 472, **473**, **474**, **475**
Masarachia, P. 239, **249**
Matera, E.M. 15, 17, 18, 19, 20, 21, 26, 27, 28, 30, **31**, **32**, **33**
Matsukawa, K. 37, **56**, 358, **376**
Matsunami, K. 411, **429**
Matthews, B.H.C. 413, **429**
Matthews, H.R. 435, **449**

Author index

Matthews, M.A. 243, **248**
Matthews, P.B.C. 79, 81, **91**, 117, 119, **120**, 128, 138, **143**, 155, **166**, 170, **172**, 179, 180, **185**, 214, **215**, **216**, 218, 225, **228**, 418, **428**, 431-49, **432**, **435**, **436**, 438, 440, 442, 443, 444, 445, **448**, **449**, 466, **474**
Maxwell, D.J. 239, **248**
Mayer, R.F. 468, **475**
Meche, F.G.A. van der 40, **55**, 357, 358, **376**
Meinck, H.M. 359, **377**
Mellon, DeF. 280, 283, 291, **295**, **296**
Melvill Jones, G. 188, **207**
Merton, P.A. 58, **75**, 79, **90**, 212, 213, **215**, **216**, 225, **228**, 412, 426, **429**, 433, **436**, **448**, 455, **464**, 466, 471, **474**
Miles, T.S. 466, **475**
Mill, P.J. 187, **207**
Miller, J.P. 22, **33**, 308, **314**
Miller, S. 40, **55**, 357, 358, **376**
Minoda, K. 358, **376**
Mirolli, M. 147, **166**
Miyan, J.A. 279, 281, 282, 283, 284, 286, 288, 289, **295**, **296**
Miyoshi, S. 329, **335**
Möhl, B. 309, 313, **314**, 337, 338, 347, 348, **350**, **352**, **354**
Mohren, W. 199, **207**
Monnier, S. 290, **297**
Moran, D.T. 188, 199, 201, 202, 203, 204, **208**, 243, **250**
Morasso, P. 424, **427**
Morielli, A. 28, **33**
Morimoto, T. 87, **91**, 118, **119**
Morin, W.A. 239, **247**
Mortin, L.I. 456, **464**
Morton, H.B. 213, **215**, 225, **228**, 426, **429**, 433, **436**, **448**, 455, **464**, 466, 471, **474**
Moss, V.A. 130, **143**
Mpitsos, G.J. 20, 21, 27, **31**, **32**, **33**
Muller, K.J. 215, **216**, 237, 245, **248**, **249**
Munson, J.B. 78, **91**
Murakami, T. 366, **376**, 461, **464**
Murphey, R.K. 93, **111**
Murphy, P.R. 118, **120**, 130, 131, 132, **143**

Murthy, K.S.K. 179, **186**

Nachtigall, W. 309, **314**, 337, **352**
Nakajima, Y. 328, **335**
Nakamura, Y. 83, 87, **91**
Nashner, L.M. 407, **409**, 451, 453, 455, 456, **463**, **464**
Neil, D.M. 277-98, **278**, 279, 280, 281, 282, **283**, 284, 288, 289, 290, 291, **294**, **295**, **296**, **297**
Neumann, L. 309, 313, **314**, 337, **352**
Newland, P. **287**
Nicholls, J.G. 245, **249**
Nijenhuis, E.D. 197, **207**
Nolen, T.G. 93, 99, 105, 109, **110**
Nordh, E. 417, **429**
Norris, B.E. 242, **247**
Norrsell, K. 35, **53**
Norrsell, U. 35, **53**, 303, **305**
North, A.G.E. 434, **449**
Noth, J. 179, 180, **185**
Novotny, M. 194, **208**
Nozaki, S. 83, **91**
Nunez, R. 239, **249**
Nyman, E. 242, **247**

O'Donovan, M.J. 37, **55**, 78, 83, **90**
O'Shea, M. 243, **249**, 344, **352**
Obeso, J.A. 424, **429**, 465, 468, 471, **474**, **475**
Oda, Y. 37, **56**
Okajima, A. 254, **258**
Olivo, R.F. 257, **258**, 283, 291, **296**
Olsson, K.A. 366, **376**
Orlovsky, G.N. 35, **37**, **53**, **55**, **56**, 169, **172**, 214, **216**, 369, **376**, 411, **429**, 462, **463**
Orr, C.A. 9, **12**
Orsal, D. 359, 364, **373**, **375**
Otto, D. 59, **75**
Ottoson, D. 152, **166**

Pabst, H. 337, **352**
Page, C.H. 277, 285, 286, 293, **296**
Pagès, P. 128, **142**
Pagni, C.A. 467, 471, **474**
Palka, J. 93, **111**
Parnas, I. 105, **111**
Parry, D.A. 343, **352**
Pasztor, V.M. 391, **400**

483

Author index

Paul, D.H. 37, 55
Pavlidis, T. 322, 355
Pearson, K.G. 9, 12, 47, 50, 51, 54, 55, 56, 60, 74, 189, 191, 192, 193, 194, 197, 199, 204, 206, 207, 208, 233, 249, 302, 303, 305, 307-15, 308, 309, 310, 311, 312, 313, 314, 337, 338, 341, 343, 344, 347, 351, 353, 357, 375, 395, 400
Pedotti, A. 451, 464
Perl, E.R. 239, 249
Perret, C. 37, 56, 118, 120, 319, 320, 325, 334, 359, 364, 373, 375, 377
Peterson, B.W. 262, 269, 271, 275
Petit, J. 132, 142, 143, 418, 428
Pflüger, H.-J. 237, 249, 250, 339, 347, 353
Philippen, J. 339, 351
Phillips, C.E. 242, 249
Phillips, C.G. 426, 427, 433, 434, 437, 449, 466, 475
Phillipson, M. 40, 56
Pi-Suner, J. 413, 428
Pinneo, J.M. 20, 32
Pinter, M.J. 37, 55
Pitman, R.M. 215, 216
Plummer, M.R. 103, 111
Polit, A. 424, 427
Pollack, A.J. 105, 107, 111
Poon, M.L.T. 319, 355
Powell, T.P.S. 434, 449
Powers, R.K. 177, 186
Precht, W. 78, 90
Preiss, R. 339, 353
Priest, T.D. 288, 289, 290, 296
Pringle, J.W.S. 197, 199, 207, 208
Prochazka, A. 82, 91, 115-21, 116, 117, 118, 119, 120, 169-72, 169, 171, 171, 172, 177, 180, 184, 186, 357, 369, 377, 411, 412, 415, 420, 428, 429
Proske, U. 118, 120, 139, 142, 180, 184, 214, 216
Provencher, J. 39, 56

Rack, P.M.H. 135, 144, 217-29, 218, 219, 220, 221, 224, 226, 227, 228, 426, 429
Rademaker, G.C.J. 254, 258
Ralston, D.D. 239, 249

Ralston, H.J. 239, 249, 451, 464
Rastad, J. 242, 249
Rechtmann, M.B. 461, 464
Redman, S. 233, 249
Reichert, H. 292, 296, 302, 337-54, 341, 343, 347, 353
Reid, S.A. 78, 91
Reingold, S. 194, 195, 208
Reinking, R.M. 177, 185
Rethelyi, M. 239, 249
Retzius, G. 325, 335
Reye, D.N. 50, 51, 56, 309, 310, 314, 337, 353
Reynolds, E.H. 468, 473
Richardson, H.C. 442, 448
Richmond, F.J.R. 177, 186
Ridge, R.M.A.P. 118, 120
Rigby, B.J. 221, 228
Ripley, S.H. 214, 216, 242, 250
Risling, M. 38, 56
Ritter, M.A. 221, 229
Ritzmann, R.E. 105, 107, 111
Roberts, A. 145, 158, 165
Roberts, B.L. 5, 12, 37, 55, 319, 329, 355
Roberts, J. 7, 11
Roberts, W.J. 233, 248
Robertson, D.G.E. 451, 464
Robertson, R.M. 50, 51, 56, 308, 309, 310, 314, 337, 344, 353
Robine, K.-P. 271, 274
Robinson, D.A. 260, 264, 274, 275
Roesler, J. 359, 377
Roll, J.P. 118, 120, 421, 429
Romnevi, L-O. 242, 245, 248
Rose, P.K. 215, 216
Rosen, T. 468, 475
Rosenthal, G.A. 338, 352
Ross, H.F. 218, 219, 220, 221, 224, 226, 227, 228
Rossignol, S. 39, 40, 42, 43, 47, 49, 50, 54, 55, 56, 175, 185, 187, 207, 302, 329, 330, 333, 334, 355-77, 356, 357, 359, 360, 366, 367, 369, 370, 375, 376, 377, 384, 400, 460, 461, 463, 464
Rothwell, J.C. 424, 429, 437, 448, 465-75, 465, 466, 467, 468, 471, 473, 474, 475
Roucoux, A. 269, 270, 271, 274, 275
Rovainen, C.M. 319, 325, 333, 355

Author index

Rowell, C.H.F. 302, 337-54, 338, 341, 343, 344, 347, 348, **352**, **353**
Rudjord, T. 413, **428**
Rudomin, P. 239, **249**
Ruit, J.B. 357, 358, **376**
Runion, H.L. 189, **208**
Russell, D.F. 370, **377**
Russell, I.J. 93, **111**
Rutkowski, S. 118, **120**
Rymer, W.Z. 78, 89, **90**, 413, 414, 415, 424, **427**, **429**, 431, 434, **448**

Sand, A. 318, **334**
Sandeman, D.C. 254, 257, **258**, 277, 280, 281, 283, **296**, **297**
Sanders, T.M. 221, **229**
Sanes, J.N. 424, **429**
Santini, M. 426, **428**
Sato, H. 461, **463**
Scapini, F. 279, 280, 281, 282, 283, 290, **296**, **297**
Schieber, M.H. 118, **121**, 411, **429**
Schmalbruch, H. 4, **11**
Schmidt, R.F. 234, 239, **247**
Schomberg, E.D. 81, 82, **91**, 177, **186**, 358, 359, **377**
Schöne, M. 279, 280, 281, 282, 283, 284, 288, 289, 290, **296**, **297**
Schwartzkopf, J. 337, **352**
Scott, J.J.A. 142, **143**
Sears, T.A. 118, **121**, 214, **216**
Sedgwick, M. 468, **475**
Séguin, J.J. 85, 87, **90**
Selverston, A.I. 22, **31**, **32**, **33**, 53, **56**, 107, **111**, 303, **305**, 307, 308, **314**
Semba, K. 239, **249**
Severin, F.V. 35, 169, **172**, 214, **216**, 411, **429**
Shaw, M.K. 237, **247**
Sherman, E. 194, **208**
Sherrington, C.S. 187, **208**, 211, 212, **216**, 356, 363, **377**
Shibasaki, H. 468, 470, **475**
Shik, M.L. 35, 37, 47, **55**, **56**, 169, **172**, 214, **216**, 411, **429**
Shimazu, H. 266, **275**
Shinoda, Y. 266, **275**, 466, **473**
Shtilkind, T.I. 451, **463**
Shuto, S. 83, **91**

Siegler, M.V.S. 20, 21, **32**, **33**, 70, **75**, 215, **215**, **216**, 234, 235, 237, 239, 241, 245, **247**, 249, **250**
Sigvardt, K. 9, **11**, 177, **185**, 319, 324, 325, **334**
Sillar, K.T. 164, **166**
Silvey, G.E. 280, **297**
Simmons, P.J. 245, **250**, 341, 342, 343, 344, 345, **353**
Sjöström, A. 456, **463**
Skinner, R.D. 36, **54**
Skoog, B. 177, **185**
Skorupski, P. 164, **166**
Skuse, N.F. 418, 425, **427**, 467, **475**
Smith, J.D. 116, **121**
Smith, J.L. 39, **56**
Smith, L. 39, **56**
Smith, M.M. 36, **54**
Smith, R.S. 123, **143**
Smola, U. 346, **353**
Snow, P.J. 215, **216**
Sojka, P. 417, **427**
Sombati, S. 61, 62, 63, 64, 66, 67, 68, 69, 70, 71, **75**, 338, **353**
Somjen, G. 78, 81, 83, **90**, **91**
Sontag, K.H. 357, 369, **377**
Sotelo, C. 243, **250**
Spikes, J.D. 221, **228**
Spira, M.E. 105, **111**, 245, **250**
Spirito, C.P. 188, 199, **206**, 379, 383, **399**
Spitzer, N.C. 60, **74**
Spray, D.C. 245, **250**
Stacey, M.J. 128, 137, 139, **142**
Stark, L. 116, **121**
Starr, A. 116, **119**, 467, **475**
Steeves, J.D. 303, **305**, 313, **314**
Steffens, H. 359, **377**
Stein, A. 278, 280, 284, 290, **297**
Stein, P.S.G. 35, 50, 52, **56**, 322, **335**, 456, **464**
Stein, R.B. 118, **120**, 182, **186**, 357, **376**
Steiner, I. 318, **335**
Stent, G.S. 9, **12**, 245, **249**
Stephens, J.A. 78, 83, 89, **90**, **91**, 174, 177, **186**, 403-9, 426, **428**, 466, **474**
Stewart, W.W. 215, **216**
Stretton, A.O.W. 215, **216**

Struppler, A. 413, **427**
Stuart, D.G. 83, **91**, 174, 177, **185**, **186**
Sugawara, K. 279, 280, **295**
Sumino, R. 83, **91**
Sutherland, F. 126, 128, 132, 138, 142, **143**
Sutton, G.G. 79, **90**, 468, **475**
Suzuki, Y. 284, **297**
Svidersky, V.L. 347, **354**
Sybirska, E. 81, **90**, 177, **185**, 303, **305**
Sypert, G.W. 78, **91**
Szumski, A.J. 413, **427**

Tagliasco, V. 270, **274**
Takahata, M. 280, 284, 285, 286, 288, **297**, **298**,
Takeuchi, A. 239, **250**
Takeuchi, N. 239, **250**
Tanaka, Y. 62, **74**
Tao-Cheng, J.-O. 328, **335**
Tasker, R.R. 436, **448**
Tatton, W.G. 225, **228**, 434, 436, 442, 445, **448**, **449**, 466, **474**
Tautz, J. 237, **249**, 339, **353**
Taxi, J. 243, **250**
Taylor, A. 9, 77-91, 78, 83, 87, **90**, **91**, 117, 118, **119**, **120**, **121**, 169, **171**, 405, **409**, 411, 419, 420, **427**, **428**
Taylor, C.P. 338, 339, 341, **353**
Taylor, J. 118, **120**
Ter Braak, J.W.G. 254, **258**
Teräväinen, H. 5, **12**
Terdiman, J. 116, **121**
Thach, W.T. 118, **121**, 411, **429**
Theophilidis, G. 214, **216**
Thilmann, A.F. 219, 221, 224, 226, **228**
Thomas, P.K. 424, **429**, 465, **475**
Thompson, K.J. 65, **75**
Toh, H. 329, **335**
Tom, W. 95, **97**, 101, 102, **110**
Torebjörk, H.E. 181, **186**, 412, 426, **429**, **430**
Tracey, D.J. 176, **186**, 466, **475**
Traub, M.M. 424, **429**, 465, **475**
Treherne, J.E. 60, **75**
Trott, J.R. 118, **120**
Tweedle, C.D. 215, **216**

Tyc-Dumont, S. 243, **248**
Tyrer, N.M. 233, 235, 237, **247**, **250**, 337, 346, 347, **350**, **354**

Uchizono, K. 239, **250**
Udo, M. 37, **56**, 358, **376**
Usherwood, P.N.R. 188, 189, 191, 204, **206**, **208**

Vallbo, A.B. 169, **172**, 181, **186**, 218, **228**, 411-30, 411, 412, 413, 414, 415, 416, 417, 418, 419, 420, 421, 422, 423, 424, **428**, **429**, **430**, **473**, **475**
Van Gisbergen, J.A.M. 264, **275**
Varanka, I. 347, **354**
Varela, F.G. 201, **208**
Varju, D. 254, **258**
Vedel, J.-P. 118, **120**, 188, **208**, 289, 290, **295**, **297**, 383, 387, 388, 399, **400**, 421, **429**
Velho, F. 413, **427**
Vernon, A. 219, 221, **228**
Verhaart, C. 270, **274**
Vidal, P.P. 260, 263, 264, 266, 267, 269, 270, 271, 272, **274**, **275**
Volman, S. 95, 97, **110**

Wagner, H. 339, 344, **354**
Waldren, I. 52, **56**, 309, **314**, 337, **354**
Wallén, P. 9, **11**, 35, 36, 38, 39, 50, **55**, **56**, 175, **185**, 301, 302, **305**, 312, 313, **314**, 317-35, 318, 319, 320, 321, 322, 323, 324, 329, 330, 331, 333, **353**, **354**, **355**, 356, **376**
Wallin, B.G. 181, **186**, 412, 417, **427**, **428**, **430**
Wallin, E.U. 435, **448**, 466, **473**
Walmsley, B. 233, **249**, 466, **475**
Walsh, J.V. 78, **90**, 424, **427**
Walters, D.K.W. 219, 221, 224, 226, **228**
Wand, P. 82, **91**, 177, 180, **184**, **186**, 357, 369, **377**, 415, 420, **429**
Ward, J. 124, 129, 132, 138, **143**
Washio, H. 62, **74**
Watkins, J.C. 319, **335**
Watson, A.H.D. 63, **75**, 231-50, 237, 243, 246, **250**

Author index

Watt, D.G.D. 188, **207**
Webb, P.W. 286, **298**
Weeks, J.C. 22, 23, **33**, 107, **111**
Wegner, U. 188, 194, **206**, 311, 312, **314**
Weis-Fogh, T. 338, 339, 346, **354**
Weiss, K.R. 26, **33**, 53, **55**
Wendler, G. 109, **110**, **111**, 197, **208**, 309, **314**, **315**, 337, 348, **352**, **354**
Westbury, D.R. 135, **144**, 224, **228**
Westerfield, M. 245, **250**
Westerman, R.A. 82, **91**, 411, **429**
Westin, J. 105, 106, 107, 109, **111**
Westling, G. 426, **430**
Westman, J. 242, **249**
Weston, B.J. 357, **375**
Wetzel, M.C. 83, **91**
Whitear, M. 145, **165**, **166**
Widen, L. 467, **474**
Wiens, T.J. 383, **400**
Wiersma, C.A.G. 214, **216**, 242, **250**, 290, **298**
Wiesendanger, M. 434, **449**, 466, **475**
Wilén, M. 319, **334**
Williams, M. 22, **32**
Williams, T.L. 9, **11**, 38, 39, **55**, **56**, 302, 317-35, 318, 319, 326, 327, **334**, **335**
Williams, V. 243, **248**
Williamson, R.M. 5, **12**, 343, 347, **354**
Willis, W.D. 234, **239**, 243, **247**, **248**
Willows, A.O.D. 8, **11**, 58, **74**
Wilson, D.M. 51, 52, **56**, 109, **111**, 199, 206, **208**, 308, 309, **315**, 338, 339, **351**, **354**
Wilson, M. 339, 341, **354**

Wilson, V.J. 269, 271, **273**, **275**
Wine, J.J. 93, 107, **110**, **111**, 239, 248, 286, 292, **296**, **298**
Winter, D.A. 451, **464**
Wirta, R. 451, **463**
Withey, T.P. 83, **91**
Wohlfart, G. 242, **247**
Wong, R.K.S. 197, 199, **207**, **208**, 233, **249**
Woo, S.L.-Y. 221, **229**
Wotherspoon, R.M. 288, 289, **296**
Wyman, R.J. 309, **315**

Yamashita, Y. 470, **475**
Yang, G. 239, **249**
Yoshida, K. 260, 263, 264, 266, 267, 269, **274**, **275**
Yoshida, M. 271, **273**
Yoshino, M. 284, 285, 286, 288, **297**, **298**
Young, D. 199, **208**
Young, R.R. 415, **428**, 466, **473**

Zajac, F.E. 370, **377**
Zangger, P. 9, **12**, 37, 38, 39, **55**, 117, 118, 119, **120**, 319, 320, **334**, 358, **376**, 456, **463**
Zarnack, W. 338, **354**
Zattara, M. 359, 364, 373, **375**
Zengel, J.E. 78, **91**
Zernicke, R.F. 39, **56**
Ziccone, S.P. 82, **91**, 411, **429**
Zill, S.N. 169, **171**, 187-208, 188, 199, 201, 202, 203, 204, **207**, **208**, 243, **250**, 388, **400**
Zomlefer, M.R. 39, **56**

SUBJECT INDEX

afferents
 flexor reflex in cat 302
 in ventral roots 37
 joint receptor 176
 muscle spindle 78, 123, 147, 169-70, 179-80, 213, 233, 403-5, 415-26, 466-67
 of leg hairs in locust 237-39
 S and T of crab TCMRO 147
 secondary spindle 405-7, 431-48
 tendon organ 169-70, 176-77, 403-4, 412-15
alpha-gamma linkage 118-19
annelids, muscle in 6
antenna, movement in lobster 289-91
Aplysia, muscle in 7
appropriateness, test for command neurone 26
arthropods
 central pattern generators and flight in 308-10, 337
 equilibrium system in 277-94
 escape behaviour in 93-110
 eye movements in 253-57, 278-83
 fast & slow muscle in 5
 generation of behaviour in 59-74
 intracellular recording from 214-15
 local reflexes in 231-47
 muscle receptors in 145-64
 walking in 187-206, 310-12, 379-98
autotomy, of leg in lobster 388-94
axon collaterals 245

bias 116, 118, 126, 128, 142, 171, 180
brainstem locomotor areas 36

buccal ganglion
 in Philine 7-8
 in Pleurobranchaea 15-21

campaniform sensilla 171, 199-204
canavaline 338
cat 9, 35-56, 77-91, 115-22, 123-44, 169-71, 173-86, 259-76, 355-79
central oscillator 308
 see also central pattern generator
  central rhythm generator
central pattern generator 7-9, 37-41, 171, 175, 304, 307-13
 & movement related feedback 9, 41-47
 definition 51-52
 in cat walking 373-75
 in cockroach walking 188, 193-97, 202-6
 in human walking 456-57
 in jaw movement of cat 83
 in lamprey and dogfish 320-25
 see also central oscillator
  central rhythm generator
central rhythm generator 307
 in locust flight 337, 343, 347
 see also central oscillator
  central pattern generator
cercal hairs, in cockroach 95, 105
cerebellum, effect of ablation 37
chordotonal organs 145, 163-64, 197-99, 280, 374, 379, 384
circadian rhythms 57
clasp knife phenomenon 413
coactivation
 alpha-gamma 181
 in crabs 161-63

489

Subject index

cockroach
  escape behaviour 93-110
  motor output pattern following deafferentation 9
  octopamine in 72
  walking 171, 187-206, 310-12
  command neurones 10, 15, 25-27, 52-53, 61, 310-12
  compartmentalised function, concept of 14-15
  compliance of tendons in man 217-28
  compensatory responses 253, 259, 277
  consensus, principle of 10, 18, 21, 26
  contact, sense of 174
  coordinating neurones 52
  corollary discharge 20
  corollary signals 259, 264
  cortex, sensory area 3a in baboon 434
  cortical lesions, long latency stretch responses following 436
  cortical myoclonus 408-9, 465, 467-73
crab
  chordotonal organ afferents in 384
  eye movements 254-57
  intraleg reflexes 163, 383-84, 387
  muscle fibre types 6
  TCMRO in 115, 117-19, 145-64, 379
crayfish
  escape swimming in 292
  eyestalk movements in 278-80
  fictive locomotion in 163
  muscles of 5-6
  righting reactions in 284, 286-88
  role of octopamine in 60
  startle responses in 293
creep in intrafusal fibre 128-30
cricket courtship 59
  song 58-59
curare paralysis 37, 42, 319
cutaneous
  mechanoreceptors & cat jaw movements 85-87
  receptors initiating phase-dependent reflexes 171, 175-76
  reflexes & human locomotion 460-61
  sensation & walking 50, 82-83, 175-76, 356-72
  transcortical reflexes, in man 80, 213, 408, 466-67
cuticular stress detectors 374, 387

deafferentation, effect on
  cockroach walking 9, 310-12
  cricket song 58-59
  fish swimming 318-19
  grasshopper courtship 59
  insect flight & walking 307-12
  lamprey swimming 9, 38, 319
  motor control in man 465
  Tritonia, swimming in 8, 58
  walking in decerebrate cat 9, 37-38
decerebrate cats
  behaviour of 35
  midbrain stimulation in 35
  walking in 9, 37-38, 357-58, 370
decorticate mammals, behaviour in 35
deefferentation of muscle spindle 180, 182
Deiters' nucleus 370
distributed function, principle of 18, 21
driving of Ia discharge 130-32
DOPA 359
DUM neurones
  in locust metathoracic ganglion 62-66
  initiating movement in locust 69-71
dynamic
  gamma motor neurones 118, 123, 129, 138, 141, 180-84
  index 128
  sensitivity of muscle spindle 128

ear, muscles of 116
economy, principle of 303
edge cells of lamprey spinal cord 9, 325-28
efferent innervation
  of crab TCMRO 148-52
  of muscle spindle 117-18, 123-24, 138-42
efference copy 37, 384
effort, sense of 174
electroencephalogram in cortical myoclonus 467-71
elemental origin of network properties 14-15
emergent network properties, principles of 18, 21
entrainment 41-47, 301, 320-25
epilepsia partialis continua 472-73
equilibrium, control of 251-98

Subject index

escape behaviour
 in cockroach 93-110
 in crustaceans 286-87, 292
 in Tritonia 8, 58
 locust jump 73-74, 243
extraocular muscles 78, 80
eye
 compound, role in flight 339, 343-46
 -head coordination 254-55, 259-73
 movements, in crustaceans 253-57, 278-83
 movements, in rabbits 255
 pupil of 116

feedback
 abnormal, in man 408, 465-66
 efferent control of 115-19
 entrainment of locomotor rhythm by 41-47, 301, 320-25
 in cockroach escape turning 95-110
 in cockroach locomotion 171, 187-206
 in control of rhythmic movement 15, 22, 355
 in equilibrium control 254, 278-83
 in fish & lamprey swimming 317-333
 in lobster walking 383-98
 in locust flight 308, 337
 in motor pattern generation 9, 51
 in saccade generation 264
 loops in Pleurobranchaea 10, 25-28
 multimodal, in behaviour 302
 negative 87, 160-61, 177, 233, 254, 278
 non-involvement of, in escape behaviours 93
 of muscle force 87, 180
 of muscle length 80, 87, 180
 positive 160
fictive locomotion
 in crayfish 164
 in spinal cat 42, 358-59, 461
 in spinal dogfish & lamprey 319-25
fish
 muscle types & their innervation 4-5
 swimming 317-33
fixed action patterns 57
flexor burst generators 194-97, 312
flexor reflex afferents in cat 302
flight
 course maintenance in locust 338-50
 pattern generation 109, 308-10, 337

fusimotor
 activity in human subjects 415-22
 control of intrafusal fibres 138-41
 end plate 126-28
 set 118-19

gain 116-19, 163, 254, 264, 278-83
 of stretch reflex 19, 214, 218, 227, 405
gastropods
 feeding behaviour in 15-22
 muscle innervation in 7
gating of sensory input 9, 50, 81, 246, 301, 330-31, 343, 347-48, 461, 462
gaze 259, 269-71
Golgi tendon organs 174, 176-77, 403-4, 412-15
 output during stepping 169-70

hair, sensory
 in cockroach escape 95, 105
 in locust 235-37
 see also wind receptive hairs
hair plates, trochanteral 197-99
head movements 255, 259, 269-73
heirarchy, concept of 10, 14-15, 52-53
Hemideina (weta) 59, 72-73
hip movements, as oscillator 42
horseradish peroxidase 215, 231, 260, 262, 263, 267, 326-27

inhibition, presynaptic 239, 382, 384
interneurones
 Ia inhibitory 81, 233
 command 10, 15, 25-27, 52-53, 61, 310-12
 flight, in locust 310, 343, 346
 giant, in cockroach 105-10
 inhibitory burst 263-64
 in stretch reflex 436, 444, 447
 multisensory 81, 348-49
 non-spiking, local 70, 215, 237, 246
 ocellar, of locust 341-43
 spiking, local 215, 235, 239-42
 thoracic, in locust 239-42, 343, 346
 tritocerebral commissure giant 347
 vestibular 266-69, 270, 273
 visual, in locust 341-46
 wind-hair, in locust 346-48

Subject index

intrafusal muscle fibre types 123, 126-32
intraspinal mechanoreceptors in lamprey 9, 325-28
iris muscles 116
isometric contractions in man 417-20

Jacksonian epilepsy 472-73
jaw movements, inputs to motor neurones during, in cat 83-89
jaw muscles, stretch reflex in 78
joint movement
 in cat locomotion 43, 360-66
 in crustacean locomotion 379-83
joint receptors 174, 176
 in cockroach 197-99
 in crustaceans 145, 379
jump, locust 60, 73-74, 243, 303, 313

labyrinth 254
lateral line organ 116
lamprey
 edge cell in 9, 325-28
 motor neurones in 5
 swimming in 9, 319-28
learning 28-29, 74
leech
 muscle and its innervation 6, 245
 swimming 9, 10, 22-23
length sensitivity of muscle spindle 126-32, 174, 404, 422-23
lidocaine 182
load compensation 58, 79-80, 199-201, 203, 384
lobster
 equilibrium 257, 277-94
 stomatogastric ganglion 22, 245, 308
 role of octopamine in 60
 walking 379-98
locomotion
 in arthropods 9
 in cats 9, 35-51, 81-83, 169-71, 175-84, 356-75
 in cockroaches 171, 187-206, 310-12
 in crustaceans 379-98
 in fish and lamprey 317-33
 in insects 307-13
 in leech 9, 22-23
 in locust 22, 51-52, 68, 109, 301-4, 308-10, 337-50
 in man 407-8, 451-62

locust
 behaviour generation in 58-74, 215
 extensor tibiae muscle 6
 flight 51-52, 68, 109, 301-4, 308-10, 337-50
 jump 60, 73-74, 243, 303, 313
 local reflexes in 231-47
 walking 22
Lucifer yellow 327
Lymnaea, muscle innervation in 7

man 217-30, 403-75
 walking in 407-8, 451-62
mastication, control of
 in cat 83-87
 in rabbit 366
mechanoreceptors
 cutaneous, in jaw movements 85-87
 cuticular stress detectors 374, 387
 intraspinal, in lamprey 38, 325-28
 Pacinian corpuscles 440
mesencephalic cat, walking in 9
mesencephalic locomotor region 36
metacerebral giant neurone 15-17, 21
metathoracic ganglion of insects 60, 63, 66, 95, 194, 233, 235, 237
microneurography 403, 411
modulator neurones 11, 59-74, 242-43
molluscs
 behavioural plasticity in 28-29
 central feedback loops in 15-22
 muscle innervation patterns in 7
monosynaptic reflex 78-80, 179, 213, 227, 231-33, 383, 431
motor end plate, fusimotor 126-28
motor neurones 242-46
 alpha 58, 118, 131, 178-79, 183, 212-14, 233
 beta 118, 179
 dynamic & static 123, 141, 142
 electrical coupling between 243-45
 fast & slow, in invertebrates 5, 242
 flight, in locust 341-43
 gamma 58, 213-14
 dynamic & static 118-19, 123-24, 126-27, 129-32, 138-42, 178-84, 419
 inhibitory in invertebrates 5, 242
 metacerebral giant 15-17, 21
 of buccal mass in Philine 7-8
 of crayfish abdominal muscles 5-6

Subject index

of lamprey 5
of locust 243
  extensor tibiae muscle 6, 243
  flexor tibiae muscle 66, 243
  of S type muscle fibres 78, 83
  pool size 242
  receptor (Rm1 & Rm2), of crab TCMRO 117-19, 148-50, 154-63
  segmental, in leeches 6, 245
motor programme generator 57
motor unit
  recruitment 4, 79
  size in mammals 4-5
multimodal interneurones & locust flight 348-49
multipolar receptors 197, 201
muscle
  abdominal, of crayfish 5-6
  accessory flexor, of crayfish 5
  extensor tibiae of locust 6
  fast & slow response types
    in invertebrates 5
    in vertebrates 4-5
  flexor pollicis longus, in man 219
  intrafusal, fibre types 123, 126-32
  iris 116
  leg, fibre types in crab 6
  middle ear 116
  obliquely striated, in gastropods 7
  receptor, of crab TCMRO 146-56
  slow type in leeches 6
  triceps surae, in man 219
  unstriated, in molluscs 7
  weight bearing & non-weight bearing, monosynaptic reflexes in 80
muscle receptor organ
  abdominal 145
  thoraco-coxal 115, 117-19, 145-64, 379
  see also myochordotonal organ
muscle spindle 78, 89, 115-19, 123-42 171, 174, 177-84, 212-14, 217-18, 223, 225-28, 405-7, 412, 431-48
  afferents 78, 123, 147, 169-70, 179-80, 213, 233, 403-5, 415-26, 466-67
  reptilian & amphibian 118
myochordotonal organ 145, 379
myoclonus
  action 472
  cortical reflex 408-9, 465, 467-73
  EEG in cortical 467-71

spontaneous cortical 472-73
myotatic reflex see stretch reflex

necessity, test for command neurone 26
neck muscles, gaze signals in 269-71
neurones
  cell C2 in Tritonia 22, 24
  cells 204 & 205 in leech 10, 22
  command 10, 15, 25-27, 52-53, 61, 310-12
  coordinating 15, 52
  corollary discharge 20
  DCMD of locust 343-44
  DUM in locusts 59-60, 62-66, 69-70
  inhibitory 6, 19, 28, 81, 87, 233, 239, 245, 263-64, 343, 383-84, 440
  metacerebral giant 15-17, 21
  modulator 11, 59-74, 242-43
  oscillator 15, 19, 52, 61, 308
  paracerebral 15-21, 28
  PD of stomatogastric ganglion 22
  pyramidal tract 434
  Renshaw 234, 245, 440
  reticular 262, 270-72
  tritocerebral giant of locust 347
  vestibular 265-70, 273
  see also interneurone
  see also motor neurone
neuropile 60, 63, 66-68, 72, 215, 237-39, 338
Nialamide 358, 359
nystagmus 256
  optokinetic 254, 266, 278,
  vestibular 254, 266, 270

ocellar sensory input in locust flight 339, 341-43, 348-49
octopamine
  in locust flight 338
  in weta arousal 59
  initiating movement in locust 66-68
  potentiating excitatory transmission 60
  producing specific postures in crustaceans 60
  promoting blood sugar release in cockroach 72
  synthesis within DUM neurones in locust 59-60, 63
olfactory bulb 116

493

Subject index

orchestration hypothesis 10-11, 60-74
orienting movements 254-55, 259, 262-64, 267
oviposition digging in locusts 65

Pacinian corpuscles 440
palatal receptors 85
Parkinson's disease 442, 443
Pecten, muscle innervation in 7
periodontal receptors 83-85
Periplaneta see cockroach
phase
  changes induced by leg autotomy in lobster walking 388-94
  control in human walking 461-62
  dependent reflex modulation 50, 175-76, 329-31, 356-75, 385-87
  dependent reflex reversal 164, 301, 329, 356
  relationships of legs in lobster walking 388, 393
  resetting in leech swimming 22
Philine, muscle innervation in 7-8
Planorbis, motor innervation in 7
plasticity 10, 28-29, 73-74, 164
Pleurobranchaea
  behavioural plasticity in 28-29
  control of feeding rhythm in 15-22
  feedback loops in 10, 25-28
Portunus, leg muscle fibre types in 6
position
  controlling servomechanisms 175, 213-14, 217-18, 226-27, 384
  sense of 174
  sensitivity of muscle spindles 128
presynaptic
  densities 237
  facilitation 239
  inhibition 239, 382, 384
  input to central processes of TCMRO afferents in crab 164
privileged access, test for command neurone 27
proprioceptors 9, 77, 175, 187-88, 302, 312, 317, 411-12
  in cockroach walking 202-6
  in control of eye movements 254, 256, 280
  in control of posture in man 451-53
  in crustacean walking 384-87
  in locust flight 337

joint receptors in cat 176
  of crustaceans 145, 254, 256, 277, 280-83, 289-90, 290-91, 379
  of jaw in cat 83
  of legs in cockroaches 197-201
  see also edge cells
  see also Golgi tendon organs
  see also muscle spindles
  see also TCMRO
pupil of eye 116
pyramidal tract 434

rabbit
  eye & head movements 255
  mastication 366
Randezellen see edge cells
reafference
  in cockroach escape behaviour 95-103
  in locust flight 109, 340-41
  receptor muscle of crab TCMRO 146-56
receptors
  central control of 113-60
  cutaneous, in cat 85, 171, 175-76
  cuticular, in Crustacea 379, 387-88
  joint, in cat 176
  joint, in cockroach 197-99
  joint, in crustaceans 145, 379
  multipolar, in cockroach 197, 201
  palatal, in cat 85
  periodontal, in cat 83
  providing feedback in flight 308-9, 337-39
  TCMRO 115, 117-19, 145-64, 379
  tension, in crustaceans 379
  wind hairs in cockroach 95-96
reciprocity, principle of 10, 14-15
redundancy 10, 27, 350
reflexes 209-47
  assistance 164, 199, 383
  autogenic, of crab TCMRO 156-61
  cervico-collic 269
  cervico-ocular 267
  compensatory 253, 259, 277
  crossed extension 363
  cutaneous 50, 82-83, 175-76, 356-72, 460-61
  flexion, in cat 356, 363
  inter-joint, in crab 163
  jaw opening in rabbit 366
  monosynaptic 78-80, 179, 213, 227, 231-33, 383, 431

Subject index

multisynaptic 78-80, 179, 233-35, 444
oculo-collic 270
of campaniform sensilla 201
of hair plates in cockroach 197
optokinetic 254-56, 267, 278-83
 phase dependent 301
 in cat 50, 356-75
 in fish 50, 329-31
 in lobster 385-87
postural, in man 451-57
resistance
 in cockroach 199
 in crustaceans 155, 164, 285, 290-91, 383-84
 see also stretch reflex
reversal 164, 188, 301, 329, 356
righting 194-97, 253, 277, 283-91
scratch, in turtle 457
stretch see stretch reflex
three neurone in locust 234-37
vestibulo-collic 259, 269-70
vestibulo-ocular 255, 259, 266-67, 270
Renshaw inhibition 234, 245, 440
resonance hypothesis 435
reticular neurones 262, 270-72
righting reactions 253, 277
 in cockroach 194-97
 in crustaceans 283-91

saccadic eye movement 254-55, 262-64
sensory
 control of cockroach walking 187-206
 control of fish & lamprey swimming 317-33
 control of steering in flight 339
 control of step cycle in cat 41-47
 control of crustacean walking 379-98
 deprivation, effect of 9, 37-39, 58-59, 307-12, 318-19, 465
 endings of crab TCMRO 147
 endings of muscle spindle 123
 input, central control of 113-66
 input during normal movements 167-208, 403-5, 411-26
 input, gating of 9, 50, 81, 246, 301, 330-31, 343, 347-48, 461, 462
 input modulating motor output in insect walking & flight 308-12, 337

inputs, convergence of 9, 77-89, 302, 348-50
 interaction in the control of equilibrium 253-57, 277-94
 perception 81
 responses of crab TCMRO 147-48
 tape 58
servocontrol 58, 87, 155, 175, 178, 213-14, 217-18, 226-27, 384, 412, 433-34
size principle 4-5
skin afferents 77, 83, 171, 174
 and phase dependent responses in cat walking 50, 175-76, 356-75
spasticity 79
specialisation in neural networks 27
speech, sensory input in 313
spindle see muscle spindle
spinalisation
 effect on locomotion in cats 39-41
 effect on locomotion in dogfish & lamprey 318-20
 persistance of long latency stretch reflexes following 436
spinocerebellar tracts 37
static
 gamma motor neurones 118, 123-24, 138-42, 180-81, 419
 sensitivity of muscle spindles 128
statocysts 254, 280-81, 286
steering
 behaviour in locust flight 338-50
 reactions in crustaceans 285-88, 292
stiffness, muscle 89, 181
stomatogastric ganglion 22, 245, 308
stretch
 activation of muscle spindles by 129
 activation of crab TCMRO by 147
stretch receptor of locust wing hinge 233, 237, 309-10, 337
stretch reflex
 gain of 79, 214, 218, 227, 405
 in crustaceans 155-56, 383-84
 in man 217-28, 405-7, 408-9, 431-48, 466-67
 load resisting in man 79
 long loop 80, 213, 405-7, 408, 433-38, 466-67
 mono- & multisynaptic pathways of 78-80, 179, 213, 233, 431
 transcortical see long loop

495

stumbling corrective reaction 50, 175, 356-57
succinyl choline 124, 132-38
sufficiency, test for command neurone 26
swimmerets 257, 284, 293, 385
swimming
  in dogfish & lamprey 317-33
  in leech 9, 22-23
  in Tritonia 8, 22, 58, 65
synapses
  of local interneurones in locust 239-41
  of motor neurones in locust 243-46
  of motor neurones in vertebrates 243, 245
  of sensory afferents in locust 237-39
  of spinal afferents 239

temporal primacy, test for command neuron 27
tendon
  Achilles, in man 224-25
  compliance 217-28
  flexor pollicis longus 220-24
  mechanical properties of 220-26
  organs, see Golgi tendon organ
tension
  of intrafusal muscle 124-26
  of receptor muscle in crab TCMRO 150-54
  receptors in crustaceans 379
thalamic cat, walking in 357
thoraco-coxal muscle receptor organ (TCMRO) 115, 117-19, 145-64, 379
training, effect on neural mechanisms of feeding in Pleurobranchaea 28-29
tremor, physiological 415
Tritonia
  escape swimming in 8, 22, 58, 65
  muscle in 7
turtle scratch reflex 457

unicellular command, concept of 15, 52-53
uropods 257, 284-88, 292

velocity sensitivity
  of crab TCMRO 148
  of muscle spindles 129-30, 174, 180

vesicles
  of inhibitory synapses 239
  synaptic, in locust 237
vestibular
  neurones 265-70, 273
  nucleus 264, 369
  reflexes 255, 259, 266-67, 269-70
vibration, reflex responses in man 405-6, 438-44
visual correction in human walking 457-60

walking
  afferent input during, in cat 169-71, 175-80
  afferent input during, in cockroach 171, 202-5
  convergence of several sensory modalities during, in cat 81-83
  development of, in infants 461
  direction changes during, in cat 47, 49-50
  effect of deprivation of sensory input during, in cat 9, 37-39
  function of proprioceptors during, in cockroach 171, 197-201
  in arthropods 9
  in crustaceans 379-98
  in decerebrate cats 9, 37-38, 357-58, 370
  in insects 310-12
  in intact cats 360-69
  in locust 22
  in low spinal cat 39-47
  in man 407-8, 451-62
  insect & cat 313
  modulation of reflex responses during, in cat 356-60
  motor activity during, in cockroach 191-93
  motor output resembling, in isolated crayfish CNS 164
  weta 59, 72-73
  threat posture in 73
wind-receptive hairs
  in cockroach escape 95-96, 103-5
  role of, in locust flight 346-48
  wing receptors in locust 309, 337

Young's modulus of tendon 222

RAYMOND H. FOGLER LIBRARY
DATE DUE

OOKS ARE SUBJECT TO
ER TWO WEEKS